50 Years of Transportation in Singapore

Achievements and Challenges

World Scientific Series on Singapore's 50 Years of Nation-Building

Published

50 Years of Social Issues in Singapore
  edited by David Chan

Our Lives to Live: Putting a Woman's Face to Change in Singapore
  edited by Kanwaljit Soin and Margaret Thomas

50 Years of Singapore–Europe Relations: Celebrating Singapore's Connections
with Europe
  edited by Yeo Lay Hwee and Barnard Turner

Perspectives on the Security of Singapore: The First 50 Years
  edited by Barry Desker and Cheng Guan Ang

50 Years of Singapore and the United Nations
  edited by Tommy Koh, Li Lin Chang and Joanna Koh

50 Years of Environment: Singapore's Journey Towards Environmental Sustainability
  edited by Tan Yong Soon

Food, Foodways and Foodscapes: Culture, Community and Consumption in
Post-Colonial Singapore
  edited by Lily Kong and Vineeta Sinha

50 Years of the Chinese Community in Singapore
  edited by Pang Cheng Lian

Singapore's Health Care System: What 50 Years Have Achieved
  edited by Chien Earn Lee and K. Satku

Singapore–China Relations: 50 Years
  edited by Zheng Yongnian and Lye Liang Fook

Singapore's Economic Development: Retrospection and Reflections
  edited by Linda Y. C. Lim

Singapore and UNICEF: Working for Children
  edited by Peggy Kek and Penny Whitworth

Singapore's Real Estate: 50 Years of Transformation
  edited by Ngee Huat Seek, Tien Foo Sing and Shi Ming Yu

The Singapore Research Story
  edited by Hang Chang Chieh, Low Teck Seng and Raj Thampuran

The complete list of titles in the series can be found at
http://www.worldscientific.com/series/wss50ynb

World Scientific Series on
Singapore's 50 Years of Nation-Building

50 YEARS OF TRANSPORTATION IN SINGAPORE

Achievements and Challenges

Editor

Fwa Tien Fang

National University of Singapore, Singapore

World Scientific

NEW JERSEY · LONDON · SINGAPORE · BEIJING · SHANGHAI · HONG KONG · TAIPEI · CHENNAI · TOKYO

Published by

World Scientific Publishing Co. Pte. Ltd.

5 Toh Tuck Link, Singapore 596224

USA office: 27 Warren Street, Suite 401-402, Hackensack, NJ 07601

UK office: 57 Shelton Street, Covent Garden, London WC2H 9HE

National Library Board, Singapore Cataloguing-in-Publication Data
Name(s): Fwa, Tien Fang, 1951–   editor.
Title: 50 years of transportation in Singapore : achievements and challenges / editor, Fwa Tien Fang.
Other title(s): Fifty years of transportation in Singapore.
Description: Singapore : World Scientific Publishing Co. Pte. Ltd., [2016]
Identifier(s): OCN 945485167 | ISBN 978-981-46-5159-2
Subject(s): LCSH: Transportation--Singapore. | Transportation –Singapore--History.
Classification: LCC HE274.A2 | DDC 388.095957--dc23

British Library Cataloguing-in-Publication Data
A catalogue record for this book is available from the British Library.

Typeset by Stallion Press
Email: enquiries@stallionpress.com

Contents

Preface

Singapore celebrated her 50 years of independence in 2015. The island city state's success in nation building, transforming from a developing nation to one with per capita GDP among the highest in the world, has no parallel in history. Among her major achievements is the development of an integrated comprehensive world class transportation system.

Singapore is now a key world transportation hub for both sea trade and air travel. The Port of Singapore is the world's largest seaport in terms of the annual tonnage of ships calling at the port; and the second largest container port based on the number of containers handled each year, next to Shanghai. The Changi International Airport has been rated the best airport by travelers for many years. The land transport system of Singapore is widely acclaimed as one of the most efficient in the world.

This book attempts to capture factually the detailed process and developments that have taken place in Singapore over the last 50 years to build a world class integrated land, sea and air transportation system that effectively serves the people and trade of Singapore as well as the whole world.

There are altogether 8 chapters in this book. Chapter 1 presents an overview of land transport development in Singapore since 1965 when Singapore gained her independence, and the principles and strategies that underpinne Singapore's land transport policies. Chapter 2 provides an in-depth account of the evolution of an efficient and eco-friendly public transport system in Singapore. Singapore is known worldwide for her successful management of road travel demand. Chapter 3 expounds the various travel demand management measures implemented by Singapore in the last 50 years, including the use of taxes to suppress car ownership in the 1960s, the implementation of the world's first congestion pricing scheme in 1975, vehicle quota system in 1990, as well as the more recently introduced 'Travel Smart' measures. Chapter 4 provides a brief history of road traffic management in Singapore and the rationale behind the technical and implementation decisions adopted. Chapter 5 elaborates how Singapore's road safety has improved and become globally benchmarked to be among the safest countries in the world.

Chapter 6 delves into the historical development of Singapore as a key logistics centre. It explains the key driving factors of the country's growth as one of the world's leading logistics hub. Chapter 7 explains how Singapore's clever and far-sighted planning of port development and expansion has made a difference. It analyzes the results of various strategies implemented, including equipment and management innovation, and application of information technology and smart technology. The plans for the future port aims to bring Singapore into the future capable of coping with diverse and critical challenges. The final chapter gives an account of Singapore's road of success as an international air-hub in operating a world-leading efficient airport combined with a premium world-class airline in the form of Singapore Airlines.

The experience of Singapore in transportation infrastructure development and successful operation of a safe and efficient transportation system is a wealth which can be shared by transportation professionals and agencies worldwide. Each of the chapters in this book is written by transportation professionals who have one way or another involved in creating the success story of Singapore. The materials presented in this book provide many valuable case studies for practitioners and researchers. It will serve as a useful reference and example of how transportation theories and ideas can be implemented in practice.

T. F. Fwa

Editor

Professor, Department of Civil & Environmental Engineering

Director, Centre for Transportation Research

Deputy Director/Head of Research, Centre for Maritime Studies

National University of Singapore

Republic of Singapore

About the Editor

Professor Fwa is currently a professor in the Department of Civil and Environmental Engineering at the National University of Singapore. He is the Director of the Centre for Transportation Research, and Deputy Director and Head of Research of the Centre for Maritime Studies. He is active academically and professionally in the areas of transportation engineering, covering land, sea and air transportation. Professor Fwa serves the international professional transportation engineering community in various capacities. He is currently the Asia Region Editor for the ASCE Journal of Transportation Engineering, and Immediate Past President of the International Society for Maintenance and Rehabilitation of Transport Infrastructure. He has chaired the Transportation Engineering Technical Committee of the Institution of Engineers, Singapore for more than 20 years since 1994.

About the Contributors

CHEW Ek Peng received his Ph.D. in Industrial Engineering from the Georgia Institute of Technology, USA. He is currently an Associate Professor and Deputy Head (Undergraduate Studies) in the Department of Industrial and Systems Engineering at the National University of Singapore. His current research areas is port logistics, maritime transportation, simulation optimization and inventory management. Some of his works are published in journals such as European Journal of Operational Research, IIE Transactions, Naval Research Logistics, and Transportation Science. He is serving as Editor-in-Chief of the Asia Pacific Journal of Operational Research and as a member in the Editorial Board of OR Spectrum, Flexible Services and Manufacturing Journal, and International Journal of Industrial and Systems Engineering. He and Dr Lee Loo Hay have recently lead a multidisciplinary team to win the Next Generation Container Port Challenge with a grand prize of US$1 mil by proposing a revolutionary double-storey container terminal.

CHIA Lin Sien MSc (McGill), PhD (Singapore), formerly Associate Professor in the Department of Geography, National University of Singapore, is currently Adjunct Senior Research Fellow of Centre for Maritime Studies (CMS). He has research interest in maritime transport including port geography, oil transportation and maritime labour. Among other activities, he was consultant to the former Division of Transport, Communications and Tourism, UNESCAP, in Bangkok. He is a member of the Advisory Panel of the Chartered Institute of Logistics and Transport (CILT) Singapore; and Advisor to the Singapore National Shippers Council (SNSC).

CHIN Kian Keong is an engineer by training, and has been involved in the planning, designing, implementing and constructing of various road, traffic management and road safety projects in Singapore. He was intimately involved in the Electronic Road Pricing (ERP) project, which started full-scale operations in September 1998 and various Intelligent Transport Systems (ITS) to better manage traffic. He is now the Group Director, Transportation & Road Operations and concurrently the Chief Engineer, Transportation at Singapore's Land Transport Authority.

CHOI Chik Cheong is Deputy Director, Knowledge Management Division with the Land Transport Authority (LTA), Singapore. He has over 15 years' experience in overall integrated land-use and transport planning, covering road and rail projects. He has directed studies on bus planning, road pricing, car-parking standards and multi-criteria evaluation of transport schemes. He worked with urban planners to integrate Light Rapid Transits into both old developments and new towns, and has had over 10 years' experience in the building and construction of the Mass Rapid Transit system in Singapore. He is currently working on the knowledge management and documentation of institutional land transport knowledge within LTA into case studies for training. He has an MSc in Transport from Imperial College, London, UK.

HO Seng Tim is the Head of Road Safety Engineering Unit of Land Transport Authority (LTA), Singapore. He graduated with a Master of Science in Transportation Planning and Engineering from the University of Southampton (UK) and has more than 30 years of experience in traffic management and road safety engineering. Mr Ho was from Public Works Department and has been with the LTA since its formation. Mr Ho played a key role in setting up Road Safety Engineering Unit in 1998 and continues to spearhead road safety initiatives including Silver Zones in Singapore. He has presented and published papers on Road Safety Engineering at several conferences and technical journals.

LEE Loo Hay is an Associate Professor in the Department of Industrial and Systems Engineering at National University of Singapore and visiting professor at the Department of Systems Engineering and Operations Research at George Mason University. He received his B.S (Electrical Engineering) degree from the National Taiwan University in 1992 and his S.M and PhD degrees in 1994 and 1997 from Harvard University. He has served as the associate editor of several journals and he is currently the co-editor for Journal of Simulation and is a member in the advisory board for OR Spectrum. He is a senior member of IEEE. His research focuses on simulation optimization, maritime logistics which includes port operations and the modeling and analysis for the logistics and supply chain system. He has co-authored the book Stochastic Simulation Optimization-Optimal Computing Budget Allocation (World Scientific) and co-edited the book Advances in Maritime Logistics and Supply Chain Systems.

LEW Yii Der is the Group Director, Corporate Planning & Development of Land Transport Authority (LTA), Singapore. He spearheads corporate development in LTA which includes amongst others expanding the research and training capacity of the land transport industry. He joined the then Public Works Department and has been with LTA since its formation in 1995. Mr Lew has held various management positions in policy, traffic management, safety and transport technology. He was involved in revamping the vehicle tax structure, developing the ERP charging policy and the comprehensive review of Singapore's land transport plans that culminated in the Land Transport Master Plan launched in 2008. He holds a first class honours degree in Civil Engineering from the National University of Singapore and a Masters in Public Management from the Lee Kuan Yew School of Public Policy. In 2015, he was awarded the Public Administration Medal (Silver).

LOOI Teik Soon is the Dean of the LTA Academy. He holds the concurrent appointment of Director, Future Mobility & Industry Development in Land Transport Authority (LTA), Singapore. His portfolios include knowledge management and capability building programmes, catering to both international and local participants. He oversees the Research as well as the Technology and Systems Divisions, including the Autonomous Vehicle Programme Office (AVPO). Prior to these, he was the Director of Policy, in charge of public transport policies, and was also former secretary of the Public Transport Council. A civil engineer by training, Mr Looi is experienced in road and rail planning and design, road operation and maintenance, etc. He holds a Masters in Public Management from the Lee Kuan Yew School of Public Policy.

A. P. Gopinath MENON was with the Public Works Department and Land Transport Authority (LTA), Singapore for over 35 years. He served as Chief Transport Engineer in LTA from 1991 to 2001. He was instrumental in introducing many new concepts in traffic management such as computerised area traffic control of traffic signals, electronic road pricing, bus priority measures and pedestrian safety programmes. He has been an Adjunct Associate Professor at the School of Civil and Environmental Engineering of the Nanyang Technological University since 1991. He retired from the public service in 2001 and is currently a Council Member of the Singapore Road Safety Council, a board member of the Public Transport Council and a member of LTA's Technical Advisory Committee on Transport Telematics Hub. In March 2013, he was elected as an Honorary Member of the Road Engineering Association of Asia and Australasia.

Giulia PEDRIELLI is currently research fellow at the Department of Industrial & Systems Engineering in National University of Singapore. She got her Master Degree at Politecnico di Milano in 2009 in Industrial Engineering. She got her Ph.D. in Mechanical Engineering in March 2013 graduating with honors. She was also research fellow for the Institute of Industrial Technologies and Automation within the National Research Council, Italy, (ITIA-CNR). She has been a visiting Ph.D. student at University of California Berkeley (USA), Department of Industrial Engineering & Operations Research (IEOR).

She develops her research activity at the in the field of simulation and simulation optimization, with a particular interest in statistical modeling and real time optimization. Her main application fields are manufacturing and logistics systems spanning from port logistics to biopharmaceutical supply chains.

Giulia will be joining Arizona State University as Assistant Professor in the fall 2016.

POON Joe Fai is Director, Policy at the Land Transport Authority (LTA), Singapore. He holds a PhD at the University of Sydney and an Intercollegiate Masters of Science in Transport from the University of London. His current work portfolio includes bus industry policies and international relations. He joined LTA in 1997, and has worked in the areas of policy, planning and public transport regulation. Prior to 2015, he was responsible for private transport ownership and usage policies.

Karmjit SINGH has been the Chairman of the Chartered Institute of Logistics and Transport, Singapore since 1992. Apart from that, Mr Singh was also the Chief Executive of SATS Airport Services from 1998 to 2004. Prior to joining SATS in 1998, Mr Singh was with Singapore Airlines for 24 years, serving in multiple senior management positions.

Mr Singh was also conferred the 2014 National Day Awards Public Service Medal by the President of Republic of Singapore.

TAN Kok Choon is associate professor of Decision Sciences at NUS Business School and Director of Degree Education at The Logistics Institute — Asia Pacific (TLIAP), overseeing the Double Master Program in Logistics & Supply Chain Management, a collaboration between NUS and Georgia Institute of Technology. He is also program co-director of the M.Sc. (Supply Chain Management) program which is jointly offered by the Department of Decision Sciences, Department of Industrial & Systems Engineering and TLIAP.

Dr Tan received a PhD in Operations Research from Massachusetts Institute of Technology. He has held appointments as Container Terminal Process Manager with PSA Singapore Terminals and Manager of Operations Research at PSA International and at Land Transport Authority. He has published in international journals including Mathematics of Operations, European Journal of Operational Research, OR Spectrum, Intermodal Transportation Research, Supply Chain Optimization, among others.

Chelsea C. WHITE III holds the Schneider National Chair of Transportation and Logistics at the Georgia Institute of Technology, where he is the former Director of the Trucking Industry Program (TIP) and the former Executive Director of The Logistics Institute. While he was a member of the University of Michigan faculty, he was the founding Engineering Co-Director of what is now the Tauber Institute for Global Operations.

He serves on the boards of directors for the Industry Studies Association and the Bobby Dodd Institute and is a former member of the board of directors for Con-way, Inc. (NYSE: CNW, 2004–2015), The Logistics Institute-Asia Pacific, ITS America (a Utilized Federal Advisory Committee), and the ITS World Congress. He is a former member of the World Economic Forum trade facilitation council.

His most recent research involves monetizing data for real-time supply chain control.

Chapter 1

Overview of Singapore's Land Transport Development 1965–2015

LEW Yii Der and CHOI Chik Cheong

This chapter paints the landscape of land transport development in Singapore since 1965, and outlines the principles and strategies that have underpinned Singapore's land transport policies.

1. Introduction

Travelling is an integral part of living. Since the formation of the earliest cities, the need for mobility has been a feature of urban living. To facilitate economic growth and ease of travel, many cities poured in vast investments in transport infrastructure and systems to cope with rising travel demand from an increasing urban population and economic activities.

Like other cities, Singapore faces its share of challenges in meeting the travel demand of its population and its economy. Everyday, over 12.5 million journeys are made on its transport network comprising: a comprehensive network of buses, roads and expressways, as well as an extensive Rapid Transit System (RTS) made up of the Mass Rapid Transit (MRT) and Light Rapid Transit (LRT) networks. Currently, six in ten households are within a ten minute walk of a RTS station. It is a densely populated and highly urbanised city, with a population of over 5.5 million residing on a land area of only 718 square kilometres.

1.1. *Circumstances and constraints*

During the early years of nationhood from 1965, Singapore was confronted with a rapidly increasing population faced many social and economic challenges. It had to address severe housing problems, high unemployment rate and a transport system that was not able to cope with the demands of a growing nation. In those early years, funding was the main constraint for expanding the transport infrastructure and facilities to meet a growing economy and population.

Hence, up till the 1970s, Singapore experienced severe traffic congestion in the city areas. Bus services were unreliable and inefficient. There were serious flooding during the monsoon periods and the roads were not well maintained. In the early 1970s, the government stepped in to restructure the public transport industry and put in place a policy to manage road use.

Land constraint has always been a major challenge in the development of modern Singapore due to the country's small land area and high population and density. In view of this, a supply side approach to maintain mobility is not sustainable without measures to manage demand. It was recognised early in the day that Singapore needed policies to restrain the growth of car population and restrict its use especially in the Central Business District.

Demand management can only be effective if there are alternative means of travel on a public transport system that serves the needs of the people. The early planners saw the need for a high quality public transport system as a crucial element in Singapore's land transport system. Planning for the Mass Rapid Transit (MRT) system started in 1971 as part of the planning process to develop Singapore's first integrated land use and transport plan. Given Singapore's funding constraints then, the question of whether to build the MRT was proposed.

The S$5 billion MRT system was vigorously argued by proponents of different camps, including members of the public. Eventually, after weighing all the pros and cons, the government decided to proceed with the building of the MRT was MRT system in1982.

1.2. *Present challenges*

Transport policies in Singapore continue to evolve in order to address the changing needs and aspirations of the population. As Singapore's transport development becomes more complex, the need for institutional integration became more apparent. This resulted in the setting up of Transit Link in 1989, an agency to integrate bus and MRT services, and the formation of the Land Transport Authority (LTA) in 1995 to integrate all aspects of land transport from planning to implementation and regulation.

Over the years, Singapore's economy and population have expanded significantly resulting in an ever-growing demand for an efficient and accessible land transport system. Singapore's population grew from 4.8 million in 2008 to 5.4 million in 2013, and the Gross Domestic Product (GDP) has grown by more than 37% in the same period. By 2030, the Population White Paper projects that Singapore's total population could range between 6.5 and 6.9 million. As such, travel demand is expected to increase by more than 50% from the 12.5 million journeys today.

The future challenges facing land transport in Singapore will still include the very ones we have grappled with over the last 50 years: the scarcity of land, the growing population, the greater propensity for car use, and the higher public expectations. The changing demographics, coupled with the responsibilities of maintaining an aging rail system while expanding the network in built territory, and the aim to be an inclusive society catering to diverse needs, are new challenges to boot and they will have an impact on our land transport policies.

2. A Brief History of Land Transport Governance

2.1. *Transport in colonial Singapore*

Singapore has come a long way in land transport development from its early days as a British colony. Under colonial rule, land transport development in Singapore took a more *laissez-faire* approach. The Public Works Department (PWD) first established by the British in 1872, was in charge of building road transport infrastructure such as roads, bridges and pedestrian walkways as well as all other public facilities such as schools, hospitals and libraries.

Given the *laissez-faire* approach, road-building was done in a somewhat 'leisurely' manner. By 1920, PWD was in charge of only 340 km of public roads which suggested that about 3.4 km of road was built every year since the British founded Singapore in 1819. Unsurprisingly, many of the public roads built during this period were often funded by impatient merchants who were anxious to promote business rather than wait for the government to build them (Sharp, 2005).

2.2. *1950s — A turbulent time for Singapore*

By the 1950s, public transport was in a state of disarray with poor regulation and lack of enforcement.

It was a turbulent time for Singapore with communist agitation in the trade unions (including bus unions) and strong anti-colonial feelings. Bus workers that were controlled by communist unions frequently initiated work stoppages to get their employers to address their low wages and poor working conditions, thus paralysing the whole bus system. There were altogether 57 strikes in the Chinese Bus Companies in 1955 alone, including the infamous Hock Lee Bus Riot.

The 'Great Singapore Traction Co. Strike' also commenced in the same year, lasting 142 days — making it the longest strike in post-war Singapore. Workers in other public bus companies also joined in for about a month, aggravating the already seriously crippled public transport system.

The tension and violence of the immediate post-World War II years seriously hampered the work of the authorities in trying to bring order to the transport system. In 1956, the government published the Hawkins Report that recommended, among other things, the 11 existing bus companies (Fig. 1) be

Source: Land Transport Authority.

Fig. 1.   Bus belonging to one of the Chinese Bus Companies, 1960s.

merged, with common ownership and common management either as a fully nationalized government-run entity or partly financed by private investment. However, this recommendation could not be realized until almost 20 years later in 1973 with the formation of the Singapore Bus Service, as it might cause severe disturbances to the political, economic and social fabric of Singapore at that time.

2.3. 1960s — Self governance and independence

The 1960s marked an era of great change for Singapore. Given the rapid population and economic growth, the government was aware of the need for a reliable and efficient public transport system to support nation building.

Early Public Bus System

However, the public transport system then had received little to no systematic transport planning. The public had to put up with a myriad of timetables, routes and fares offered by the different bus companies. The lack of integration often resulted in long and inconvenient journeys with multiple transfers, especially for commuters who wanted to travel between territories served by different bus companies. In order to maximise profits, buses were not regularly maintained by the operators, resulting in frequent breakdowns.

Pirate Taxis

As an alternative to the inadequate public transport system, pirate taxis were rampant. These were private and usually old vehicles that transported people without a taxi licence (Chin, 1998). During this time, about 5000[a] illegal or pirate taxis were reported to be plying the roads. Such taxis were not metered, requiring passengers to negotiate their fares at the start of the journey and often, sharing the taxi with strangers picked up along the way. As they were not regulated, drivers often tried to take as many passengers as they could to increase their takings, paying little attention to service quality and safety.

However given the disorganized state of the public bus system that was often paralysed by labour strikes, the government did not strictly enforce the taxi industry until the 1970s, as the pirate taxis supplemented the public

---

[a] Commission of Inquiry into Public Passenger Transport System of Singapore Report, 1956.

transport system at the time and was a source of employment in a time of high unemployment and job shortage.

Dawn of a New Era

After separation from Malaysia in 1965, Singapore found itself on the cusp of a new era, one filled with uncertainty and unrest. These circumstances fostered a sense of urgency and vulnerability which prompted the government to take decisive actions and a strong interventionist approach to secure economic development and social stability to ensure Singapore's survival.

Thus, the circumstances behind Singapore's independence have shaped land transport governance to a large degree; from no systematic planning to a long-term, forward-looking approach of land-use and transport planning. Next is the application of novel and radical policies like using road pricing to manage congestion (first implemented as a manual scheme under the Area Licensing Scheme (ALS) in 1975 and then converted to the Electronic Road Pricing (ERP) scheme in 1998). Another major initiative was the building of the Mass Rapid Transit (MRT) system, as well as the practice of studying examples from other cities and adopting their policies to suit Singapore's context such as the public transport financing framework.

2.4. 1970s — The beginning of the interventionist approach

Long-term Integrated Planning

Long-term, forward-looking planning is firmly entrenched in Singapore's land-use and transport development process. Singapore's growing economy and land constraints necessitated sound long-term city planning. Today, Singapore uses the Concept Plan, which is a strategic land-use and transport plan that guides long-term physical development (40–50 year planning horizon) to ensure that there is sufficient land to meet long-term population, economic growth and mobility needs while providing a good quality living environment. This strategy goes beyond merely planning the physical land use and transport network holistically. It underscores the importance of the state envisioning an overall concept plan for the long-term future of itself, from which a development master plan and a supporting transport master plan are then developed. The government needs to take the long view in order to ensure the continuity of this vision and process. It also requires various government agencies to work hand-in-hand as a networked entity in developing the different plans and coordinating each other's actions to ensure maximum chance of realising them.

Rapid developments in the 1960s called for a comprehensive land use and transport plan as the 1958 Master Plan left by the colonial government had its limitations and was not geared towards high population growth. As such, the government sought technical assistance, and the first concept plan (Fig. 2) was developed in 1971 as the outcome of a four-year State and City Planning (SCP) Project. This project was commissioned by the government in 1967 and completed in 1971, with the help of the United Nations Development Programme (UNDP) to address issues of redevelopment. The 1971 Concept Plan laid the foundation for Singapore's growth for a better quality of life with new towns, transport infrastructure and access to recreation.

Within this 1971 Concept Plan was the Strategic Transport Plan which proposed a comprehensive island-wide network of expressways and arterial roads as well as an MRT system. At that time, the planning horizon was only 20 years ahead. By 1989, much of the infrastructure envisioned in the 1971 Concept Plan had been put in place. The Concept Plan was subsequently reviewed in 1991, 2001 and in 2011 to factor in changes in local and global trends, and ensure that Singapore's plans remain relevant to address future challenges and meet needs.

The Concept Plan Review is a whole-of-government exercise which mobilizes all land-use agencies from the economic, social, environmental and infrastructure sectors as well as extensive public consultation. Updated

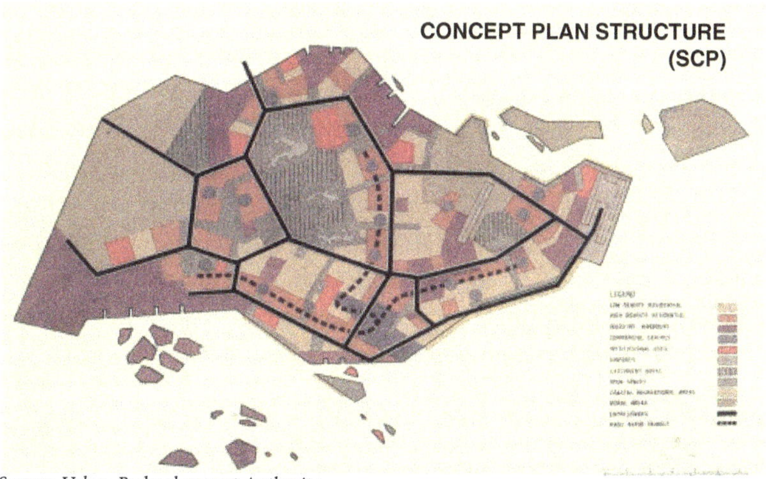

Source: Urban Redevelopment Authority.

Fig. 2.   Singapore's first integrated land-use and transport development plan 1971 Concept Plan.

Planning Process

Road and RTS Master Plan	Concept Plan	Long term plans 30-40 years
Road and RTS Safeguarding Plan	Master Plan	Medium term plans 10-15 years
Implementation Plans	Planning Feasibility Studies	Near term plans 5-10 years
5 Year Road Development Programme	Rail Lines	Bus Routes & Infrastructure Development

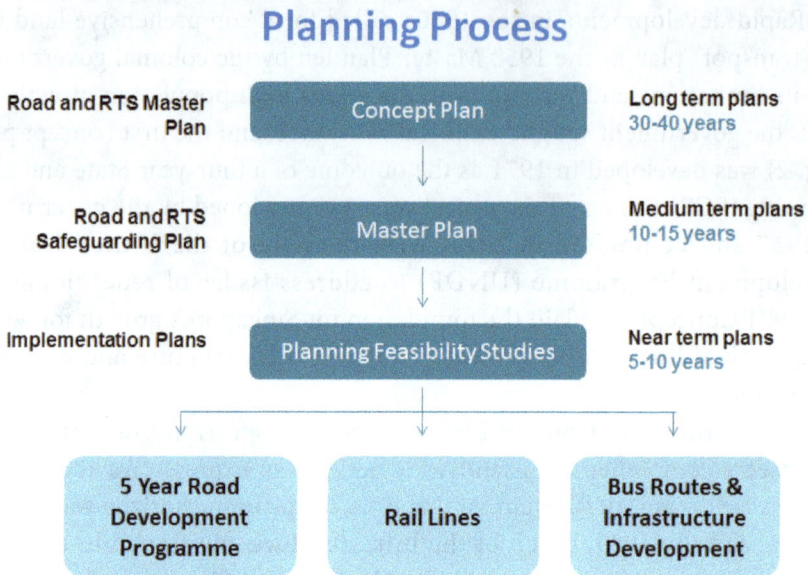

Source: Land Transport Authority.

Fig. 3.   Concept Plan Process in Singapore.

economics trends and projected population parameters are also taken into consideration during the reviews. As illustrated in Fig. 3, this multi-agency review effort comprehensively looks at all land use needs such as housing, businesses, transport, infrastructure, recreation, and community needs and balances between different demands for land to ensure that there is sufficient land to meet the needs of future generations.

The Concept Plan (Fig. 4) reflects the need to integrate the planning of various land uses and major transport infrastructure such as road and rail networks. This facilitates advanced planning and safeguarding of transport corridors and systematic expansion of transport infrastructure in tandem with land development and traffic growth. Where needed, the state would step in with policy measures like the Land Acquisition Act, which might be draconian to some, but for the larger public good, to ensure order in the physical developments and to retain the flexibility of the state to build the necessary infrastructure. Figure 4 shows the Concept Plans that have been formulated over the years.

Reorganisation of Public Transport Industry

The 1970s ushered in an era of stronger government intervention in the public transport sector. The government implemented new regulations to ensure better bus and taxi services. In 1970, the government published a White Paper on

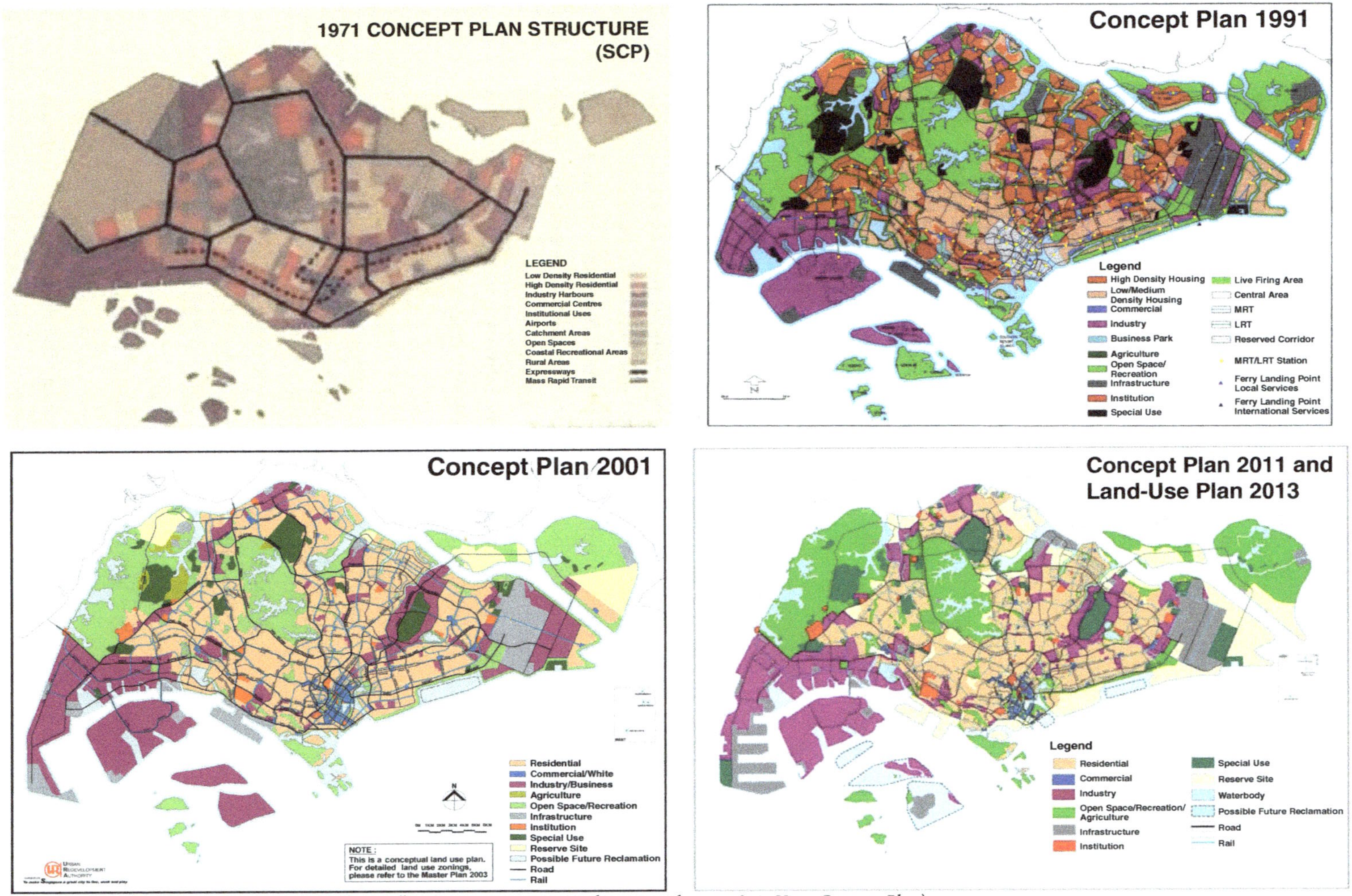

Image Source from Urban Redevelopment Authority (https://www.ura.gov.sg/uol/concept-plan.aspx?p1=View-Concept-Plan)

Fig. 4.   Concept Plans across the years.

the Reorganisation of the Motor Transport Service of Singapore to address the shortcomings of the public transport service which was described as owing to: the lack of a definite policy coupled with ad-hoc decisions being made as and when problems arose, the lack of continuity in the Registrar of Vehicles post, as well as insufficient financial and manpower support for the Registrar to perform and enforce its multifarious duties.

The 1970 White Paper addressed important aspects of road transport which needed improvement. It laid the foundation for the overhaul of the overall public transport service — the merger of 10 Chinese Bus Companies into three major bus companies with clear territorial demarcations; and the reorganisation of the Singapore Traction Company (STC) — the main bus company providing the main routes within the Central Business District at the time. It also provided the mandate for the government to eliminate the illegal pirate taxi industry with the proposed plan to restructure the public transport industry.

Bus Reforms

At the same time, the government commissioned a study on bus services and operations. The outcome of this study was the Wilson Report in November 1970, which advised the government on making the bold steps to amalgamate the Chinese bus companies into three, to rationalise bus services and to intro-duce a new unified fare structure in April 1971. Implementation in mid-April was planned to coincide with the school vacation period when the public transport services were not taxed to capacity. This would make the change of such magnitude less difficult to cope with, especially with any teething trouble that may be encountered during the first few days of operations.

In spite of the first set of reforms in mid-1971, the disorganized situation deteriorated further. This was accompanied by lower profits (the STC was making huge losses and eventually ceased operation on 5 December 1971), frequent bus breakdowns, absenteeism and industrial unrest.[b] This prompted the government to intervene again in 1973 with the merger of the three remaining bus companies into a single private entity called the Singapore Bus Service (SBS) as first envisioned by the Hawkins Report in 1956.

The merger saw the integration of all bus routes and the standardisation of fares. In 1974, the Government Team of Officials (GTO) was sent in to the amalgamated Singapore Bus Service (SBS) company to improve organisational

---

[b] A. P. Gopinanth Menon and Loh Chow Kuang, Lessons From Bus Operations, Land Transport Authority, March 2006.

efficiency and provide reliable services to the public. The GTO was an inter-departmental team of 100 civil servants, including army engineers, mechanics and the Police. The move also created economies of scale and ultimately improved bus services.

Taxi Reforms

New regulations were implemented to weed out the illegal taxi industry in accordance with the 1970 White Paper. These included:

i. Raising diesel taxes on private diesel vehicles (majority of which operated as pirate taxis) by 100%;
ii. Suspension of pirate taxi drivers' driving licences for one year if caught; and
iii. Making pirate taxi operations a seizable offence where offenders could be arrested on the spot and charged the following day. Other new regulations included establishing an age limit of seven years for taxis and a maximum age of 63 for taxi drivers.

Furthermore in 1970, the NTUC Comfort (cooperative commonwealth for transport) was formed to manage the taxi drivers, freeing them from unscrupulous taxi barons that controlled clusters of up to 100 pirate taxis and drivers. It helped former pirate taxi drivers to become licensed taxi owner-drivers — members of the cooperative could take up vehicle loans allowing them a four-year period to repay the loans. Taxi operations also improved as drivers took personal interest in maintaining their own vehicles (*Chin, 1998*).

Rising Traffic Congestion

As the economy grew and incomes rose, Singapore experienced rapid growth in car ownership. The effect of traffic congestion was most severe within the Central Business District (CBD). Furthermore, The SCP studies projected massive traffic congestion by 1992 without car restraints. Hence, the Road Transport Action Committee (RTAC) was set up in the 1970s to look into pre-empting traffic jams, which led to the implementation of the world's first congestion pricing scheme, the Area Licensing Scheme (ALS). This era saw a radical shift in policy towards heavier government intervention and demand management policies.

The ALS was a usage restraint measure to control traffic congestion in the CBD during peak hours. An imaginary cordon was set up around the most congested parts of the CBD. The cordoned area, called the Restricted Zone, was

Source: Land Transport Authority.

Fig. 5.   Area Licensing Scheme in operation.

demarcated by overhead gantry signs. During the Restricted Hours, a paper licence had to be purchased and displayed on the vehicle windscreen to enter the Restricted Zone. Enforcement was manual (Fig. 5), requiring enforcement personnel at all gantry points and sales staff at roadside sales booths located at the approach roads to the Restricted Zone. Licences could also be purchased at petrol stations, post offices and convenience stores.

Parking Policy

In 1965, a policy of stipulating minimum parking provision on zonal and use-mix basis was introduced to ensure that sufficient parking is provided within each development. The rationale was to prevent illegal parking or unnecessary cruising in search of parking spaces, and queuing on roads while waiting for available parking spaces.

Along with the ALS, parking controls were applied to discourage trips by car, especially work trips into the city in 1975. This was done in two ways: raising parking charges differentiated between CBD and non-CBD areas and imposing a charge on each lot owned by the private car park operators. However, this scheme was suspended in 1998 with the introduction of the Electronic Road Pricing (ERP) Scheme, which was a more efficient and flexible traffic management tool.

Today, Singapore's parking policy requires that each privately-owned building or facility provide a minimum level of parking based on the space developed, type of use and location.

Over the last two decades, parking standards have been tightened progressively in the city centre, and in and around transport nodes. In 2005, the Range-based Car Parking Standard implemented jointly by LTA and the Urban Redevelopment Authority (URA) gives developers the flexibility to provide up to 20% fewer parking spaces than the prevailing standard. This enabled developers to better match parking provisions with their assessment of demand based on operational and business considerations.

2.5. *1980s*

The Era of Expressways

The 1980s saw the completion of several major transport infrastructure projects inked in the first Concept Plan such as, the first expressways that provided faster islandwide connectivity such as the Bukit Timah Expressway (completed in 1986) which connects the Pan Island Expressway (completed in 1981) at Bukit Timah to the Woodlands Causeway. Table 1 lists the major events in the era of expressways, and Figs. 6 and 7 shows the expressway network in 1989 and 2014 respectively.

The Great MRT Debate

Despite all the reforms and changes to the transport system, a burgeoning Singapore required more to be done to ensure efficient and convenient travel for the masses. First mooted in the 1970s during the SCP project, the study envisioned the need for a rail transit system to cater to Singapore's growing transport needs. Thus began 'The Great MRT Debate'. The decision to build a S$5 billion solution did not come easy. While some argued for a bus-rail system, others preferred an all-bus system.

The decision took a long time primarily due to fiscal concerns. The government had to weigh the benefits of the investment against the great cost of building it at a time when Singapore needed funds for urgent housing development, education and defence. In the end, the argument for the MRT won as it was felt that an all-bus system would impose considerable traffic externalities to all users; and developments in the CBD would have to be scaled back to give enough space and priority for buses. Furthermore, it was felt the MRT would be an investment to the economy by improving Singapore's competitiveness in attracting the kind of higher value-added investments

Table 1. Chronology of major events in the era of expressways.

Year	Event
1964	• Construction of Singapore's first expressway — Pan-Island Expressway (PIE) begins. The PIE was built in several phases over a period of almost 30 years.
1981	• East Coast Parkway (ECP) expressway completed • Benjamin Sheares Bridge completed
1982	• Green light finally given to build MRT
1985	• Bukit Timah Expressway (BKE) completed
1987	• Incorporation of Singapore Mass Rapid Transit Ltd (SMRT), the private operator of the MRT system. • Public Transport Council (PTC) set up to regulate bus services and public transport fares. • 1st section of the MRT system, from Yio Chu Kang to Toa Payoh, opened for service. Within three weeks, the millionth ride is recorded
1988	• Ayer Rajah Expressway (AYE) completed.
1989	• Central Expressway (CTE) opened to motorists • Tampines Expressway (TPE) opened to motorists • Transit Link set up by SBS, SMRT and TIBS to develop an integrated bus-rail public transport system.
1994	• Kranji Expressway (KJE) completed • Start of on-going upgrading works of the Outer Ring Road System (ORRS), a network of major roads that forms a 'ring' along the outer areas of the city. ORRS links motorists to the expressways and other major roads allows travel between east and west of Singapore without going through the city.
1998	• Seletar Expressway (SLE) completed
2008	• Opening of the Kallang — Paya Lebar Expressway (KPE)
2013	• Opening of the Marina Coastal Expressway (MCE)

desired, especially in the financial and business sector (Phang, 2002). The MRT would also boost investor confidence and have a multiplier effect on real estate values.

In 1987, the Singapore Mass Rapid Transit (SMRT) Pte Ltd was incorporated to run the MRT system as a private operator. As explained by Mr. Ong Teng Cheong, (Fig. 8) then-Minister for Communications, "The MRT is much more than a transport investment, and must be viewed in its wider economic perspective. The boost it'll provide to long term investor confidence, the multiplier effect and how the MRT will lead to the enhancement of the intrinsic value of Singapore's real estate, are spin-offs that cannot be ignored."

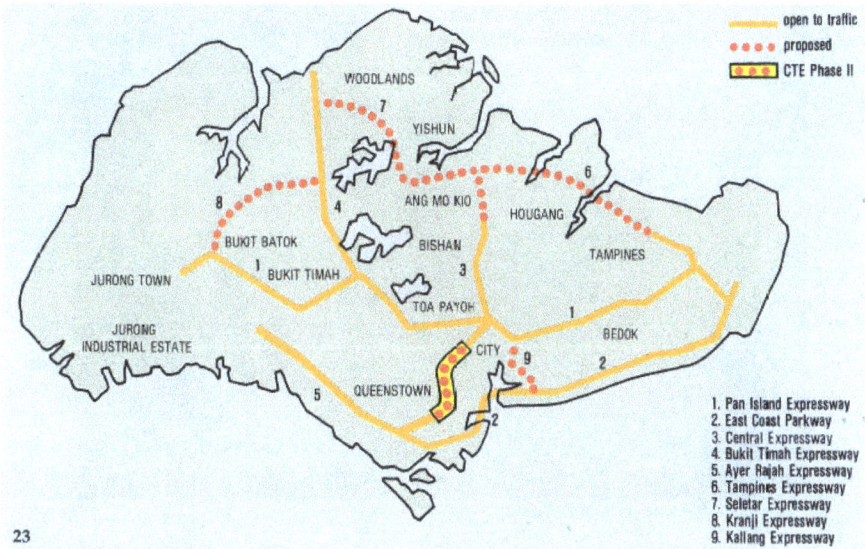

Source: Public Works Department. Annual Report 1988/1989.

Fig. 6.   Expressways Network in 1989.

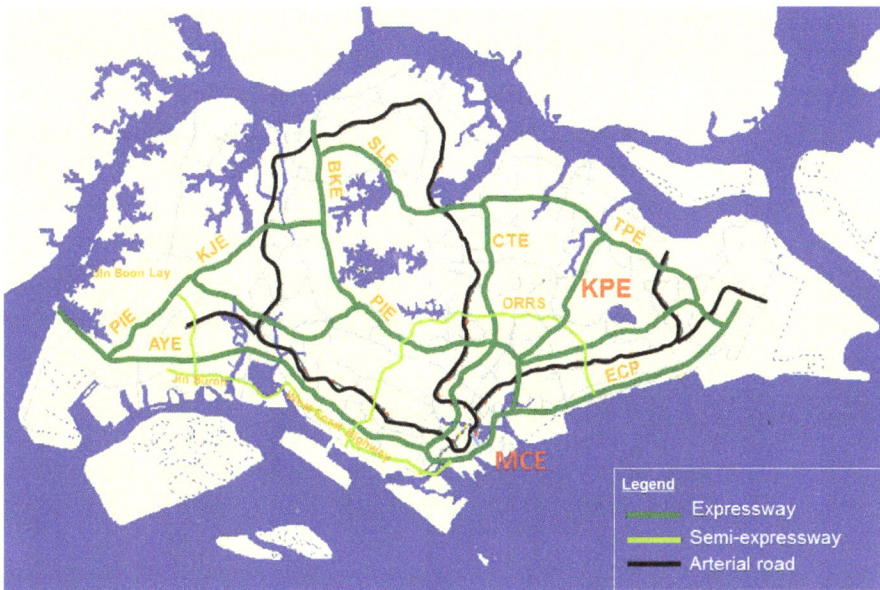

Source: Land Transport Authority.

Fig. 7.   Expressways in Singapore up to 2014.

Source: Land Transport Authority.

Fig. 8.   Deputy Prime Minister, the late Mr Ong Teng Cheong, opening the initial section of the MRT at Toa Payoh MRT Station on 7 November 1987.

Need for competition

By the 1980s, there was increasing public dissent against the SBS as a monopoly given its strong profits and fare increases. To address this, the government mooted the idea of a second bus company to provide some degree of competition to the SBS.[c] In 1982, the Trans-Island Bus Services Pte Ltd (TIBS) was formed. and started operations in April 1983. It was also hoped that the introduction of TIBS would serve as a catalyst for improving bus services. Mr Ong Teng Cheong, then-Minister for Communications, remarked: "Each company will act as a natural impetus to enhance the performance and efficiency of the other in the spirit of healthy competition, and in the process, help bring about a better level of service."[d]

Formation of the Public Transport Council

In response to public concerns for a wider representation on fares and services, the Public Transport Council (PTC) was established under the Public Transport

---

[c] In 2001, Singapore bus service changed its name to SBS Transit Limited. This was to reflect its status as a multi-modal transport operator, as it had won the tender to operate the new North-East Line and the Punggol LRT Line and Sengkang LRT Line.

[d] Speech by Mr Ong Teng Cheong, Minister for Communications and Labour at the Inauguration of Trans-Island Bus Services (PTE) LTD (TIBS, at Anson Road Bus Terminus, 3 April 1983. Source: National Archives, Singapore. http://www.nas.gov.sg/archivesonline/data/pdfdoc/otc19830403s.pdf Retrieved 9 Mar 2015.

Council Act to regulate the public transport industry in 1987. The PTC took over the Bus Services Licensing Authority in regulating bus services as well as the additional responsibility of regulating fares, which was formerly the responsibility of the Ministry of Communications. The main aim was to create greater public representation (council members are appointed by the Minister for Transport from both the private and public sector) in determining bus routes as well as to serve as a "curator" on behalf of the public in matters that affected commuters such as fares. The launch of the newly-built MRT system in the same year also created a need to integrate public transport fares across different transport modes.

Formation of Transit Link

To facilitate the multi-modal integration between the newly-built MRT system and buses, Transit Link was jointly established by then-public transport operators (SMRT, SBS Transit and TIBS) in 1987 to integrate fare collection system for bus and rail network.

Following the introduction of the MRT in 1987, an integrated ticketing system was introduced by Transit Link in 1990 to provide a common cashless fare payment system on both bus and rail services. This allowed commuters to switch from bus to rail and vice versa, without the fuss of having to change payment methods.

The launch of the integrated cashless ticketing system also allowed for the introduction of transport fare rebates in January 1991. Without the hassle of reimbursement through cash, the system enabled rebates to be automatically credited into travelling cards when commuters made transfers. Rebates were introduced to offset subsequent boarding charges associated with bus-to-rail or bus-to-bus transfers.

Transit Link also played an essential role in the network integration of Singapore's public transport. It took on the role of central planning and coordination of the bus network before LTA took over as the central bus planner at the end of 2009.

On 30 April 2010, the LTA acquired Transit Link from the PTOs to enhance the integration and efficiency of public transport services. As the central bus network planner as well as the agency in charge of land transport matters, LTA would be in a better position to plan the public transport network from the commuter's perspective and improve the journey quality while balancing system costs.

Today, Transit Link provides integrated services and solutions to the authorities, the public transport operators and the Card Managers (EZ-Link and NETS). It plays an important role in Singapore's public transport system as the Transit Acquirer which provides integrated services and solutions to the PTC, LTA, the public transport operators and the Card Managers (EZ-Link and NETS). It processes transit transactions and apportions revenue to the public transport operators, provide card sales, refunds and replacements and top-ups of stored value smartcards for commuters.

Liberalisation of the Taxi Industry[e]

In 1998 and 2003, the government took steps to liberalise the taxi industry. The rationale for the liberalisation of the industry was to allow the forces of demand and supply to determine the pricing of taxi services. This would optimise the utilisation of the taxi fleet in order to improve supply and demand matching during peak hours and improve service standards. With deregulation, taxi companies had to carefully consider and individually decide on fare revisions based on their pricing strategy and supply-demand considerations. The move resulted in a more competitive market that proved beneficial to cabbies in terms of more employment choices and attractive terms.

The government deliberately adopted a phased approach for liberalisation instead of a big bang approach (to deregulate both fares and taxi supply simultaneously). This allowed taxi companies, taxi drivers and commuters some time to adjust to the deregulation of fares,[e] and gave the government time to monitor the impact of fare deregulation before the "next obvious step" to deregulate supply.[f]

In 1998, taxi fares were deregulated and taxi companies were allowed to set their own fares without having to seek the PTC's approval, provided that they gave advance notice of fare adjustments to PTC and the public.[g]

The second phase of liberalisation came in June 2003 with the lifting of controls on the number of taxi companies in the market and the fleet quota imposed on taxi companies. A new Taxi Operator Licence (TOL) framework was also established to set out how new taxi companies could be registered and ensure that taxi service standards would not be compromised with the increased competition.

(Continued)

---

[e] The deregulation of fares in September 1998 coincided with the revision of the vehicular tax structure and implementation of Electronic Road Pricing (ERP) which might impact taxi fares. Taxes for taxis were more closely aligned with those for private cars. The registration fee and additional registration fee rate for taxis were increased to that for private cars. However, road and diesel taxes for taxis were reduced. Taxis are subject to ERP charges, like other private and public transport vehicles using the roads. Commuters pay for ERP charges incurred on taxi trips.

[f] Kaur, Karamjit. Quota on Taxis May be Lifted. *The Straits Times.* 30 July 1998.

[g] PTC had earlier deregulated surcharges that taxi operators levy for phone bookings and premier services in April 1997.

2.6. *1990s — Applying more demand management levers*

The 1990s saw the completion of the first MRT network consisting of the 67 km radial North–South and East–West Lines. The government also ventured into the greater use of transport demand management policies that targeted both vehicle ownership and road usage charges.

(Continued)

However, the complaints of poor service and taxi availability especially during peak hours still persist. Today, there are six taxi companies and the size of the taxi fleet has increased by 46% since liberalisation in 2003. On the other hand, taxi ridership has increased by over 20% from 805,000 in 2003 to 967,000 daily passenger trips in 2013. Furthermore, the varied surcharges levied by different taxi companies result in complaints about confusion over taxi fares and complicated fare structures. Today, there are close to 10 different flag-down fares, three metered fare structures, more than 10 types of surcharges and eight kinds of phone booking charges.

To address these concerns, LTA conducted a review in 2012 and implemented several measures to improve service levels and taxi availability. This led to the revision of TOL requirements for all new taxi companies to equip their entire taxi fleet with the capability to identify taxis within the vicinity of the caller and assign call booking jobs to taxis in the vicinity, among others. The minimum fleet size requirement was also increased from 400 to 800 taxis to ensure that individual taxi companies have sufficient critical mass to provide an adequate level of service to the commuters

A new set of Taxi Availability (TA) standards were also introduced in January 2013 to increase the utilisation of the taxi fleet. It requires taxi operators to maintain a set percentage of their fleet on the roads during peak periods, and a percentage of taxis clocking in a minimum daily mileage of 250 km. Taxi companies need to meet TA requirements to expand their fleet and financial penalties will be imposed for taxi companies who do not meet standards.

2.6.1. *Extension of the ALS*

Initially, the ALS scheme was restricted to passenger cars and taxis with less than four occupants during the morning peak hours from 7.30 am to 9.30 am. All other vehicles, such as motorcycles and goods vehicles, were exempted.

As vehicle population and car ownership grew, several refinements were made over the years to better manage congestion. For example, the ALS was extended to cover all vehicles except emergency vehicles (e.g. ambulances, fire engines) and public buses in 1989. Car and taxi pools were no longer exempted. This was deemed to be more equitable as all vehicles contributed to the congestion in the area and all should therefore be subject to the same restraints. Restraint on all also meant a milder levy was sufficient. The ALS fees were thus reduced by 40% to S$3 per day and S$60 per month. The Restricted Hours were extended to include the evening peak hours between 4.30 pm and 7.00 pm due to worsening traffic conditions.

In 1994, the Restricted Hours were further extended to full days, with peak and off-peak differential pricing. The objective of the changes was to spread out traffic flow more evenly throughout the day, so that the roads were better utilized to reduce congestion. The Restricted Zone also expanded over the next 14 years from an area of 610 hectares to 725 hectares with the expansion of the CBD.

In 1995, manual road pricing scheme was extended to expressways. Known as the Road Pricing Scheme (RPS), it was meant as a pilot scheme to introduce the idea of road pricing at other congested points outside the Restricted Zone. It also served to familiarize motorists with passage pricing as opposed to area or cordon pricing of the ALS.

When the ALS was introduced, it had a dramatic effect on traffic entering the CBD. Traffic volume dropped by 44% initially, but it tapered to 31% drop by 1988. This was despite growth by a third in employment in the city and a 77% growth in vehicle population during the same period. The drop in traffic was caused by the decanting of motorists whose destinations were not the CBD but had merely been using the city roads as a bypass, as well as by those who changed their journey start time to avoid paying the ALS fees.

Despite various fiscal measures to curb growth in vehicle population, the growth rate of the car population was as high as 12% per annum between 1975 and 1990, before the recession of 1985. Taking into account the limited land available for development, it became clear that some stricter form of vehicle restraint was necessary.

Shortcomings of the Manual Schemes

Being manual schemes, the ALS and RPS had certain limitations. Firstly, they were labour intensive and extending the schemes to other points would have

Fig. 9.   ALS Coupons.

required even more people. Enforcement was also difficult because over the years, numerous types of licences were introduced bearing different shapes and colours for easier differentiation of different areas (e.g. on expressways), timings (peak/off-peak) and types of vehicles, throughout the day. (Fig. 9) This made the extension of and any changes to the ALS very difficult.

The ALS' flat licence fee for unlimited entry made it inequitable to motorists because the charges were not commensurate with the congestion caused (e.g. if a vehicle entered the Restricted Zone many times a day). Furthermore, there was always a rush to enter the Restricted Zone just before or after the restricted hours, resulting in short and sharp peaks of traffic volume entering the Restricted Zone. "Shoulder charging", or having intermediate rates, would smoothen out the peaks, but this was difficult to implement in a manual system. Having more categories of licences would make enforcement more difficult and more prone to mistakes.

2.6.2. *Electronic Road Pricing (ERP)*

With the shortcomings of the manual road pricing schemes, the search for a more efficient technology began in earnest before 1990. In July 1989, the Government announced that it would implement the ERP. Study missions were sent to USA and Europe where some form of electronic pricing had been implemented for road toll systems. However, although the relevant technology required for ERP had been developed, there were no ready off-the-shelf ERP systems which met Singapore's requirements to charge vehicles travelling in a multi-lane environment. Singapore's ERP system was therefore the first of its kind in the world.

The ERP system was finally implemented in 1998 after a prolonged period of technology trials and re-tenders. It was essentially an automated version of the ALS, with the flexibility to vary charges at different times and places according to traffic conditions. This provided for more efficient control of congestion, and presented the opportunity to use road pricing much more extensively as a traffic management tool. When the ERP system was implemented, it was to replace the manual road pricing systems like the ALS in the first instance, and then gradually to extend further to other areas of congestion.

The ERP system was designed to be simple to use. All that it required the user to do is to insert the smart-card with a stored value into the in-vehicle unit (IU). When a vehicle passes through the ERP gantry, the appropriate ERP charge is deducted from the smart-card. If there is insufficient cash in the smart-card or no smart-card in the IU, the enforcement cameras in the gantry would take a picture of the rear of the vehicle. The ERP was designed as a pay-as-you-use system, where motorists pay each time they pass a gantry. Charges are set in 30-minute blocks and the rate for each block at each location is adjusted periodically based on the traffic conditions in the vicinity of the road-pricing point for each time period. This is a more refined and equitable form of usage charging than the ALS, which had encouraged a buffet syndrome by allowing a motorist to make as many trips as he or she wanted using the Area Licence for the day.

Packaging the Introduction of ERP with Vehicle Tax Rationalisation

When the ERP was implemented in 1998, the government rationalised the vehicle tax structure to reflect the shift towards higher reliance on usage measures to control traffic congestion. This was achieved by reducing upfront vehicle taxes such as the registration fee, Additional Registration Fee (ARF) and the road taxes of vehicles. The changes also aimed for a more consistent vehicle tax structure, based on road space occupied (and hence contribution to congestion) after giving proper regard to social and equity considerations. With the introduction of ERP, the government sought to strike a good balance between ownership and usage restraint measures.

Another objective of the vehicle tax changes was to win over public confidence and assure the public there was no fiscal agenda in the introduction of the ERP, such as revenue objectives. Therefore, the government bore the entire upfront development and implementation cost, and on top of providing tax incentives, the IUs (including installation) were provided free to owners of all existing vehicles.

All vehicle owners benefited from the reduction of vehicle ownership taxes. In addition, a package of road tax rebates was also given to them for a period of five years. Together with the reduction in ownership taxes, these measures significantly eased the introduction of ERP by helping motorists adjust to the new usage-based ERP charges.

Since the implementation of ERP, the government has continued to lower the ARF for cars, in line with the policy to lower upfront taxes as the government moved to a more usage-based congestion management regime. The ARF today is 100% of Open Market Value (OMV), down from 140% when ERP was implemented in 1998.

Managing the Impact on Businesses

In implementing the ERP, the government also took into consideration the potential impact on businesses. Therefore, in addition to the tax rationalisation and rebate measures, the ERP charges for commercial vehicles, i.e., buses and goods vehicles, were phased in over a period of four years, while that for taxis were phased in over a period of three years. These measures served to enable businesses to better adjust to the usage-based charging system.

2.6.3. Managing vehicle ownership

From the outset, the government has used fiscal disincentives to control vehicle ownership and usage. An excise duty (ED) is levied by Customs on each vehicle imported into the country. Owners must pay a registration fee (RF) and additional registration fee (ARF) to register the vehicle in addition to an annual road tax.

Over the years, heavier restraints were imposed primarily on cars due to the strong demand for cars. From 1975, excise duty, ARF and road taxes were raised periodically. This worked successfully for many years. From 1975 to 1989, the car population grew on average by 4.4% per annum, or by a cumulative total of 80% from 141,875 to 258,537 vehicles.

The Vehicle Quota System (VQS)

As the economy recovered from the economic recession of the mid-1980s, the growth rate of cars picked up and reached 8.2% in 1989. At this rate, it was projected that the car population would double to 570,000 in nine years. In August 1989, a Parliamentary Select Committee was formed to examine the need for measures to curb road usage, to review existing government policies on the measures and to make recommendations. The Committee held public hearings to hear views of members of the public and transport experts. The Committee eventually recommended a quota system as a long-term solution to the problem of rapid growth in the car population.

The Ministry of Communications accepted the proposal and in May 1990, implemented the VQS to control vehicle population. As the VQS had no precedent, it started out as a simple system. The objective was to allow vehicle population to grow in tandem with road capacity. This system controls the growth of the vehicle population at a sustainable rate. The key features of the system are:

a. **Universality**: the scheme applies to all vehicles except scheduled buses, school buses and emergency vehicles such as ambulances and fire engines. Including almost all vehicles under the scheme avoids distorting economic behaviour, e.g. buying a minibus instead of a car to avoid paying for a COE.

b. **Rationing by market-based pricing**: anyone who intends to register a vehicle must first bid for a Certificate of Entitlement (COE) in an open tender. The tender method was chosen because pricing is a more efficient way (than alternatives such as balloting or queuing) to ration a scarce resource, in this case, the right to own a car. Successful bidders need only to pay the lowest successful bid price.

c. **Time-limited rights**: The COE is valid for 10 years. Beyond that, the owner must either de-register his vehicle, or renew the COE for another 10 years by paying the prevailing quota premium. If he de-registers his vehicle's COE, LTA will then recycle his COE and put it up for bidding by new buyers. This ensures that nobody holds perpetual rights to the ownership of a vehicle.

d. **Separate COE categories[h]**: With the merger of the small (1000cc & below) and medium (1001–1600cc) car categories, as well as the big (1601–2000cc) and luxury (2001cc & above) car categories, there are currently five categories for the COE quota: Category A (cars up to 1600 cc & 97 kW and taxis), Category B (cars above 1600cc or 97kW and above), Category C (goods vehicles and buses), Category D (motorcycles) and Category E (open). To provide flexibility to cope with higher demand in some categories, there are open COEs (Cat E) which can be used to register any type of vehicle.

---

[h] Based on cylinder capacity (cc) and kilowatts (kW).

In managing private transport demand, the VQS and ERP illustrate how Singapore has recognised the potential of market mechanisms to allocate scarce resources, i.e., road space. By using auctions, the VQS allocates the right to own a vehicle in an economically efficient manner. Car owners who value the ownership of vehicles more are willing to pay higher prices for the COEs. Similarly, road pricing mechanisms allocate road space to motorists who are willing to pay for using it. ERP charges and fuel taxes have the added benefit of promoting sustainability based on the principle that the polluter pays for the pollution he causes.

2.6.4. *Towards a more integrated approach in land transport*

Formation of a Single Integrated Land Transport Agency, LTA

The mid-1990s welcomed a new era of integrated transport management with the merger of all land transport functions, both private and public transport into a single integrated land transport agency. In 1995, the Land Transport Authority of Singapore (LTA) was formed with the merger of (i) the Roads and Transport Department of the PWD which was responsible for the planning, design, construction and maintenance of roads, pedestrian and commuter facilities; (ii) the Mass Rapid Transit Corporation (MRTC) which was responsible for the planning, building and regulation of the MRT system; (iii) the Land Transport Division of the Ministry of Communications (MINCOM) which was in charge of land transport strategies and policies; and (iv) the Registry of Vehicles (ROV), which was in charge of the administration, regulation and enforcement of land transport policies and rules.

At the macro level, the Concept Plan reflected the need to integrate the land-use planning and major transport infrastructures, as well as to safeguard transport corridors. At the micro level, land-use integration can only be achieved if there is advanced planning for high density commercial/residential developments at and around major transport nodes such as MRT stations and bus interchanges. This helps to facilitate seamless travel by commuters and encourage public transport usage. Commuters can walk through the inter-changes in greater comfort and convenience.

The formation of LTA was intended to facilitate land-use and transport integration at both the macro and micro-level. It heralded the official shift in its approach to look at land transport issues in totality by the integration of operations in both private and public transport. This also ensured that the policy for land-use for transport purposes could be rationalised in a more holistic fashion, i.e., road space for buses *vis-à-vis* road space for car.

A White Paper for a World Class Land Transport System, 1996

In 1996, LTA came up with a landmark document, the White Paper for a World Class Land Transport System which spelt out the transport vision of Singapore and the operating philosophy of the vision. It introduced the current rail financing framework that allowed the building of the North East Line (NEL), the first line to be constructed after the main MRT network was completed in 1990. The NEL was built to serve commuters from Harbour Front in the south through the Serangoon corridor to Sengkang and Punggol in the northeast.

The rail financing framework introduced in 1996 set out a new funding criterion for the rail system: the second set of operating assets would be paid through a combination of farebox revenue covering the historical cost of the first set of operating assets and government co-financing for the balance, i.e., the inflationary component. In this way, each generation would pay for the operating assets they consume (Lew and Choy, 2009).

The adoption of this framework allowed for the construction of the NEL, which would not have attracted enough ridership during the initial years to become viable, albeit with the caveat of a need for fare premium. The NEL began its constructions in 1997 and was opened in 2003. Other subsequent MRT lines, the Circle Line, the Changi Airport Extension, and the Boon Lay Extension, were also justified on the same basis.

Apart from the adoption of the rail financing framework, the White Paper crystallised the concepts of transport planning that was already in practice by Singapore planners since the late 1980s. It underlined the importance of integrating land use, town and transport planning, as well as managing the demand of road usage through ownership measures such as the VQS and usage measures such as the ALS.

The White Paper presented a more integrated and holistic approach towards land transport governance in Singapore which involved the four major policies of:

- Integrating land use, town and transport planning so as to reduce the need for people to travel and at the same time, maximise the accessibility to public transport services.
- Developing a comprehensive road network that takes into consideration practical concerns and economic principles, as well as applying technology to maximise its use.
- Managing demand of road usage through ownership and usage measures.
- Improving public transport to attract more commuters to use public transport.

2.6.5 *Harnessing Technology to Maximise Capacity*

In addition to spelling out initiatives to develop a more comprehensive road network, the White Paper highlighted the need to harness technology to maximise network capacity and optimise the available road space. This led to the adoption of the Intelligent Transport Systems (ITS) to improve the operational efficiency and road safety of Singapore's road network. More components were also added to the Green Link Determining (GLIDE) system that computerised traffic signal control, first implemented in 1988.

Intelligent Transport Systems to Maximise Road Capacity

To maximise road network capacity, LTA continued to invest in various innovative ITS. The Expressway Monitoring and Advisory System (EMAS) launched in 1998, for example, performs live-video traffic surveillance, incident detection and traffic advisory functions. EMAS also informs motorists of prevailing traffic conditions through strategically located electronic signboards and radio broadcasting. ITS also provides motorists with timely traffic information, thus enabling them to plan their journeys more effectively. To illustrate, the GPS-enabled Traffic Scan system implemented in 1999, collects information on traffic speeds on major roads by relying on taxis equipped with GPS readers which capture information on their locations and speeds. This information would then be disseminated to motorists, therefore helping them decide the best routes to their destinations. Other ITS devices that provide timely quality information to motorists are the Junction-eyes (J-Eyes), i-Transport System, Traffic Prediction Analytics and the Parking Guidance System.

Much was also done to provide greater convenience to motorists and public transport users in the form of e-services. LTA's One.Motoring Portal launched in 2001 provides a one-stop channel for the public to access motoring-related information anywhere and anytime. Among other things, the portal highlights latest news releases from LTA as well as real-time traffic news. The launch of e-Services@One.Motoring makes available a series of self-help online services, such as the renewal of road tax, payment of vehicle transfer fees and bidding for vehicle registration numbers, thereby reducing the need for the public to physically queue at LTA premises (*Lew and Ang, 2010*).

Applying technology for making more informed travel decisions

To enhance the travel experience, people need accurate information to plan their journeys. LTA has been working to provide real-time travel information for commuters, motorists and cyclists. This has manifested in the form of

MyTransport.SG, a portal that groups transport information to better serve travel profiles of the public transport user, students, cyclists, motorists and commuters with special mobility needs such as wheelchair-accessible buses, which was launched in 2010.

MytTransport.SG has a website and mobile application format that Smartphone users can access on-the-go. This includes:

- Real-time bus arrival information
- Search and locate nearby bus stops, bus services, metro stations and taxi stands using location based services
- Real-time traffic news updates & live traffic images
- Receive notification on expressways traffic news during peak hours and train service delay information
- Explore cycling towns, routes, and parking facilities

On top of the information for the commuting and motoring public, third-party developers can tap on Singapore transport, traffic and geospatial data available in the portal's 'Data Mall' section. The objective is to make it easier for developers to incorporate such information into their product and services for the benefit of all commuters and motorists. Today, over 20 mobile applications are available for download under the 'Apps Zone' section of the portal. Applications available include travel planning tools, parking availability, and viewing of real-time, island-wide images of traffic conditions.

Applying technology in making better planning decisions

Data analytics plays a key role in LTA's efforts to meet the ever-growing challenges and aspirations of the people. The core of LTA's data analytics capability is Planning for Land TrAnsport NETwork (PLANET), an enterprise data warehouse project. PLANET was developed with information from Public Transport, Traffic, Vehicle and Geographical Information System (GIS) systems. First rolled out in July 2010, PLANET enables LTA to analyse fare-card data and provide a better understanding of travel patterns to enable effective introduction of public transport improvement measures.

2.7. *2000s to present*

In recent years, the focus of land transport governance has shifted from providing quality transport infrastructure to providing quality transport experience. Given Singapore's land constraints, promoting public transport as the main mode of travel became a priority in light of increasing travel demand and population growth. This is evident in the posturing of the Land Transport Master Plan 2008 and 2013.

Apart from doubling the rail network, increasing public bus capacity and bus routes, the two master plans reflect the government's commitment to building a more comprehensive transport network as well as taking a greater hand in managing public transport. This is reflected by the government's adoption of the new rail financing framework in 2010, and the transition towards a Government Contracting Model for the bus industry by 2016, where all operating assets and infrastructure will be owned by the government and leased to operators to ensure sufficient capacity and service standards to meet future needs.

2.7.1. A new transport strategy, the land transport master plan

In 2008, LTA updated its land transport strategy with the launch of the Land Transport Master Plan (LTMP) 2008. The LTMP shifted the focus of land transport towards building a more 'people-centred' land transport system that will meet the diverse needs of an inclusive, liveable and vibrant global city. This was in response to the challenges faced by the transport system in Singapore: increasing travel demand, limited land, declining public transport modal share, changing demographics and rising expectations of the public. The three strategies to attain this goal are firstly to make public transport a choice mode for commuters; secondly to manage road usage; and lastly to meet the diverse needs of the people.

The "New Normal"

The face of governance in Singapore has changed vastly since 2011. The hotly-contested General Elections in May 2011 and the Presidential Elections of August 2011 "are indications that a more sophisticated and vocal citizenry desires not merely to be governed, but to be heard, informed and engaged, and to participate in the business of the nation."[i]

Prior to the May 2011 General Election, much debate was generated by the public regarding the organization of the public transport and how to improve services. December 2011 also saw two major rail service disruptions which led to the government calling a Committee of Inquiry (COI) to investigate the causes of the disruption. Coupled with the public's general dissatisfaction with increasingly crowded public transport services, this led to a dent in public sentiment over government's credibility to provide reliable services.

To address these new challenges in Singapore's land transport system, LTA undertook a review of the LTMP 2008 and consulted the public to gather views

---

[i] Lena Leong, *Developing Our Approach to Public Engagement*, Ethos, Issue 10, October 2011, Civil Service College.

on how to improve their travel experience. This resulted in a revised master plan, the LTMP 2013 (Fig. 10), which reaffirms Singapore's commitment to achieve its long term vision of a 'People-centred land transport system', and recognises the need to put the commuter at the heart of the land transport system and to listen to their views.

The LTMP 2013 puts a greater emphasis on improving public transport and reducing reliance on private transport. Some measures under LTMP 2013 include:

- The rail network will be expanded so that eight in 10 households will be within a 10-minute walk from a train station. This means that one MRT line or extension will be opened each year from 2013 till 2025, and a 360 km rail network to be ready by 2030. To shorten waiting times, the signalling systems of the existing North–South and East–West lines will be upgraded so that trains can run at 100-second intervals instead of the current 120-seconds. More trains will also be added to reduce crowding (Table 2). Together with the new rail lines, additional trains and upgraded signalling system, our peak hour capacity will be increased by 110% for travel into the city.

Source: Land Transport Master Plan 2013 Report, Land Transport Authority.

Fig. 10.   Areas to be addressed by the Land Transport Master Plan 2013.

- Public bus capacity will also be increased with the introduction of 80 new bus services, and over 1,000 buses will be under the Bus Service Enhancement Programme (BSEP) by end of 2017.[j] In addition, greater priority to buses will be given on the roads so that commuters can travel faster to their destinations on public transport.
- More than 200 km of sheltered walkways, over 700 km of island-wide cycling path networks and more cycling infrastructure and facilities will be built to facilitate walking and cycling trips. More integrated transport hubs will also be built to enable commuters to switch between different types of transport easily, on top of convenient access to retail, dining and other lifestyle services.

2.7.2. *Central bus network planning*

In making public transport a choice mode, Singapore recognised the importance of enhancing the integration and efficiency of its public transport services. In the past, public transport operators planned the routes for their bus services which had to be approved by the PTC, while Transit Link coordinated the central planning of the bus network. LTA took over the role of central bus planner at the end of 2009 with the aim of having a clear overview of the entire transport network and the needs of the commuters. The focus was to identify service gaps and optimising bus services with the new rail lines being rolled out. It also paved the way for the government to move towards bus contracting so as to increase contestability in the industry.

Table 2. Upcoming Train purchases for Rail Lines.

Rail	Current no. of train/train-cars	No. of additional trains	Year	Percentage increase from today's train fleet size
North-South and East-West Lines	128 trains	13 trains 28 trains	In 2014 From 2016	30%
North East Line	25 trains	18 trains	From 2015	70%
Circle Line	40 trains	24 trains	From 2015	60%
Bukit Panjang LRT	19 trains	13 trains-cars	From 2014	70%
Sengkang and Punggol LRTs	41 trains	16 trains-cars	From 2016	40%

Source: LTMP 2013.

---

[j] On 11 March 2014, LTA announced that the doubling of Government funded BSEP buses from 550 to 1,000. Together with the buses injected by the PTOs, as well as the City Direct Services (CDS) and Peak Period Short Services (PPSS) run by private bus operators, it will increase Singapore's total bus fleet available for public bus services by 35%.

In addition, LTA as the central bus network planner would enable better optimisation of resources in the bus industry. In Singapore, bus service rationalization is undertaken whenever a major train line is introduced. This was done by Transit Link to ensure that bus services are well-integrated with new train lines and reduce duplication of the rail system. This began when the North–South and East–West lines commenced in the late 1980s and early 1990s and later when the Woodlands extension and North–East line began operations. Since its formation, LTA has played the role of the central transport network planner as it is in charge of all rail, bus, and road networks as well as commuter facilities like bus stops, pedestrian walkways and integrated transport hubs. This would facilitate a more holistic approach to planning the bus networks, taking into consideration the rail networks comprising the LRT and MRT systems and other transport infrastructure, to focus on the 'total journey' experience. It is therefore a natural extension for LTA to plan the bus routes as well.

Bus Service Enhancement Programme

In recent years, there has been much public dissatisfaction over the crowdedness and reliability of rail and bus services in Singapore. The significant increase in population, without an in-tandem expansion of public transport capacity, is deemed as the main cause. Commuters' main grouses for bus services were the long waiting times and overcrowding on buses during peak hours.

In September 2012, the government announced that it was establishing a Bus Service Enhancement Programme (BSEP) to expand the bus fleet and to help the public transport operators with running costs of the new buses. Under the BSEP, the government has partnered the bus operators to grow the bus fleet by 35% within five years, with the government contributing 1,000 buses by 2017. This would help to increase public bus capacity, which would translate to less crowding and shorter waiting times for buses. In 2012, the government also announced that it would review and enhance structural assistance to the bus industry.

The BSEP would be financed by the government through a S$1.1 billion Bus Service Enhancement Fund (BSEF). While the decision received much criticism from the public at the start, the government made a decisive move and went ahead to improve bus services for commuters, ahead of the roll-out of new rail lines and capacity enhancements to existing lines in the next few years. It was also recognized that while both SMRT and SBS Transit reported yearly overall profits (through revenues from its other businesses such as rail and advertising), they had been suffering operating losses in their public bus operations for several years; and would take about 20 years to increase bus capacity by a similar size without government support.

2.7.3 *Increasing contestability of public transport industry to improve services*

Transition to a Government Contracting Model for the Public Bus Industry

In a bid to strengthen the government's ability to respond more quickly to changes in travel demand and service level expectations, the government announced the restructuring of the bus industry from the privatised model to a "Government Contracting Model" on 21 May 2014. Bus contracting will be implemented in phases over several years to ensure a smooth transition for all stakeholders.

Under the new model, LTA will contract operators to run bus services through a competitive tendering process. LTA will determine the bus services to be provided and the service standards, and bus operators will bid for the right to operate these services. Operators who are awarded the contracts will be paid the fees to operate the services, while fare revenue will be retained by the government.

The government will also own all bus infrastructures such as depots, as well as operating assets such as buses and the fleet management system. The existing bus financing framework[k] was reviewed to ensure that operators would remain solvent in light of the changes to bus policies. In addition to waiving COEs and the requirement for public buses to pay the 5% Additional Registration Fee (ARF), the government will take over the leasing and land-related costs for all bus depots and bus parks.

Additionally, a portion of the revenue currently received by the government from bus shelter advertisements will be channelled to the operators. These measures will help operators to cope with rising costs associated with bus operations. This will lower the barriers of entry to the market, attract more bus operators, and facilitate any transitions from incumbent operators to new operators. This brings the bus financing framework in alignment to the rail financing framework.

In recognition of the challenges of retaining and hiring bus industry workers, the government also works closely with the National Transport Workers Union (NTWU) and bus operators, to help safeguard the job security and welfare of bus industry workers. This is to ensure that there is sufficient manpower to run the bus industry.

By introducing more contestability to the bus industry, it is envisioned that the new model will also promote greater efficiency among operators as they now have to compete for the right to run the services. Over time, these will lead to the provision of better bus services in a cost-competitive manner.

---

[k]For Bus Service Operator Licences expiring on 31 August 2016.

Commuters stand to benefit from more responsive bus services and higher service levels under the new model.

New Rail Financing Framework

In 2010, a new rail financing framework was put in place to facilitate the future expansion of the rail network in a financially sustainable manner; and inject greater contestability into the rail industry. Firstly, the license period for new rail lines starting with the Downtown Line was shortened from the existing licenses of 30–40 years to about 15 years. A shorter licence enhances the level of contestability as the operator faces the prospect of competition at the end of its licence term. It also allows LTA to refresh licence conditions at the end of each licence term to take into account changes to the operating and business environment.

Secondly, the government instead of the operator, will own the rail operating assets to allow the government to make the decisions on replacing existing trains and operating assets, as well as investing in new trains and operating assets to keep pace with growing ridership demand. Under this new arrangement, operators will be required to follow a rigorous set of Asset Management Requirements (AMR) for maintaining the assets. The takeover of ownership of operating assets also represents a move to reduce the barriers to entry into the rail industry and facilitates the injection of greater contestability in the market, thus allowing operators to concentrate on the provision of rail services.

Lastly, the new framework enables future lines to be introduced as long as the rail network as a whole is viable. Prior to the adoption of the new framework, new rail lines were evaluated on a line-by-line basis and had to be financially viable on its own before it could be implemented. The rationale for adopting this new method of assessment is to adopt a more holistic approach towards accounting the benefits that will accrue to the existing line as a result of the new line. The new framework takes into account the "network effect" of existing bus commuters and private transport owners who reside in less densely populated areas, switching to rail due to the travel time savings arising from using the new rail lines which are connected to the existing lines.

As the RTS coverage becomes more comprehensive over time, the network effect will become more significant as new lines add to the ridership of existing lines. Thus, new lines will be assessed based on the financial viability of the overall network, and not just that of the new line under this framework. This is reflective of Singapore's long-term view and responsiveness in refining existing policies to meet changing demands. There is a heightened awareness of the need to accelerate the build-up of capacity in light of unexpected increases in travel demand. The Downtown Line and the 43-kilometre long Thomson-East Coast Line which, when completed in 2024 will be Singapore's sixth line, were justified according to the new framework.

2.7.4 Fare Regulation in Singapore

Public transport services are provided on a commercial basis, within the maximum fares approved by the PTC. To keep public transport fares affordable to the general public, public transport infrastructures such as MRT/LRT lines and bus interchanges are funded entirely by the Government. However, no direct subsidies are provided for public transport operations.

In regulating bus and train fares, the PTC carries out its statutory mandate to safeguard public interest by keeping fares affordable while ensuring the long-term financial viability of the public transport operators.

In 1998, the PTC introduced a formula (also known as the Fare Review Mechanism) to calculate the annual transport fare adjustment cap. This fare cap calculated, in percentage, would be the maximum allowable increase for public transport fares for that year. Nevertheless, these fare adjustments are not automatic in nature and operators have to submit applications for fare adjustments annually which are subject to the PTC's deliberation.

Over the years, the fare review mechanism been reviewed several times to better determine the appropriate price index and level of productivity extraction to share the productivity gains with commuters. The current formula adopted follows the recommendation of the 2013 Fare Review Mechanism Committee which takes into account the changes in the Wage Index, Energy Index and Core CPI — which excludes items that are not relevant to public transport. The factor "X" was retained as productivity extraction based on an equal sharing of the operators' past average annual productivity gains. Currently it is set at 0.5%, valid for five years from 2013 to 2017.

Targeted help for the lower income

Targeted help is given for the lower income group to address concerns with affordability of public transport. Needy families can apply for public transport vouchers from the local community centres. These vouchers are funded by the Public Transport Fund which was set up in 2006. As part of the fare review exercises, operators are required to contribute a portion of the increased fare revenue to the fund to help those affected by fare increases. Over 250,000 Public Transport Vouchers worth $30 each were distributed to low-income households from the proceeds of this Fund in 2014. Apart from this, operators also offer concessionary fares for students, National Servicemen, low-wage workers and persons with disabilities.

Distance-based through-fares

Public transport also has to be integrated, reliable and efficient for it to be a mode of choice for commuters. Singapore's public transport system is based on a hub and-spoke design. This is much more efficient than having many direct services as this model allows for more frequent and faster services to bring commuters to a transport hub and then onwards to their destinations, with the same amount of resources. However, transfers are an integral part of our hub-and-spoke system and they have to be made seamless for public transport to be attractive.

In July 2010, the distance-based through-fare structure was implemented to better integrate the fare system and facilitate seamless transfers between different modes and operators. In the past, commuters were levied a fare penalty when transferring between buses and/or rail rides serviced by different public transport operators. Under the Distance Fare scheme, all commuters are charged a fare based on their total distance travelled, regardless of whether they travelled by bus or train.

The new scheme brings about a more integrated fare structure that allows commuters to make transfers without incurring additional costs. Fares are computed on a journey basis, without a boarding charge being imposed for every transfer trip that makes up the journey so as to improve connectivity and facilitate transfers. Gradual steps were taken to prepare for the fare restructuring so as to minimize the impact on commuters. This was done by reducing the transfer fare penalty progressively over two years in 2008 and 2009. Figure 11 shows that commuters (in particular the 2nd quintile income group) are spending less on public transport as a proportion of income, even as incomes increases.

Before the changes, commuters making transfers were cross-subsidizing those who take direct journeys. For the same distance travelled, they paid more than commuters making direct journeys. The distance fares thus provides a fairer system where commuters travelling the same distance on the same transport mode will pay the same fare, regardless of whether or not transfers are made.

The distance fares system also helps to integrate the two public transport modes in our hub-and-spoke system. It creates a more seamless journey for

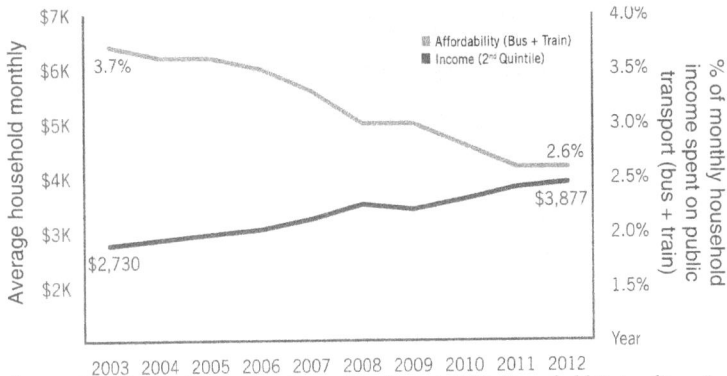

Source: Fare Review Mechanism Committee Report 2013, Household Expenditure Survey 2008, Department of Statistics.

Fig. 11.   Income level and fare affordability of the 2nd quintile (21st–40th percentile) income group households have improved.

commuters by providing more choices and greater flexibility in choosing routes to their destination.

The shift towards distance-based through fares represents a long-term approach to enhance integration of the public transport system. It facilitates transfers by enabling commuters to make seamless journeys without having to incur transfer penalties. Coupled with infrastructure development and improvements to public transport service levels, the new fare structure being more equitable in nature will help to maintain reasonably priced public transport and attract more ridership in future.

2.7.5 Meeting diverse needs and managing travel demand

In meeting the diverse needs of the society, the first approach was to ensure physical accessibility for all, in particular the less mobile such as the elderly and handicapped with prams, as well as cyclists. LTA has been investing in accessibility for many years by purchasing modern vehicles and upgraded public transport facilities to improve access to mobility and public transport, including retrofitting lifts in rail stations, replacing the public bus fleet with low-floor wheelchair accessible buses and providing barrier free access to all public transport nodes. Lifts will also be installed to another 40 pedestrian overhead bridges (POBs) to improve walking access between homes and amenities, especially in areas located near MRT stations, bus hubs, health instructions, and schools for special-needs children.

Today, all MRT stations have barrier free access, more than 80% of the MRT stations have at least two barrier free access routes; 40% of the existing stations are fitted with extra lifts to improve accessibility for less mobile commuters like the elderly; 95% of the pedestrian walkways, taxi and bus shelters are barrier free; 86% of the pedestrian overhead bridges are fitted with shelters; and about 85% of the public buses are wheelchair accessible, with a plan to achieve 100% coverage by 2020.

Facilitating Walking and Cycling

When the first MRT line was opened in 1987, cyclists were observed to cycle to and from these stations daily. In 1991, the authorities constructed between 20 and 80 bicycle parking stands per station at 24 MRT stations. By 1997, there were 869 bicycle stands at 38 MRT stations (*Koh and Wong, 2012*). This reflects the popularity of cycling as the first-mile and last-mile connection between homes and transport nodes.

Given the increasing popularity of cycling and its potential as a sustainable travel alternative especially for short trips, the government is committed to facilitate this by building intra-town cycling infrastructure and better bicycle parking facilities around rail stations and bus interchanges, as well as at Housing Development Board (HDB) blocks, amenities, schools and other places where there is demand.

A National Cycling Plan was also developed to build cycling routes for recreational and short commuting purposes. The aim is to provide all housing towns across Singapore with comprehensive intra-town cycling networks for residents to cycle to and from MRT stations and neighbourhood centres. By 2030, we aim to have a comprehensive island-wide cycling path network reaching over 700 km long that will link intra-town and inter-town routes, the Park Connector Network (PCN) and round-island routes to allow people to cycle safely and comfortably within and between towns.

In addition to this, the Walk2Ride Scheme was launched in 2013 to expand the existing sheltered walkway network. It is aimed at improving the walking experience and to provide alternatives to motorized trips between transport nodes such as MRT stations, and to trip-generating hubs, such as schools, healthcare facilities, public amenities, offices and residential developments within 400m of the stations. Over S$330 million has been committed to this. By 2018, the network will increase more than four times from current 46 km to about 200 km.

Influencing Travel Behaviour

While building up the capacity of our MRT system, the government has also been exploring measures to manage commuter demand with the focus on shifting travelling commuters to off-peak periods. 'Travel Smart' is a broad initiative to influence travel behaviour and reduce travel demand altogether.

Employers play a pivotal role in enabling shifts in the travel patterns of their employees. Travel Smart started in October 2012, with 12 companies participating in a pilot to implement various initiatives such as telecommuting and allowing staff to work from home before travelling to the office after the morning peak period.

The pilot scheme has shown encouraging results and enabled more employees of the pilot companies to shift their travel patterns (See Chapter 3 on Travel Demand Management).

In July 2014, LTA announced that it would partner organisations with over 200 employees in major employment centres located near MRT stations to help fund the flexible travel plans companies developed for staff. Companies can receive as much as just over half a million in funding from the LTA over three years.

Travel Smart also includes free pre-peak period travel in the morning as well as a Travel Rewards Scheme, formerly known as the 'Incentives for Singapore's Commuters' (INSINC) study first launched in August 2012. The scheme rewards commuters who travel on public transport, especially during the off-peak periods. Apart from spreading out peak hour travel demand, Travel Smart aims to encourage a switch to more sustainable modes of travel such as public transport, walking and cycling, car-pooling, car sharing, or to reduce travel demand altogether.

2.8. *Future of land transport: A car-lite Singapore*

The future of transport in Singapore will be a greener one with greater emphasis on more sustainable transport modes such as public transport and active mobility. Firstly, with the addition of new and upcoming MRT lines, eight in ten households will be within a 10-minute walk from a train station by 2030.

Apart from expanding the cycling path network, new innovative features and designs will be piloted in towns, starting with Ang Mo Kio, to provide a better walking and cycling environment. The Walk2Ride programme will expand the sheltered walk-way network linking transport nodes to homes and key amenities, making it more convenient for people to walk to their destinations.

More car-free zones in the city will also be created through road closures for public activities like at Haji Lane and Ann Siang Road. New HDB towns in the future will be transit-oriented and designed with less roads and fewer car parks. Walking, cycling and perhaps, even the use of Personal Mobility Devices (PMDs) will be the main feeder modes to the public transport nodes and other amenities in the town or precinct level. To allow people convenient access to car use without having to own one, more car sharing schemes will appear in the future, together with autonomous vehicles to cater to the first mile–last mile legs of a journey.

Parking policy has been under-utilised as a demand management tool since the implementation of ERP in 1998. More can be done to understand motorist behaviors towards parking management tools and how they would supplement the two prime levers of VQS and ERP. Parking policy will thus need to be enhanced with the view to restrain demand as well as to restrict supply in order to further discourage car usage.

More research is also needed to better understand commuters' travel motivation and their mode and time choices, especially that of the future where the economic structure, demographics and social setting will be very different from those of the present and past generations. The Singapore economy will be more knowledge-based and the free lancing workforce in the informal sector may be considerable. The population will be ageing with a higher proportion of less mobile individuals who still want to make the same trips on the transport system.

Socially, income disparity and dissenting voices will be more significant. The concept-planning and master-planning will have to take all these trends into account so that we can plan for and thus shape the future needs in time. At the strategic level, employment and services must continue to be brought closer to homes through decentralization. At the detailed planning level, compatible land uses should be co-located within the same precinct to minimize the need to travel. Urban structures and transport systems must in combination in order to nudge residents to select sustainable travel options. At the building and development control level, building designers and owners must henceforth play a larger role in transport planning and management for the occupants to minimize external impact on the surroundings. They should adopt a life cycle approach towards ensuring sustainability of the building operations. For example, a building should have excellent connectivity to public transport nodes through walkways, cycling tracks and shuttle buses; it should provide facilities to encourage active mobility like bicycle parks and shower rooms; and it should discourage private car use by limiting cheap parking spaces.

3. Principles & Strategies

3.1 *From 'world-class' to 'people-centred'*

To the transport planners in Singapore, like their counterparts around the world, they walk a tightrope of balancing competing objectives and optimising outcomes amidst constraints. The constraints faced by Singapore are perhaps a tad more severe than in other cities because of its limited size and rapid pace of development.

The Singapore of the 1960s had few roads, totaling less than 1,000 km in length. Since then, Singapore's total population has more than doubled and its vehicle population almost quadrupled. Its transport system today comprises a comprehensive network of more than 3,000 km of roads and highways and a high-quality public transport system. About 12% of the total usable land area is taken up by roads alone, close to the 15% by housing. The number of daily motorised trips has grown exponentially from a mere 2.7 million in 1981 to over 12.5 million today.

Yet, Singapore has avoided the many negative consequences of rapid urbanisation and rising societal affluence typical in major cities, such as traffic congestion and air pollution. Its land transport system has received many accolades from international transport institutions and experts. For example, the International Association of Public Transport (UITP) rated Singapore's public transport system amongst the best in the world, together with Helsinki and Vienna (UITP 2006). Its transport infrastructure consistently ranks among the top 10 in the World Economic Forum's Global Competitiveness Report. Hans Rat, the former Secretary-General of the UITP,[1] once said "If public transport is a religion, then Singapore must be a holy city".

Today, motorists in Singapore enjoy one of the highest urban traffic speeds of 25 km/h. The average speeds on its expressways exceed 45 km/h even during the peak hours. About 66% of public transport commuters are able to reach their destinations within 60 minutes from the time they leave their homes. Accessibility to public transport is universal and fares are one of the lowest in the developed world, even though there is no direct state subsidy on operating cost. Singapore is also one of three cities in the world where public transport fare-box revenue exceeds operating costs (*UITP, 2006*).

·In 2014, Singapore ranked overall third among the 30 cities studied by PricewaterhouseCoopers (PwC) in the sixth edition of its Cities of

---

[1]Union Internationale des Transports Public (International Association of Public Transport).

Opportunity report. Singapore ranked highest among Asian cities, and is one of only four cities, along with London, Stockholm and Sydney, to finish first in more than one indicator, taking the top spot in the ease of doing business and transport and infrastructure categories. In terms of transport alone, Singapore repeated its top ranking for the quality of its transport and infrastructure. The PwC report states that "Singapore blazes the trail in urban mobility" scoring third overall in terms of public transport systems. In addition, the World Economic Forum (WEF) ranked Singapore as the second most competitive economy in the world in the same year. One of the key factors attributing to the competitiveness and productivity of the economy is that it possesses world-class infrastructure. Singapore was ranked second overall in terms of infrastructure; it is world-class with excellent roads, ports and air transport facilities.

3.2. *Five principles for urban governance*

The story of land transport in Singapore over the past 50 years reads like those of the other aspects of urban development in the city state. Its governance is underpinned by the very same principles that the Singapore government has applied in infrastructure sectors like airport, seaport, water, electricity and telecommunications. The transport policies have sound theoretical foundations, and are quintessentially textbook solutions that transport academics and professionals would love to cite for examples. More importantly, Singapore has been blessed with the political will and the resources to implement those policies, often hard-nosed ones, which other developed cities could only wish to emulate for themselves.

The authors would argue that Singapore's achievements are attributed to five principles of governance generally and four urban transport strategies specifically. However, they would also admit that they do not have the final word on this, nor are those principles or strategies exhaustive. Many writers have quoted the same and other contributing factors to the success, but scholars and practitioners alike would agree on those mentioned below in this section.

a.   Strong Governance Structure

The principle of strong governance structure is a *de rigeur* in good urban governance. It is present in every major city around the world that has been recognised as economically and culturally vibrant, environmentally and socially sustainable, and providing safe and secure living conditions. Strong structure does not rely on a single charismatic leader or mayor, as seen in cities such as New York and Seoul, although they did help to bring about the tough reforms. Rather, the institutional structures and ethos of the administrators are the crux to create an environment conducive to changes and attractive to economic agents in order to generate the wealth and the buzz. In recent years, many mega-cities have appointed their own mayors who are given more direct administrative powers to set up structures to support their grand visions. This development has helped those cities to make rapid changes at the municipal level that are independent of the rest of their respective nations, and thus allowed them to progress at a much faster rate. The likes of London, Beijing and Jakarta come to mind: these capital cities have implemented bold traffic restraint measures within their city boundaries to solve their own problems. Such measures include congestion pricing, vehicle growth quota, restricting vehicles with certain registration numbers, and the banning of some vehicle types on highways.

As a city–state, Singapore has enjoyed the unique advantages of both a city and a nation. The transport policies for the city also apply to the country as a whole, thus eliminating any leakage or externality that may undermine their efficacy. For example, the VQS can effectively control the vehicle population on the island without concern for vehicles from outside the city limits. It also helps that Singapore is an island with only two border control points for vehicles, and therefore the entry and exit of foreign vehicles can be easily managed. Being small in physical area and having a unitary level of government facilitated Singapore to effect its urban plans quickly and comprehensively. Its efficient public service that has attracted the top brains from each student cohort ensures that policies are well thought through before they are announced and implemented. Every government agency imbibes the pursuit of excellence and quality customer service as its mantra, and is accountable to its overseeing minister for proper delivery of policies and expected results.

A strong governance structure requires not only the leadership of politicians but also the support of highly educated and competent, morally upright and responsible, and deeply motivated professionals. The good institutional trust, respect and cooperation between the politicians and professionals have stood Singapore well since the days of independence. This was evident in the commissioning of the State & City Planning (SCP) study in 1967, and ultimately the completion of Singapore's first Concept Plan in 1971, to address inadequacy of the 1958 Master Plan in addressing the economic and housing expansion of 1960s. The government took the SCP's recommendations very seriously, and since October 1972, the government started introducing a series of tax measures on car ownership growth in the form of very steep import duties on cars and stiff road taxes.[m]

Good governance leadership also means the ability to make quick, determined and effective decisions to resolve problems without delay and, at the same time, garner ground support for those decisions, which may not always be popular to everyone, at least not for the short term. Leadership is crucial in gaining trust and securing a social compact from the populace that the government knows what it is doing and will bring good to the society for the long term. The importance of engaging the public effectively so that they can come to a general consensus and accept hard-nosed policies cannot be over-emphasised. The earliest examples in Singapore's land transport history were the bus industry reforms, the heavy tax measures on car ownership, and the implementation of the Area Licensing Scheme (ALS) for the Central Business District (CBD) in the 1970s. Later examples include the decision to develop the Mass Rapid Transit (MRT) in 1982 after more than three years of public debate, and the implementation of the VQS in 1990 after hearings and recommendations by a Parliamentary Select Committee.

Fortuitously, the successful implementation of those difficult policies in the early years of transport development has helped, to a large extent, their continuation and subsequent modifications as the public has over time come to accept them as integral features of Singapore's transport landscape over first. Hence, the evolution of the ALS to the Electronic Road Pricing (ERP) system and that of VQS to its present form were made easier because there was no further need to reopen debate on their *raison d'etre* and policy makers could move on quickly to engage stakeholders on the details of the policy design.

b. 'Whole-of-Government' Approach

Being a city–state, Singapore inherently enjoys the efficiency and convenience that a unitary government and a unicameral parliament can bring, allowing her to integrate her policies and plans across different agencies in a 'Whole-of-Government' (WOG) approach. Conflicts of interest or competition for scarce resources such as land allocations between different ministries can be resolved quickly at the highest level i.e., the Cabinet, without going through too many layers. Decisions made by the Cabinet can be swiftly implemented at the street level without intervening bureaucracies. This approach is seen replicated in the

---

[m] Quote by Mr Joseph Yee, Director of Planning & Transport at LTA from 1999 to 2003, lifted from Centre for Liveable Cities, *Transport Overcoming Constraints, Sustaining Mobility Singapore Urban Systems Studies Booklet Series*. 2013: Singapore.

administrations of cities that have a fully empowered mayor who can make direct executive decisions on material matters concerning their cities; essentially, "the buck stops at the mayor".

This unifying ethos runs through the public service of Singapore from the day of its independence to this present day. Numerous inter-ministerial and inter-agency committees as well as cross-cutting structures have been formed in the past 50 years to study, recommend or implement solutions to solve sectorial problems, not just cross-sectorial ones. The solutions often require multi-agency coordination and cooperation to effect optimal results. The first recorded inter-ministerial committee for transport was the Road Transport Action Committee (RTAC), which was set up in 1973 to coordinate the transport planning efforts and formulate transport policies to manage the traffic demand, particularly within the city. The committee comprised staff from the then Public Works Department of the Ministry of National Development, and others from several other ministries. The 1991, 2001 and 2011concept plan reviews also involved various agencies working together to plan the long term integrated land use plans for the years 2010, 2020 and 2030 respectively. These exercises ensure that the long term requirements of housing, transport, recreational, industrial and other land uses are safeguarded for a sustainable and liveable urban environment. For example, the land space (above and underground) for MRT lines and roads for 2030 and beyond is already protected from encroachment by building developments, parks, channels and pipelines, present or to be built.

There are also permanent institutions that have been set up to ensure integration of policies and plans. The Master Planning Committee (MPC) led by the Chief Planner at the Urban Redevelopment Authority (URA) puts all land use agencies together to holistically coordinate their separate land use requirements. The merger of four agencies under two ministries to form the Land Transport Authority (LTA) in 1995 is yet another example of a new structure that was formed to engender coherence in objectives and policies of different sectors to seek overall effectiveness in transport policy outcomes.

c. Think Long-term for Good Urban Governance

Twinning the principle to think holistically is to think long-term. Long-term urban planning is good governance because it allows planners to anticipate future demands and cater to them through the supply of infrastructure capacity and/or control them through demand management measures, all implemented in good time so as to achieve maximum benefits for all. The State & City Plan

(SCP) project in the 1960s best illustrated this in integrating transport and land use planning in the early years. The basic infrastructure and long-term policies were laid down, recognising the need for future growth. Later, the 1971 Concept Plan had an outlook of 20 years that mapped out major transport developments.

The importance of forward-planning and good foresight in building infra-structure in a land scarce city state could not be over-stated. The Great MRT Debate in the late 1970s and early 1980s was essentially one over long-term thinking against short-term solutions i.e., building the MRT lines which would cost the nation a hefty bill of almost S$5 billion in those days compared to an all-bus system which would have been expedient and cost a lot less. The MRT network would also create other benefits that were not intuitively foreseeable at that time: fostering economic and social growth in the city; and MRT stations becoming lifestyle nodes for modern living. Once the first two MRT lines were built and operated, it became all too obvious to expand to form a dense net-work that would eventually cover the entire island. As explained by former Chief Executive of LTA (1995–1998), Mr Liew Heng San: "We should have moved faster. It's very hard to retrofit (MRT lines) and it's very expensive. Examples were that of tunneling works and infrastructure building of the Circle and Downtown lines. If those were done 10 years earlier, we could have done it at half the price."

Long term plans when made known and set in place will thus help to reduce the cost of future infrastructure developments as and when they are needed. They provide the time flexibility for capital investments in order to reap maximum returns. They also guide short-term developments such that the long-term developments are not encumbered in time or in space. For example, short leases of vacant land plots can be issued for them to expire just before the start of physical construction of public infrastructure that requires occupation of those plots.

d.   Sound Economic Principles for Sustainability

Another key factor for Singapore's success in urban governance, and particularly of transport solutions, has been its steadfast, almost dogmatic, application of market-based principles to its policy design. Economic sustainability in the form of financial prudence is an overriding consideration. Even as policies change over the years in response to shifting social and other imperatives, it remains an important consideration for new plans to ensure that future generations are not burdened by debts incurred by this generation, and that competition is used to keep quality high and costs low. Admittedly, such relentless pursuit of sound economic principles has not been without political ramification, as many of the transport policies and schemes are famously effective but infamously unpopular with the people.

The most notable examples are the vehicle restraint schemes. As land is scarce in this city state, car ownership and usage have to be 'rationed' through the market. The VQS and the road pricing schemes (i.e., the ALS and ERP) have long been admired by many other cities, some of which have subsequently emulated for themselves. London, Sweden and Milan are others that have since implemented a congestion scheme, while New York and several others are seriously contemplating to have one. Shanghai, Beijing and Guangzhou, all mega-cities of China, have developed a variant of the VQS for their residents.

The VQS is a remarkable public policy innovation. By setting an annual vehicle quota that the road system can take in comfortably and allocating them to those who most value the right to own a car for 10 years through an auction, the government effectively reaps the economic rent instead of allowing it to become a windfall for individuals (who are lucky to win in a ballot system); yet at the same time, the traffic congestion problem is contained. The pricing for the limited Certificates of Entitlement (COE) is thus determined by the market, i.e., by the bidders themselves. COE prices have seen many ups and downs in the 25 years since its implementation in 1990, but there has only been one clear determinant — the interplay of the COE supply based on the allowable growth rate and of the COE demand, which is influenced by the prevailing economic conditions amongst other factors.

Similarly, road pricing mechanisms allocate road space to motorists who are willing to pay for using it. In both cases, the state makes use of the revenue collected to benefit the society at large. Charges for road usage (ERP) and energy use (fuel tax) have the added benefit of promoting environmental sustainability based on the "polluter-pays" principle. Road pricing charges are also determined by the congestion levels on the roads, i.e., by the road-users themselves. A self-adjusting mechanism is deployed whereby traffic speeds are monitored quarterly and the charges increased, decreased or remained the same to keep the speeds within desired bands for optimal throughput in the road network.

On the public transport infrastructure front, economic principles are all the more applied due to the high costs of building and operating the MRT and bus systems. The main principle of capital investment by the government and self-sustaining operations and maintenance by the operator and users was established for the MRT network when the first lines were built. The 1996

White Paper further enshrined this principle and worked in a framework of Economic & Financial Viability Tests for deciding when and where to build a new rail line as well as a funding framework to ensure each generation of MRT users pay for the assets they use. The underlying rationale is that as a natural monopoly, mass public transport investments have positive economic and social externalities that cannot be fully captured financially by any non-state actor; and only the state has the wherewithal to absorb the finances and risks behind the very costly investments. Hence, it is incumbent on the state to put up the capital investments and even continue with the operational subventions (partially or fully) where it makes economic and social sense. The Viability Tests are project appraisal exercises done prior to the investment decision to first demonstrate social benefits exceed monetary cost, and next, the life-cycle operations can be sustained by fare-box revenue with the inflationary component of the operating asset[n] replacement cost subsidised by the government. The infrastructure[o] upgrading is the ambit of the government under the rail financing framework, which aims for prudence in expenditure and assurance of sustainability and minimal wastage.

No other operating subsidy is given to the MRT operators, which are private companies left to run the system on their own, subject to the government's concession and regulation. There is concession because the North-South-East-West MRT line (NSEWL), the North–East Line (NEL) and the Circle Line (CCL) were handed over to their respective operators to generate fare-box revenue without encumbrance or debt for a period of 30–40 years. In return, the operators are to maintain the systems in good conditions and replace parts and assets such that they can be returned to the state at the end of the said period in the same condition as when they were first taken over. The operators may generate retail and advertising revenues out of the spaces in the MRT stations after paying the state the necessary market land premiums. There is regulation because LTA as the regulatory body makes sure that the operator meets the stipulated operating and maintenance standards in delivering the MRT services. Failure to upkeep the standards will subject the operators to censures and financial penalties. As a result, the operators face both market and regulatory pressures to be more efficient, as well as to provide quality services to raise ridership, in order to eke out a decent return on its capital.

---

[n] Operating assets are generally railway systems with typical lifespans of less than 30 years, notably the rolling stocks, the computer systems and software, the E&M systems in an MRT tunnel and in a station, etc.

[o] Infrastructure refers to those systems that have much longer lifespans e.g. railway track, tunnels, viaducts, station shells and structures, etc.

A similar market-based framework is also applied to the financing and regulation of the public bus industry: the government constructs infrastructure such as roads, bus interchanges, terminals, depots and bus-stops, but it does not subsidise the operating cost of the bus operators including the purchase of the buses. Like the MRT operators, the public bus operators pay a nominal rent of $12 per annum for the use of the bus infrastructure. They are allowed to retain the revenue collected from commercial facilities in the bus interchanges, such as shop spaces and advertising panels to help defray costs of maintaining the interchanges.

Recognising that public transport infrastructure and services are typical natural monopolies, the government goes about regulating their prices and service levels to prevent operators from reaping supernormal profits by raising fares and cutting back outputs. This is Public Economics 101. For fare regulation, Singapore has adopted the price-cap model with productivity extraction such that efficiency gains are shared with the commuters. The price-cap formula takes in cost drivers like wages, energy prices and general inflation, using CPI as the proxy.

The Public Transport Council (PTC) is a uniquely Singapore institution that was set up in 1987 under the Ministry of Transport (MOT) to carry out this, sometimes thankless, job of determining the fare cap and approving the fare adjustment packages of the Public Transport Operators (PTOs) in accordance with the cap of each year. The PTC's mission is to weigh the economic situation of each year, and hence, the general affordability of the fare increases, against the commercial viability of the PTOs. It is fully cognisant of the need to incentivise the PTOs to stay in the business and continue to invest for quality and productivity improvements. This is in keeping with the 'user-pay' principle so that wastage is minimised and government subsidies can go to where they are most needed, e.g., low-income households can apply for Public Transport Vouchers to defray the fare increases, and students and senior citizens enjoy lower fares and can apply for concessionary season passes.

On the regulation of the Rapid Transit System (RTS), which covers both MRT and LRT systems, LTA is the authority on service and technical standards for historical reason. All service breakdown incidents are investigated by LTA and, where justified, financial penalties are imposed on the PTOs. Since the major MRT breakdowns in late 2011 which led to a public inquiry by a committee (COI) in 2012, LTA has enhanced its regulation regime from performance-based oversight to a more interventionist approach towards what the PTOs should and should not do in their daily operations as well as in their maintenance routines. Such tightening of the regulatory hand is not out of line with the need to 'tame' a monopoly in the absence of credible market competition.

It is well established in the public policy literature that public transport like other infrastructure-based services is a natural monopoly, and unbridled competition for a share of the market will lead to undesirable consequences. An unregulated bus market will see buses rushing to pick up passengers at the bus stops or no bus serving remote areas, as we saw in the pre-independence days of Singapore. Without protection of the ridership along the MRT corridor through transit-oriented housing and commercial developments or through bus route re-routing to serve as feeders, no MRT operator would have taken up the licence to run the network for 30–40 years under the previous rail financing framework. Under such a monopolistic condition, there is little room for 'competition in the market' to succeed without a level-playing field for new entrants, nor certainty of revenue stream for the incumbents.

In the post-2000 era, there merges a new trend of 'contracting-out' of public transport services by public authorities or state-owned enterprises in European cities in compliance with laws passed by the European Parliament to foster competition. 'Competition for the market' allows open competition to take place in the provision of public services that were previously incontestable due to either natural monopoly or public good conditions where the fees paid by those directly enjoying the services are not, or cannot be, commensurate with the true public value of those services. Even though some of the services may be excludable to those not paying the fees like in public transport or public utilities, the upfront and recurring capital investments are so tremendous that the payback period has to be much longer or the fees much higher than what the public could tolerate. Competition space is thus carved out by the state to allow participation by the private sector in order to reduce the capex and opex subsidies through efficiency gains. The city of London has for many years let out its bus routes for competition, so has the city of Stockholm for its metro service. Some other cities like Paris and Berlin have signed Service-Level Agreements (SLAs) with their corporatized transport entities to run their services, laying out financial incentives/penalties for over-/under-performance of standards to instill market discipline. Such SLAs are often seen as an intermediate step towards open market competition. Under 'competition for the market', the right to operate the contested services is protected for the duration of the contract, and the revenue risk is either fully taken by the government (as in Gross-Cost Contracts) or mitigated through government interventions such as regular fare increases (as in Net-Cost Contracts).

In Singapore, the government has taken the decision since the 2008 Land Transport Master Plan (LTMP) to enhance competition in the public transport sphere. The RTS market was the first to open up through the new rail financing framework (NRFF), which has been applied to the Downtown Line (DTL). Different from the previous tenders for the right to operate NEL and CCL, the competition for DTL operation was a financial one in which bidders offered their own formulae to share revenue and profits with the government. The rail operating licence is a 15–year Net-Cost Contract whereby the LTA retains ownership and control of the infrastructure and operating assets and directs their maintenance and upgrading activities. The revenue and profit contributions from the operator will go into a sinking fund to be used for upgrading and replacement of assets.

On the bus front, LTA has taken over the role of the master bus route planner from the incumbent regulated bus operators since 2008. In the 2013 LTMP, the government made a further move towards market competition by announcing the Government Contracting Model (GCM) under which all bus routes will be packaged into 12 parcels (unlike the London model where each route is subject to bidding). The first three parcels will be awarded to the successful bidders from 2015, and the remaining nine will be contracted to the incumbents via negotiation. The 5+2 year licence for each parcel is a Gross-Cost contract and new buses will be provided by the government in order to level the playing field for new entrants. Hence, the need for competition is real and the government is doing what it can to ensure that competition does happen and will produce the desired results. In the run-up to the GCM, the government has also launched the Bus Service Enhancement Programme (BSEP) whereby 1000 new and replacement buses are pumped in by the state and put into operation either by the incumbent public bus operators or new operators through tendering. This helps to prepare the local bus industry for the eventual competition as well as the government in the form of price discovery and contract specifications.

For the taxi industry, Singapore has gone a full circle from almost no regulation to strict regulation and then back to deregulation of fares and liberalisation of supply. The taxi industry in the 1950s and 1960s was a freewheeling one, operated by both legally licensed taxis and illegal 'pirate-taxis', which were private cars driven by unlicensed taxi drivers out to make a quick buck. These pirate taxis had poor service standards, posed road safety problems and charged irregular fares. The 1970 White Paper made pirate taxi operations a seizable offence, among others, to properly regulate the taxi industry. To regularise them under the new framework employment, opportunities were provided to the former pirate taxi drivers and a loan given to the National Trades Union Congress (NTUC) Workers Cooperative Commonwealth For Transport Limited (NTUC Comfort) so as to bring them under its umbrella by offering vehicle loans. Pirate taxis thus perished quickly, and in their place were several new licensed taxi companies. Over time, other taxi companies were formed to compete in the market. A regulated competitive industry was thus formed where taxi companies owned and rented out their fleets to individual hirers, who are subject to the company's service standards and government's regulatory discipline. Taxi fares and fare increases were controlled by the PTC in the same way as for bus and MRT fares.

However, taxi services continued to be a hot-button issue throughout the 1990s and the first decade of the 21st century. The then Minister for Transport, Mr Yeo Cheow Tong, addressed Parliament on this issue in 2003: "A major source of dissatisfaction with the taxi industry… is the mismatch between supply and demand during peak periods, and poor service standards… Once we have put in place the rules on how taxi companies can be set up and how they are to be accountable for their service standards, we will lift controls on the number of taxi companies, and the fleet quota imposed on each company. Such a move will allow the market to respond freely to demand, thereby providing a better match between supply and demand."

In 1998, Singapore decided to take the unconventional road of liberalising the industry in two measured phases: taxi fares were first deregulated where taxi companies were allowed to set their own fares without having to seek PTC's approval, and later in 2003, control on the number of taxi companies and the fleet quota imposed on them was lifted. To keep taxi companies and drivers in check, taxi operator licences are issued to those meeting minimum entry requirements and their performance in call-booking services is monitored. Those who fail to meet stipulated standards are penalised financially. Otherwise, taxi companies are very much left to their own devices to compete with one another for taxi rentals through better vehicle and booking offerings.

This industry regime has persisted to this day, with further tweaks in recent years to forcibly increase taxi availability on the streets. The jury is still out on the efficacy of this 'regulated free market' policy because the peak period demand-and-supply mismatch is not yet fully resolved, but competition has led to a plethora of different taxi services in terms of pricing structures[p], vehicle types and service packages, which some feel are rather confusing.

---

[p] At the time of writing, PTC and LTA announced in March 2015 that some parts of the taxi fare structure will be standardised, to prevent taxi fares from becoming even more complex for com-

e. Flexible, Responsive & Adaptive Plans to Meet Changing Circumstances

In spite of the above principles for good urban governance, one other key principle stands out apparently in contradiction: that no policy is too sacred to never be shifted or adjusted in response to changing circumstances. Cities have to constantly innovate and be bold to try new measures if old ones do not work or cease to work well. The world is more complex now and fresh challenges abound; cities cannot always rely on the same old strategies and ignore what is happening elsewhere that may have references to themselves. Principles may stay intact but flexible, responsive and adaptive strategies and plans are necessary to deliver the outcomes that resonate with the populace of the day. Even though these outcomes are clear, there has been no predetermined path to them. Policy makers also do not have all the answers and they can be blind-sighted to whatever that challenges their assumptions and mindsets.

The history of land transport in Singapore has had plentiful examples of such responsiveness and pragmatism. Policies are tweaked and sometimes 'U-turned' whenever the need arises to solve implementation issues while preserving their intents and the larger principles underlying them. The VQS is one of the most tweaked schemes since its launch in 1990, not least due to its political sensitivity. In the early days, there were unhappiness about the dominance of the quota market by the rich and the manipulation by the motor agents. Successive committees and task forces were thus formed to fine-tune the scheme to make it more palatable. The current scheme is quite dissimilar from its first version not just in terms of the quota categories but also in the way the auction is conducted, for example. Despite the differences, the policy intent of limiting vehicle population growth to a sustainable pace and of using the market mechanism to allocate the quota is still very much valid today as it was 25 years ago. Similarly, the evolution of another controversial policy — road pricing schemes — from the ALS in 1975 to the RPS in 1995, to the ERP in 1998, is filled with adjustments of coverage and rate formulation and phasing-in tax packages. All for engendering public acceptance.

Other vehicle restraint policies like the Weekend Car/Off-Peak Car Schemes, the vehicle tax structure rationalisation of the ARF/PARF, excise duty and road licence fee, and the petrol duty and tax on diesel vehicles, are tweaked from time to time for different purposes. Sometimes they are meant to restore the effectiveness of the policies, other times they helped to ease in the signifi-

---

muters in the future. This has been published in the Gazette under the Public Transport Council (Taxi Fare Pricing Policy) Order 2016 which came into operation on 22 January 2016.

cant changes of the VQS and ERP schemes. Not least the growing aspirations towards car ownership in Singapore, as she develops into a mature economy, have turned these policies into visceral issues which require careful and calibrated responses.

Whereas we see incremental changes being made on private transport policies, the changes on public transport have been epochal and bold, albeit less frequent. The bus reform story is one prime illustration. The government set up the Bus Service Reorganisation Committee (BSRC) in 1973 to resolve problems associated with the dissatisfaction of bus workers, and to more fundamentally change the bus industry from a free market of many players into a centrally planned one operated by a regulated monopoly firm that was formed from a merger of the different companies to bring about new efficiencies. However, the new entity still suffered from mismanagement due to the pre-existing management methods for small bus companies. This led the government to appoint a team of government officials to study and revamp its operations that consequently gave rise to productivity and profitability improvements. Nine years later, in what seemed like a policy reversal, a second bus company — TransIsland Bus Services Pte Ltd (TIBS) — was formed when it was felt that more competition was needed within the public bus sector to raise service standards. However, this was a competition neither in the market nor for the market, but a benchmarking competition; for what was really at stake was only the reputation of the two bus companies. The industry remained a regulated dual-monopolies.

The next wave of epoch-making changes began in 2008 when LTA took on the responsibility of bus planning, and in 2012 with the start of the BSEP and later the announcement of the GCM. From 2006 onwards, Singapore started to see rapid increases in its resident population as its economy recovered from the doldrums of earlier years and more foreigners were employed to meet its manpower needs. Public buses and trains were becoming more congested as the infrastructure capacity fell behind in growth. The government had to act quickly to bump up the capacity gaps: the DTL was announced in 2007 and the 2020 RTS master plan in 2008. Each RTS line takes almost 10 years to complete and therefore cannot solve near-term problems. Quick and decisive measures had to be implemented, notwithstanding the ideologies. The BSEP is a very significant change in the long-standing position of not giving operating subsidy to public transport operators for fear of wastage and inefficiencies. The government bought new buses for the operators to beef up existing service frequencies and to run new routes. Government funding is given to raise the wages of bus drivers in order to attract and train more of them to drive those buses. The GCM which works on Gross-Cost Contracts shields the operators from revenue risks and obviates them from high upfront capex for bus fleet procurement and bus-related facilities. To minimise the potential of wastage and moral hazards on the part of the operators, strict financial controls are instituted to track expenditures, and regular reviews were conducted to monitor efficiency. Market mechanism in the form of tendering is nevertheless employed to secure the best value for money.

In the same vein, the new Rail Financing Framework sets the stage for enhanced competition in the RTS market to ensure that operators focus on delivering quality service. The worldwide trend for such competition has also given us the impetus to try out this new market-based solution.

To ease the crowdedness on the RTS network, bold and immediate measures like the free pre-peak MRT travel for the selected CBD stations and the Travel Smart initiatives, including the Travel Smart Rewards (formerly called the INSINC), have been put in place. Singapore is probably the first major city in the world to have implemented these incentives to encourage commuters to start and end their journeys before the morning peak period, rather than for deferring infrastructure investments. This is another score for Singapore for policy innovation, not unlike the VQS, ERP and the public transport fare cap formula adopted by the PTC since 2005 and revised in 2013. Hong Kong also adopted a similar formulation soon after Singapore had first published its revised Fare Revision Mechanism 10 years ago. The crux of this mechanism is to ascertain small and regular fare adjustments (either upwards or downwards) in tandem with exogenous cost variations experienced by the PTOs. In response to the widening income gap and the social inequity faced by the ageing, low-income and physically-challenged segments of the population, the public transport fare structure was revamped under the 2013 edition of the Fare Review Mechanism. Together with the revised formula, new concessionary fares and season passes for those segments were introduced; some at the expense of the PTOs, others subsidised by the government. 'Sugar-coating' of policy packages is often used to obtain buy-in while achieving the larger objective of instilling market discipline.

3.3. *Four key strategies for sustainable transport*

Underpinned by the aforesaid five principles of good urban governance, Singapore's long-standing transport strategies can be summarised in four key thrusts. The ultimate objective of these strategies is to build and maintain a sustainable transport system for higher liveability and economic growth.

a. Integrated Master Planning for Transit-Oriented Development

Faced with scarce land space and growing population on an island, Singapore has had no choice but to guard its land use jealously. The allocation of land space must be carefully planned for the long-term such that future uses for, say infrastructure development, are not constrained or that future growth is limited due to a lack of supporting amenities. From the earliest Master Plan of 1958 to the State & City Planning (SCP) project in 1967 to the first Concept Plan of 1971, Singapore's city planners had started to integrate transport and landuse planning in the early years before Singapore became more urbanised. Hence, basic infrastructure were laid down and longterm transport policies implemented in coalition, recognising the need for future growth. Integrated master planning links the movement of people to their residences and places for work, education, community development and recreation. If suitably located, these places would have a very strong influence on the need to travel long distances, to travel in a particular direction, or to travel at all. Therefore, by organising land use into high-density, self-contained satellite towns linked by an island-wide transport network running through their centres, travel demand can be better managed at source in space and in time, and not over-strain the infrastructure. Keeping a healthy job places-worker residence ratio for each geographic sector would reduce peak hour commute trips across the sectors. Satellite towns are planned to have a full range of facilities for non-work purposes so that there are fewer reasons to travel out of them. The transport infrastructure comprises RTS lines, bus corridors, feeder buses and roads to enable efficient traveling within and between the towns, and the commercial and industrial hubs. The expressway network connects the major origins (i.e., satellite towns) and destinations (the Central Business District in the south, the Jurong/Tuas Industrial Estate to the west, and the Changi Airport to the east).

To further reduce travel needs and distances and to improve traveling convenience, housing and land use densities are usually enhanced around key transport nodes i.e., MRT/LRT stations, bus interchanges and terminals. Land allocation for these nodes and the developments around them have to be planned decades ahead of their realisation. The 1991 Concept Plan further enshrined the integrated master planning for transit-oriented development strategy by adopting a "constellation concept" of setting up regional centres, aimed at decentralising commercial activities, and thus reducing the conges-tion in the city and the travel demand towards it. The four regional centres planned were Tampines in the East, Seletar in the North East, Jurong East in the West and Woodlands in the North. The subsequent Concept Plan Reviews in 2001 and 2013 sought to plan for heavier investments in transport infra-structure, higher densities around transport nodes, and more even job-worker ratios, in order to accommodate anticipated larger populations in the future without causing a drop in liveability and efficiency.

Correspondingly, the 1996 White Paper for a World-Class Land Transport System and the 2008 Land Transport Master Plan (and its 2013 edition), both published by the LTA, put the integration of land-use planning with large-scale RTS network expansion as key to solving the transport challenges that Singapore would face with increasing travel demand. Taking a long-term view about future needs and setting aside land space for future infrastructure to cater to those needs make for a sustainable transport system, where future generations of Singaporeans will not be strait-jacketed by less options than we have today.

b. Build a Quality, Efficient and Sustainable Public Transport System

Likewise, the key to solving our challenges and ensuring sustainability is to have majority of the commuters in a dense city like Singapore use public transport. Hence, the corollary to the first strategy is to build an efficient public transport system that provides convenient and affordable services. A quality public transport system will meet the needs of individuals' economic and social aspirations, as well as reduce the reliance on private passenger vehicles, which often give rise to negative environmental externalities of traffic congestion and pollution. The 1996 White Paper laid this out in no uncertain terms that its first priority was to provide a world-class public transport system to cater for mass movement. The two editions of the LTMP reinforced the Government's commitment to make public transport the chosen mode of travel and even set the mode share targets of 70% and 75% for the years 2020 and 2030 respectively.

Throughout the history of public transport developments in Singapore, the primary efforts of the Government have been firstly to set aside land for the public transport system, as in MRT/LRT corridors and stations, road space for buses, bus stops and interchanges, etc., and to integrate them within the CBD, residential towns and commercial and industrial centres. Secondly, the government has been determined to ensure adequate public transport service provision at affordable prices and to develop a commercially viable private sector for it, all through harnessing the market mechanism for eco-nomic sustainability and exercising the regulatory power for social inclusion. From the early bus reforms, to the Great MRT Debate and to the recent pol-icy changes to bring about competition for the market and in conjunction operating subsidies, the public transport industry has gone from free markets to central planning to monopolies, and now back to market competition again, albeit with more control and financial support from the government. These changes were made in response to changing needs and to political demand for more and better services, as unpopular policies on restricting car growth and road usage kicked in and were tightened over time.

Iterative reviews and new institutions are put in place to attain and balance economic and social goals of the public transport industry. Top on the government's mind has always been the concern of whether there are adequate resources to expand the costly RTS network, and to sustain its operation and maintenance without causing an undue toll on the finances of future governments and people of the nation. To maximise efficiency for Singapore's public transport system, a 'hub and spoke' model has been adopted: the MRT network forms the backbone and serves the heavy transit corridors primarily for long-haul travel, with buses and the LRT serving the lighter corridors and providing intratown feeder services to connect commuters to the MRT and long-haul bus networks. When new MRT lines are opened, pre-existing bus routes that were serving the same corridor are re-routed to provide better connections to the MRT stations and to reduce duplication of capacity that may be under-utilised as a result of the new lines.

Other means to attract more people to use public transport include providing walkways and cycling tracks to connect to the MRT stations and bus interchanges, and installing more bicycle parking facilities. Integrated ticketing and fare integration are also implemented across the network and for different travel modes to enhance convenience. From magnetic farecards to contactless CEPAS cards, Singapore has adopted a common standard of stored-value card for use on buses, trains, taxis and small value retail transactions. Since 2010, the public transport fares have migrated to a distance-based through-fare regime where the total fare for a door-to-door journey is computed based on the total distance that is added up from each leg on a different mode, without paying the boarding charge more than once. Such a unified fare system allows for a better integration of transport modes and encourages commuters to switch amongst them for faster and more efficient travel.

To further aid public transport efficiency and to emphasise the importance given to it, various bus priority measures have been a signature on Singapore roads from the early days. Bus priority lanes were implemented in the 1970s, the Full-Day bus lanes in 2005, and the Mandatory Give Way to Bus scheme in 2008. Since June 2008, the network of normal bus lanes were extended from 120 km to 150 km, and Full-Day bus lanes trebled from 7 km to

23 km. These measures have improved bus speeds by an average of 7% and as much as 16% on some roads.

With advancements in information and communication technologies, information integration has improved bus services through better planning of their routes, and widespread availability of their real-time arrival-at-bus-stop information on mobile devices and at the bus stops themselves. On-time information assists commuters in managing their waiting time and making informed travel decisions. LTA is also launching a new bus monitoring and control system progressively from 2015, to enhance its capabilities to manage incidents and to respond to travel demands in a timely manner, as well as to provide more comprehensive and accurate information about the locations of moving buses and the movement of bus passengers.

c. Managing Road Usage

Concomitant to the 'pull strategy' of good public transport services is the 'push factor' of restraining road usage. Singapore has been a pioneer in this field starting with the fiscal measures since the late 1950s to the ALS in 1975, then the VQS in 1990 and the ERP in 1998. The 1971 SCP study, the 1973 RTAC, the Concept Plan Reviews of 1991 and 2001, the 1996 White Paper, and the two LTMPs in 2008 and 2013, all advocated managing road usage as a key plank of urban transport governance. Given the limited land to build more roads and car parks, this is the natural conclusion not just for Singapore but also for many global cities facing traffic congestion and pollution issues caused by increasing motorisation. This has become a necessary and urgent thing to do to rehabilitate streets choked with cars and smog.

Whether it is imposing high taxes on car purchase or fuel, or charging for the use of roads and parking spaces, demand restraint measures are just as unpopular and are often baulked at by politicians. While many countries have managed to implement fuel and vehicle taxes, parking tariffs and toll charges, these are usually meant to raise revenue for general taxation or for specific infrastructure investment and maintenance. Seldom are they publicly declared as congestion management strategies. Hence, road pricing has been deemed a holy grail by many transport professionals, especially as other measures fail to grapple with the immense urge of car-owners to drive. Road pricing in Singapore, or congestion charging in other jurisdictions, is backed by the economic principle of a Pigou Tax such that those who cause a negative externality to society (in this case traffic jams, noise and air pollution by their acts of driving) should be made to pay a tax to benefit society at large (in Singapore's case through the government coffers). This transfer will minimise the overall welfare loss and improve efficiency of the system. Congestion charging became technically feasible only towards the end of the 20th century although the concept had come about much earlier. London and Stockholm implemented it using video recordings, while Singapore uses Dedicated Short Range Communication (DSRC) technology for its ERP gantries. All three cities adopt a gateway or point charging concept whereby identification of the vehicles is made at fixed locations.

To manage road usage in Singapore, the government has adopted a delicate balance of ownership and usage restraint schemes, while continuing to expand the road network in a calibrated manner and where it makes economic sense. Before 1990, the constantly increasing registration taxes could not keep up with the burgeoning demand to own cars and the growing wealth of the population. ALS and parking surcharge were introduced in the mid-1970s for the CBD to control the inflow of vehicle traffic. While these measures regulated traffic congestion downtown, it did not help the rest of the island or at other times of the day. By 1989, evening ALS had to be introduced. In 1990, the implementation of the VQS put demand management strategy on a new page because upfront taxes could henceforth be cut without affecting the demand for cars. The COE premium is a reflection of the market demand for cars and not a direct policy intent of the government, as the supply of COEs is fixed and known upfront. However, COE premiums remain a sensitive political issue to this day.

The high COE premiums in the first half of the 1990s added pressure to do something on the road usage front, and hence the expansion of the manual ALS to the RPS on the expressways and later the automation of the manual system to an electronic one — the ERP in 1998. The advent of ERP was another watershed event, for henceforth the government would have the two key levers to manage road usage demand. To demonstrate the promise of the new lever upfront, registration taxes were cut and recurrent vehicle licences were rationalised and reduced, but the vehicle quota remained at 3% per annum, which had been the rate since 1990. To further assure the public that ERP is not a revenue-generating ploy, its charges are reviewed quarterly and adjusted up or down according to actual traffic conditions in order to maximise throughput on the roads. The criteria and the annual collection are made transparent.

For almost 20 years, the vehicle population had been growing at 3% per annum. With a growing base, the absolute increase became larger and larger every year. This called to question the sustainability of the growth rate and its impact on road use for the long term. The rate was thus halved in 2009 to 1.5% per annum. From 2012 to 2015, the annual rate was further cut in quick succession to 1.0%, then 0.5% and finally to 0.25%. It was also announced that the long term rate would be zero, and the COE quota in each quarter would just be the recycled number of deregistered vehicles in the previous quarter. This rapid reduction in growth rate is a reflection of the urgent need to curb the use of passenger cars against the backdrop of economic expansion and increasing population on an island of almost capped land area and road network. The conditions for these reductions were right: their actual impact was not really felt in the available quota because the recycled numbers formed a larger proportion and were more so in the last couple of years, due to a bumper crop of COEs that had been released 10 years earlier and had now expired.

On road usage control, the next generation of the ERP System slated to be implemented by 2020 will be another milestone, not least the first in the world to be able to levy a charge based on distance travelled by the vehicles in an urban setting. Conceptually, such a system would obviate the need for any form of upfront or recurrent fixed taxes, and charge motorists on actual usage, much like traveling in a taxi. This will open up new policy innovation opportunities for Singapore as it works assiduously towards moving more people to use space-efficient forms of transport.

Like the other cities which have implemented congestion charging, Singapore has faced the same political challenges in the introduction and upholding of unpopular vehicle restraint policies. It takes a lot of willpower, political acumen, stakeholder engagement and 'sweeteners' to pull through each of those schemes at the right timing. Often, existing taxes had to be cut or eliminated to give way to new ERP gantries. Rebates were also given to specific groups that were worst hit to ease in the new schemes. In the final analysis, both ownership and usage restraint measures are necessary and complementary to each other, and the government has to use them judiciously to achieve the outcome because over-reliance on any one of them will create ill effects. For example, if we stopped using ERP, there would be localised congestion around high-density areas even if there was zero or negative vehicle population growth. On the other hand, if we scrapped VQS and relied on ERP alone, the daily odium of high congestion charges felt by motorists would be tremendous. To garner buy-in for these hard-nosed policies, the schemes and their purposes should be easily understood by the public. Motorists or car-buyers should feel that the schemes are fair and that they have a choice not to pay. The schemes must also show results so that cynicism does not set in about their purported intent.

d. Ensure Environmental & Social Sustainability

A sustainable transport strategy in the general sense cannot be complete if it considers only the economic and financial dimensions. Environmental and social sustainability are just as critical to the making of a liveable city. Both have been very much embedded in Singapore's urban transport policies although they are less highlighted as key objectives than the economic need to ensure free-flowing roads and efficient mass mobility. All the same, free-flowing roads and car usage restraint give rise to a cleaner and greener environment, mitigating carbon emission and reducing other externalities like noise and air pollution. The roads are planned and built with copious space for trees and bushes. Buildings are set back from roads to provide a noise buffer for their occupants. The vehicle tax regime favours younger vehicles and those with fuel-efficient or cleaner engines. Stringent vehicle emission standards and mandatory periodic inspections provide the assurance of their road worthiness. The strategy of making public transport the predominant and desired travel mode in a city is, in fact, the most cost-effective way to shrink carbon footprint; far more than electrifying the vehicle fleet or introducing new emission standards which only make the vehicle more costly but no less used.

Social sustainability refers to social inclusion, social equity and social involvement in policy making. In terms of urban transport, universal access to basic public transport services and a just taxation system are prime features. Institutions like the PTC and the fare cap formula have been put in place for ensuring universal access to mass public transport as well as affordable fares to the majority of the commuters. Special fare concessions and public transport vouchers help the low-income families, students, senior citizens and physically disabled groups with their transport expenditures. The VQS has a separate category for lower cost cars even though the Government does not guarantee that everyone in Singapore can own one. The vehicle taxation system is a progressive one whereby luxurious cars attract proportionately more tax than low-end cars do.

Involving the communities and the general public in consultation is an indispensable step in the policy process. Depending on the nature of the policies or schemes, views of appropriate stakeholder groups should be sought before and after the implementation. Transport policies in Singapore are no exception: from municipal matters like parking restrictions and traffic schemes for which local communities were consulted, to fare revision mechanism and VQS changes where academics, trade associations and Members of Parliament formed focus groups for discussion. Such

engagements would go a long way towards garnering support and ownership for the outcome of the process.

4. Next 50 Years

The experience of land transport governance over the last 50 years in Singapore
can be summarised as transitioning from a system of chaos with a fairly unregulated market in the 1960s; to one where the government took full control to
reorganise the industry in the 1970s; to market-based structures with light-touch
outcome-based regulation against market failure in 1980s to 2000s; and, since
2010, back to more state interventions (including operating subsidies and micro-
management of operations and maintenance) to ensure adequate provision of
transport services and quality travel experience for all. At each juncture, the
government took measured and deliberate steps in response to the challenges of
the day without worry or fear of reversing past policies, as long as the problems
could be resolved in a sustainable manner.

Cautious of the experience of cities with nationalised public transport
industry, Singapore has tried to minimize the inefficiencies by aligning incentives to ensure that public transport operators remain viable and commuters
are provided with good service. The bus and rail financing scheme have been
aligned with government taking ownership of all infrastructure and operating
assets and leasing it to public transport operators. This reflects the change in
policy for the government to shoulder more responsibility and impose greater
control over maintenance and operational decisions to ensure that service
standards remain satisfactory.

What will Singapore's urban transport be like in the next 50 years? More
contestability will be added to the public transport market as the government transitions to a Government Contracting Model for the bus industry
and as operating licenses become shorter under the new rail financing
scheme adopted in 2010. As a centralised network planner, the government
will be better placed to plan the entire public transport network in a more
holistic manner and to enhance connectivity and mobility to meet diverse
needs.

With the focus on people as the centre of the land transport system, the
government is prioritizing individual mobility needs and meeting expectations for a more diverse array of travel options. The commitment to facilitate
walking and cycling as first- and last-mile modes is apparent with measures
like Walk2Ride and the National Cycling Plan. To ensure that people will be
more likely to benefit from the new travel demand management schemes and

effectively influence travel behaviour, the government has also decided to partner employers to facilitate more flexi-travel arrangements. In encouraging people to make more informed travel decisions, the government is committed to sharing data and providing more real-time travel information for people to access on-the-go.

Underpinning this experience is the understanding that ensuring traffic remains smooth-flowing for both people and goods is vital to Singapore's economic growth and sustainability. Hence, more reliance on road pricing and parking demand management can be expected in the future as the allowable vehicle population growth rate tends towards stagnating at zero. Advanced technologies will also be harnessed to make way for smart mobility to optimise the 12% of our land area that has been used for transport. Some pundits have come up with a "no privately-owned car" scenario whereby all vehicles are shared and transport is but a service of moving an individual or goods from one place to another on demand. Different combinations of mode can be used depending on the price one chooses.

This chapter has related a brief story of Singapore's land transport policies and developments in the first 50 years of her independence. It has also laid out the underlying principles and strategies for the urban transport governance that has produced a world-class transport system that many cities can only admire. The following chapters will provide more details and in-depth analyses on those policies and schemes. The same challenges — limited land and increasing travel demand — that have confronted Singapore in the past half a century will continue to confront her in the next five decades. The principles and strategies have hitherto served Singapore well. The authors are confident that they will continue to be relevant for guiding future generations of transport professionals towards maintaining sustainability of the transport system and raising liveability of the city–state.

References

Centre for Liveable Cities, *Transport Overcoming Constraints, Sustaining Mobility Singapore Urban Systems Studies Booklet Series*. 2013: Ministry of National Development, Singapore Land Transport Authority, Ministry of Transport, Singapore.

Chin, Hoong Chor, "Urban Transport Planning in Singapore." In *Planning Singapore: From Plan to Implementation*, edited by Belinda Yuen, p. 133–168. Singapore: Singapore Institute of Planners, 1998.

Chin, Kian-Keong, Congestion Pricing Experiences in Singapore. Land Transport Authority, Singapore, 2008.

Committee on the Fare Review Mechanism. 2005. Report of the Committee on the Fare Review Mechanism. Singapore.

Committee on the Fare Review Mechanism. 2013. Report of the Committee on the Fare Review Mechanism. Singapore.

Ilsa Sharp, *The Journey — Singapore's Land Transport Story* (Singapore : Published for the Land Transport Authority by SNP Editions, 2005)

Land Transport Authority of Singapore, "A World Class Land Transport System, White Paper." Singapore: LTA, 1996.

Land Transport Authority of Singapore, "LT Master Plan: A People-Centred Land Transport System." Singapore: LTA, 2008.

Lew Yii Der and Ang Chor Ing, "The LTA Journey: 15 Years On," JOURNEYS November 2010, LTA Academy.

Lew, Yii Der and Choy, Maria "An Overview of Singapore's Key Land Transport Policies: Optimising under Constraints." Singapore: LTA Academy, 2009.

Looi, T. S. and Tan, K. H. 2007. Striking a fare deal — Singapore's Experience in introducing a fare review mechanism. Paper presented at 11th World Conference on Transport Research, 2007.

Looi, T. S. and Tan, K. H. *Instituting Fare Regulation*. JOURNEYS November 2009. LTA Academy.

Minister of Transport (1999–2006) Mr Yeo Cheow Tong in Parliament on 25 January 2003. Speech. URL: http://sprs.parl.gov.sg/search/topic.jsp?currentTopicID=00068846-ZZ&currentPubID=00069930-ZZ&topicKey=00069930-ZZ.00068846-ZZ_1%2Bid004_20030125_S0002_T00022-bill%2B

Mr. Liew Heng San, interviewed by Grace, Leong Ching, 28 March 2011.

Phang, Sock Yong. *Strategic Development of Airport and Rail Infrastructure: The Case of Singapore*, Singapore Management University Economics & Statistics Working Papers, June 2002.

PricewaterhouseCoopers (PwC). *Cities of Opportunity 6th Edition*. 2014. URL: http://www.pwc.com/us/en/cities-of-opportunity/

Union Internationale des Transports Public, UITP. *Mobility in Cities Database Analysis and Recommendations Report*. July 2006.

World Economic Forum, The Global Competitiveness Report 2014–2015: Full Data Edition., Switzerland: 2014. URL: http://www3.weforum.org/docs/WEF_GlobalCompetitivenessReport_2014-15.pdf

Chapter 2

An Evolving Public Transport Eco-System

LOOI Teik Soon and CHOI Chik Cheong

Compiling and consolidating from various narratives and sources, this chapter dwells an some key changes in the evolution of the public transport ecosystem over the last 50 years. It is not meant to comprehensively trace all aspects of changes but the fundamental developments that have brought about notable changes, specifically in the provision of bus, mass rapid transit and taxi services.

1. Relentless Pursuit to Improve Bus Services[a]

The transformation of Singapore's bus industry illustrates a unique public transport evolution which started from a free-wheeling unfettered market to one with regulated competition and restricted players. This journey of evolution owes its present state to the concerted effort of decades-long initiatives in transport planning and restructuring, adaptations, and government interventions. Key milestones on bus reforms in Singapore are summarised in Annex A.

1.1. *Early years of trams and trolleys up to 1930s*

The earliest form of public transport in Singapore started with the establishment of the Singapore Electric Tramways Company (London) in 1902. The electric tram system carried an average of 11,000 passengers daily at its peak, but its services failed to make profit and soon became inadequate to meet public demand.[b]

[a] LTA Academy, *The Singapore Bus Reform Story*, (2012).
[b] Sharp, I., *The Journey — Singapore's Land Transport Story*, (SNP Editions, 2005).

During the First World War, the maintenance of the track and equipment fell into arrears, and their condition deteriorated rapidly. After the war, rising public dissatisfaction with the tram service led to the governor-in-council appointing a commission in 1920 to inquire into the state of the electric tram services. Soon after, the operation of the tram service was taken over by the Shanghai Electric Construction Company Limited, which did improve the efficiency of the tram service. Nevertheless, the track and equipment were still in a derelict condition.

As such, in preference to reconstruct and re-equip the tram network, it was decided that the electric trams be converted to trolley bus operations instead. A government-owned new company — the Singapore Traction Company (STC) Limited was formed and operated under powers, and conditions laid down in the Singapore Traction Ordinance of 1925.[c] This probably marked the first known attempt to enact a legislation specifically to provide public transport in Singapore.

Trolley buses started operating by 1929, with the first trial run on a west–east route between Johnston's Pier at Collyer Quay and Geylang. Like the electric tram, trolley buses received power from overhead electric wires but had the advantage of not being restricted by tracks laid on the road. Thus, trolley buses were able to manoeuvre around obstacles.

The rising popularity of motor vehicles in the early 1930s had the STC's motorised omnibuses running alongside the trolley buses. The STC was given a 30-year monopoly to operate trolley and motor buses within the city limits.

Informal "mosquito buses"

However, weak regulations for enforcement in the 1930s had led to privately-owned bus companies providing bus services outside of the STC routes. These buses were nicknamed "mosquito buses" for the way they weaved in and out of traffic, competing fervently to be the first to arrive at bus stops to pick up passengers. The "mosquito buses" were first built by attaching a body with seats for seven passengers to a motor car chassis.

With the lack of institutional oversight and regulations, the bus industry was a relatively free market which was imperfect in many ways. It was a profit-driven system, and when coupled with the benign neglect of the government, it resulted in a high sense of dissatisfaction among bus commuters.

In spite of that, the Registrar of Vehicles (ROV) did acknowledge that the "mosquito buses" had a useful informal role to transport people efficiently.

---

[c] A management agreement was formed between the Traction Company and the Shanghai Electric Construction Company Limited for the operation of the trolleybuses, and later the buses, which continued until 1935 when it was terminated by mutual agreement.

Table 1.   The 10 chinese bus companies.

The 10 Chinese bus companies
Changi Bus Company
Easy Bus Company
Green Bus Company
Hock Lee Amalgamated Bus Company
Kampong Bahru Bus Service
Katong-Bedok Bus Company
Keppel Bus Company
Paya Lebar Bus Service
Ponggol Bus Service
Tay Koh Yat Bus Company

As reported by the ROV in 1939: "The mosquito bus companies have established a very useful form of transport and are very popular with the travelling public. An attempt has been made to establish definite time schedules on a properly organised basis and this is proving successful".[d]

An amalgamation of these many small "mosquito bus" operators was deemed necessary by the ROV for order to be instilled in bus schedules and routes. In 1935, the ROV managed to have the many operators consolidated to 10 companies (also known as the Chinese bus companies) with 144 buses (see Table 1). These buses ran regular services in the different parts of Singapore. This probably marked the first government's attempt to centrally organise private bus operators, in addition to operating a government-owned STC.

1.2. The Chaotic years of bus system from 1940s to 1960s

After the Japanese Occupation in 1945, the STC was left with only 51 serviceable vehicles, comprising 29 trolleybuses and 22 buses. The STC started a long process to re-build its operation in tandem with post-war economic rebuilding. Between 1951 and 1954, bus ridership increased by 49%, which was a huge growth in a short period of time.[e]

The Chinese bus companies also managed to recoup and salvage what was left of their "mosquito buses" after the war. Within four years, 249 buses were

---

[d] Hawkins, L.C., *The Report of the Commission of Inquiry into the Public Passenger Transport System of Singapore*, (1956).
[e] Ibid.

Table 2.   Services provided by STC and Chinese bus companies.

	STC		Chinese bus companies	
	Number	Per cent of total	Number	Per cent of total
Passengers carried per year	85,842,000	48%	94,111,000	52%
Miles worked per year	14,694,000	38%	23,691,000	62%
Vehicles owned	404	47%	461	53%
Seats	13,509	49%	14,222	51%

Source: Hawkins, L. C. The Report of the Commission of Inquiry into the Public Passenger Transport System of Singapore, January 1956.

already operating on approved routes. So the Chinese bus companies once again proved to be popular among commuters during the 1950s.

Although government-owned STC was the largest among the 11 bus companies and owned 47% of the bus fleet, it only worked 38% of the miles which was considerably less than those worked by the other 10 Chinese bus companies (see Table 2). This coexistence of both a government-owned bus company and 10 privately-owned bus companies was a problematic arrangement.

This arrangement created intense competition from the Chinese bus companies, which severely affected STC's operations resulting in serious labour and financial problems. This led to poor wage for the workers and high union agitation. There was a clear lack of transport planning. Thus, bus companies simply cherry-picked the lucrative routes and ran buses in high demand areas, leaving many low demand areas uncovered. This greatly affected the travelling population living at rural areas. In a nutshell, providing a comprehensive route coverage and accessibility were not the considerations at all, other than commercial profits. There were no universal service obligations that the bus companies had to deliver.

First landmark Hawkins report for bus reform

By 1955, both the STC and Chinese bus companies were already plagued with operational difficulties, poor management and labour unrest. The bus industry was so bad that the government found it necessary to intervene and overhaul the entire bus system.

In October 1955, it set up a commission of inquiry under the chairmanship of L.C Hawkins of the London transport executive to review the public passenger transport system of Singapore and its future developments.[f] On 12 January 1956, a report titled *Report of the Commission of Inquiry into the Public Passenger Transport System of Singapore*, also known as the Hawkins Report, was published. This was the first landmark report that aimed at reforming the bus system.

The report recommended that the 11 existing bus companies be consolidated into a single undertaking, with common ownership and common management. This single company could be a nationalised government-run company or a statutory limited liability company, financed partly by government and partly by private investment. It also recommended for the establishment of a single licensing authority to replace the existing dual system of licensing provided under:

(i) Section 13 of the STC Ordinance, 1925 for STC routes; and
(ii) Section 326 of the Municipal Ordinance, 1913 for routes applicable to the 10 Chinese bus companies.

Unfortunately, during that period from the mid-1950s to early 1960s, Singapore suffered many worker strikes and police lock-outs, and this badly handicapped the Commission in its task to implement its key recommendations to improve the bus system.

Bus workers that were controlled by communist unions frequently resorted to work stoppages, paralysing the whole bus system. There were 57 strikes in the Chinese bus companies in 1955 alone.

The workers of the Hock Lee Bus Company (one of the 10 Chinese bus companies) went on strike in April 1955, which was supported by Chinese language medium school students. This was followed by the "Black Thursday" riot in May 1955 which ended with the death of four people — two police officers, a student and an American press correspondent. Thirty others were seriously injured. The next year, in January 1956, the 146-day "Great STC Strike" happened and this further aggravated the already seriously crippled bus system.[g] Given such a disorderly context, the undertaking of bus reforms was not only pressing but also more politically challenging for the government.

---

[f] *White Paper on Reorganisation of the Motor Transport Service of Singapore*, (1970).
[g] Centre for Liveable Cities & LTA Academy, *Crossroads and Nexus: Land Transport Planning and Development*, (2011).

First institutionalisation of an authority to license bus services

Although it was not possible to implement in full, the efforts of the Hawkins Report were nevertheless not entirely in vain. The recommendation of enacting the omnibus services licensing authority (OSLA) Ordinance No. 13 of 1956 was implemented by the labour front government.[h] In terms of institutional design, this probably marked the first setting up of a single authority to put all bus service licensing under one-roof. This was a significant attempt in moving towards integrated provision of bus services.

In 1966, after a decade-long hiatus due to civil unrest and riots, the OSLA resurfaced the recommendations of the Hawkins Report. It suggested for the amalgamation of the 11 existing bus companies in order for bus services to be more efficient. However, after many months of discussion, the plan for amalgamation once again fell through.

A major contributing factor to the difficulty in materialising the amalgamation was an absence of a disciplined task force to effectively check on public transport matters. The ROV's inspectorate[i], with a staff strength of 45, was disbanded during that time due to public allegations and complaints that staff was abusing its wide powers.[j] There was general distrust between the public and the authority, and this had made the situation worse for the much needed bus reform. In short, OSLA failed to deliver its mandate as intended.

In 1968, three years after Singapore's independence, the government stepped in to establish a 15-member transport advisory board (TAB) with two main goals in mind. First, to study ways and means of improving the public transport service, Second, to reduce congestion on the streets with the possibility of introducing staggered working hours in offices and schools in the city area. The TAB was made up of representatives from four sectors, namely bus workers, owners of bus companies and taxis, organisations and individuals interested in the improvement of safety and road users, and the government (see Annex B for members of TAB).

By January 1969, TAB submitted its first interim report, and seven months later, the government announced its acceptance of TAB's recommendations. The main recommendations for the improvement of the public bus service

---

[h] The labour front won the 1955 legislative elections. It formed a coalition government for Singapore which, at that time, was a separate crown colony.

[i] Established in 1945, the ROV was an influential player in Singapore's land transport development. It was sometimes known as the 'custodian of roads' and played a big part in every motorist's life. Prominent duties of the ROV were the implementation of policies relating to vehicle ownership and usage control, enforcement of vehicle safety, and the regulation of public transport services.

[j] *White Paper on Reorganisation of the Motor Transport Service of Singapore*, 1970.

embraced simple but effective quick fixes such as queuing up at bus stops, and the printing of schedules showing bus routes and, service schedules.

Such simple quick-fixes did show some tangible results of service improvements, both optically and in terms of operational efficiency. From this point onwards, the public began to slowly appreciate the deliverables. In a way, this pragmatism helped to enable the government to regain the much-needed public trust that was lost in the past decades.

1.3. *Overhauling bus system in 1970s*

The TAB was the initial step in a series of major milestones in the reorganisation of the public bus system in the 1970s, transforming it from one which was *laissez faire* with a free market of many players, to a bus industry that had a centrally-planned approach with a restricted number of operators[k] (Fig. 1).

First white paper (1970) to reorganise bus system

Following the recommendations of the TAB, the government initiated the publication of the 1970 White Paper on *The Reorganisation of the Motor Transport Service of Singapore* as a continual effort to improve the bus system in Singapore. The 1970 White Paper explicitly stated the tumultuous situation of the public transport service and the three main reasons for its unsatisfactory state. First, there had been a lack of a defined policy in the past, coupled with ad-hoc decisions being made as and when problems arise. Second, there was also a lack of continuity in the important post of Registrar of Vehicles (see Annex C for ROV incumbents from 1959 to 1970) to ensure continuity and consistency in the ROV's action plans. Third, there was insufficient financial support that resulted in an inadequate number of qualified staff, as well as the lack of proper facilities and equipment for the ROV to perform, implement and enforce its multifarious

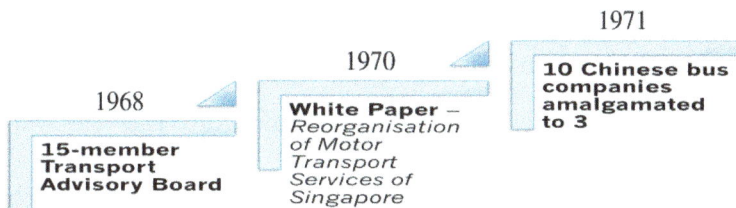

Fig. 1.   Transforming the bus industry.

[k] Centre for Liveable Cities & LTA Academy, *Crossroads and Nexus: Land Transport Planning and Development*, (2011).

duties. These three reasons underscored the critical factor of having a ready institution set-up when considering bus reform.

The White Paper also addressed some important aspects of road transport which needed improvements, namely:

- Early eradication of smoky exhausts from motor vehicles which was in line with the then anti-pollution campaign for Singapore;
- The amalgamation of the 11 existing bus companies into fewer but larger companies, and the necessity to streamline bus routes for better bus services;
- Phasing out of pirate taxis (i.e., unlicensed taxis competing with legal taxis and buses); and
- Job opportunities to alleviate the repercussions affecting drivers of pirate taxis.[1]

By tackling head-on the critical and fundamental issues concerning institution design, policy and planning, and root causes of operation problems, the 1970 White Paper had effectively laid a firm foundation for the overhaul of the overall public transport service, where the 10 Chinese bus companies finally merged into three major bus companies with clear territorial demarcations.

First bus planning study for Singapore — Wilson report 1970

In the preparation of the 1970 White Paper, the government decided to conduct a technical study on bus service and operations under the auspices of the then Ministry of Communications (MOC) and the State and City Planning (SCP) department under the Ministry of National Development (MND). The cost of the study was borne by the 11 bus companies (i.e., the STC and the 10 Chinese bus companies).

A transport consultant, R.P. Wilson, who was a traffic manager of the municipal tramways trust of Adelaide, South Australia, was appointed to conduct the study. Referred to as the Wilson Report, the study was completed in November 1970. The main aims of the study were as follows:

1. To reorganise the 117 bus routes to provide more direct services with minimum transfers;
2. To make recommendations on the allocation of bus services to each of the four operating companies (STC and three merged companies);

---

[1] Statement by the Minister for Communications, *Reorganisation of Motor Transport Service of Singapore*, (1970).

3. To review the existing bus fare structure and make recommendations on a new fare structure; and
4. To advise on any other specific features connected with bus operations, as and when necessary.

The study concurred that the amalgamation of the 10 Chinese bus companies into three regional operating groups was a step in the right direction. This would enable the reorganisation of bus services and planning for future expansions to take place on a much sounder basis. However, even with four operating companies (i.e., the STC and three merged companies) in place, the study noted that there might be complications in determining the routes of the services and the allocation of these services. It therefore recommended to ultimately having the four operating companies brought under a single operating entity, thereby also creating greater economies of scale.

The Wilson Report suggested that the reorganisation of the whole of the bus services could take place in mid-April 1971 instead of the originally intended date at the beginning of 1971. This would give the bus companies a reasonable amount of time to prepare for the changeover once the plan had been officially approved and released to them. According to the Wilson Report, "such hasty preparations of a major reorganisation of this kind could seriously jeopardise the success of the scheme."

Furthermore, the implementation in mid-April would coincide with the school vacation period when the public transport services were not stretched to capacity. This would make the change of such magnitude less difficult to cope with, especially with any teething problems that may be encountered during the first few days of operations.

Genesis of an integrated bus network

Through the analysis of data, the Wilson Report noted that there was a strong demand for public transport services to and from the following locations within the central city area:

- The commercial area south of the Singapore River in the vicinity of Collyer Quay, Shenton Way and Robinson Road;
- The Chinatown area in the vicinity of South Bridge Road, Neil Road and New Bridge Road;
- The administrative and shopping complex near North Bridge Road and High Street; and
- Orchard Road.

The report thus recommended that a series of radial bus services towards these areas, which operated on main arterial roads and connected the major areas of urban development with each of the abovementioned areas, form the basis of the integrated bus network. It further recommended that the radial services were to be supplemented by a series of ring routes and inter-urban services so as to provide connections with the radial services and direct public transport links between major residential and industrial areas.

To avoid overlapping or duplication of services in the central city area and to minimise the amount of space required at city bus terminals, the Wilson Report recommended for the services to be linked to provide through city operation between areas to the east and north-east of the city, and those to the west and north-west. This arrangement of services enabled the majority of bus travellers to complete their journeys without changing vehicles; and in cases where this arrangement was not possible, only one transfer would be necessary.

Some considerations were also given to the establishment of express or limited stop services in areas of high density population. The Wilson Report noted that such services were only successful where large numbers of people required transport between one specific area and another, and where express services provided a significant reduction in travelling time.

Furthermore, analysis of bus passenger data showed that the very high demand for bus travel during peak periods resulted in the inefficient use of buses — i.e., many buses could be used effectively for only one trip during each of the peak periods. The Wilson Report noted that this was an uneconomic form of operation because neither the buses nor the bus crews could be employed profitably at other times. For this reason, many bus companies were reluctant to purchase buses and employ bus crews to improve peak hour services, thus contributing to the overcrowding of buses during peak periods.[m]

Eliminating duplicated services, as well as arranging for more direct routes (instead of circuitous routes that went through sparsely developed areas) through densely populated areas were also proposed as a step forward in improving the standard of bus services along major arterial roadways in more developed areas. There were many outer areas which were over-serviced, especially during off-peak periods. The report suggested that bus service coverage could be widened to release additional buses for use in the more heavily populated inner areas.[n]

---

[m] Wilson, R. P., *A Study of the Public Bus Transport System of Singapore*, (1970).
[n] Ibid.

From transport planning perspective, this genesis of an integrated network of bus routes became a lasting building block that could explain much of the legacy bus routes that exist till today. It could also be said that bus network planning had its inception planted as early as at Wilson Report in 1970.

Improving frequencies to meet demands

In determining bus frequencies, the Wilson Report stated two factors that had to be taken into account. They were:

- The number of people requiring transport during any given period; and
- The carrying capacity of the buses in use.

The report recommended that bus frequencies to be arranged so that the average load during peak periods amounted to no more than 80% of average bus capacity. This would provide sufficient flexibility to meet fluctuations in demand and avoid overcrowding as well as delays to passengers during peak hours. However, the report noted that there were insufficient buses to meet this requirement at the time and any deficiencies in the standard of service arising from this could not be remedied until more buses were made available.

To better address fluctuations in demand during peak hours, the report highlighted the need to prepare schedule specifications in relation to demand over each 10 or 15-minute interval during peak hours instead of maintaining a set frequency of service through the whole of each peak period as practiced by the bus companies at the time. Detailed studies of the flow of bus passengers through peak periods were necessary to provide the data needed to establish precise peak hour schedule specifications. However, due to the dearth of necessary data available at the time, the team was unable to prepare detailed specifications.

This underscored the importance of systematically capturing and gathering timely data, without which transport planning and service specifications would not be able to be carried out.

Industry restructuring and direct allocation of bus services

The Wilson Report allocated bus services to the four proposed operating companies based on the following guiding principles:

- The number of buses each company had available by the end of 1970:
- The location of depot sites owned by the respective companies;

- The distribution of services in as equitable manner as possible; and
- The avoidance of two or more companies sharing a common route.

The number of buses required by each company to operate the services in accordance with the recommended allocation, compared with the number of buses that each company had available by the end of 1970 was as shown in Table 3 below:

Table 3.   Allocation of buses in each company, Year 1970.

Company	Buses Available	Buses Required (including spares)	Deficiency
Associated Bus Services Pte. Ltd.	461	467	6
United Bus Co. Pte. Ltd.	384	387	3
Amalgamated Bus Co. Ltd.	271	277	6
Singapore Traction Co (STC).	426	429	3
Total	1,542	1,560	18

Source: Wilson, R.P., A Study of the Public Bus Transport System of Singapore, November 1970.

The team evaluated the estimated deficiency of 18 buses as of November 1970 and concluded that it should not cause any serious difficulties as all companies already had additional buses on order. While waiting for the orders to arrive, the Wilson Report suggested that careful planning of servicing and maintenance procedures would help alleviate the interim deficiency.

A common fare structure to eliminate anomalies

The Wilson Report recommended that a common fare structure be adopted for all bus services. As all four companies provided services over common routes within the central area and in some suburban areas, differing fare levels would create grave anomalies which may ultimately affect bus services as a whole.

Before the rationalisation of bus routes and services, fares charged by STC were higher than those charged by the 10 Chinese bus companies. Despite the higher fares, STC sustained losses while the Chinese bus companies were profitable on the whole. The Wilson Report recommended that uniformity could be achieved by adopting a simplified 10 cent fare scale throughout the system, rather than a 5 cent fare scale which was in effect then. The scale provided for five fare values — 10, 20, 30, 40 and 50 cents. A flat journey fare of 10 cents for children travelling any distance on any one bus service was also recommended.° This common fare structure could be described as the first integration of bus fares that aimed to eliminate the legacy of anomalies.

Concerted publicity to minimise public confusion

The Wilson Report recommended that detailed publicity through the press, radio and television stations be used to reach all sections of the community so as to readily inform them of the new routes, fares and stopping places in the central city area. This was to avoid confusion during the major reorganisation of whole bus system. It suggested that publicity materials include route maps, route descriptions, fare schedules and a plan showing the location of bus stops for each service in the central city area. It also recommended that training should be given to the bus crews (i.e., drivers and conductors) so they could be thoroughly conversant with the routes of the new services and with the new fares.

In summary, the Wilson Report recommended specifically for:

- Low fares;
- Fast journey times;
- Frequent, regular and reliable services;
- Conveniently located routes and stopping places;
- Modern attractive, clean and comfortable vehicles;
- Adequate information services as to routes, timetable and stopping places; and
- Courteous and efficient crews.

Supporting measures to ensure success

Besides giving its recommendations for the improvement of bus services, the Wilson Report also observed that bus passengers in Singapore were regarded as second class citizens. For example, car parks were found at prestige locations whereas city bus terminals were inconveniently located for both passengers and the operating companies. The road space provided for bus stops were also grossly inadequate, with priority given to car users. Furthermore, one way streets which were designed to facilitate general vehicular traffic resulted in many circuitous bus routes which were not only inconvenient, but also resulted in increased journey times and operating costs. It was little wonder that Singaporeans chose to become car owners, or chose pirate taxis to meet their transport needs. Considerable improvements were necessary to create a better image for the public transport system, which would encourage its greater use.

° Ibid.

In addition, the report found it imperative to restrict or control the access of vehicular traffic into the city area in order for bus services to be effective as a means of mass transportation for the travelling public. The report opined that buses should be given some form of priority on roadways and intersections, or even the allocation of road space for the exclusive use of buses.[p]

With the aim of providing direct and efficient service for bus commuters, the government accepted the advice of the Wilson Report and made bold steps of rationalising the existing 117 bus services and introducing the new unified fare structure of a 10-cent fare scale on 11 April 1971.

Looking at the thinking behind the comprehensive recommendations made, the Wilson Report was indeed a turning point in the history of bus transport in Singapore. In fact, most of the insights provided by the recommendations remain relevant and applicable even today.

Implementation of the amalgamation of Chinese bus companies

The first step taken by the government to rationalise the bus service was to urge the 11 bus companies to amalgamate into four bus companies (STC and three merged private companies) by 1971. During the preparation of the 1970 White Paper, the managements of the 10 Chinese bus companies had agreed to amalgamate among themselves into three regional groupings by August 1970. The failure to do so would result in the government's intervention in order to attain the "national objective" of reorganising the public transport system before the end of 1970.

As it turned out, on 16 November 1970, the Hock Lee Amalgamated Bus Co. Pte Ltd, Kampong Bahru Bus Service Pte Ltd, and Keppel Bus Co. Pte Ltd came together to form the Amalgamated Bus Co. Ltd, which served the western part of Singapore. Soon after, on 10 December 1970, the Changi Bus Co. Pte Ltd, Katong Bedok Bus Pte Ltd, Paya Lebar Bus Service Pte Ltd, and Ponggol Bus Service Pte Ltd also came together to form the Associated Bus Services Pte Ltd which served the eastern region. However, the northern group comprising the Easy Bus Co. Pte Ltd, Green Bus Co. Pte Ltd, and Tay Koh Yat Bus Co. Pte Ltd failed to meet the deadline of 1 January 1971.[q] The government therefore intervened and they eventually merged to form the United Bus Co. by April 1971. The amalgamation of the Chinese bus companies proceeded as shown in Table 4.

---

[p] Ibid.
[q] Parliament No. 2, *Bus Services Licensing Authority Bill*, (1971).

Table 4. Amalgamation of the 10 Chinese bus companies.

| | | No. of | No. of buses | | Total |
		Services	Operating	Spares	Fleet
(A) ASSOCIATED BUS SERVICES PTE LTD					
1. Paya Lebar Bus Service		7	91	10	101
2. Changi Bus Co.		6	82	11	93
3. Katong-Bedok Bus Co.		4	48	6	54
4. Ponggol Bus Service		1	17	2	19
	Total:	18	238	29	267
(B) UNITED BUS CO.					
1. Tay Koh Yat Bus Co.		21	147	16	163
2. Green Bus Co.		13	95	10	105
3. Easy Bus Co.		2	14	4	18
	Total:	36	256	30	286
(C) AMALGAMATED BUS CO.					
1. Hock Lee Amalgamated Bus Co.		15	134	18	152
2. Keppel Bus Co.		5	43	7	50
3. Kampong Bahru Bus Service		2	12	1	13
	Total:	22	189	26	215

Source: White Paper on Reorganisation of the Motor Transport Service of Singapore, 1970.

Setting up of a new bus services licensing authority (BSLA)

Following the 1970 White Paper, the Singapore Traction Ordinance and the OSLA Ordinance were repealed. Replacing the outmoded OSLA was the bus services licensing authority (BSLA) in 1971.

Although the OSLA was established with the objective of providing adequate and better public bus transport services, its efforts were greatly hampered by the provisions of the 1956 OSLA Ordinance, which *inter alia* imposed upon OSLA the obligation to consider the interests of the STC when granting new routes, including those within the city area. In addition, the OSLA could not withdraw any of the STC's routes, even if the STC was inefficiently managed. No condition could be attached to any of the STC's authorised routes. The STC was empowered to licence its own drivers and conductors so that even the ROV had no authority to withdraw or suspend their vocational licences for any breach of conduct.[r] In other words, OSLA was in fact hamstringed to protect the STC's privileged interests.

Under the new legislation, BSLA had wider powers to regulate the bus industry including coordination and rationalisation of bus routes. BSLA comprised six members (instead of the previous three in OSLA), with representatives from different government departments involved in land transport such as the Public Works Department (PWD), Housing and Development Board (HDB), and ROV.[s] The then Ministry of Communications, the predecessor of the Ministry of Transport, was responsible for approving requests for fare adjustments from the bus operators.

Formation of a single bus company in 1973 — Singapore Bus Service

Even with the amalgamation of the 10 Chinese Bus Companies into three and the rationalisation of existing bus services in 1971, the disorganised situation continued to deteriorate. This was accompanied by lower profits, frequent bus breakdowns, absenteeism and industrial unrest.[t] At the same time, the STC continued to be in financial difficulties, which was largely caused by the introduction of the common fare structure, and it eventually had to cease operations. This prompted the government to elicit a solution for the future of Singapore's bus transport, and in 1973, the three bus companies merged to form a single private entity — Singapore Bus Service (SBS).[u] See Fig. 2 for events leading to the formation of SBS.

SBS was a privately owned company. The government decided that nationalisation was not an option at that point in time, largely in view of the prevailing political and racial volatility situation then. As explained by a former Minister for Communications and Works, Mr Francis Thomas, "The problem is to get a take over of industry in which you will not have a head-on collision

Fig. 2.   Events leading to the formation of SBS.

---

[r] Ibid.
[s] Ibid.
[t] Menon, A.P.G, and Low, C.K. *Lessons From Bus Operations*, (Land Transport Authority, 2006).
[u] Centre for Liveable Cities & LTA Academy, *Crossroads and Nexus: Land Transport Planning and Development*, (2011).

between the workers and the employer who, in this case, will be the government. When you had the recent Hock Lee trade dispute, it was possible for the government to intervene and mediate. It would not be so easy for them to do so if the government was, in fact, the employer"[v]. The government wanted to retain the upper-hand flexibility to act should similar situation arise that would warrant its intervention.

The merger however did not quite meet the expectations of the public, and reports of the time described SBS as "an outsized body with an underdeveloped brain". In 1974, it was common for SBS to have a total of 400 buses out-of-service and breakdowns to average about 800 a day.[w]

Further intervention by government team of officials in 1974

Despite the merger of the three bus companies, the old inefficient way of management persisted. None of the companies attempted to establish a unified management structure or introduce a uniform system of operation, accounting and maintenance. Internal feuds and squabbles were known to exist as each company effectively continued to operate independently of the others.

In addition, the managers had vested interest. They had their own businesses or petrol stations and were making purchases for SBS without any proper tendering. In short, the amalgamation of the three bus companies was more of a paper exercise. The owners continued to run the company their ways. There was no proper accounting and tickets were piled up all over in the bus depots. Anyone could pick up bus tickets and sell them. SBS was in a big mess.

Being frustrated by the incompetent management of the SBS, the government felt that there was no other option but to assist SBS by seconding a government team of officials (GTO) to SBS in May 1974 (see Annex D).

The GTO was initially tasked to put together a consultancy report on SBS. Subsequent to the report, this inter-departmental team of about 100 civil servants, police officers and military personnel were mandated the important task of restructuring SBS's operations, and improving the company's productivity and profitability.

It was in the government's view that if SBS were to be left to run on its own, the quality of Singapore's bus service would deteriorate even further. SBS simply lacked sufficient personnel of the right calibre and experience to turn-around and improve the company's performance. The GTO was the

---

[v] Parliament No. 0, *Public Passenger Transport Commission*, (1955).

[w] Sharp, I., *The Journey — Singapore's Land Transport Story*, (SNP Editions, 2005).

government's direct and hands-on intervention in the affairs of a privately-owned company.

The GTO was tasked to identify the weaknesses in the systems of management, operational and accounting procedures, personnel management and training, provisioning and procurement systems for vehicle repairs and maintenance as well as to make recommendations for improvements in these areas.[x]

Initially, the GTO encountered a lot of resistance from the management and staff of SBS. The GTO had to stop illegal practices, enforce staff discipline and worked with the unions to bring order and discipline back to the workforce.[y]

Nevertheless, with much effort and hard work, the GTO succeeded in digging out the root causes of SBS's inefficiencies, one of them being the lack of employee welfare. "We found a lot of the breakdowns were caused by the drivers themselves, because their pay was low and they needed to work overtime… when they needed rest, they would take an iron rod and puncture the tyres so that while they waited for repair crews to come, they could get some rest!" said Mr Wong Hung Khim, who was seconded from the Ministry of Labour to head the GTO.

Besides reorganising the management structure of SBS, the GTO kept records, computerised the takings and monitored the staff. It also had discussions with the Urban Redevelopment Authority (URA) and the Housing Development Board (HDB) to acquire land to build new bus depots. Within six months, Mr Wong was able to report that conditions in SBS had improved as well as achieving a remarkable turnaround with profits.

SBS was finally free from the old inefficient, parochial ways of management. The GTO's swift intervention, together with the cooperation of SBS's management laid the foundation for SBS to grow into what it is today. The year 1974 also saw public buses being given greater priority on the roads with the introduction of dedicated bus lanes.[z]

By 1978, SBS was doing well enough to be listed on the Stock Exchange of Singapore, with a public issue of 20 million shares at S$1 each.

---

[x] Report of Government Team of Officials, *Management and Operations of Singapore Bus Service Ltd.*, (1974).

[y] Oral History Interview with Mr Mah Bow Tan, 24 February 2012. Mah Bow Tan was a member of the GTO. He was a member of the Cabinet from 1991 to 2011, serving as Minister for Communications (1991–99), Minister for the Environment (1993–95) and Minister for National Development (1999–2011).

[z] Sharp, I. *The Journey — Singapore's Land Transport Story*, (SNP Editions, 2005).

Stop-gap measures — supplementary public transport services

Insufficient peak hour capacity remained a major public transport problem in the mid-1970s. SBS had just been formed and immediate solutions to the acute public transport situation could not be expected of the newly established company. Besides introducing bus lanes in 1974, two supplementary public transport services, schemes A and B, were therefore introduced to ameliorate the insufficient peak hour capacity.[aa]

Scheme A bus services to carry adult workers

The Scheme A system was introduced in February 1974 to allow lorries, goods vehicles, private hire and school buses to transport workers between their homes or designated points and places of work. The use of lorries (lorry-bus, as they were called) was phased out after a few years due to their unpopularity with passengers.[ab]

Scheme A operators had to apply for an adult workers contract (AWC) permit and a public service vehicle permit issued by the ROV. Monthly rates were negotiated between the employers and operators. No restriction was set on the hours of operation, but the operators were not allowed to pick up and charge passengers at bus stops along the way.

Scheme B bus services to carry peak hour passengers

In March 1974, the government introduced the Scheme B system to cater to peak hour demand for public bus transport between the residential estates, the city and the industrial areas. The scheme allowed school and other buses to ply for passengers along authorised routes during peak hours. The scheme lightened the load of the then overcrowded SBS buses.

There were two scheduled services under Scheme B. Schedule I buses were recognised with their red plates and plied between housing estates and the city. Schedule II buses bore blue plates and plied between housing estates and industrial areas. The Scheme B licence was issued by the BSLA and the vehicles required a PSV issued by ROV.

In July 1978, the government extended the Scheme B system to serve some rural areas in Singapore. Known as the Scheme B rural services, it catered to residents who needed to walk a long way to reach the main roads. Scheme B system was expanded in 1982 to integrate school and private hire buses into SBS's scheduled services. For better operational control, the pilot project was

---

[aa] Menon, G., Low, C.K., *Lessons From Bus Operations*, (Land Transport Authority, 2006).
[ab] Ibid.

restricted only to fleet owners who had more than seven buses. Operators were required to run each cluster of service, which was a mixture of popular and unpopular SBS's routes. Buses of these services carried similar SBS service numbers and had to keep to schedules controlled by SBS's time-keepers.[ac]

New city shuttle services to support Park-and-Ride

In conjunction with the Park-and-Ride (P&R) and Area Licensing Scheme (ALS), the City Shuttle Service (CSS) was implemented in May 1975 to cater to the expected increase in demand for public bus transport at the fringe of the city, which was marked as the Restricted Zone (RZ). The ALS required motorists to purchase a paper licence if they wished to drive into the RZ during restricted hours. The P&R and CSS were introduced to position the ALS with something more attractive.

The P&R was seen as a good alternative for those who wanted to drive but did not wish to purchase the ALS licences. It encouraged motorists to park their cars at the outskirts of the city and switch to public transport for their trips into the RZ. The CSS was thus provided to commute motorists who parked at the P&R fringe car parks to destinations within the RZ.

Two companies, Singapore Shuttle Bus (Pte) Limited and National Trade Union Congress (NTUC), were involved in the CSS. When the CSS first began, it had a total of 11 services, seven were operated by the Singapore Shuttle Bus (Pte) Ltd and four by NTUC. It had a combined fleet of 88 medium-sized buses and initially plied between fringe car parks and the destinations within the RZ. However, the CSS was not well received both before and after the ALS was implemented. Its routes were thus adjusted to improve patronage. CSS expanded its operations between housing estates and the city. Headways of CSS services ranged from six to 15 minutes.[ad]

First limited-stop direct bus services — SBS's blue arrow services

In 1975, with the introduction of the ALS, SBS also introduced six "blue arrow services" (service 301 to 306) in an effort to encourage motorists to use buses. These buses were called "blue arrow services" because the buses carried a blue plate with two horizontal arrows in opposite directions. Such services mainly served the private residential estates and were one-man operated. It charged a slightly higher flat fare of 50 cents each way and made limited stops so that buses

---

[ac] Ibid.

[ad] Land Transport Division of the Ministry of Communications and Information, Brief on Bus Services in Singapore, (1985).

ran express to the CBD. Bus timings were from 7.00am to 9.30am and 4.30pm to 7.00pm on weekdays. The blue arrow services were indeed the forerunners of today's premium bus services[ae].

In summary, the 1970s began with the government's clear intention and intervention to overhaul the bus industry structure by consolidating the bus companies and rationalising bus routes. Implementation of policy directions and integrated planning of bus network took centre stage. To a large extent, such direct interventions had laid a strong foundation for a bus industry that could grow to meet the travel demands brought about by economic development. Specific bus services were also introduced to better meet peak demands and complement the basic bus services.

1.4. *Instituting regulatory oversight and integration from 1980s to 2000s*

In early 1980s, there were increasing public dissent against SBS for being a monopoly. The government saw that the introduction of a second bus company would be a catalyst for improving bus services.

A second bus operator to rival SBS

In March 1982, following the government's announcement on the need of competition within the bus industry to raise service standards, the Singapore Shuttle Bus (Pte) Ltd was given a licence to operate the second bus company which would rival SBS on an equal footing. Trans-Island Bus Services (TIBS) was subsequently incorporated on 31 May 1982. The new bus company started operations in April 1983 with 37 buses plying on two services — 160 and 167. The establishment of TIBS was a catalyst for creating competition for the bus market, with the aim of enhancing bus services.

Initially, TIBS outsourced management and administration, repair and maintenance of its buses to Trans-Island Management and Engineering Services (TIMES). In March 1987, TIBS decided to end its outsourcing arrangement with TIMES and took over the role of managing its own administration, repair and maintenance. By 1991, TIBS had become a company which was publicly listed on the SESDAQ[af]. It also launched its taxi division and underwent a restructuring of business units. Together with other subsidiary

---

[ae] The premium bus service is a seating-room only bus service aimed at bringing customers from their homes to the city in comfort.
[af] Stock Exchange of Singapore Dealing and Automated Quotation (SESDAQ) was established in 1987 to meet the fundraising needs of local small and medium enterprises (SMEs).

companies such as TIBS Motors (Pte) Ltd and TIBS Taxis (Pte) Ltd, TIBS came under the umbrella company of TIBS Holdings Ltd.

In order to improve their competitive advantage, both SBS and TIBS introduced a series of cost "rivalry" measures to offer superior services. To increase capacity and patronage, double-decker and bendy buses were introduced in the 1980s and mid-1990s respectively. TIBS was the first bus operator to introduce the bendy buses in 1996 and SBS followed suit in the same year. In addition, both operators converted many of their services to one-man operations (also known as driver-only operation) where buses are operated by the driver alone without a conductor. There were some degree of rivalry between SBS and TIBS.

With the second bus operator coming onboard and the launch of Singapore's mass rapid transit (MRT) system, it became evident that there would be a need to institute an independent agency to regulate public transport in Singapore.

Instituting regulatory oversight — Public Transport Council 1987

The Public Transport Council (PTC) was formed on 14 August 1987 to replace BSLA. The set-up was to create greater representation in determining bus routes as well as to serve as a watchdog on behalf of the public in matters that affected commuters such as fares. The launch of the new MRT system[ag] also created a need to integrate public transport services and fares across different modes of transport, namely buses and MRT trains. The PTC operated as a non-profit making statutory board with a small supporting staff, and was funded by a government grant.[ah] (see Annex E for the PTC members as at 31 March 1988).

The PTC's key statutory mandate was the twin objective of looking after the interests of the commuters and at the same time, ensuring the viability of the public transport operators. From the start, the PTC had to deal with three public transport operators (PTOs), namely SBS and TIBS, which were providing bus services, and the Singapore Mass Rapid Transit (SMRT) Ltd, which was operating the MRT system.

The appointed PTC members were selected from the public and private sectors, academia, unions, and grassroots organisations. Initially, the PTC also

---

[ag] While PTC regulated bus service standards, the MRT system was regulated by the former Mass Rapid Transit Corporation (now part of the Land Transport Authority).

[ah] Parliament No. 6, *Public Transport Council Bill*, (1987).

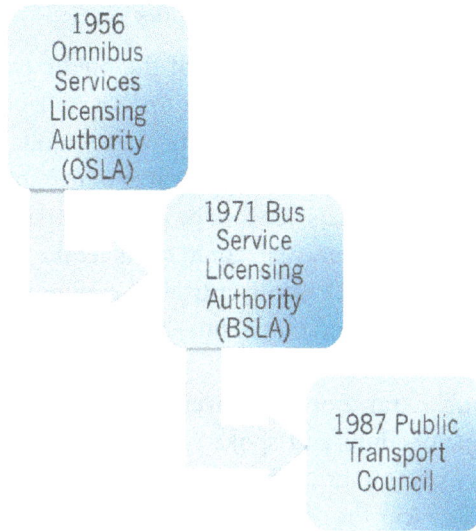

Fig. 3.   Evolution of public transport regulation.

included representatives from the three PTOs so as to achieve a more well-integrated public transport system so that their views could be included in the decision process for matters such as the determination of bus routes and fares. Representatives from the PTOs were nevertheless not allowed to vote on matters in which they had or were deemed to have an interest.

The PTC continued the work of BSLA such as considering applications for the issue of bus service licences for bus routes and ensuring bus service standards, but it also took on the additional responsibility of regulating public transport fares[ai] which was formerly the responsibility of the then Ministry of Communications.[aj]

From an institutional viewpoint, the regulation of bus services had morphed from the first OSLA in 1956 to BSLA in 1971 and to PTC in 1987 (see Fig. 3 on evolution of public transport regulation). With each change, the statutory functions were strengthened to support policy and regulatory changes.

[ai] Bus, MRT and taxi fares. Taxi fares were deregulated in 1998. The taxi industry was fully liberalised in 2003, with liberalisation of supply where taxi operators are allowed to determine their own fares and fleet sizes.

[aj] Parliament No. 6, *Public Transport Council Bill*, (1987).

Integrating bus and MRT services — Transit Link

With the advent of the MRT, the capacity of the public transport system in Singapore had significantly increased. To ensure that transport resources were optimally utilised so that the operating costs of the public transport system would not be significantly increased and the fares were kept affordable, it was considered imperative that the bus network be rationalised so as to complement the MRT system.[ak] A fare system which could facilitate the commuters to make transfers between buses and MRT would critical. Therefore, the PTC tasked itself to oversee the smooth implementation of bus–MRT integration and fare system for the public transport system.

As integrated public transport, the government had planned for the MRT to form the backbone of a hub-and-spoke system by serving the major urban corridors with heavy commuter traffic. Buses were envisaged to play the role of providing trunk services for commuters on major corridors not served by the MRT and feeder services to channel commuter traffic from neighbouring precincts to the nearest transport hub — a bus interchange or an MRT station.

Following the introduction of the MRT in 1987, the three PTOs, SBS, TIBS and SMRT Ltd[al] came together to establish Transit Link Pte Ltd (Transit Link)[am] to take care of bus–MRT network integration, information integration and fare integration. Transit Link's immediate task was to assume the essential role of network integration of Singapore's public transport.

First major bus-MRT integration exercise, 1991–1994

Transit Link took on the role of central planning and coordination of the bus network and it rolled out a series of rationalisation of bus services after MRT started revenue service. Transit Link would conduct a detailed study on the travel patterns the affected commuters, which would be subsequently reviewed and approved by the PTC. The first major bus–MRT integration exercise started in January 1991, and the PTOs had until December 1993 to complete 14 phases of bus service changes involving 94 amendments, 37 withdrawals and 19 new services.[an] Even after the completion, there were further fine-tunings of the

---

[ak] Public Transport Council, *Annual Report 1987/88*, (1988).
[al] SMRT Ltd. was formed in 1987 to operate the MRT network.
[am] Transit Link eventually became a subsidiary of the Land Transport Authority (LTA) on 30 April 2010.
[an] Public Transport Council, *Annual Report 1993/1994*, (1994).

bus network in 1994 involving 29 amendments, two withdrawals and 16 new services.[ao]

Transit Link played a critical role of fronting the public and engaging the grassroots, local communities and constituents. This was also probably the first time such an extensive public engagement programme on bus route rationalisation had been carried in the public transport industry.

Second major bus–MRT integration exercise, 2003

Bus service rationalisation continued from when the North-South East-West lines (NSEWL) were launched to the North-East line (NEL) in 2003. The exercise undertaken for NEL in 2003 was the second major bus–MRT integration exercise. It was completed in three phases. The first phase was done in June 2003, immediately upon the opening of the NEL, to divert bus services for better linkages with NEL stations. The second phase, which was done in July/August 2003, was to cutback or withdraw bus services which were directly duplicating the NEL. The third phase was completed in December 2003 and it was to amend/withdraw bus services complete full integration with NEL. There were a total 27 services, including three Scheme B bus services, were affected. Out of these, 19 were for service amendments and there were six withdrawals.

From the bus commuters' point of view, the NEL bus rationalisation was notably adverse as bus commuters were not only affected by the inconvenience caused by the withdrawal of bus services, but they were also made to take NEL, which charged a higher fares than bus. The unhappiness subsided after refinements were made to the plan and new bus services, service 119 and service 84, were introduced to serve the various schools in Upper Serangoon/Sengkang and Punggol residents respectively. There were also some lessons from the episode on grassroots and public engagement.

Integration of bus–MRT ticketing system in 1990

In 1990, Transit Link introduced the integrated ticketing system that provided a common cashless fare payment system on both bus and MRT services. The farecards were magnetic strip plastic tickets that came in various forms to

---

[ao] Public Transport Council Annual Report 1994/1995.

cater to the different needs of commuters. There were stored value tickets that came in three values, namely: the S$10 blue ticket for adults, the S$10 magenta ticket for senior citizens, and the S$5 red ticket for children. Commuters would insert their magnetic farecards into an on-board bus validator and choose the appropriate fare, after which their farecards would be returned together with a ticket to evidence the fare paid. (see Annex F for picture of farecards)

The launch of the integrated cashless ticketing system also allowed for the introduction of transport fare rebates in January 1991. Without the hassle of reimbursement through cash, the system enabled rebates to be automatically given in a form of discount in fares for the second leg of a journey involving a transfer[ap]. The objective was to offset subsequent boarding charges associated with bus-rail or bus-bus transfers.

By the late 1990s, 85% of public transport trips were paid using the magnetic fare cards. They were deemed as a more convenient payment method by most commuters as compared to the hassle of carrying sufficient cash. The reduction in the management of cash transactions resulted in a significant saving of about 7% in operation costs for the PTOs.

Harnessing ticketing technology for seamless travel

After 10 years of using magnetic farecard technology, Singapore's public transport ticketing system was upgraded to the enhanced integrated fare system (EIFS), a contactless smart card (CSC) ticketing system in 2002, in time for the opening of the NEL and Sengkang-Punggol Light Rapid Transit (SP LRT) system. With EIFS, for the first time, a minimum one-cent fare adjustment increase quantum was possible as compared to 5-cent cash. The system also incorporated multi-fare structure, multi-mode and multi-operator functionalities.

Subsequently, EIFS was replaced by Symphony e-Payment (SeP) in 2009, which used a contactless e-purse application (CEPAS) for the payment of fares. This was to further enhance commuters' travel experience by speeding up boarding times on buses and trains with the 'tap-and-go' payments of the contactless EZ-Link smart cards. For the first time, CEPAS

[ap] Initially set at 25 cents per transfer for adult rides (15 cents for student rides), transfer fare rebates were increased to 40 cents and 50 cents on 1 October 2008 and 1 April 2009 respectively for adult EZ-Link fares before the implementation of the distance-based through-fares on 3 July 2010.

could enable card users to have a single multi-purpose stored value card for micro-payments in Singapore, under the brand name of EZ-Link cards, issued by a card company set up by the Land Transport Authority (LTA). This comprised payments not just for public transport but also for retail services.

In addition to implementing an integrated ticketing system, Transit Link also provided coordinated and convenient information on almost all aspects of travelling on buses and rail. The earlier Singapore bus guide book published by SBS was replaced by Transit Link's guide book. It proved to be a popular information source on MRT and bus services for the commuters.

Instituting fare regulation

In the early years before the introduction of MRT services in 1987, bus fares were regulated by the government. Whenever there were cost increases, public transport operators would apply to the government to adjust fares to cover their costs. In 1987, the Public Transport Council (PTC) was set up as an independent body to regulate public transport fares. The PTC's roles was to safeguard commuters' interest by ensuring adequate public transport services at affordable fares, and at the same time ensure the long term viability of public transport operations. It comprised members from a wide cross-section of the society: union representatives, academia, grassroots leaders, and professionals from the public and private sectors. This facilitated a wide representation of views aimed at making PTC's decisions more acceptable to the public.

Mandating PTC as a public agency with independent decision making power was a unique feature in Singapore's fare regulatory framework. The underlying assumption was that commercial public transport operators would be well placed to optimise their services and earnings if they were to bear full fare-box revenue risks, as long as the minimum standards for service delivery and universal service obligations continued to be regulated by authorities. If revenue risks were to be borne by the government, as in the case of government contracting model in other cities, the PTC's role as formulated in 1987, would become less relevant (Looi and Tan 2007). This was because the government, rather than the PTC, would have to be the party to reconcile between revenue and cost of service provision.

1998 Fare Review Mechanism

From inception in 1987 to 2014, the PTC carried out its statutory mandate in fare regulation to safeguard public interest by keeping fares affordable while ensuring the long-term financial viability of the public transport operators. Public transport operators were required to apply to the PTC for approval to adjust fares. Post-2014, the role of PTC is expected to evolve with the implementation of the government contracting model for bus services in Singapore.

In 1998, the PTC started to introduce a formula to calculate the annual transport fare adjustment cap. This fare cap calculated, in percentage, would be the maximum allowable increase for public transport fares for that year.

The Fare Review Mechanism did not allow automatic fare adjustments as the public would find it hard to accept (Looi and Tan, 2007). The operators had to submit annual fare revision applications to the PTC before 1st May each year. The PTC would deliberate the application and announced the decision in June, with the fare revision taking effect in July in the same year. The fare formula was:

$$1998 \; Fare \; Review \; Formula: \quad CPI + x$$

CPI was the change in the Consumer Price Index (CPI) over the preceding year, representing the cost increases of the PTOs. The value of "x" was set based on net effect of wage increases after accounting for productivity gains.

In 2002, there was a heated public debate centered on the formula after the fare adjustments that year. The public felt that the fare adjustments formula lacked transparency[1] and was not responsive to economic conditions. There was also a perception that it favoured the public transport operators, as it seemed to be a cost-plus formula, which guaranteed the operators a fixed rate of returns without the need to keep costs low. There was also the misconception that fares were not allowed to decrease and could only increase under this formula (Looi and Tan 2009).

There were also questions surrounding the profitability of the operators. It was raised that the operators were making healthy profits and the public questioned why there was a need to ensure that they maintained relatively high profits amid the economic downturn then. There was also a lack of clarity on whether the PTC had set an acceptable rate of return on equity for the operators. This controversy led to the stepping down of the operators' representatives from the PTC and

---

[1] Due to the complexity in deriving x, the actual derivation of x was not made public and was set at 2% for 1998 to 2000 and 1.5% for 2001 to 2005 (Looi and Tan, 2007).

(Continued)

(Continued)

the setting up of the Committee on the Fare Review Mechanism (FRMC) in 2004, which brought about many changes in 2005, including a new fare formula.

2005 Fare Review Mechanism

Responding to the criticisms in the 2002 debate, as well as the recommendations by the 2005 FRMC report, the fare cap formula was revised to:

$$Price\ Index\ = 0.5(\Delta CPI) + 0.5(\Delta WI)$$
$$Fare\ Cap = Price\ Index - 0.3\%$$

CPI remained as the changes in the CPI in the preceding year and the WI[2] was the changes in the Wage Index in the preceding year. The weights for CPI and WI was set to reflect the operating cost structure then. The CPI and WI formed the Price Index and this would improve the responsiveness of the fare adjustment formula to changes in CPI and wage that were macro-economic indicators.

The value of "x" was changed to a productivity extraction that was based on an equal sharing of the operators' past average annual productivity gains. This gave the operators more incentive to reduce cost and improve efficiency and balanced the need to motivate the operators to be productive and allow the commuters to benefit from the productivity gains. The value of "x" was set at 0.3% in 2005 to 2007.[3] Subsequently, the validity period of the revised formula was lengthened from three years to five years to provide greater certainty to both the commuters and operators.

Key Changes in the Mechanism

In addition to the formula, there was a new mandate that stipulated two extenuating circumstances where the PTC should intervene and exercise its powers when deciding of fare adjustment: when there were adverse economic conditions or when there was significant deterioration in the overall affordability of public transport fares. The new formula also allowed for negative fare adjustments depending on the annual changes in CPI and WI. When the formula yielded a negative value, the PTC could consider a downward adjustment of fares or give a rebate.

---

[2] Wage Index is the national average monthly earnings adjusted for any change in the employer's contribution to the government's Central Provident Fund.
[3] After the review in 2008, "x" was recalibrated and set at 1.5% from 2008 to 2012.

(Continued)

(Continued)

The PTC started using the Return-on-Total-Assets (ROTA)[4] values as a reality check for fare revision. The operators' ROTA would be benchmarked against companies in a similar industry with similar risk profiles. The ROTA values should be enough to make capital investments to sustain their services but not excessive.

New Affordability Indicator

As fare affordability was one of the key factors the PTC would consider in exercising its flexibility to intervene, a robust affordability indicator was required. Prior to 2005, fare affordability was monitored through the Household Expenditure Survey (HES).[5] The HES findings were only published every 5 years for calculating two indicators: the average monthly household expenditure on PT as a percentage of average monthly household income, and the average monthly household expenditure on public transport as a proportion of total household expenditure.

The PTC implemented a new fare affordability indicator to track trends more closely on an annual basis. The new indicator used a characteristic family that would correspond with the second quintile (20% to 40%) of household income. This characteristic family would then be representative of the average public transport users, who were the median of the bottom 60% group. The indicator tracked the percentage of household expenditure on public transport on an annual basis in between the two available data points of HES.

Public Transport Fund

In 2003, the PTOs, Singapore Labour Foundation, NTUC Club and Community Development Councils contributed $6 million to set up the Public Transport Fund to cushion the impact of the cut in the CPF rate. The Fund helped the union members who were low-income families facing financial hardship to meet the public transport expenses of their school-going children. The FRMC 2005 recommended that help should continue to be rendered through such targeted means as the Public Transport Fund.[6]

---

[4] ROTA measures the net profit after tax divided by total assets.
[5] HES was a survey done every 5 years by the Department of Statistics in Singapore.
[6] FRMC 2005 report

(Continued)

(Continued)

2013 Fare Review Mechanism

After five years of using the 2005 fare review mechanism, the PTC led another review in 2012. A new Fare Review Mechanism Committee (FRMC) was set up to review fare concessions, the fare adjustment formula and the fare review mechanism. The PTC adopted a new formula recommended by the FRMC:

$$\textit{Fare Adjustment} = \textit{Price Index} - \textit{Productivity Extraction}$$

where **Price Index** $= 0.4\ cCPI + 0.4\ WI + 0.2\ EI$
and **Productivity Extraction** $= 0.5\%$ (*valid for* 2013 *to* 2017)

The cCPI[7] tracked the changes in the Core CPI, the Wage Index (WI) tracked the changes in wage costs and the Energy Index (EI)[8] tracked changes in the energy costs in the preceding year. The value of "x" was kept as a productivity extraction that was based on an equal sharing of the operators' past average annual productivity gains. It was set at 0.5%, valid for five years from 2013 to 2017.

The weights for cCPI and WI were revised to reflect the prevailing operating cost structure. This was to allow the formula to be more responsive to the economic environment. The usage of cCPI in exchange for the CPI allowed for the exclusion of influence of irrelevant items on public transport fares such as the housing prices and cost of private transport. A new component, the Energy Index (EI) that would capture the cost of fuel and electricity, was newly included. The weight for EI was also set to reflect the operating cost structure. This revised Price Index that comprised cCPI, WI and EI would allow for a more representative macro-economic calculation in the cost increases in provision of the public transport services.

Key Changes in the Mechanism

In addition to the changes in the formula, the fare review mechanism also adopted a rollover system that gave the PTC a flexibility to vary the fare adjustment quantum granted or to defer it, partly or wholly, to the next fare review exercise.

---

[7] cCPI excludes items from the CPI that are not relevant to public transport, including items accounted for in the Energy Index.
[8] EI is a composite index created for the tracking of energy cost changes, with a 50:50 split between fuel and electricity.

(Continued)

(*Continued*)

This created a more flexible mechanism for the PTC to balance between its twin responsibilities — to safeguard commuter interests and to ensure overall system viability in the long run.

The mechanism also allowed the PTC to vary the weightage of fare adjustments to be given to the bus and train modes. The equal weightage throughout the years had been to the disadvantage of the bus services due to the hub-and-spoke transport system where trains were a more attractive transport option and where more bus routes were restructured to shorter routes feeding the train system. Under Distance Fares, this hub-and spoke arrangement had resulted in a structural shift of fare revenue from bus to train mode. In the new mechanism, the PTC was able to tweak the weightage for fare revenue adjustments so that, in terms of fare revenue allocation, it could benefit the bus mode more as compared to the train mode so as to bolster the long-term sustainability of bus services. Commuters, however, would not see any change in fares paid as this tweaking of weightage would merely affect the back-end allocation of revenue collected.

The PTC had also removed the ROTA reality check. With the new rail financing framework introduced in 2010 by the government, the operators would not own operating assets and the ROTA would not be meaningful as a reality check. The PTC would instead focus on the fare review mechanism and monitoring of fare affordability.

Improved Monitoring of Fare Affordability

The previous indicator tracked the percentage of monthly household expenditure on public transport via a characteristic 2nd quintile income family. The PTC changed the affordability indicator to the monthly household expenditure of the household on public transport as a percentage of the monthly household income of the 2nd quintile household group. This way, there would no longer be a need to rely on a constructed characteristic family like was done in 2005. Using available and government-published statistics on household data also improved the robustness of the indicator and ensures consistency with such published information.

The expenditure of the household on public transport also excluded taxis. This would be more accurate to capture the expenditure on bus and train fares.

In addition, to ensure that fares would remain affordable for the low-income group and not just the majority of users, the PTC would also track the same affordability indicator for the 2nd decile income group (10% to 20%). Figure 12 shows the improvement on affordability over the years.

(*Continued*)

(Continued)

Monthly Public Transport Expenditure as a Percentage of Household Income

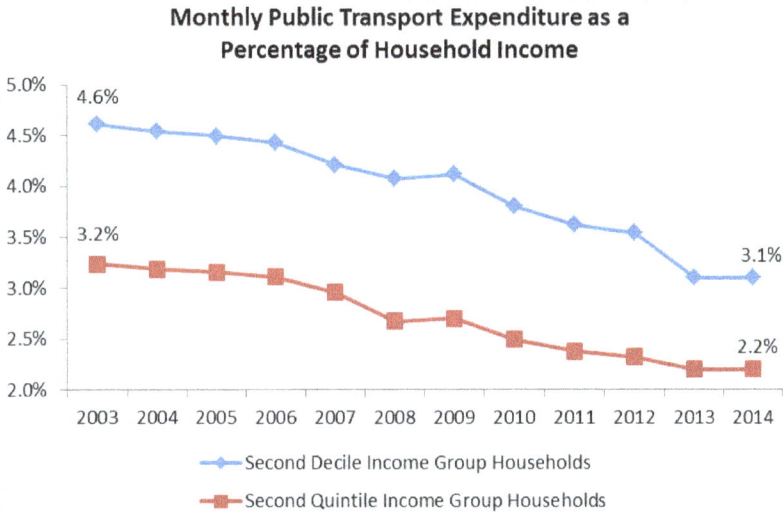

Fig. 12. Affordability of public transport over the years has improved as monthly public transport expenditure as a percentage of household income has declined. Source: Public Transport Council

More Concessions

The PTC also rolled out new concession schemes and improvements to existing concession schemes. The concession schemes for persons with disabilities and the low-income group, which would be funded by the government, were introduced. Adult monthly passes were made more affordable. Concession schemes for young children (under 7 years old), senior citizens, and polytechnic and university students were improved and the concessions were made available to a broader group of students who are Singapore citizens. The funding of the improvements to the existing concessions will come from the full fare paying commuters.

Public Transport Fund

Previously, the government largely funded the Public Transport Fund, with small voluntary contributions by the operators. To ensure affordable fares for the low-income group, the PTC mandated the operators to contribute to the fund a minimum of 20% of the expected increase in fare revenue when there is a fare adjustment. This could increase up to 50% depending on the prevailing operators' profitability. In addition, to close loop with commuters, proceeds from financial penalties imposed on service lapses of the operators would also be channelled to the fund. This way, there would be more resources to sustain the fund.

An integrated agency to deliver a world class land transport

As the land transport system grew more complex, conscious efforts were put in to simplify regulatory and governance framework. The move towards more integration went beyond instituting Transit Link. In 1995, the government established the Land Transport Authority (LTA) to integrate different areas of land transportation, including planning and regulatory functions for both public and private transport. The LTA was formed through the merger of four government departments, namely, the former roads and transportation division of PWD[aq], the ROV[ar], the former mass rapid transit corporation (MRTC) and the former land division of the Ministry of Communications (MOC). (see Annex G for the first LTA Board Members in September 1995).

This merger formed a single agency focusing on land transportation to coordinate planning for private and public transport infrastructure, the formulation of vehicle restraint policies, as well as the evaluation of trade-offs between alternative infrastructure and policy options. The rationale is simple. Mr Low Tien Sio, former Executive Director of MRTC, explained, "In Singapore, [the government] puts together land policy, vehicle registration, road transport, rail transport, all into one organisation and that is a significant move because it means that they have brought together a lot of the thinking that were quite diverse into a central body that is tasked to envision Singapore's land transportation in the next decade."[as] Taking a broader perspective, he added, "Institutional integration is what differentiates us from the rest of the world. It provides a drive, it provides the initiative, and it influences perspective and eventually the end delivery."[at]

At the launch of the LTA on 15 September 1995, the government set a challenge for the newly-formed LTA to build a world-class land transport system in Singapore. Barely four months later, the LTA completed its blueprint to meet

---

[aq] Former Public Works Department (PWD) took charge of most public works in Singapore. Its Roads and Transportation Division was in charge of building transport infrastructure, which included roads, bridges, bus interchanges and shelters.

[ar] The Registry of Vehicles (ROV) took charge of implementing policies relating to vehicle ownership and usage control, enforcement of vehicle safety and the regulation of public transport services. ROV was also the licensing authority and regulator of the taxi industry until it was subsumed into LTA in 1995.

[as] Oral history interview with Low Tien Sio by Centre for Liveable Cities and LTA Academy, April 2011.

[at] Ibid.

this challenge. The White Paper on *A World Class Land Transport System* was presented to the Parliament and accepted by the government.[au] The White Paper 1996 set out the LTA's bold transport vision for Singapore, its operating approach and the initiatives. It also laid down the overarching philosophy governing public transport to forge a social contract with the people, in which all these parties, namely the government, public transport operators and commuters, share the cost of public transport in a sustainable manner.[av] In terms of bus–MRT integration, two key concepts could be teased out from the 1996 White Paper — integrated development for value capture and multi-modal public transport operators.

Integrating developments at transport hubs for value capture

In the past, large bus interchanges were provided by the HDB[aw] in the centre of new towns close to commercial hubs. As many of these HDB estates were built before the construction of the MRT system in 1980s, integrated bus–MRT interchanges were not planned for in these estates. Commuters transferring between buses and MRT would have to walk a fair distance due to the bus interchanges being a few hundred metres away from the MRT stations.

In order to increase the convenience of commuters and capture the value of land-use, the government made a concerted effort to integrate bus interchanges with MRT stations at the planning stage. HDB towns were designed to support public transport. Existing MRT stations which already had bus interchanges in the area would go through a comprehensive redevelopment plan. Better design for the physical integration of bus and MRT stations ultimately reduces potential loss in land-use development and improves transportation and pedestrian flows.[ax]

Integrated developments at bus–MRT interchange became more common after the formation of the LTA. Among the first few examples were at Toa Payoh (2002), which was the first air-conditioned bus interchange in Singapore, and Sengkang (2003), which was the transport hub for bus, NEL and SP LRT. Commuters' journey experiences were enhanced through the provision of integrated transport hubs which provided seamless and comfortable transfers

[au] Land Transport Authority, *Annual Report 1996/1997*, (1997).

[av] Ibid.

[aw] HDB is Singapore's public housing authority and a statutory board under the Ministry of National Development.

[ax] Report of the *Inter-Ministry Committee on MRT/Bus Integration*, (1995).

between bus interchanges and MRT stations. In fact, the provision of easily accessible amenities had made such integrated developments to be lifestyle hubs, thus maximising the value capture from land-use viewpoint.

Moving towards multi-modal public transport operation

The concept of multi-modal operation of bus and train services was mooted in 1998 when the Ministry of Transport (MOT) and LTA were concerned with two longer term industry structure issues. First, bus operators would be faced with a shrinking market share as the expanding MRT system would inevitably take away more of the lucrative bus routes from them. Buses would then play a more supportive role as feeders to the MRT network. If this happened, bus operators would have little incentives to continue to invest in improving their bus service. Second, transport operators were better able to integrate their bus and MRT services as they had the incentive make their services efficient in serving the commuters of entire the entire journey instead of by bus or MRT mode.

In December 1998, the LTA invited a closed tender to select the operator for the NEL and SP LRT. It only allowed the two incumbent bus operators, SBS and TIBS to take part in the tender exercise and explicitly precluded the SMRT from doing so. This marked a clear intention by the government to bring about a multi-modal public transport operation for Singapore. When the SBS won the NEL operating contract in May 1999, it was already known that SBS would become the first multi-modal public transport operator in Singapore. To reflect its new multi-modal operations, SBS rebranded itself to be SBS Transit (SBST) in 2003 when it started operating NEL.

SMRT Corporation, the public listed holding company of SMRT Ltd., became the second multi-modal operator in 2001 when it acquired TIBS to form SMRT Buses Ltd., which was a sister company of SMRT Trains Ltd. Concurrently, TIBS, TIBS Taxis and TIBS Motors were rebranded as SMRT Buses Ltd., SMRT Taxis Pte. Ltd. and SMRT Automotives Pte. Ltd. respectively.

This concept of multi-modal operation was an important turning point that gave rise to the current duopoly structure of Singapore's public trans-port industry. The intent was to enable the multi-modal operators to better integrate their bus services for feeding into the MRT network, creating con-venience for commuters and greater efficiency for the public transport as a whole. This also provided a base for peer benchmarking in the areas of ser-vice standards and cost efficiency between the two multi-modal public transport operators (PTOs) — SBS Transit and SMRT. This in turn could prevent regulatory capture, and the government would be in a stronger position to tender and award of contracts for operation of bus routes and new MRT lines.

1.5. *Strengthening regulations and rethinking bus services from 2005 to present*

From 2005 onwards, the bus industry went through even more changes. It started with the Public Transport Council (Amendment) Bill that was passed on 17 October 2005. The PTC's statutory jurisdiction was enlarged to cover the licensing of PTOs and not just bus routes; the licensing of ticket payment services in public transport, and the tighter regulation of bus and MRT fares.

The PTC was previously only allowed to licence individual buses. Under the amended PTC Act, all public bus services must be licensed. In addition, any bus operator running 10 or more such bus services must have a bus service operator licence (BSOL), unless exempted by the PTC. The two largest bus operators — SBS Transit Ltd. and SMRT Buses Ltd. — were each granted a 10-year BSOL with effect from 1 September 2006. Other smaller bus operators were exempted from BSOL.

Under the BSOL regulatory regime, the two bus operators were assigned with their respective areas of responsibilities (AoRs) designated by the PTC. The AoR allocation was as enshrined in their respective BSOL.[ay] SBS Transit's AoR was in the eastern and south-western part of Singapore, while SMRT's AoR was in north-western Singapore. Both SBST Transit and SMRT buses were allowed to run bus services from their AoRs to the central area. Although there were some overlaps of the bus networks, both operators were in a good position to provide integrated multi-modal transport services within their designated territories (Fig. 4[az]).

In return for the exclusive rights to run bus services in their AoRs, the two bus operators would have to comply with universal service obligations (USOs), in which the PTC would require them to operate bus routes in their AoRs with a minimum level of demand even if they were unprofitable.

Tightening of bus service obligations through licensing

As early as in 1994, the PTC had put in place a set of specifications on bus services and carried out audits on the bus service performance against the specifications. To ensure that the specifications remained relevant to the

---

[ay] Until BSOL was put in place under the PTC in 2006, SBS Transit's and SMRT Busses' (also the former TIBS) areas of responsibility were informally allocated to and observed by the PTC and the two bus operators.

[az] Public Transport Council, *Annual Report 2006/2007*, (2007).

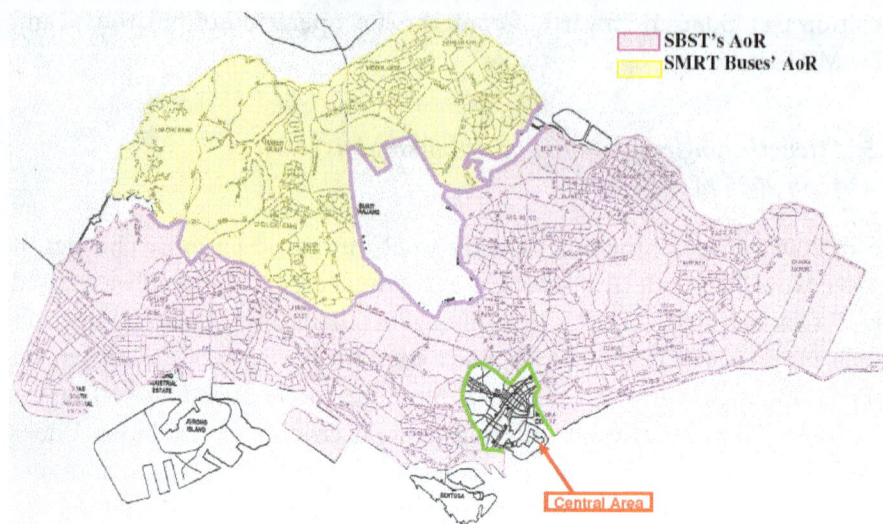

Fig. 4.   PTO's Areas of Responsibility (AoR).

changing industry environment and reflected the prevailing commuters' needs and expectations, the PTC reviewed and introduced a new set of quality of service (QoS) standards in 2006. The new QoS standards were incorporated in the new BSOL that was granted to each of two operators in the same year. To enforce the standards and licensing conditions of BSOL, PTC could impose financial penalties of up to S$100,000 or revoke licences for breaches.

The bus QoS standards comprised two categories — Operating Performance Standards (OPS) and Service Provision Standards (SPS) and covered 19 parameters. OPS measured minimum daily or monthly operational deliverables either at the bus network or route levels, such as bus reliability, loading and safety, by using indicators on adherence to scheduled bus trips and headways, bus loading during peak periods, and accident rates. SPS measured overall bus route planning and provision of services which covers service availability, integration and provision of information, using indicators which include adherence to scheduled bus trips and headways, bus loading during peak periods, and accident rates.

In addition, the PTOs were required to submit monthly performance reports to PTC for assessment of compliance with the QoS standards. PTC also regularly audited the PTO's reporting processes and systems to ensure that the performance reports were accurate and complete, as well as conduct on-site audits or spot checks every month. A review of the PTO's performance was conducted by

PTC every six months in the months of April and October. The results of such reviews were announced by PTC and published in its annual report.

From 2006 onwards, the bus QoS had been progressively tightened by the PTC through exercising its licensing power under the PTC Act. The regulation of bus services had since taken on a more transparent mode in terms of public accountability by the PTC.

Master planning public transport by the LTA

In October 2006, about 10 years after the 1996 White Paper, the government embarked on a comprehensive land transport review. It consulted a broad spectrum of the public, believed to be a first major public exercise done on land transport since the start of MRT services in 1987, to solicit views and suggestions. More than 4,500 people contributed their time, energy and ideas in this review.[ba] Supporting this land transport review was a fundamental study of the public transport industry structure that would be needed in relation to the regulatory and institutional arrangements. This study made specific recommendations on the role of government, network integration, competitive procurement of services, industry structure, number of PTOs, funding and investment criteria.

The land transport review culminated in the launch of a land transport master plan (LTMP) in 2008— *A People-Centred Land Transport System*. This visionary and initiative-specific LTMP 2008 strived to make Singapore a great city to live, work and play in, over the next 10 to 15 years. One of the three key strategic thrust in the LTMP 2008 was to make public transport a choice mode. It called for the LTA to take on the role of central bus network planner for bus services so as to better integrate bus and MRT networks from a total journey experience of public transport commuters. Buses would be given greater priority on the roads to improve bus service reliability. It also called for the gradual opening up of the bus industry to allow for competition in the market, i.e., competitive tendering for the right to operate a package of bus services.

Five years after LTMP 2008, the LTA launched a refreshed version of the plan and launched the LTMP 2013 to address three key challenges of changing commuter expectations and norms, growing population and economy, and tighter land constraint and trade-off. The LTMP 2013 aimed to enhance travel experience through more connections, better service and liveable and inclusive community. It set a long-term measurable vision 2030 of Singapore having eight in 10 households living within a 10-minute walk from an MRT station, 85% of public transport journeys (less than 20 kilometres) completed within 60

[ba] Land Transport Authority, *Land Transport Master Plan: A People-Centred Land Transport System*, (2008).

minutes, and 75% of all journeys during peak hours undertaken on public transport.

The LTMP 2013 focused on improving public transport. For the bus industry, QoS standards were tightened further, a bus service enhancement programme (BSEP) was pursued, more bus priority measures were planned, better bus travel information were being put in place, more integrated bus–MRT transport hubs were planned, and, for the first time, a bus service reliability framework (BSRF) that aimed to incentivise bus operators to ensure reliable bus services was being introduced.

In a nutshell, the LTMP 2013 had repositioned improving public transport as enhancing travel experience for commuters, truly reflective of the same vision of a people-centred land transport system as articulated in the earlier LTMP 2008.[bb]

Policy shift towards centralised bus network planning and greater contestability

Being the agency in charge of planning and developing the rail and road networks, LTA took over the bus planning function from the two operators in 2009 in order to plan Singapore's land transport network more holistically. The intention was to have one dedicated agency that looks into the needs of commuters in drawing up land transport plans for buses, MRT and roads from the planning stages for better optimisation of bus and MRT services, and a more holistic integration for the land transport system. As the central bus planner, the LTA is also responsible for planning and developing the necessary supporting transport infrastructure such as bus interchanges and bus depots. This move to retake the central bus planning function from bus operators represented a major policy shift in the bus industry not seen since the 1996 White Paper.

Dovetailing the taking over of the new bus planning function in 2009, the LTA started to work on the appropriate model and approach to gradually open up the bus industry for greater competition and eventually package bus services routes planned by the LTA for competitive tendering. Competition *in* the market, like the bus industry situation in the 1960s, where bus operators competed head-to-head for market share was not adopted. This was because such head-on competition in the market would destroy integration and lead to undesirable consequences such as wasteful duplication of services and danger-

---

[bb] Ibid.

ous bus driving to compete for passengers. There was also the risk that less profitable routes would be neglected.

Instead, the government adopted regulated competition *for* the market where bus operators compete periodically for the right to operate the bus services. The idea was to inject a real threat of competition to spur efficiency and innovation. As stated by the former Minister for Transport (MOT), Mr Raymond Lim, "...the public interest is best served not by simply having a single public transport operator...but by ensuring there is a threat of competition to keep these dominant market players on their toes".[bc] This public articulation on making the bus industry a more contestable market represented a significant shift in the government's thinking that rivalry in the duopolistic bus market was deemed not adequately intense for greater efficiency and service innovation.

With the LTA as a central bus network planner, the PTC continued to retain oversight of bus service regulation, ticket payment services, and bus and rapid transit system fares. With this new function in 2009, the LTA aimed to take on a more integrated approach to planning the bus network with the MRT system and making the entire public transport more commuter-centric.[bd]

Bold move by government to fund bus service enhancement programme (BSEP)

Two years after the launch of LTMP 2008, buses continued to experience crowdedness despite further tightening of bus QoS standards, continual service improvements and periodic review of bus operations between the LTA and the two bus operators. This was primarily driven by the rapid increase in population and economic growth. Singapore's population grew from 4.8 million in 2008 to 5.3 million in 2012, more than 10% increase in just a short span of four years. In the same period, Singapore's gross domestic products (GDP) grew by 28%.[be] Such a significant growth had brought a new urgency to roll-out the bus and MRT plans that were charted out in LTMP 2008.

As the planned MRT projects would take time to be completed, some form of short-term measure to mitigate the insufficient peak hour bus capacity was necessary. This was as buses were still carrying the main bulk

---

[bc] Mr Raymond Lim, *Minister for Transport Speech at Committee of Supply Debate (Land Transport — Part 1)*, (2009).
[bd] Land Transport Authority, *Land Transport Master Plan: A People-Centred Land Transport System*, (2008).
[be] Land Transport Authority, Land Transport Master Plan, (2013).

of daily public transport trips given that the MRT system then was only serving areas of high population density. This new urgency led to a proposal to have the government partner the two bus operators to quickly increase the bus capacity. On 17 February 2012, the government announced to implement a bus service enhancement programme (BSEP) to expand the bus fleet quickly and help the two bus operators with the running costs of its new buses. The BSEP would be financed by the government through a S$1.1 billion bus service enhancement fund (BSEF) which covered a period of 10 years.

The objectives of BSEP was three-fold — to reduce crowding level of buses, improve frequency of bus services and enhance connectivity of bus services. To achieve these objectives, it would ramp up bus capacity and pre-vailing bus service levels by injecting 800 new buses into the bus fleet within the next five years, i.e. by 2017. The government decided to fund the purchase and operation of 550 of these buses with the S$1.1 billion BSEF, while the two bus operators would pay for the other 250 buses to meet the projected rider-ship increase. These 800 new buses would account for a 20% increase in fleet size over the five years.

The S$1.1 billion funded by the government included the salaries of new bus drivers to drive the additional buses, and all other running costs for these buses such as fuel and maintenance. Government funding was not expected to result in a profit for the bus operators as the revenue derived from operating the 550 buses would be recovered. Disbursement of funds would commensurate actual expenditure incurred. In return, the two bus operators were required to deliver agreed service level improvements. LTA was entrusted by the government to scrutinise the actual costs for the pur-chase and running of the buses. Should the cost turn out to be lower than expected, government funding would be reduced correspondingly.[bf] The LTA launched the BSEP in 2012 after a series of intense negotiations with the two bus operators on the agreed mechanisms and payment terms to roll out the programme.

There were some initial adverse public reactions to the BSEP announce-ment in 2012. The BSEP was seen by some critics as a subsidy for the two bus operators which were run commercially. The government took pains to explain the rationale for BSEP and reassured the public that the BSEP was meant to be a subsidy for bus commuters, not for the PTOs. The government also had to

---

[bf]Mr Lui Tuck Yew, *Minister for Transport Speech at Committee of Supply, Part 1 — Public Transport*, (2012).

ensure that BSEP would deliver tangible bus improvements on the ground as experienced by commuters.

The initial public criticism subsided and the BSEP was proven to be popular with the commuters who benefitted from the improved bus services. This prompted the government to decide in 2013 to enhance the BSEP by bringing forward the planned bus buys and also increasing the government funded buses from 550 to 1,000 by 2017.

The enhanced BSEP introduced more bus services to serve new areas and to improve connectivity, new peak hour short service (PPSS) to improve peak period bus frequency, new full day short services (FDSS) to strengthen bus connections to transport hubs, and more city direct services (CD) to link up housing areas with the city centre. Other private bus operators were also roped in to run some of the CD and PPSS services via competitive tenders based on bus routes.

Such involvement of private bus operators was firstly aimed at tapping on the available resources to augment the speedy rollout of BSEP and secondly, gathering more information on operating costs via competitive price discovery. In a way, this was a precursor to the eventual opening up of the bus market for competitive tendering.

This bold government's funding of buses and bus operation was clearly a departure from the long-standing government policy on bus financing. In a way, it represented a significant policy shift by the government to step in to subsidise the bus industry with a decisive objective to immediately bring about better bus services for the benefits of commuters, while the appropriate model and mechanism for competitive tendering of bus services were being worked out by the LTA.

Structural shift and first competitive tendering of bus services in 2014

Two years after the launch of the BSEP in 2012, the government announced on 21 May 2014[bg] that it would restructure the bus industry to a government contracting model (GCM). Firstly, GCM aimed to make bus services more responsive to changes in ridership and commuter needs, and secondly, to inject greater contestability so as to bring about better bus services for commuters. Under GCM, bus operators would be contracted by the LTA to run bus services via competitive tendering. The LTA would determine and specify the requirements of the bus services to be provided and the bus operators would bid for

---

[bg] Land Transport Authority. "Transition to a Government Contracting Model for the Public Bus Industry." Press Release, 21 May 2014.

the right to operate them over a fixed period. Contracted bus operators would be paid a fee while the revenue would be retained by the government.

One major change was that the government also decided to own all bus infrastructure such as depots, buses and bus fleet management system. The rationale was that this would lower the barriers of entry to the market and thereby attracting more prospective bus operators to participate in the tender exercise, and this would also facilitate the transition from an incumbent to a new operator should the incumbent not win the tender upon contract expiry.

A phased approach was adopted to ensure the transition to a full GCM would not risk disrupting the continuity of bus services to the commuters. All the bus services island-wide would be bundled into 12 bus packages with about 300 to 500 buses each. Three packages of bus services would be tendered out first. The contract period would be five years with an option to extend it by two years on good performance. In total, the three packages would comprise about 20% of the existing buses.

The remaining nine bus packages would continue to be operated by the two incumbent operators. The LTA would negotiate with them to run the nine packages under the contracting model for durations of about five years. These negotiated contracts would replace the BSOLs which were due to expire on 31 August 2016. Upon the progressive expiry of these negotiated contracts, the bus packages would then be competitively tendered out.

Following the announcement of the decision to adopt GCM on 3 Oct 2014, the LTA invited tenders for the first bus package, comprising 26 existing bus services originating from the Jurong East, Bukit Batok and Clementi bus interchanges, and to be supported by the new Bulim bus depot, were announced by the LTA. The contract award was expected to be in early 2015. The bus services of the first contracted bus package would be progressively rolled out, starting from second half of 2016. It would start with about 380 buses before growing to about 500 buses in 2021.

1.6. *Once over lightly and some observations*

In retrospect, the evolution of the Singapore's public transport ecosystem could be described as almost synonymous with the reform story of the bus industry. Since the days of trolley buses in the 1925, there had been two landmark reports — 1956 Hawkins Report and 1970 Wilson Report, two White Papers — in 1970 and 1996, two master plans — LTMP 2008 and LTMP 2013, and countless other unreported committees' work and studies done that had characterised the bus story. Over a period of almost 60 years, Singapore had found its own way to reform the bus industry and adapt to changes that suited the prevailing transportation peculiarities. The bus industry has indeed come a long way in a relatively short period.

A few notable observations can be teased out of this 60-year bus story and, perhaps, public transport in general. First, Singapore's vision had been consistent to make public transport a priority among the delivery of public programmes. It had not wavered in improving public transport to benefit the commuters, primarily based on sound socio-economic considerations.

Second, the government tended to take a long-term view for planning public transport. Integration to enhance network and capture value had been consistently carried out.

Third, there were decisive interventions by the government to put things right when or even before things went wrong. This results-driven action was evident in the manner that such interventions were carried out. The execution could be described as both pragmatic and effective.

Fourth, sound institutions were put in place to ensure governance and execution. Fundamental structural issues were dealt with upfront through the restructuring of such institutions. At the same time, there was a general sense of pragmatic engagement to solicit timely buy-ins of initiatives and tough measures.

Fifth, working with markets, e.g., competition and benchmarking, and harnessing technology, e.g., ticketing systems, were two common characteristics. The government would leverage on market incentives and discipline, and well as technology innovation in its design of interventions, in addition to technical solutions. Looking ahead, Singapore's public transport ecosystem would continue to evolve. This bus story would not end.

2. Delivering Sustainable Rapid Transit System (RTS)

Briefly, Singapore's urban mass transit development started in 1967, two years after Singapore's independence. In 1971, the state and city planning (SCP) project was completed and it recommended an efficient public transport system by means of a rail transit system (RTS). This study roughly coincided with the bus industry's 1970 White Paper on *the Reorganisation of the Motor Transport Services of Singapore*. Detailed feasibility studies ensued in the 1970s to look at the needs, alignments, and costs and benefits of the RTS which was then budgeted at $5 billion.

As the budget was a hefty sum for Singapore at that time, the government held a MRT debate in the media in 1980. The public debate compared the merits of building an MRT system supported by buses with that of an all-bus system. The government appointed teams of consultants to examine both options, including a team led by Professor Hansen of Harvard University (known as the Hansen Review team) which advocated that an all-bus system deploying a large number of high-capacity, high-speed buses would suffice to meet travel demand well into the 1990s.

In 1982, the government decided to proceed with the building of the RTS. The first RTS revenue service started in 1987. Since then, the RTS industry as a whole had also undergone some notable changes in industry structure, regulatory and licensing regime, financing framework and competitive tendering. Key milestones on RTS development in Singapore were summarised in Annex H.

2.1. *A financially sustainable RTS operation from the outset*

Singapore's RTS operation started in 1987 with the opening of the first RTS lines, i.e., north–south and east–west (NSEW) lines. SMRT Ltd was set up by the government to operate and maintain the NSEW lines through a licence and operating agreement (LOA) for a period of 10 years. The 1987 LOA was structured with the aim to assist SMRT Ltd. to achieve self-sufficiency during the initial start-up of commercial operation. The RTS trains and infrastructure were leased to the operator at a nominal rate. But the operator had to bear all revenue risks even though it had no control over the setting of RTS fares which were regulated by the Public Transport Council (PTC). The PTC was also instituted in 1987, just prior to the start of RTS revenue service in the same year.

In addition, the operator was also obligated to set aside a portion of their post-tax surpluses annually, equivalent to the depreciating rate of the operating assets, in a reserve fund for future asset replacement. However, to help defray the operating cost, some RTS stations in the city had some commercial spaces that the operator could lease out to generate some non-fare revenue. Operating assets which were replaced would become the property of the operator.

From the outset, the government had required two criteria, i.e. economic evaluation and financial viability, to be met before proceeding with an MRT project. First, the economic cost–benefit analysis (CBA) had to be done to determine whether to invest in a potential RTS project. Both the benefits and costs of a project were evaluated over the project lifecycle. The government would require that the ratio of benefits to costs was at least one for investment prudence. Second, the government would require that the projected operating revenue to at least cover the projected operating costs, including depreciation, for financial viability. As a whole, such fiscal prudence aimed to ensure that the operations of RTS network would not require operating subsidies in the future, and the operator would be able to fund future asset replacements from the depreciation that had been set aside (Fig. 6).

Rail financing framework in 1996 white paper

The government also recognised that RTS operating assets were costly and should the costs increase over time, the operator might not have enough funds to finance asset replacements even though they had prudently set aside funds in their asset replacement reserves. To ensure that the investment on infrastructure would not be wasted due to the inability of operators to replace obsolete operating assets, the government was prepared to co-share the financial burden of asset replacement.

So, in 1996, the government spelt out the rail financing framework in the White Paper on *A World Class Land Transport System*. In it, it articulated that an overarching financing policy for transport that was based on the concept of partnership. The government would provide for the infrastructure, the commuter would pay for the operating costs and the operators would extract efficiency dividends within the service standards and fares set by the regulators.[bh]

The 1996 White Paper also redefined the financing policy for assets replacement. While the first set of operating assets would be handed over to the operator, subsequent sets of operating assets purchased would have to be co-funded by the operator and government. The operator would fund the subsequent assets at a cost equivalent to the historical cost of the first set of operating assets, while the government would fund the remainder, i.e., the inflationary component, through a replacement grant (RG) (Fig. 5).

Economic Evaluation Criteria: Total Benefits > or = Total Costs
Financial Viability Criteria: Total Revenue > or = Total Operating Costs + Depreciation

Source: Land Transport Authority.

Fig. 5.   Economic evaluation and financial viability criteria.

[bh] Land Transport Authority, *Land Transport Master Plan: A People-Centred Land Transport System*, (2008).

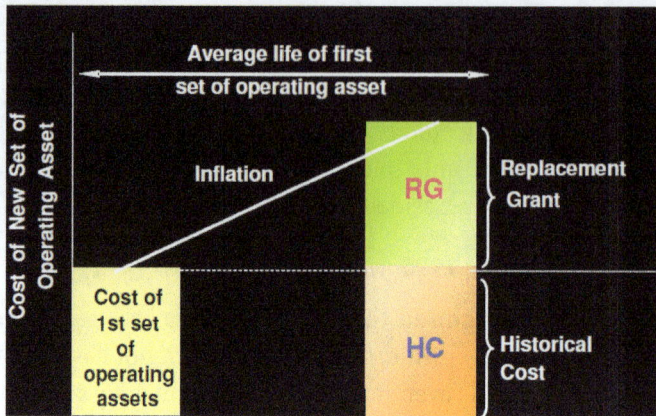

Source: Land Transport Authority.

Fig. 6.   Funding of assets replacement.

This meant that, in effect, the total investment injected by the government were not only limited to the long-term infrastructure and initial operating assets but also included the inflationary costs of subsequent set of operating assets. This effectively transferred the inflationary risk from the operator to the government.

Assets ownership passed over to SMRT in 1998

Two years after the 1996 White Paper, a new licence and operating agreement (LOA) was entered into between SMRT and LTA. The goal then was to set up a financial and licensing regime that would facilitate the flotation of SMRT. To achieve this goal, the 1998 LOA was more comprehensive than the first one done in 1987. Besides the incorporating the financing arrangement for SMRT to buy over and own the operating assets henceforth, it also spelt out the obligations of the operator and LTA. At the same time, a lease and maintenance agreement (LMA) was established to make clear the maintenance obligations of the operator with regards to infrastructure such as stations, viaducts, and depots. This LMA also made clear the lease titles, such as the station premises that the operator would be responsible for use in generating non-fare revenue.

With the 1998 LOA and LMA, the roles of operator and LTA became clearly delineated. The operator's role was to own, operate and maintain the operating assets. Its decisions on maintenance, replacement and procurements of such assets were to be based on benefits and costs, while complying with the service and safety standards set by the LTA. The LTA's role was to build the

MRT system, regulate the operation and maintenance and administer the replacement grants. LTA would own the infrastructure and fares system software.

With this clarity of roles, LTA further strengthened its regulatory and licensing regime to ensure that the operator would abide by its obligations in the 1998 LOA. This also placed LTA on a firm foundation of instituting a strong regulatory oversight after SMRT's flotation which eventually took place two years later in 2000.

2.2. *Instituting regulatory oversight on RTS operation*

A year before the formation of the LTA in 1995, the former MRTC had set up a small regulating unit, in the form of a committee, to monitor, audit and benchmark SMRT's operating performance against overseas best practices. When the LTA was formed, it set up a new transit services division under its Vehicle and Transit Licensing (VTL) group to take on a dedicated role for regulating and licensing rapid transit system (RTS) operators. This division took over the work done by the regulating unit. In 1996, the transit services division established a new performance regulatory framework. For the first time, a set of operating performance standards (OPS) was incorporated by the LTA in the 1998 LOA with SMRT. A general penalty clause was also incorporated into the 1998 to make any breach of the conditions of the LOA enforceable. This enabled the OPS to be used for the LTA to gauge, monitor and enforce SMRT's service provision.

Following the same regulatory framework and similar terms of 1998 LOA, a separate LOA for Bukit Panjang (BP) light rapid transit (LRT) system was also signed with Singapore LRT Ltd., a subsidiary of SMRT. Bukit Panjang LRT was offered to SMRT to operate as it was a *de facto* feeder system that would serve the Bukit Panjang new town and connect it with the NSEW lines.

First competitive tendering of RTS operation in 1999

Separately, with the rail financing arrangement settled for the 1998 LOA with SMRT, the development of the North-East Line (NEL), Sengkang LRT and Punggol LRT (SPLRT) presented an opportunity for the LTA to introduce a second RTS operator. It was recognised that a second RTS operator would facilitate the benchmarking of operators' service levels, performance and operational efficiency. However, the LTA also recognised that having too many operators would lead to difficulties to reap the economies of scale and benefits of service integration. Given the then foreseeable size of the RTS network, it was reckoned that the RTS industry structure could potentially support about two to three RTS operators.

So, in 1999, the LTA conducted the first competitive tender for the operation of the North-East Line (NEL), Sengkang LRT and Punggol LRT (SPLRT). This tender exercise was deliberately limited to the two incumbent bus operators, namely SBS Transit and TIBS. This decision to go for a limited tender was premised on two objectives; firstly, to have a second home-grown RTS operator, and secondly, to have a multi-modal operator that would run both RTS and bus services. SMRT Ltd. was not allowed to participate in the tender because it was deemed that it would have a natural advantage, given its track record on RTS operation then.

The NEL and SPLRT operating tender was structured as a quality tender with the objective to select the better of the two operators. It was not a financial tender. Both tenderers did not bid on licence fee or fares, but on quality aspects such as operation and maintenance, marketing and customer service, safety management, recruitment and training, and capital structure.

The tender also adopted the same regulatory framework and similar terms as that developed in 1998 LOA. In 1999, the LTA announced that SBS Transit won the licence to operate NEL and SPLRT. SBS Transit became the first multi-modal operator in Singapore, as earlier alluded to in the bus story.

In 2001, the LTA conducted the second competitive tendering for the operation of the Circle Line (CCL). The format and evaluation remained largely the same as that of the first operator tender, as the fundamental regulatory and licensing regimes had remain unchanged since the appointment of SBS Transit as the operator for NEL and SPLRT. However, for this second round, SMRT was allowed to compete with SBS Transit and TIBS as the earlier twin objectives of having a second RTS operator and a multi-modal operator had already been achieved. SMRT won the tender and was appointed as the CCL operator in 2001.

Linking operating performance to penalty

In early 2000, the 1998 LOA was further modified to prepare for the impending flotation of SMRT in the Singapore stock exchange (SGX). One major modification was the incorporation of a new penalty framework pegging the penalty against the OPS that would measure the impact of a service disruption or failure. Henceforth, the LTA could impose a penalty that was explicitly linked to a failure to comply with the OPS standards. This 2000 LOA paved the way for LTA to rely on the OPS standards as an added punitive mechanism to regulate SMRT's service provision.

But relying on LOA arrangement to regulate the RTS operator had its limitations as the LOA was, in essence, a contract between the operator and the

LTA. Any modifications to the terms of the LOA would have to be agreed between the operator and the LTA before it could be incorporated in the LOA and for it to be enforceable. Clearly, this was not an ideal arrangement as the operator could choose not to agree with the modifications proposed by the LTA. So, to overcome these limitations, the LTA reviewed its regulatory model and legislative powers, soon after the 2000 LOA was completed. It looked at various licensing models and best practices such as those adopted by the tele-communication industry then.

Regulating RTS operation with new licensing powers

The year 2002 was a turning point in the regulation of RTS operation. That year, it was decided that a licensing model would be adopted instead of relying on LOA mechanism. The RTS (Amendment) Bill was passed in 2002, giving the LTA explicit new powers to grant RTS licences, impose licence conditions, modify or amend licence terms, issue and modify code of practices and give directions for compliance by the licensees.

The powers also provided for punitive measures for financial penalty, forfeiture of security deposits, suspension and revocation of licence. To assure that the interests of the licencee were not unduly compromised, mandatory due process was included for the LTA to comply should it decide to lawfully modify the licence terms or code of practice. There was also an appeal process that the licencees could rely on to make representations for review if they were aggrieved by such modifications and the directions issued by the LTA.

In 2003, the NEL and SPLRT licence became the first RTS licence granted by the LTA under the amended legislation. The next one was the CCL operating licence which was issued in 2009.

This licensing model marked the departure from the earlier LOA contractual arrangement that had been in existence since the start of RTS operation in 1987. Subsequently, the LTA also relied on the same legislative powers to modify the existing LOAs it had with SMRT.

Henceforth, the roles of RTS operators and the LTA as the regulator and licensor were not only clearly delineated, but also formalised by law. This enabled clear public accountability.

2.3. *Master planning and new rail financing framework*

In 2008, following a comprehensive review of the land transport in Singapore, the LTA published the Land Transport Master Plan (LTMP) on *A People-Centred Land Transport System*. In the master plan, the LTA aimed to double Singapore's RTS length from 138 km in 2008 to 278 km by 2020. To facilitate this expansion of RTS network, the LTMP 2008 charted out bold directions to shape the future RTS industry and the financing framework needed to support it, while ensuring fare affordability.

Recognising network benefits for future RTS lines

As the RTS network was being expanded, future RTS lines would mainly have to be built underground in a land-scarce and urban environment. Invariably, they would be more costly to be built, operated and maintained as compared to the existing lines that were built earlier. In addition, such future lines would also be built in less mature corridors with lower urban density and ridership. Such higher cost and lower ridership would make such lines less financially viable initially than the existing lines.

However, as new lines were added, the RTS network would also become more interconnected and comprehensive. As such, they would generate positive externalities which would benefit the rest of the existing RTS network. Network benefits from new rail lines could arise from commuters who would shift their preference from buses to rail, and motorists who would find public transport more attractive with the completion of new rail lines.

Therefore, to keep pace with future expansion of the RTS network in a financially sustainable manner, the LTMP 2008 envisioned that the financial evaluation of future lines would allow for a network approach to be adopted. This way, the construction and operation of new lines could potentially be brought forward by a few years, since the financial viability of the new rail lines would not be evaluated on a line-by-line basis but on a network basis.

Injecting greater contestability with shorter RTS licences

In order to introduce greater contestability for the RTS industry, LTMP 2008 called for the RTS operating licences to be restructured. The licence period would be shortened from 30–40 years to about 15 years. The shorter period aimed, firstly, to provide a more effective threat of contestibility to the incumbent operator. Secondly, it would provide LTA with the flexibility to review licence conditions as well as to appoint a new operator should the incumbent fail to maintain its good performance. The overall goal was to keep the operators on their toes to run RTS efficiently and improve services deliveries to benefit commuters.

With such shorter licences, operators would not be able to continue to invest in the rail operating assets which could have useful lifespans beyond their licence period. As such, the government decided that the LTA, instead of the operators, would own the operating assets and be responsible for replacing the assets. A new payment mechanism was put in place whereby potential operators would competitively bid for the right to operate the rail lines and to pay the LTA for the use of the operating assets through a licence charge (LC).

The design of the LC was based on the sensible sharing of risk between the operator and the LTA. The LC comprised fixed and variable components, with a certain portion to be determined through competitive bidding. The fixed charge would be pre-determined and would reflect a schedule of charges, whereas the variable charge could be pegged to a percentage of revenue or profit. The variable components would allow for some risk-sharing. For instance, in a case where ridership fell due to poor economic conditions, the LTA would partially shoulder the fall in revenue through a lower amount of LC collected.

New railway sinking fund for asset replacement

Under the new RTS financing framework, monies collected from the LC would be paid into a newly created railway sinking fund (RSF). The use of this fund would be ring-fenced for the purpose of procuring assets for replacement. Effectively, the RSF broadly replaced the asset replacement reserve requirement that was imposed on the existing lines under the previous RTS financing framework. The LTA would take control of the sources and use of the RSF, and it would ensure its adequacy to meet future obligations to replace assets.

With the set up of RSF and the transfer of asset replacement obligations from the operator to the LTA, the government's replacement grant (i.e., the subsidy needed to cover the inflationary component of the historical cost of the replaced assets) would correspondingly be made to the RSF instead of the operator (Table 6).

2.4. *First financial tender for RTS Licence — Downtown Line*

In 2010, legislative amendments were made to the RTS Act, firstly, to support a new rail financing framework using a financial evaluation approach that would enable new RTS lines to competitively tendered out, and secondly, to provide LTA with more regulatory powers to enhance network integration, safeguard the continuity of RTS operation and facilitate the transfers of undertakings between outgoing and incoming operators, as well as enhance passenger security. Consequential amendments were also made to the LTA Act to enable the setup of RSF under LTA's jurisdiction.

Table 6.   Comparison of RTS financing frameworks.

Items	Whose Responsibility?	
	Previous Framework (After 1996 White Paper)	New Framework (After LTMP 2008)
Fare and Non-fare revenue risks	Operator	Operator
Payment of licence charge (LC) for use of assets	Not applicable	Operator
Infrastructure (stations, tunnels, etc.) cost	Government	Government
1st set of operating assets (trains, signalling, etc.) cost	Government	Government
Replacement of operating assets at historical cost	Operator	*Government*
Inflation risk (which is covered by Replacement Grant)	Government	Government
Purchase of additional operating assets	Operator	*Government*
Operating (manpower, energy, maintenance, etc.) cost	Operator	Operator

Source: Land Transport Authority.

Retaining licensing model in legislative amendments

At the point when preparing for the legislative amendments, it was decided that the existing licensing model be retained rather than switching to a franchising model. Under a franchising model, the legal relationship between operators and the LTA would be based on a contractual agreement and this would be a departure from the existing regulatory and licensing oversight that the LTA was entrusted with since the last RTS (Amendment) Bill was passed in 2002.

Had the government adopted the franchising model in 2010, it would have been a significant policy change which could entail challenging transitions to move the existing licences to over to franchises. It could also risk inadvertently fettering the existing statutory powers of the LTA and as a result, compromising its regulatory oversight to ensure safe, efficient and good quality RTS services. As it turned out, the final decision made was to amend the legislations to augment the licensing powers of the LTA, after the government had given due consideration to the alternative franchising model. The legislative amendments were completed in 2010, in time for the invitation of tender for operating the DTL.

Start of new approach for appointing RTS operators

The legislative amendments stipulated that the all new RTS lines henceforth would have to be competitively tendered out based on the new financing framework, in which the operator would also put in a cash-bid in addition to their tender proposal. The cash-bid would be the bidded portion of the licence charge (LC) payable by the operator to the LTA for the grant of the licence. Unlike the earlier operating tenders done for NEL and SP LRT as well as the CCL, operators of future lines, starting from DTL, would not be appointed solely based on qualitative criteria, but also on financial criteria which included the cash-bid.

As the DTL operating tender would be the first tender to be carried out, the government decided that it would adopt a gradual and calibrated approach, and limit the tender to the two existing operators then, i.e., SBS Transit and SMRT, mainly because the government wanted to phase in and implement the changes before it would consider opening the sector to new RTS operators.

The operating tender for the DTL was also called towards the end of 2010. In 2011, SBS Transit clinched the rights to operate and maintain the DTL under the new financing framework. The licence period was about 15 years from full completion of the DTL in 2017. It was estimated that the SBS Transit would pay a total licence charge of S$1.6 billion over the licence period.[bi]

With the new DTL operating licence granted, the LTA went on to negotiate with the two incumbent RTS operators to transition the exiting licences for NSEW lines, NEL and SP LRT, CCL as well as BP LRT, to the new RTS financing framework. The transition of such existing licences to the new licences would not be straightforward, as it would entail negotiating the appropriate levels of licence charge and financial terms to be incorporated. Alternatively, the existing licences would be allowed to run-out till licence expiry dates, after which, the new licences under the new rail financing framework would be granted via competitive tenders. Until the transition was fully completed, the LTA would have to grapple with two financing frameworks in its licensing and regulatory regimes for the time being.

2.5. *Once over lightly and some observations*

In retrospect, starting from the state and city planning (SCP) study in 1967–1971, an RTS network had already been envisioned as the backbone of public

---

[bi] Land Transport Authority. "LTA Appoints SBS Transit Limited To Operate Downtown Line Under New Rail Financing Framework." Press Release, 2 August 2011.

transport to provide an efficient and quality mass transport for urban mobility. In a relatively short period of less than 30 years from the start of operating the first RTS line in 1987, many RTS projects had already been rolled out. The latest series of RTS initiatives was charted out in the LTMP 2013. A few observations could be teased out from this RTS story.

First, Singapore had consistently placed strong emphasis on the need for an integrated public transport system with the RTS network as the backbone. This was adhered to despite the many measures that had been taken to build roads and keep them relatively congestion free for the movement of people and goods.

Second, to ensure sustainability, RTS lines were implemented and operated based on economic and financial prudence. The sound principle of appropriate risk-sharing between the LTA and operators was adopted in the RTS financing framework. User-pay principle had also been applied.

Third, strong regulatory and licensing regimes had been instituted to ensure that the operational performance would be up to par to meet the evolving needs of commuters.

Fourth, the approach to governance had been to harness market forces to where they could help to improve efficiency, while ensuring the need to retain sufficient economies of scale in structuring the industry. There was also a clear sense of what the government's roles were *vis-à-vis* the commercial operators and how these could be effectively performed with minimal risk and interventions.

3. Facilitating Convenient Taxis Services

3.1. *Brief evolution of taxi industry*

The taxi industry in Singapore had come a long way since the 1950s and 1960s (see Annex G for key milestone in taxi industry). At that time, the taxi industry was not tightly regulated. Due to high unemployment, there were hardly any restrictions on people registering their cars as taxis and being taxi drivers. The industry was fragmented, comprising mostly taxi companies with small fleets and some taxi owner-drivers.

Prior to a 1966 reform which allowed only non-transferrable new licences, some 3,200 transferrable taxi licences had already been issued by the Registry of Vehicles (ROV). These remained transferrable after the 1966 reform and fetched a high price in the black market.[b]

---

[b] Sharp, I., *The Journey — Singapore's Land Transport Story,* (SNP Editions, 2005).

There was also the serious problem of pirate taxis — private cars operating as taxis without licences. These pirate taxis were not metered, had poor service standards and posed road safety problems.

Eliminating pirate taxis

The change came in 1970 when the government published a White Paper on *Reorganisation of Motor Transport Service of Singapore*. A key strategy of the White Paper was to eliminate the problem of pirate taxis. While recognising the informal role of pirate taxis as they served as an alternative to the inadequate bus transport then, such activities would not be tolerated nor worthwhile,[bk] with the proposed plan to restructure the bus services.

New regulations arising from the 1970 White Paper included: (i) raising diesel taxes on private diesel vehicles (majority of which operated as pirate taxis) by 100%; (ii) suspension of pirate taxi drivers' driving licences for one year if caught; and (iii) making pirate taxi operations a seizable offence where offenders could be arrested on the spot and charged the following day.

The government also supported the formation of a National Trade Union Congress (NTUC) Transport Cooperative. NTUC Comfort was established in October 1970. It helped pirate taxi drivers become licensed taxi owner-drivers. A thousand new taxi licences were issued to NTUC Comfort when it was formed. It was also given existing licences from deregistered over-aged vehicles if their owners were unable to replace them. For the first time, there was a concerted effort to partner with the union to orgnanise taxi driving licences.

Consolidating taxi operations

Pirate taxis were quickly eliminated. Besides NTUC Comfort, several new licensed taxi companies emerged, namely Singapore Airport Bus Services (SABS), Singapore Bus Service taxi (SBS Taxi), Singapore Commuter (SC) and TIBS taxis. The emergence of new technologies, such as radiophone call booking in 1978 and global positioning systems (GPS) in the 1990s, led to a consolidation of the taxi industry. SABS, SBS Taxi and SC merged in 1995 to form CityCab, becoming a sizeable competitor to TIBS Taxis and NTUC Comfort.

Separately, NTUC Comfort's taxi owner-drivers proved difficult to manage, and their service and maintenance standards varied. In a move for greater control, NTUC Comfort introduced a taxi rental scheme in 1983 and began

---

[bk] Mr Yong Nyuk Lin, Minister for Communications, *Speech on Reorganisation of Motor Transport Service of Singapore in Parliament*, (1970).

converting its taxi owner-driver shareholders into taxi hirers by purchasing their taxi operator licences.

The three major taxi companies, namely, NTUC Comfort, CityCab and TIBS Taxis, owned fleets of taxis which they purchased, paid taxes and insurance for, and maintained. They rented their taxis to licensed taxi drivers at fixed daily rates. The taxi drivers paid for the rental and fuel costs but kept all their fare revenues. There was also a relatively smaller group of Yellow-Top taxi owner-drivers left which did not belong to any taxi companies.

In 1996, soon after the LTA was formed, NTUC Comfort bought over 797 out of 2,159 licenses from the individual Yellow-Top taxi owner-drivers at S$10,000 each. By 1996, NTUC Comfort dominated the taxi industry with its fleet of 10,000 while CityCab was half its size with 5,000 taxis and TIBS Taxis had less than 2,000 taxis.[bl]

The ROV was the licensing authority and regulator of the taxi industry until 1995. Being the issuer of taxi licenses to taxi companies, the ROV was able to control the number of taxi companies in the market and the size of their taxi fleets. However, ROV's function of taxi fare regulation was taken over by the Public Transport Council (PTC) when it was formed in 1987.

In 1995, the ROV was subsumed into the newly-established Land Transport Authority (LTA) which then became the regulator of the taxi industry.

3.2. Deregulation of taxi industry

1998 liberalisation of taxi fares

The first phase of the liberalisation involved the deregulation of taxi fares where taxi companies would be allowed to set their own fares without having to seek PTC's approval, provided that they gave advance notice of fare adjustments to PTC and the public.[bm] The government felt that the taxi companies could do a better job at setting optimal taxi fares than PTC, which had to balance the interests of commuters and taxi companies.

When the deregulation of taxi fares was announced in the parliament in 1998, there were some concerns that the three incumbent taxi companies might form a cartel to fix and raise taxi fares. However, the thinking then was that the taxi industry could potentially be further deregulated, if necessary, to allow for more new entrants into the market so as to increase competition.

[bl] Sharp, I., *The Journey — Singapore's Land Transport Story*, (SNP Editions, 2005).
[bm] PTC had earlier deregulated surcharges that taxi operators levy for phone bookings and premier services in 1997.

The government deliberately adopted a phased approach for liberalisation instead of a big bang approach (i.e., to deregulate both fares and taxi supply simultaneously). This allowed taxi companies, taxi drivers and commuters some time to adjust to the deregulation of fares,[bn] and gave the government time to monitor the impact of fare deregulation before moving on to deregulate supply.

The deregulation of fares took effect on 1 September 1998. During the initial period after deregulation, taxi companies did not intensify price and service competition. There were minor price differentiations among the taxi companies and their fare structures converged after a short period. It was usually the case that NTUC Comfort, being the market leader, would increase the fares and surcharges first, followed closely by CityCab, TIBS Taxis and Yellow-Top.

2003 liberalisation of taxi supply

The second phase of liberalisation came in June 2003 with the lifting of control on the number of taxi companies in the market and the fleet quota imposed on taxi companies.

Under the regulatory framework then, anyone wanting to set up a taxi company had to apply for a taxi operator licence (TOL) from the LTA. The application would be assessed on a comprehensive set of criteria, including financial resources and plans for providing a satisfactory taxi service. To ensure sufficient economies of scale to provide good services, such as investing in GPS-based taxi despatch system, new taxi companies were required to operate a minimum fleet of 400 taxis within four years of commencing operation but subject to a fleet quota of 100 taxis for the first year. The TOL was valid for 10 years.

All taxi companies also had to comply with a set of taxi Quality of Service (QoS) standards which specified performance targets in three areas — availability of taxis via radiophone booking; safety records; and customer satisfaction. There were two tiers to the QoS standards such that companies with more than 1,000 taxis had to comply with the higher standards. Taxi companies that consistently failed to meet the QoS standards could be fined up to S$100,000 or have their TOLs revoked.

[bn] Fare deregulation coincided with the revision of the vehicular tax structure and implementation of Electronic Road Pricing (ERP). Upfront taxes for taxis were made more aligned with those for private cars. However, usage related road and diesel taxes were reduced. Taxis are subject to ERP charges and commuters had to pay for ERP charges incurred on taxi trips.

In 2003, a revamped vocational licence points system (VLPS) was launched to complement the TOL framework in maintaining the standards of taxi services. The VLPS was a demerit point system that specified the penalties for the more serious breaches of conduct such as overcharging and touting. The demerit points varied with the seriousness of the offences.

Also in 2003, NTUC Comfort merged with DelGro Corporation, which owned CityCab. The new company ComfortDelGro (CDG) had a combined fleet of 16,500 taxis or about 85% of the entire taxi fleet in Singapore. The remaining taxis were owned by the SMRT Taxis[bo] and individual Yellow-Top owner-drivers. As a result, the taxi industry effectively had three big taxi companies, namely CGD (running both Comfort and CityCab taxis), SMRT Taxis and Yellow-Top taxis.

3.3. *Outcome of liberalisation*

By the end of 2003, three new taxi companies Trans-Cab, SMART Cabs, and Premier Taxis had entered the industry, followed by Prime Taxis in 2007.[bp] As of end 2009, there were seven taxi companies (including Comfort, CityCab and SMRT taxis) and a small group of owner-driver Yellow-Top taxis in the industry, with a total fleet of about 25,000 taxis. This was an increase of about 30% from May 2003 before the liberalisation of market entry.

Prior to liberalisation in 2003, Singapore had 4.8 taxis per 1,000 population, compared to 2.8 in Hong Kong, 2.9 in New York, 2.6 in London and 2.3 in Tokyo.[bq] In 2009, six years after libreralisation, there were 4.95 taxis per 1,000 population in Singapore.

The 2003 liberalisation led to more visible outcomes and significant impacts on the taxi industry. There was greater differentiation in taxi services offered at differentiated fares from the standard taxi service, including premium services using luxury vehicles for business travellers, and the use of more spacious vehicles such as space wagons or "London Cabs" to cater to bigger groups of passengers and/or luggage.

Taxi companies also offered innovative value-added services such as tour guides, medical chaperons and drink jockey services. In addition, there were

---

[bo] SMRT merged with TIBS in December 2001, and TIBS Taxis was rebranded as SMRT Taxis in May 2004.

[bp] LTA rejected an application for a TOL in November 2009.

[bq] Mr Yeo Cheow Tong, Minister for Transport reply to Parliamentary question on Road Traffic (Amendment) Bill, (2003).

services that catered to members of voluntary welfare organisations and the mobility-impaired.

The increase in taxi companies, with their own distinct liveries and different car models, was very conspicuous. There were some 34 different taxi car models in 2009, including CNG-powered and hybrid-electric powered taxis, compared to just 13 in 2006.[br]

Although fare rates had increased after liberalisation, taxi services in Singapore today remained less costly than in other major cities, and are affordable to most public transport commuters in Singapore.

3.4. *Challenges of the taxi industry*

Matching Demand and Supply

Despite the relative abundance of taxis, there were complaints of difficulty in getting taxis during peak hours and rainy days, especially in the CBD. To better match demand and supply, taxi companies implemented several surcharges, such as peak-hour, late night, city area and airport surcharges, to dampen excess demand while giving some incentives for taxi drivers to serve high demand times and locations.

For example, the dearth of taxis in the CBD during peak hours was primarily attributed to taxi drivers' unwillingness to pay ERP charges to enter the CBD without passengers. Hence, besides peak-hour surcharge, another surcharge was imposed on passengers boarding taxis within the CBD during peak hours to help defray the ERP charges and encourage taxi drivers to enter the CBD.

The opposite situation was observed at Changi Airport, Resorts World Sentosa and Marina Bay Sands where surcharges were also imposed on passengers boarding taxis. Commuters at these locations did not have to wait too long for taxis, even during peak hours.

In terms of taxi drivers' earnings, there were some improvements. Speaking in the parliament in March 2008, former Minister for Transport, Mr Raymond Lim, said that the average daily gross earnings of 1-shift and 2-shift taxis for the month of January 2008 (one month after the fare increase in December 2007) had increased from $206, to $216, and from $306 to $308 respectively.[bs]

[br] Almenoar, M., Cabs Now Come In All Shapes and Sizes, *The Straits Times*, 8 January 2010.
[bs] Mr Raymond Lim, Minister for Transport, *Speech at the Committee of Supply Debate*, (2008).

Waiting times for taxi commuters in the city had also fallen from five to 22 minutes to up to six minutes after the increase in CBD surcharges in December 2007.[bt]

Complicated taxi fare structure

On another front, the slew of surcharges led to complaints about confusing taxi fares. Responding to a parliamentary question on surcharges in January 2010, Minister Lim said that taxi surcharges accounted for less than 20% of total taxi fares.[bu] Taxi companies tried to address this concern by providing detailed breakdowns of the various fare items in taxi receipts. In addition, all the taxis are required to display the taxi fare chart at the rear seat window.

On taxi booking charges, the taxi companies' point of view was that setting up taxi booking and despatch systems required significant capital expenditure and operating costs, and it was justified that the costs be passed on to users. Commuters paid booking surcharges to taxi drivers, and some taxi companies recovered part of the call-booking cost from their drivers for each successful call-booking job.

Impact of liberalisation on taxi drivers

With increased supply and fares and declining ridership, the average daily number of trips, average mileage per trip, and average utilisation rates of taxis, appeared to have fallen in recent years. To cope with high rental fees, more than half of the taxis were driven by two drivers on a two-shift system. One of the drivers hired the taxi and rented it out to the other "relief driver" for part of the time. Taxi driving was commonly perceived as a temporary job during economic downturns when many people were retrenched. There was a report in 2006 that the average time a taxi driver stayed on the job before leaving was only three years.[bv]

There were taxi drivers who lamented on the hardship of earning a livelihood as a taxi driver, with fluctuating fuel prices, high taxi rentals and increased competition. To increase the earnings, some taxi drivers would rather wait for call bookings than pick up street hail or taxi stand commuters.

[bt] Chow, J., Drivers' Earnings Up, Cab Waiting Times Down: Raymond Lim. *The Straits Times*, 22 January 2008.

[bu] Mr Raymond Lim, *Minister for Transport's reply to Parliamentary question on Taxi Surcharges*, (2010).

[bv] Chang, R., Bumpy Ride for Cabbies, *The Straits Times*, 23 September 2006.

Others would flock to Changi Airport, Resorts World Sentosa or Marina Bay Sands to earn the additional location surcharges.

There were also taxi drivers who lamented that taxi rental fees had remained high despite falling car prices, and that a reduction of rentals would help them significantly. For instance, it was reported in 2006 that the cost of a Japanese taxi had dropped from around S$100,000 in the mid-1990s to S$60,000 in 2006[bw] and taxi companies could recover their capital outlay from taxi rental in three years or less.[bx]

From the taxi companies' perspective, although prices of taxis had fallen and they could expand their fleets without quota restrictions, the increased competition meant that they had to compete to attract taxi drivers to drive their taxis. Taxi drivers now had more choices of which companies to rent the taxis from. Taxi companies thus had to keep taxi rental rates competitive and introduce other initiatives such as cash bonuses and incentives for loyalty, good performance and being accident-free to attract and retain taxi drivers.

Taxi companies also faced difficulties in recruiting taxi drivers given the strict criteria imposed by the LTA on who can be a taxi driver. Currently, only Singaporeans who were above the age of 30 with at least one year of driving experience and clean driving records could apply to be taxi drivers. They also needed to be able to understand and speak English.

3.5. *Further interventions*

Looking back, the liberalisation of the taxi industry in Singapore had its successes. Taxi numbers grew from 19,000 in 2003 to about 28,000 in 2013. Taxi fares were relatively low by international standards. Taxis were generally a safe and convenient way to travel.

In spite of this, one key operational issue was commuters' difficulty to get taxis during peak hours. This prompted the government to make several interventions to improve the availability of taxis when the demand is highest.

Enhancing quality of taxi service

In 2008, the LTA launched a new central booking number (6342 5222) in an effort to integrate taxi operators' booking systems to make it more convenient

---

[bw] Tan, C., Taxi Woes Likely to Remain for Commuters, *The Straits Times*, 7 July 2006.
[bx] Chang. *Op Cit.*

to route the booking to available taxis. In 2012, the taxi quality of service (QoS) standards were ratcheted up to shorten call and booking wait times. At the same time, a new licence obligations was introduced requiring taxi operators to:

- Equip their entire fleet with the capability to identify taxis within the vicinity of the caller and assign call bookings accordingly;
- Equip their fleet with a centrally controlled 'on-call' display to prevent misuse; and
- Increase their fleet size to at least 800, from a previous minimum of 400.

Taxi availability standards

In addition, a major initiative was the introduction in 2013 — the new taxi availability standards, which required taxi operators to ensure that by 2015, at least 85% of their fleet clock a daily mileage of 250 km, and at least 85% of their fleet is on the road during the main weekday morning and evening peaks.

If taxi operators failed to meet these new standards, they would face financial penalties under the terms of their licences. Early indications had been that the availability standards have made some improvements to the taxi supply, eventhough not all operators were able to meet the standards.

Simplifying the fare structure

With a deregulated market, taxi operators were allowed to set their own fares based on commercial considerations. However, in practice, fare revisions had come in distinct clusters in 2006, 2007 and 2011. This deregulation of fares had resulted in a complex taxi fare structure that was confusing for commuters. The government acknowledged the unintended outcome, and on 11 November 2013, the Senior Minister of State for Transport and Finance, Mrs Josephine Teo, made the following comments in the parliament:

"Taxi fares were deregulated in September 1998 to allow taxi companies to set their own fares, so that they can be more responsive to market conditions. The taxi fare structure has since evolved with different surcharges to better match taxi supply with demand by giving incentives to taxi drivers to serve locations and time periods where demand is high."

"On the other hand, the recent proliferation of different taxi fares for different types of taxis is driven largely by the different cost and therefore different rental rates for different models of taxis. That said, we recognise that the current taxi fare structure is complex and confusing for commuters. The LTA will work with the Public Transport Council (PTC) and the taxi companies to study if and how it could be made simpler and more easily comparable across different taxi companies, taking into consideration also the impact on taxi drivers, ultimately to have a taxi fare structure that best serves commuters' interests."

The challenge remained that simplifying the complicated fare structure would not be easy, given that the fares had been deregulated and the cost structure of each taxi companies would differ, largely depending on the costs of procuring and replacing their fleets. Work was still ongoing in this area and the LTA and PTC were expected to make some announcements on taxi fare structures in early 2015.

3.6. New booking methods

Over the last couple of years, third-party taxi booking mobile applications (or apps) had made their appearance in the taxi market. Most of these apps were run independently from taxi operators. They sought to link passengers directly to taxi drivers in their vicinity, by passing the taxi companies' booking systems. Some of the more prominent apps included Easy Taxi, Hailo, GrabTaxi, Uber and MoobiTaxi. They had become hugely popular among commuters and cabbies as they effectively and efficiently match commuters' demand with taxi supply.

Taxi drivers were generally supportive of the new apps. They would receive sign-up incentives and most pay little to no commissions for each job they took. Passengers too had embraced the apps readily and used them for their taxi bookings. The taxi companies however had mixed reactions.

There were also other concerns about these app services. Since they were not owned by the taxi operators, it was possible that unlicensed drivers could be using the apps. In addition, there was little avenue to safeguard passengers on service quality and fairness in pricing.

In response to such concerns, the LTA announced in late 2014 that it would look into applying new regulations to third-party booking service providers. Under this framework, all third-party taxi booking services would be required to comply with the following conditions:

- Registration of services with LTA;
- Available only on label licensed taxis and drivers holding valid taxi driver's vocational licences. This ensured that commuters are served by taxis and taxi drivers who were operating legally in Singapore;
- Put in safeguards for commuters. All information on the fare rates, surcharges and fees payable for the journey must be specified to commuters upfront, before commuters accept the dispatched taxi;

- Taxi booking services cannot require commuters to specify their destinations before they can make bookings; and
- Such booking services were required to provide some customer support services, such as lost and found services, and avenues for commuters to raise queries and complaints.

3.7. *Once over lightly and some observations*

In retrospect, the early taxi industry attracted strong government intervention in rooting out the pirate taxi industry. To resolve the problem, the government firstly reduced the demand for pirate taxis by simultaneously improving bus services and reorganising the bus industry, and also reduced the supply of pirate taxis by providing alternative forms of employment to taxi drivers. Many pirate taxi drivers chose to become bus drivers and conductors, or joined the newly-formed NTUC Comfort taxi cooperative. The move modernised taxi operations and converted these drivers into responsible owner-drivers which, in turn, helped to improve service standards. Furthermore, the NTUC Income — the union's insurance cooperative — provided insurance coverage for these drivers which consequently benefited passenger safety. Secondly, the use of cooperatives to build a sound institution to deliver taxi services and insurance coverage had helped to reform the taxi industry, and by 1971, Singapore was officially pirate taxi free.

The deregulation of taxi fares and liberalisation of the industry in the late 1990s and early 2000s demonstrated the government's commitment to work with markets to deliver efficient public transport outcomes. The rationale for liberalisation of the industry was to allow the market forces to determine the pricing of taxi services that would better match taxi supply and demand, and optimise the utilisation of the taxi fleet.

However, there was room to further improve on taxi availability especially during peak hours. As of end 2014, there were six taxi companies and the size of the taxi fleet had increased by over 48% since 2003 to 28,736 taxis as of December 2014. Taxi ridership had increased by over 20% from 805, 000 in 2003 to 967, 000 daily passenger trips in 2013.

The slew of surcharges levied by different taxi companies, and accumulated over the years, had resulted in complicated taxi fare structures. As of end 2014, there were close to 10 different flag-down fares, three metered fare structures, more than 10 types of surcharges and eight kinds of phone booking charges. The LTA and PTC would be expected to address this issue in early 2015.

References

Almenoar, M. *Cabs Now Come In All Shapes and Sizes*. The Straits Times. 8 January 2010. (Singapore).

Centre for Liveable Cities & LTA Academy (2011) *Crossroads and Nexus: Land Transport Planning and Development*. Case Study. (Singapore).

Chang, R. (2006) *Bumpy Ride for Cabbies*. The Straits Times. 23 September 2006. (Singapore).

Chow, J. (2008) *Drivers' Earnings Up, Cab Waiting Times Down: Raymond Lim*. The Straits Times. 22 January 2008. (Singapore: The Straits Times).

Government Team of Officials. (1974) *Management and Operations of Singapore Bus Service Ltd: Report of Government Team of Officials*. (Singapore: Singapore National Printers).

Hawkins, L.C. (1956) *The Report of the Commission of Inquiry into the Public Passenger Transport System of Singapore*. (Singapore).

Land Transport Authority (1996), *White Paper for a World Class Transport System*. (Singapore).

Land Transport Authority (1997), *Annual Report 1996/1997*. (Singapore).

Land Transport Authority (2011), *LTA Appoints SBS Transit Limited to Operate Downtown Line under New Rail Financing Framework*. Press Release. (Singapore). Retrieved from http://www.lta.gov.sg/apps/news/page.aspx?c=2&id=659z82u5joc nrr4j4it759812yw2etknbsr66ucn2jd67avxjm.

Land Transport Authority. (2008) *Land Transport Master Plan: A People-Centred Land Transport System*. (Singapore).

Land Transport Authority. (2014) *Transition to a Government Contracting Model for the Public Bus Industry*. Press Release. (Singapore) Retrieved from http://www.lta.gov.sg/apps/news/page.aspx?c=2&id=28fca09a-bed6-48f4-99d4-18eeb8c496bd.

Lim, Raymond (2008) *Minister for Transport Speech at Committee of Supply Debate (Head W — Ministry of Transport) at Parliament*. (Singapore).

Lim, Raymond (2009) *Minister for Transport Speech at Committee of Supply Debate (Land Transport — Part1) at Parliament*. (Singapore).

Low, Tien Sio, interview by Centre For Liveable Cities and LTA Academy. *Oral History Interview*. (April 2011).

LTA Academy (2010), *Singapore's Integrated Public Transport System — Connecting the City-state*. Case Study. (Singapore).

LTA Academy (2010), *Taxi, Taxi, Everywhere but Not One to Stop For Me*. Case study. (Singapore).

Lui, Tuck Yew. (2012) *Minister for Transport Speech at Committee of Supply (Part 1 — Public Transport) at Parliament*. (Singapore).

Mah, Bow Tan, interview by LTA Academy. *Oral History Interview*. (February 2012).

Menon, A.P.G. & Loh, C.K. (2006) *Lessons from Bus Operations*. (Singapore: Land Transport Authority).

Parliament No. 0 (1955) *Public Passenger Transport Commission*. Session 1, Volume 1. (Singapore).

Parliament No. 2 (1970) *Reorganisation of Motor Transport Service of Singapore*. Session 1, Volume 30. (Singapore).

Parliament No. 2 (1971) *Bus Services Licensing Authority Bill*. Session 1, Volume 30. (Singapore).

Parliament No. 6. (1987) *Public Transport Council Bill*. Session 2, Volume 49. (Singapore).

Public Transport Council. (1988) *Annual Report 1987/1988*. (Singapore).

Public Transport Council. (1994) *Annual Report 1993/1994*. (Singapore).

Public Transport Council. (1995) *Annual Report 1994/1995*. (Singapore).

Public Transport Council. (2007) *Annual Report 2006/2007*. (Singapore).

Sharp. Ilsa, The Journey: Singapore's Land Transport Story. Singapore: SNP Editions, 2005.

Singapore Government (1955) *Report of the Inter-Ministry Committee on MRT/Bus Integration*. (Singapore).

Singapore Government (1970) *White Paper on Reorganisation of the Motor Transport Service of Singapore*. (Singapore).

Tan, C. (2006) *Taxi Woes Likely to Remain for Commuters*. The Straits Times, 7 July 2006. (Singapore).

Wilson, R.P. (1970) A Study of the Public Bus Transport System of Singapore. (Singapore).

ANNEX A

KEY MILSTONES ON BUS REFORM IN SINGAPORE

Timeframe	Key Milestones
1902	Singapore Electric Tramways Company (London) established.
1925	Singapore Traction Company (STC) was formed to operate trolley buses.
Early 1930s	Weak regulations led to rampant "mosquito bus" operation.
1935	"Mosquito bus" operators consolidated to 10 Chinese Bus Companies.
1955	Commission of Inquiry set up to review the public passenger transport system.
1956	Hawkins Report published and the Omnibus Services Licensing Authority (OSLA) formed.
1968	The 15-member Transport Advisory Board established.
1970	White Paper on *the Reorganisation of the Motor Transport Service of Singapore* and the Wilson Report published.
1970–1971	Amalgamation of the 10 bus companies to 4, and rationalisation of bus routes.
1971	Bus Service Licensing Authority (BSLA) replaced OSLA.
1973	Formation of Singapore Bus Service (SBS).
1974	Government Team of Officials seconded to SBS.
1974	Introduction of dedicated bus lanes and supplementary public transport service.
1975	Introduction of SBS's Blue Arrow Services (Forerunner of today's Premium Bus Services).
1978	SBS listed on the Singapore's stock exchange.
1982	Formation of Trans-Island Bus Services Pte Ltd (TIBS).
1987	The Public Transport Council (PTC) replaced BSLA.
1989	Transit Link Pte Ltd formed.
1990	Integrated Ticketing System (ITS) launched.
1994	PTC stipulated bus service specifications.
1996	White Paper on *A World Class Land Transport System* published.
2000	First comprehensive bus service audits by the PTC.
2001	Merger of TIBS and SMRT, forming Singapore's first multi-modal public transport operator.
2002	Enhanced Integrated Fare System (EIFS) launched and the PTC introduced the first feeder bus service competition framework.
2003	SBS rebranded as SBS Transit and became Singapore's second multi-modal public transport operator.
2005	First annual bus passenger satisfaction survey by the PTC.

(Continued)

(*Continued*)

Timeframe	Key Milestones
2006	First 10-year Bus Service Operator Licences (BSOLs) issued by the PTC, incorporating the mandated Area of Responsibility and QoS standards on licensed bus operators.
2007	First bus QoS penalty regime launched by the PTC.
2008	Land Transport Maser Plan (LTMP) 2008 launched.
2009	Symphony e-Payment (SeP) replaced EIFS.
2009–2010	LTA took over the role of the central bus planner.
2012	Bus Service Enhancement Programme (BSEP) launched.
2013	Land Transport Maser Plan (LTMP) 2013 launched.
2014	Government contracting model (GCM) for bus services announced and first competitive tender of bus services launched.

ANNEX B

MEMBERS OF TRANSPORT ADVISORY BOARD
(As of November 1970)

Chairman:

1. Phua Bah Lee Parliamentary Secretary (Communications)

Members:

(a) *Workers' Representatives*

2. Hashim bin Idris (President, Singapore Traction Co. Employees' Union)
3. Yeo See Bah (Secretary, Singapore Bus Employees Union) — Up till October 1969 as union was deregistered in May 1970. Vacancy would be filled by a representative from the Singapore Industrial Labour Organisation.
4. Robin Sim Boon Woo (Organising Secretary, Singapore Taxi Drivers' Association)

(b) *Bus Owners/ Taxi Owners' Representatives*

5. H. M. J. Jensen (Managing Director, Singapore Traction Co.)
6. Tan Kong Eng (President, Singapore Chinese Bus Owners' Association) Tay Soo Yong — Up till April 1970
7. Vacant (to be filled by a representative from the NTUC Workers' Transport Co-operative, when formed).
   Lim Chuan Huat (President, Taxi Transport Association) — Up till June 1970
   Low Kim Kee (Secretary, Taxi Transport Association) — Up till October 1969

(c) *Interested Organisations etc.*

8. Milton Tan (Chairman, National Safety First Council)
9. William S. W. Lim (Chairman, Singapore Planning and Urban Research Group)
10. Miss Phyllis Tan (Advocate and Solicitor)

(d) Government Departments
11. Goh Yong Hong (Registrar of Vehicles)
     Lim Kuan Ming (Registrar of Vehicles) — Up till May 1970
12. Koh Lian Wah (Officer-in-charge, Traffic Police)
13. J. T. Nallaiah (Senior Executive Engineer, Roads Division, Public Works Department)
14. Lim Leong Geok (Assistant Coordinator Traffic Transportation, State and City Planning Department)

(e) Secretary
15. Ling Teck Luke (PAS, Commmunication)
     Yap Soon Hoe (AA, Communication) — Up till January 1970
     Wong Meng Voon (AA, Communication) — Up till January 1969

Source: White Paper on Reorganisation of the Motor Transport Service of Singapore. 1970.

ANNEX C

REGISTRARS OF VEHICLES (1959–1970)

S/N	Name	Period	Length of Service
1.	Goh Kee Song (Acting)	13 September 1958–2 July 1961	2 years 9 months
2.	M. C. Schubert (Acting)	3 July 1961–30 October 1962	1 year 3 months
3.	Wong Keng Sam	1 November 1962–22 January 1965	2 years 2 months
4.	Kwa Soon Chuan	23 January 1965–8 January 1967	1 year 11 months
5.	H. F. G. Leembruggen	9 January 1967–13 June 1967	5 months
6.	Lim Kuan Ming (Acting)	14 June 1967–16 July 1967	1 month
7.	Lim Phai Som	17 July 1967–10 January 1968	5 months
8.	Lim Kuan Ming (Acting)	11 January 1968–5 May 1970	2 years 3 months
9.	Goh Yong Hong (Acting)	6 May 1970 onwards	

Source: White Paper on Reorganisation of the Motor Transport Service of Singapore. 1970.

ANNEX D

COMPOSITION OF GOVERNMENT TEAM OF OFFICIALS (GTO)

S/N	Name	Designation
1	Wong Hung Khim (Team Leader)	Deputy Secretary, Ministry of Labour
2	Yap Boon Keng	Assistant Commissioner of Police
3	Ang Teck Leong	Deputy Accountant-General
4	Yeo Seng Teck	Deputy Director, Economic Development Board

(Continued)

		(Continued)
S/N	Name	Designation
5	Lo Wing Fai	Head Industrial Engineering Unit, National Productivity Board
6	Ong Chuan Tat	Accountant, Accountant–General's Office
7	Mah Bow Tan	Industrial Engineer, Ministry of Defence

Source: Report of Government Team of Officials, Management and Operations of Singapore Bus Service Ltd, 20 July 1974.

ANNEX E

THE PTC MEMBERS AS AT 31ST MARCH 1987

Chairman:

1. Mr Michael Fam

Deputy Chairman:

2. Mr Lock Sai Hong

Members

3. Mr Lim Leong Geok — representing Singapore MRT Ltd
4. Mr Tan Kong Eng — representing Singapore Bus Services (1978) Ltd
5. Mr Ng Ser Miang — representing Trans Island Bus Services Ltd
6. Mr Sam Chong Keen — representing NTUC Workers' Co-operative Commonwealth for Transport Limited *(up to 30 September 1987)*
7. Mr Lew Syn Pau — representing NTUC Workers' Co-operative Commonwealth for Transport Limited *(from 1 October 1987)*
8. Lee Yew Kim *(up to 31 March 1988)*
9. Mr Phua Tin How *(From 1 April 1988)*
10. Mr Bunno Hylkema
11. Mr Chia Lin Sien
12. Mr Victor Ng
13. Mr Lew Seng Huat
14. Mr Ong Cheng Sng
15. Mrs Vasantha Kumaree Siva

Source: Public Transport Council Annual Report 1987/1988.

ANNEX F

PICTURES OF OLD MAGNETIC FARECARDS

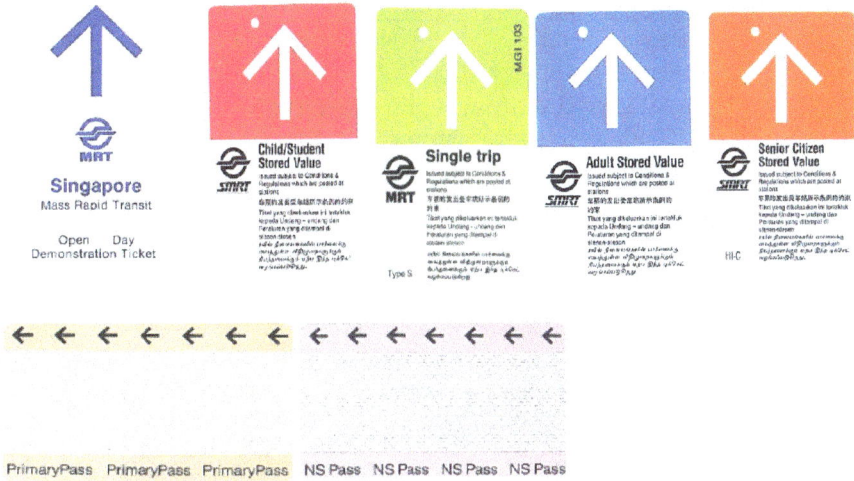

ANNEX G

THE FRIST LTA BOARD MEMBERS IN SEPTEMBER 1995

Chairman:

1. Mr Fock Siew Wah

Members

2. Dr Tan Cheng Bock
3. Mrs Yu-Foo Yee Sharon
4. Dr Cheong Siew Keong
5. BG Wesley D'Aranjo
6. Mr Michael Lim
7. Mr Peter Chen
8. Mr Khoo Teng Chye
9. Mr Han Fook Kwang
10. Prof Henry Fan
11. Dr Han Cheng Fong
12. Mr Darke M Sani
13. Mr Liew Heng San

Source: LTA Annual Report 1996.

ANNEX H

TIMELINE OF RAPID TRANSIT SYSTEM DEVELOPMENTS

Timeframe	Key Milestones
1967–1971	The State and City Planning (SCP) Project recommended an efficient public transport system by means of a rail transit system by 1992.
1980	Provisional Mass Rapid Transit Authority (PMRTA) was set up. The MRT debate on an MRT system supported by buses versus an all-bus system.
1982	Decision to build the MRT system (North-South and East-West Lines) at an estimated cost of S$5 billion.
1983	Formation of the Mass Rapid Transit Corporation (MRTC).
1987	Singapore Mass Rapid Transit Ltd or SMRT was incorporated.
1990	MRTC explored various options for Woodlands Extension.
1994	Announcement to build two LRT systems — Bukit Panjang and Buona Vista. The latter did not take off as it was deemed not financially viable.
1995	Formation of Land Transport Authority (LTA).
1996	White Paper on *A World Class Land Transport System* and rail financing framework. Opening of Woodlands Extension.
1998	A new Licence and Operating Agreement (LOA) signed between LTA and SMRT Ltd.
1999	SBS Transit was awarded operating licences for North-East Line, and Sengkang and Punggol LRTs. Opening of Bukit Panjang LRT.
2000	SMRT Corporation Ltd was fully incorporated.
2001	Circle Line operating licence was awarded to SMRT Corporation. Opening of Dover MRT station.
2002	Amendment of Rapid Transit System (RTS) Act to strengthen licensing power.
2003	Opening of North-East Line (NEL) and Sengkang LRT.
2004	Existing East-West Line would be extended to meet growing demand from commuters in Jurong Industrial Estate.
2005	Opening of Punggol LRT.
2008	Land Transport Master Plan 2008 published.
2009	Opening of Stage 3 of the Circle Line. Opening of Boon Lay Extension
2010	Amendment of RTS Act for new rail financing framework and greater contestability. Opening of Circle Line stages 1 and 2.
2011	Downtown Line operating licence was awarded to SBS Transit. Opening of Circle Line stages 4 and 5.

(*Continued*)

(*Continued*)

Timeframe	Key Milestones
2012	Opening of Circle Line Extension to Marina Bay.
2013	Land Transport Master Plan 2013 published. Opening of Downtown Line stage 1.
2014	Opening of North-South Line Extension to Marina South Pier.

Source: Land Transport Authority.

ANNEX G

KEY MILESTONES IN SINGAPORE'S TAXI INDUSTRY

Date	Key Milestones
1950s	• Taximeters introduced to provide a fairer service to the public. Singapore Taxi-Owners Co-operative Motor Garage and Stores Society Limited, a major taxi company in Singapore, became one of the first taxi companies to adopt the meters. By the end of 1953, all taxis in Singapore were required to install the taximeters.
1960s	• Heyday of "pirate" taxis in Singapore due to paralysed bus system as a result of frequent labour and union strikes in the 1950s and 1960s. • Police launched a campaign in Singapore to clamp down on "pirate" taxis following complaints that these "pirate" taxis were frequently overcrowded and endangered the safety of its passengers. Furthermore, passengers were not covered by insurance should an accident take place.
1966	• STC released an annual report citing the "pirate" taxi problem as a cause for its operating losses of nearly $1 million in revenue. STC estimated that there were 12,000 "pirate" taxis illegally plying the roads for hire and had taken away about 6,732,000 fares from STC in 1965.
1970	• The National Trade Union Congress (NTUC) Workers' Co-operative Commonwealth for Transport was established with a fleet of 1,000 taxis. It would later become NTUC-Comfort, the largest player in the local taxi-cab industry for decades. • 1970 White Paper on the Reorganisation of Motor Transport Service of Singapore was published recommending new regulations to weed out "pirate" taxis implemented such as raising diesel taxes on private diesel vehicles by 100 per cent (majority of which operated as taxis), suspension of driving licences for one year if caught; and making pirate taxi operations a seizable offence where offenders would be arrested on the spot and charged the following day.
1971	• "Pirate" taxis in Singapore were officially "eradicated". Facing uncertainty and unemployment, many pirate taxi drivers decided to switch to licensed taxis with NTUC-Comfort, or became bus drivers or conductors.

(*Continued*)

Date	Key Milestones
1983	• NTUC Comfort introduced a taxi rental scheme and began converting its taxi owner-driver shareholders into taxi hirers by purchasing their taxi operating licences.
1995	• CityCab was formed by the merging of Singapore Airport Bus Services (SABS), Singapore Bus Service Taxis (SBS Taxi) and Singapore Commuters taxi companies.
1997	• PTC deregulated surcharges levied by taxi operators for phone bookings and premier services.
1998	• Deregulation of taxi fares. • Revision of vehicular tax structure. • Introduction of ERP.
End 1999 to early 2000	• NTUC Comfort raised CBD surcharge and adjusted peak period hours. • Other operators followed.
2000	• NTUC Comfort raised its distance-based rate. • Other operators followed.
2002	• Road Traffic Act amended. • Deregulation of taxi supply and introduction of TOL framework.
2003	• TIBS Taxis raised distance-based rate to match NTUC Comfort, CityCab and Yellow-Top. • NTUC Comfort and DelGro merged to form ComfortDelGro (CDG). • TOL regime implemented with taxi QoS standards. • Quota restrictions on fleet size of taxi operators were lifted. • Singapore Taxi Academy (STA) formed and was tasked by the LTA to train, test and certify taxi drivers vocational licences. • Trans-Cab, Premier Taxis and SMART Cabs were granted TOLs.
2006	• ComfortDelGro raised standard taxi fares. • Other operators followed.
2007	• TOL's QoS standards on radiophone booking services were enhanced. • Prime Taxis was granted TOL. • LTA introduced Mystery Customer Audits of taxi services.
End 2007 to early 2008	• ComfortDelGro raised taxi fares. • Other operators followed.
2008	• Street hail of taxis in the CBD was prohibited. • Common taxi number to facilitate phone bookings launched.
2009	• Government grants a waiver of the Diesel Tax for taxis that are not hired out in the period from 1 March 2009 to 28 February 2010 to provide relief to the taxi industry during the economic downturn.

(*Continued*)

(Continued)

Date	Key Milestones
2012	• QoS standards for taxis hourly call booking and safety performance indicators raised from October 2012. • Age limit for taxi drivers raised from 73 to 75 years from June 2012. • CBD taxi rules eased from 1 March 2012 for more convenient pick-up and drop-off of passengers in the CBD.
2013	• Introduction of new Taxi Availability (TA) standards from January 2013 to increase utilisation of the taxi fleet. • New online matching portal to facilitate hirers in finding relief drivers and vice versa launched by LTA.
2014	• New regulatory framework for third-party taxi booking services to protect the safety and interests of commuters announced with effect from 2nd quarter 2015. • Release of taxi customer satisfaction survey 2013 results — 5.6% of respondents said they were satisfied with taxi services in Singapore.

Source: Land Transport Authority.

Chapter 3

Travel Demand Management

POON Joe Fai

This chapter expounds the various travel demand management measures undertaken by Singapore in the last 50 years. Beginning with the use of taxes to suppress car ownership in the 1960s, to the implementation of the world's first congestion pricing scheme (1975) and vehicle quota system (1990) as well as the more recent 'Travel Smart' measures which seeks to influence travel behaviour change and smoothen peak-hour public transport demand, it recounts the changes made to each scheme with the rationale and impact of each measure. Organised chronologically, the writer briefly describes each policy measure together with the transport paradigm of the period.

1. Introduction

Like most developing cities then, Singapore in the 1970s was already facing road congestion, arising from increasing number of vehicles on the roads. As a small island city-state, she was particularly vulnerable. With limited potential for road capacity expansion, Singapore introduced constraints on travel demand much earlier than her counterparts, and has now decades of application in travel demand management.

Since the 1970s, Singapore has progressively implemented a variety of vehicle ownership and usage restraint measures. Vehicle ownership restraints are mainly in the form of fiscal disincentives such as import/excise duties (ID), Additional Registration Fees (ARF) and road taxes. Usage restraints include the Area Licensing Scheme (ALS) and Electronic Road Pricing (ERP).

2. Early Years — Suppressing Car Ownership Demand through High upfront Taxes

In the early years, the primary approach to curbing growth of the vehicle population, particularly of cars, was to discourage vehicle ownership through tax measures. ARF and ID were levied as a percentage of the Open Market Value (OMV)[1] of the vehicle in a bid to raise car ownership costs and restrain vehicle ownership.

In the 1960s, import duties (ID) were imposed on cars at 10% of the Open Market Value (OMV), before they were raised to 30% in 1968. An Additional Registration Fee (ARF) was also imposed at different rates, with locally assembled cars attracting the lowest rate of 10% of the OMV, imported Commonwealth cars at 15%, and non-Commonwealth cars at 25%. This was before the ARF on all newly registered cars in 1972 was standardised at a rate of 25% of the OMV. In the same year, ID on cars was further raised by 15 percentage points to 45%, just four years after the last revision, and the road tax was revised to a sliding scale upwards.

As shown in Table 1, the ARF for cars continued to be revised upwards in the 10 years since, a reflection of the challenges of, and limited success in, curbing the rapid growth in the vehicle population, especially of cars. In the

Table 1.   Rates of Additional Registration Fees (ARF) implemented in Singapore.

Year	Additional Registration Fee of Cars (as percentage of Open Market Value)
Before 1972	10%–locally assembled 15%–imported Commonwealth 25%–imported non-Commonwealth
1972	25%
1974	55%
1975	100% (Preferential ARF Scheme was also introduced)
1978	125%
1980	150%
1983	175%

Source: Land Transport Authority.

---

[1] The OMV of a vehicle is assessed by Singapore Customs, based on the price actually paid or payable for the imported vehicle which includes purchase price, freight, insurance and all other charges incidental to the sale and delivery of the car to Singapore.

intervening period, the ARF was raised every few years. By 1983, the ARF reached a peak of 175% of the OMV.

The annual road taxes, or the fees to renew the vehicle licence, were also increased several times in the same period. They were increased in 1975 together with the revision of ARF rates. Another round of increase was implemented in 1980, and was accompanied by a Registration Fee rate of $1,000 for private cars (from a low rate of $15), and $5,000 for company cars.

The road tax structure was also revised in 1975 to serve another purpose — to keep the vehicle population young. A 10% surcharge on the road tax was imposed on vehicles that were more than 10 years old. This was because older vehicles tended to break down more often and cause traffic congestion.

Another fiscal tool to encourage car owners to replace their cars early was the Preferential Additional Registration Fee (PARF) scheme, which was also introduced in 1975. The scheme allowed the car owner to pay a lower ARF rate on a new replacement car, if he de-registered his old car before it reached the tenth year of age. This was introduced in consideration of the increasing cost of cars, particularly with the large hike in the ARF in the same year that could lead to car owners prolonging the use of their cars. The PARF rate was set according to the engine capacity (Table 2), and ranged between 35% and 55%. The next revision in 1983 saw the PARF rates increased by 10 percentage points across the board.

For more than a decade since the early 1970s, ownership taxes were raised incessantly. By the mid-1980s, car buyers had to pay substantially more than the OMV of a car to own one. Such a strategy proved to be necessary, but not sufficient, to address the challenges posed by increasing travel demands in Singapore.

Table 2.   Preferential additional registration rates implemented in 1975 and 1983.

Preferential Additional Registration Fee Rates (As % of ARF of new car)		
Engine Capacity	1975	1983
Less than 1,000 cc	35%	45%
1,001–1,600 cc	40%	50%
1,601–2,000 cc	45%	55%
2,001–3,000 cc	50%	60%
3,001 cc and above	55%	65%

3. Birth of Road Pricing — Area Licensing Scheme (1975)

The high upfront taxes turned out to be a blunt restraint measure that had limited impact on congestion in local areas. Addressing neither temporal nor geographical dimensions of travel demands, they could not ensure the optimal utilisation of our roads. In particular, the roads in the city area faced increasing traffic problems.

So, in 1973, a high level inter-ministerial committee was set up to look into measures to improve the transport situation. It formulated a scheme called the Area Licensing Scheme (ALS), put forth to the public for feedback. Following a one-year dialogue on the details and some subsequent modifications, Singapore introduced her first road pricing scheme in June 1975. The ALS was also the world's first road pricing scheme and an unprecedented policy innovation at that time which heralded Singapore's road usage restraint measures.

An area of 725 hectares of the city, or slightly more than 1% of the total area of Singapore, was designated as the Restricted Zone (RZ), around which a cordon was established. Thirty-three overhead gantries were erected along the approach roads to the RZ.

The ALS started with the restricted hours of 7.30am to 9.30am daily, except on Sundays and Public Holidays. During this period, all non-exempted vehicles entering the RZ were required to have an area licence displayed on their windscreens (or handlebars, in the case of motorcycles). The drivers of these vehicles were to buy the area licences in advance, before entering the RZ, at sales booths along the approach roads, and other sales outlets such as petrol stations, post offices or convenience stores (Fig. 1).

There were different licence fees set for various types of vehicles. There were also licences for daily and monthly use. The paper licences came in different colours and shapes to enable the different types of licences to be easily distinguished. This was because enforcement was carried out by police officers (Figure 2), who were stationed at the entry gantries, and the licences needed to be quickly identified by sight, typically within a few seconds. Vehicles that did not have the valid licences would have their details recorded by the enforcement officers. A summons to pay a fine would then be sent to the registered owners.

Initially, taxis, public transport buses, goods vehicles, motorcycles, and passenger cars involving in car-pooling, i.e., with four or more passengers, were exempted from the scheme, although taxis were subsequently included

Source: Land Transport Authority.

Fig. 1.   ALS Licence sale booth.

Source: Land Transport Authority.

Fig. 2.   Enforcement officers standing at an ALS Gantry.

and treated as normal cars two months later. Car-pooling was allowed initially to optimise vehicle usage and to counter the perception that ALS favoured only the rich. A park-and-ride scheme was also implemented at 13 fringe car parks to provide alternatives to motorists entering the RZ by parking their cars there for a low fixed fee and taking a shuttle bus into the RZ at a flat fare.

Economics had long established the case for the ALS, or road pricing in general. It was a classic example of the use of market mechanism as instruments of governance to allocate scarce resources, for which Singapore was well known. The scarce resource in this case was road space, which was severely limited in this small island state, especially in the city area. The ALS allocated road space within the RZ to motorists who were willing to pay, through the licence fees, for using it. Those who were not willing to pay would need to seek out alternative modes, such as public transport, park-and-ride or car-pooling, or travel after the restricted hours.

The licence fee could also be seen as charging for the negative externalities of traffic congestion, such as pollution from vehicle emissions, noise and other disamenities to society, on a 'user-pays' or 'polluter-pays' principle. By reducing the number of vehicles and minimising traffic congestion on the city's roads, ALS helped to achieve environmental objectives.

Since its implementation, ALS was subject to adjustments and fine-tuning over time. The first came within just three weeks, with the restricted hours extended to 10.15am. This extension was to restrain the surge in traffic immediately after the restricted hours ended at 9.30am. Nonetheless, on the whole, the ALS had achieved what it was meant to do. Traffic into the RZ was reduced by 44% in the three months after implementation.

To sustain its effectiveness in the face of growing vehicle population and the travel demand generated, the ALS fees were increased over the years. The fees for cars started with $3 for the daily licence and $60 for the monthly licence. They were increased to $4 and $80 respectively in January 1976, and then $5 and $100 in March 1980.

In June 1989, all classes of vehicles were subjected to ALS, including motorcycles and goods vehicles which were initially excluded. The only exceptions were scheduled public buses and emergency vehicles, such as fire engines and ambulances. Exemptions for car pools were also abolished to prevent public bus commuters from switching to taking private cars. The restriction period was also extended to the evening peak hours between 4.30pm and 7.00pm on weekdays, although the end time was later cut back by half an hour to address requests from residents living inside the RZ but worked outside.

With these changes, the fees (for cars) were reverted back to $3 and $60, for the daily and monthly licences, respectively.

In January 1994, the restricted period was further extended to cover the period between 10.15am and 4.30pm (the inter-peak period) on weekdays, and between 10.15am and 3.00pm (post-peak period) on Saturdays. The Saturday restriction period was subsequently shortened to end at 2.00pm, as traffic conditions within the RZ improved. To cater to motorists who chose to enter the RZ during the inter-peak and post-peak periods only, a new category of 'part-day licences' was introduced. The fees for such licences were set at either 70% or two-thirds of those for the original hours. For cars, they were $2 and $40 for daily and monthly licences. The original licences became 'whole-day licences' that allowed the motorist to enter the RZ throughout the entire restriction period.

4. Advent of New Ownership Measures

4.1. *The vehicle quota system*

The aggressive increases in taxes on vehicles from the 1960s to 1980s reflected the government's determination to restrain vehicle growth, but also its limited success thus far. Even as the ALS managed to restrain the growth of traffic into the city area, the vehicle population growth threatened to surpass the capacity of the road network elsewhere on the island in the long run, even as it was being expanded.

The growth of the vehicle population, particularly of the private car, remained unabated in spite of all the tax disincentives to vehicle ownership. The car fleet in Singapore grew from about 142,000 in 1975 to 257,000 in 1989. This 80% growth far exceeded the growth rate of the Singapore population, which expanded about 30% growth (from 2.26 million to 2.93 million) over the same period. The high costs of car ownership had limited effectiveness in restraining its growth at a sustainable pace was not surprising. Demand for car ownership was driven by rising affluence of society; per capita income had risen 3.4 times from around $6,000 to over $20,000. New approaches to curb vehicle ownership were needed.

In 1989, a Select Committee of Members of Parliament, led by Dr Hong Hai, Chairman of the Government Parliamentary Committee (GPC) on Transport, was appointed. The Select Committee was tasked to examine the prevailing policies and measures to control the vehicle population and managing road usage, to review their effectiveness, and to consider any other policy. It invited feedback from various professional organisations, academic institutions, and members of the public, and convened public hearings.

After a few months, the Select Committee presented its findings to Parliament in January 1990. The Committee made recommendations in several areas, such as car ownership and usage measures, PARF, parking, public transport, and infrastructure. In the key area of car ownership, it put forth the observation that the vehicle population was the single most important determining factor of road congestion. It concluded that "the number of vehicles on Singapore roads must be controlled and limited below the level that a free market in vehicles would create. Middle class aspirations for car ownership must therefore be tempered by the realities of our nation's resource limitations." Nonetheless, the Committee recognised that ownership restraint measures alone could not ensure the optimal utilisation of the road network, and required usage restraint measures to complement them for maximum effectiveness.

The most significant contribution of the Committee was its recommendation of a quota system to control the vehicle population. The Committee observed that the current approach of using high punitive ownership taxes had been inadequate in controlling car population, with high car growth during economic booms and decline in car ownership during leaner times. For greater control over vehicle growth rates to ensure that the vehicle population could be maintained within target in the long run, a quota system would be more effective than a pricing system. This was despite the former being more difficult to understand and implement than a simple tax system.

The recommended quota system would enable the annual allowable car growth to be set at a rate commensurate with the expansion of the road network capacity, and allow the market to find the appropriate price for a licence. Market demand would determine car prices: when market demand for cars was higher, licence prices would be higher.

The government swiftly accepted the Committee's recommendation to introduce a quota system to control vehicle population. Two months later, in April 1990, it was implemented, after extensive refinement of the scheme's details. Thus, another innovative policy instrument — the Vehicle Quota System (VQS) — was introduced. Like the ALS, the VQS was then the first of its kind in the world.

The VQS required a person interested in buying and registering a new vehicle to first obtain a Certificate of Entitlement (COE) issued by the government. The price of the COE is known as the Quota Premium (QP). With the COE, its owner was granted the right to own and use a vehicle in Singapore for

10 years, which was aligned with the deadline for PARF benefits. Upon the COE's expiry, the vehicle had to be de-registered (either through scrapping the vehicle body or exporting it) unless the owner revalidated its COE by paying a fee called the Prevailing Quota Premium (PQP), which was the moving average of the QPs for the previous 12 months. Alternatively, he could obtain a new COE to register for a replacement vehicle. The COE was tied to the registered vehicle such that it was transferred to the new owner when the vehicle was sold.

With a few exemptions,[2] the VQS was applied to all vehicles, even those registered before VQS was implemented.[3] However, if a vehicle was scrapped or exported before the 10-year validity of its COE was up, its owner received a rebate on the QP, pro-rated to the number of months left on the entitlement. The rebate could be used to offset the QP for a new COE.

With a the VQS, the Ministry of Communications and Information (MCI), then in charge of land transport policies, was given an effective instrument to control the growth rate of the vehicle population by fixing the quantity of new COEs to be issued each year. In the first year from May 1990 to April 1991, or the first quota year, MCI set a quota to allow about 4.3% vehicle growth. This was based on the average annual vehicle growth of 4.2% over the previous 15 years. For the second quota year onward, an initial 2.5 to 3% vehicle growth target range was set, depending on traffic conditions. Subsequently, the target annual vehicle growth rate was fixed at 3%, a rate that was maintained for almost two decades.

To ensure that the vehicle population was kept at a moderate growth path even as older vehicles were deregistered, COEs of vehicles that were de-registered in one quota year would be added back to the pool of COEs to be issued out in the following quota year. Hence, the quota in each quota year would be determined by both the number of new COEs issued that year (to allow for growth in population) and the number of de-registered vehicles that were de-registered in the previous quota year. The number of COEs to be issued in year n was

---

[2] Exemptions were given to scheduled and school buses, emergency vehicles, engineering plants primarily for off-the-road use, trailers, diplomatic vehicles, and vehicles belonging to the disabled.
[3] Owners of vehicles registered before VQS was implemented were given a grace period to revalidate their vehicle entitlements. The grace period was determined by (10-X) years, where X was the age of the vehicle (calculated from the first day of the month in which it was first registered). The minimum grace period was two years. At the end of the grace period, owners had to revalidate their entitlements for another 10 years by paying the PQPs. If they decided to dispose of their vehicles registered before May 1990, they would not receive the COE rebates for new vehicles because they did not pay for COEs when they first registered.

$$COE_n = x \ Vehicle \ Population_{(n-1)} + Deregistrations_{(n-1)}$$
where x was the target vehicle growth rate

The government decided to use the market mechanism to allocate the COEs, whose numbers were surely fewer than the demand for them. The then Registry of Vehicles (ROV) would conduct auctions, in which persons interested in obtaining the COEs would engage in competitive bidding. Thus, the market would decide the appropriate prices for limited COEs. It was deemed an economically efficient way to allocate scarce resources — those who valued them most, and were willing and able to pay for them would get the COE. The revenue from the auctions could then be used by the state to benefit the society at large.

A person interested in obtaining a COE had to make a bid during a tender. When VQS started, a closed bidding system was used; each bidder had to submit his bid in a tender form, with his bid amount unknown to other bidders. At the first tender in April 1990, which was for the quarter of April to June 1990, each individual was only allowed one bid, and each authorised motor distributor, 100 bids. From July 1990 onwards, the tenders were conducted monthly to make it more convenient for car buyers. (Motor distributors were then allowed 30 bids each, while companies could make unlimited bids but their COEs were not transferable). The tender opened at the start of each tender month and closed at the end of the month. Tender results were announced after the closing of the tender. When they submitted their bids, bidders had to make monetary deposits of half the amount of their bid price.

Each successful bidder only needed to pay for the COE a price equal to the lowest successful bid (LSB), which was the Quota Premium (QP).[4] Deposits would be returned to unsuccessful bidders. Successful bidders whose deposits exceeded the QP would be refunded and those whose deposits were lesser than the QP would pay the remaining sum.

An individual or a motor distributor who was successful in the tender would receive a Temporary COE (TCOE). It was transferable once,[5] either between individuals, from an individual to a company, or from a motor distributor to its customer. Allowing transferability improved market efficiency and social welfare, because it allowed transactions between willing buyers and sellers in the secondary market, and to allow some unsuccessful bidders to

---

[4] Companies paid twice the QP because these were tax-deductible.
[5] TCOEs of companies were not transferable.

purchase vehicles without waiting for the next tender. It also served to signal the latest market valuation of COE, which would benefit participants of future tenders.

The VQS provided the government an effective policy instrument to have absolute control over the growth of vehicle population. Given the longstanding aspirations for car ownership, the key challenge of implementing VQS was to ensure its public acceptance by addressing the perceptions of its fairness and equity. This challenge proved to be a persistent one, even after more than two decades of implementation.

While it was decided that the COEs were to be allocated through the market mechanism, it was recognised that there was a need to moderate its full effects. If the allocation of COEs was left fully to the market mechanism, it would lead only to those with the most financial means obtaining the vehicles they desired. The design of the VQS had included several elements that attempted to address the perceptions of fairness and equity from various stakeholders, including existing vehicle owners, prospective vehicle buyers, and non-owners.

First, it was decided that to separate the COEs into different categories by the type of vehicles. Initially, the VQS had seven categories by vehicle type, each of which had a quota of new COEs that could be issued each year. These categories were shown in Table 3.

The COEs in Category 7, or the Open Category, could be used to register vehicles of any type. The purpose of the Open Category was to provide bidders flexibility over the choice of vehicles to register, and to allow the composition of the entire vehicle population to evolve over time in response to trends in the vehicle market, else it would be unchanged from that in 1990.

Table 3. Categories of Vehicle Quota System (VQS).

Category	Vehicle type
1	Small Cars — Cars with engine capacity below 1,000cc
2	Medium Cars — Cars with engine capacity between 1,001cc and 1,600cc
3	Big Cars — Cars with engine capacity between 1,601cc and 2,000cc
4	Luxury Cars — Cars with engine capacity above 2,001cc
5	Commercial — Goods vehicles and buses
6	Motorcycles
7	Open Category

Source: Land Transport Authority.

For a start, 10% of new COEs (or 1,400 COEs) for the first tender in April 1990 were allocated to the Category 7. In subsequent years, the quota of this category would be made up of 20% of COEs from vehicles deregistered in the previous quota year, with the remaining 80% allocated back to their own respective COE categories.

Second, the COE validity was restricted to 10 years. Such a limit was imposed to prevent the granting of perpetual rights to own vehicles, as that would raise the price of COEs, making it extremely expensive for prospective vehicle owners. At the end of the validity period, existing owners who wished to continue owning a vehicle would either have to compete with all other interested bidders for a new COE (for a new vehicle) or pay the Prevailing Quota Premium (PQP), which reflected the prevailing market valuation of ownership rights, to revalidate the existing COE to continue using the current vehicle. This way, existing owners would not be advantaged over prospective ones.

Third, the market-based mechanism was also a fair method to allocate COEs. While there were suggestions for the COEs to be allocated by other non-market-based methods, such as by balloting or on the basis of need, these could be seen as inequitable for many. Allocation through a ballot was essentially by the luck of the draw, and fully ignored how COEs were valued differently by different individuals. Some who valued COEs highly would end up not getting them, while opportunists could obtain one, even if they did not intend to buy a vehicle. These opportunists could then sell COEs at a higher price to those who valued the COEs highly but were not lucky enough in the ballot. These opportunists would then make windfall gains at the expense of the general society, including non-vehicle owners, who would benefit from public projects and schemes that could be funded from COE revenues.

While it seemed reasonable to allocate COEs on the basis of need, it was difficult in practice to determine need and assess who was in need of a COE, in the absence of price signals. A large administrative set-up would have to be set to determine the qualifying criteria, process applications, and set the COE prices. However, there would be those who would be unhappy for being ineligible for concessionary treatment. It would also not prevent those who were eligible who might exploit the system by transferring their rights to those ineligible for monetary gains, as in the case for balloting.

There were also refinements to the VQS shortly after its implementation, partly to assuage unhappiness over certain quarters making speculative and unfair gains in the bidding exercises. Prospective car buyers and motor distributors expressed unhappiness over speculation of TCOEs, which was blamed for excessively high bids and profiteering from transfers of TCOEs. In response, after the very first quarterly tender, the frequency of tenders was changed to monthly. The validity of TCOEs was also reduced from six months to three months, and the bidding limits for motor distributors, cut down from 100 to 30.

With the implementation of VQS, the government also adopted recommendations from the same Select Committee to adjust the ARF and PARF schemes. With an effective tool to control the vehicle population in VQS, the reliance on ARF could be less. Therefore, the ARF was reduced by 25% of OMV in two steps, from 175% to 160% from 1 November 1990, and from 160% to 150% from 1 February 1991. This marked the start of a progressive lowering of ARF rates in the next two decades.

The PARF scheme was also changed substantially. Instead of tying the PARF benefit to the ARF of new car, it was converted to one that was tied to the car that was to be replaced. A rebate of 80% to 130% of the OMV (depending on the age of the car when de-registered) was assigned upfront to the car at the point of its registration, and to be used to offset the ARF of its future replacement. The maximum PARF rate of 130% was applied only when the car was de-registered before five years of age, and was reduced linearly for each subsequent year until the car reached 10 years old. This change applied to new cars registered from 1 November 1990 onwards.

The PARF benefits of existing cars were also changed. As the PARF rates were then pegged to the ARF of new cars, the reduction of ARF rates would result in existing car owners losing a significant amount of PARF benefits overnight. To avoid such an undesirable outcome, cars registered before 1 November 1990 were instead assigned a lump sum PARF benefit that replaced the PARF rates pegged to the ARF paid on new cars (Table 4). Subsequently, the PARF rate structure was refined in 2002 as shown in Table 5, such that the PARFs rate were based on the ARF paid, instead of the OMV.

Due to the significant increase in the number of vehicles older than 10 years (70% over three years to 1989), the road tax surcharge for such vehicles was increased to discourage owners from keeping them. A road tax surcharge of 10% was imposed for each year above 10 years, up to a maximum of 50%. This was increased in steps of 10% per year for five years starting from 1 May 1990, in a phased implementation. This surcharge structure remained till today.

Table 4.   Preferential additional registration fee rates for cars.

PARF for Cars registered before 1 November 1990		PARF for Cars registered on or after 1 November 1990	
Engine Capacity	PARF	Age at Deregistration (years)	PARF
Less than 1,000 cc	$9,200	≤5	130% of OMV
1,001 — 1,600 cc	$11,200	5 < age ≤ 6	120% of OMV
1,601 — 2,000 cc	$29,000	6 < age ≤ 7	110% of OMV
2,001 — 3,000 cc	$43,700	7 < age ≤ 8	100% of OMV
3,001 cc and above	$49,300	8 < age ≤ 9	90% of OMV
		9 < age ≤10	80% of OMV
		> 10	Nil

Table 5.   Preferential additional registration fee rates for cars.

Age at Deregistration (years)	PARF for cars registered with COEs obtained before May 2002 tender	PARF for cars registered with COEs obtained from May 2002 tender onwards
≤5	130% of OMV	75% of ARF paid
5 < age ≤ 6	120% of OMV	70% of ARF paid
6 < age ≤ 7	110% of OMV	65% of ARF paid
7 < age ≤ 8	100% of OMV	60% of ARF paid
8 < age ≤ 9	90% of OMV	55% of ARF paid
9 < age ≤10	80% of OMV	50% of ARF paid
> 10	Nil	Nil

Weekend Car (WEC) and Off-Peak Car (OPC) schemes

Despite much attention to social equity and fairness considerations in the design and implementation of the VQS in 1990, the public continued to have strong reservations about it. Public sentiments were certainly not helped by the large increase in COE prices shortly after implementation. In Category 2 (cars with 1,001cc to 1,600cc engine capacity), the prices increased substantially from less than $3,000 in June 1990 to more than $9,000 in August. There were considerable concerns from the ground that car ownership was already out of reach for many. To partially address such concerns, the Weekend Car

(WEC) Scheme was introduced in May 1991, just a year after the VQS was implemented. Then Minister of State for Communications Mr Mah Bow Tan acknowledged at the Second Reading of the Road Traffic (Amendment) Bill in 1991, that Singaporeans had a strong desire to own cars, and the WEC scheme was introduced to accommodate this desire, without having to compromise on the land transport objectives. In other words, this was to allow more people to own cars without causing congestion on the roads.

The WEC scheme provided generous vehicle tax benefits to a buyer of a new WEC, giving him full COE, ARF (less PARF) and import duty rebates, subject to a combined rebate cap of $15,000. The owner could also enjoy a discounted flat $100 annual road tax. These would substantially lower the cost of owning a car, especially when COE prices were rising. In return for these tax concessions, a WEC owner could use the car only between 7pm and 7am during weekdays, after 3pm on Saturdays and on the eves of selected public holidays, and on the whole of Sundays and public holidays. Beyond these periods, i.e., during the "restricted hours", he or she would have to pay a fee of $20 that took into account the estimates of the cost of marginal congestion caused by an additional car on the road, as well as the administrative and enforcement costs. The fee was paid through the purchase of paper day-licences that were required to be displayed on the car's windscreen when used during restricted periods. A normal car could also be converted to a WEC, but the benefits would be restricted to a 95% rebate of the annual road tax.

To implement the WEC scheme, a new Category 8 was introduced in the VQS. For its quota, 3,000 COEs were initially provided, with a corresponding reduction of 600 COEs in the quota of each of Categories 1 to 4, or 2,400 in total. This assumed, quite reasonably, that WEC would be used less frequently than a normal car. It also reflected the view that more could own cars as long as they were used during the non-restricted hours and did not contribute to the traffic during the congested periods. This scheme turned out to be a popular one in the public discourse on travel demand management for the next two decades, before it was finally given up in more recent years.

The WEC scheme was positioned publicly as an alternative avenue for car ownership for the more marginal car owner-aspirants by lowering the upfront cost barriers. Nonetheless, it could also be seen as a usage management measure that used incentives, rather than charges, to reduce car usage. The government then promised that the WECs would be converted to normal cars, if and when "island-wide Electronic Road Pricing" was introduced, and the

WEC scheme would serve in the interim to provide tax concessions to those who did not contribute to congestion during the peak periods.

One issue with the scheme was that many owners of high value cars found it financially more advantageous to register their cars as WEC, and pay the $20 daily fee and use them during the restricted hours. So in 1994, the WEC was replaced by the Off Peak Car (OPC) Scheme (Figs. 3 and 4). The COE Category 8 for Weekend Cars was abolished, with its quota returned to the four car categories (1 to 4) in proportion to the number of cars in the respective categories. The tax concessions were changed to an upfront rebate of $17,000 on the Quota Premium, ARF and Excise Duty, and an annual road tax rebate of $800 (subject to a minimum payment of $50 road tax). As with the WEC scheme, it allowed normal cars to be converted and enrolled in the scheme.[6]

As there were no longer any COEs reserved for OPCs, buyers of these cars would be required to bid and use COEs from the normal car categories to register them. Such a change meant that for every OPC that was registered, another person would be deprived of the opportunity to buy a normal car. Although by design, the OPC scheme did not make car ownership more widely accessible, the government indicated that if the response to the scheme

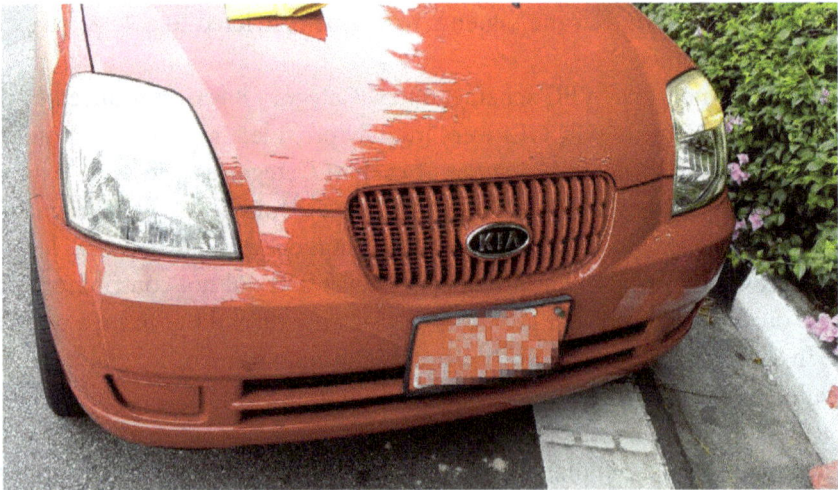

Fig. 3.   Red-plate of a revised Off-Peak Car.

---

[6] The tax concessions enjoyed by the converted OPC were $800 discount on road tax, and additional PARF rebate of $2,200 per year it remained as a converted OPC.

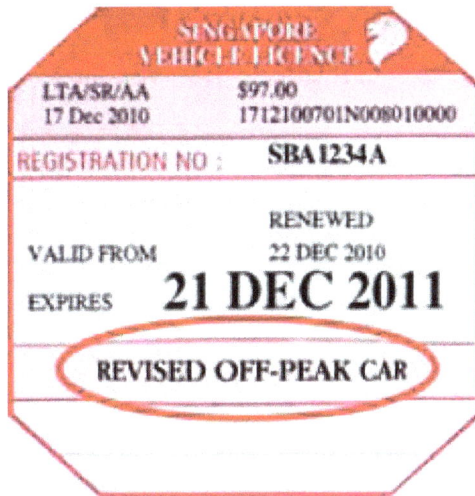

Source: Land Transport Authority.

Fig. 4.   Revised Off-Peak Car Road Tax Disc.

was good and cars were used sparingly, there would be scope to release more COEs to enable more to own cars. The government also announced that the scheme would be a permanent feature of the land transport policy, even after ERP was implemented. As for WECs, the promise to convert them to normal cars eventually was reaffirmed.

Road Pricing Scheme

Following the success of the Area Licensing Scheme (ALS), road pricing was extended beyond the city area to the expressways. While the ALS proved to be effective in suppressing the traffic into the city, congestion had built up outside. Hence, a manual pricing scheme similar to ALS, called the Road Pricing Scheme (RPS) was introduced progressively in 1995 along congested sections on three major expressways. These expressways were the East Coast Parkway (ECP), Central Expressway (CTE) and Pan-Island Expressway (PIE).

Unlike the full-day operation of the ALS, the operation of RPS was limited to the morning peak hours of 7.30am to 9.30am on weekdays only. As with the ALS, gantries were erected (Fig. 5) and vehicles passing them during the hours of operation were required to buy and display valid licences. The licences for RPS were of a different category from the ALS', although full-day ALS licences were also accepted for RPS.

Source: Land Transport Authority.

Fig. 5.   Road Pricing Scheme Gantry.

Electronic Road Pricing

The Area Licensing Scheme (ALS) and its later derivative, Road Pricing Scheme (RPS), had several limitations as manual schemes. First, large numbers of enforcement personnel was needed, given the number of entry gantries and the long hours of operation. About 60 enforcement officers were needed at all the gantry points, and around the same number were required at the licence sale booths. Expansion of the schemes would be constrained because they would require even more manpower.

Enforcement was also tedious and tiring for the enforcement officers who subject themselves to long hours under the elements and the discomfort of the roadside environment. Such a mode of enforcement which relied on visual means was particularly prone to human errors, and not all infringing vehicles were identified. This was not surprising with the enforcement personnel having to grapple with the large number of licences. After RPS was implemented, there were 16 types of licences for both ALS and RPS, divided into daily/monthly, peak/inter-peak and the different vehicle classes. The ALS/RPS licences also offered unlimited number of entries to the RZ or through RPS gantry points, leading to a 'buffet' syndrome that diluted the effect of pricing.

There was also the phenomenon of motorists rushing to enter the RZ or driving through RPS gantries, just before or after the restricted hours, due to the sharp change of licence fee from $0 to $3, or vice versa. This resulted in short but sharp surges in traffic. To smooth out such surges, having intermediate rates, or "shoulder-charging", would help smooth out the peaks. However, this would entail having even more categories of licences.

With all these shortcomings of the manual system, the government therefore decided in 1995 to automate both ALS and RPS with an Electronic Road Pricing (ERP) system. The ERP system could charge per pass, and cater to charge differentiation. Automation would also bring benefits in the form of fewer human errors in enforcement. It should also eliminate the need for motorists to queue at the sales outlets to purchase physical ALS/RPS licences. There would also be significant savings in personnel, who would no longer be needed at the gantries and sales booths anymore.

To test the feasibility of available technologies for the ERP system, three international consortia were invited in April 1993 to conduct demonstration projects to set up and test their respective solutions for an ERP prototype system for Singapore. Field tests were conducted successfully in 1994 and 1995. Development of the actual system proceeded in earnest from October after the contract was awarded to a consortium of both local and Japanese companies, employing technologies from the Mitsubishi Heavy Industries (MHI) of Japan.

The first ERP charging point was deployed in April 1998 at the RPS gantry along East Coast Parkway (ECP). After six months of pilot operations at that point, and several years of preparation, the ERP system was finally put into full operation in September 1998, and replaced both the ALS and RPS.

In the ERP system, each vehicle passing under an ERP pricing point during the operating hours would pay a fee using a pre-paid stored value card (Cashcard) inserted in a device fitted on windscreen (Fig. 6), called the In-vehicle Unit (IU). The IU would interact through short range radio communications with the equipment set up on ERP gantries erected at the pricing points to enable the electronic fee payment (Fig. 7). If there was insufficient cash balance in the Cashcard, or the IU was faulty or not present,[7] the equipment on the gantry would capture photographs of the vehicle, and notification

---

[7] It was not mandatory to install IUs. A vehicle owner could choose not to install one if he decided that he would not be driving through ERP gantries during their operating hours. However, if he did so without an IU, he would have committed an offence and could be fined. Foreign vehicles that needed to use ERP priced roads could either install or rent an IU.

Source: Land Transport Authority.

Fig. 6. In-Vehicle Unit (IU).

Source: Land Transport Authority.

Fig. 7. ERP Gantry.

would be sent to the owner to pay up the outstanding fee (with additional administrative charge) or fine (if there was no IU), or to have the IU inspected (if faulty). As the ERP system was fully automated, the need for a large army of enforcement personnel in the ALS and RPS was removed, and errors associated with manual enforcement eliminated.

The ERP system continued the practice of ALS and RPS in applying differentiated pricing according to vehicle type. Nonetheless, its introduction also provided the opportunity to refine the pricing structure. Under the ERP, the rates charged for each type of vehicles were based objectively on the Passenger Car Unit, or PCU. The PCU measured the amount of road space taken up, and hence the amount of congestion contributed, by the vehicle relative to a passenger car, which had a PCU of 1.0.[8] Larger vehicles were assigned either a PCU of 1.5 (heavy goods vehicles and small buses) or 2.0 (very heavy goods vehicles of more than 2 axles and big buses), and motorcycles, 0.5. The ERP rates were set according to the PCU. Therefore, a heavy goods vehicle would always pay 1.5 times the ERP rate set for passenger cars, and motorcycles, half of it.

The ERP system also provided flexibility to vary the operational hours and the charges that were to be applied in different time windows. This allowed road pricing to be much more targeted to the varying traffic conditions and levels of congestion across the day. For example, the highest ERP rate could be imposed during the peak-within-the-peak period when the traffic is the heaviest, slightly lower before and after, and the lowest during other times. Across the day, the rates were varied across multiple time periods that were as short as 30 minutes. Such rate variations would have been almost impossible to implement under the manual ALS or RPS systems — the number of colours and shapes of the paper licences (Fig. 8) would be too numerous.

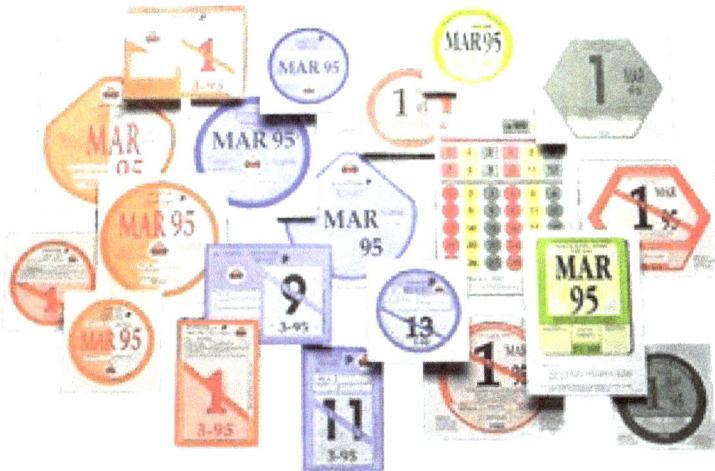

Source: Land Transport Authority.

Fig. 8. Different types of ALS Licence.

---

[8] The PCU of 1.0 also applied to light goods vehicles and taxis.

Not only did the ERP system allow for different road pricing rates to be applied across vehicle types and time periods of the day, it could cater to their periodic adjustments over a year in response to changes in traffic patterns. The Land Transport Authority (LTA) reviewed the ERP rates based on the traffic speeds along stretches of priced roads. In each review, the speeds of each monitored road stretch in every time window were measured using Global Positioning System (GPS) data from taxis and cameras used for monitoring of traffic conditions of expressways, and the average speeds were computed. To set the rates for the Restricted Zone (RZ), the traffic speeds of a basket of routes, or a collection of main road corridors, inside the RZ were measured. This was because the RZ cordon was to control the total traffic volume within it, and therefore it was necessary to use the speeds of representative routes inside the RZ rather than the speeds along the roads leading into it.

The ERP rates were to be adjusted to ensure that the average traffic speeds were maintained within the range of 45 to 65 km/h for expressways, and 20 to 30 km/h for arterial roads and roads within the Restricted Zone (RZ). These speed ranges were considered optimal, and within which the highest number of vehicles could travel along the road without exceeding its capacity. These threshold numbers were determined in a study by the Nanyang Technological University on local traffic characteristics in 1995, before ERP was implemented. If the measured speed at a particular time window were to drop below the lower threshold, indicating an unstable 'start-stop' condition, the ERP rate for that period would be increased to discourage vehicles from using the road and adding to the traffic. Conversely, if the speed exceeded the upper limit, it indicated that the road could still accommodate more traffic, and the ERP rate would be reduced, even down to zero. The ERP rate changes were implemented on a quarterly basis in February, May, August and November of each year, with a notice of change given to the motoring public one week in advance. From May 2000, there were also rate adjustments for the June and December school holiday periods, during which traffic volumes tended to decrease and the rates were correspondingly reduced temporarily.

In addition to the rates, LTA also reviewed the temporal and geographical coverage of road pricing periodically. When the speeds of priced roads outside of the prevailing ERP operating hours fell below the minimum threshold of the relevant optimal range persistently, the operating hours would be adjusted to extend to the congested window, i.e., a zero-rated window would see an introduction of a new charge. Similarly, traffic speeds of major non-priced roads were also closely monitored, and if traffic conditions along these roads had deteriorated over a sustained period, the introduction of road pricing would be triggered and a new ERP gantry would be installed.

Vehicle tax rationalisation exercise (1998)

With the introduction of the ERP system, the vehicle tax structure was also revised extensively. Among the many changes implemented, the fixed statutory taxes[9] of different categories of cars (e.g., company, rental, tuition cars) were standardised with those for private cars. As part of the rationalisation exercise, the ARF rate for cars were reduced from 150% of OMV to 140%, a second downward revision since it reached its peak between 1983 and 1990. The road tax for cars was also reduced by about $280 per year on average. The reduction of both ARF and road tax was to balance the increase in usage costs with the introduction of ERP, thus achieving a better balance between ownership and usage costs, as greater reliance was to be placed on usage measures to manage travel demand. In the same year, the Excise Duty for cars was also reduced from 41%[10] to 31% of OMV. This was part of a separate package of measures to support businesses.

Review of VQS

In February 1998, in the midst of preparing for the implementation of the ERP system, and after just eight years after it started, another review of the VQS was announced by the Minister for Communications Mr Mah Bow Tan. This time, the Government Parliamentary Committee (GPC) on Communications, headed by Mr Chay Wai Chuen, was appointed in October 1998 to carry out this review. The Committee was asked to review VQS on its effectiveness as part of Singapore's traffic management strategy, given the implementation of ERP and the new vehicle tax structure in the same year, and to recommend improvements. The intent was for the VQS to achieve its transport objectives more effectively, but with greater fairness and objectivity.

The emphasis placed on "fairness and objectivity" was not surprising. It reflected the ongoing concern of vehicle owners, both existing and prospective, of the fairness of VQS, as well as the continuing issue of how Singaporeans could fulfill their aspirations of owning a car. Despite several measures to

---

[9] Registration Fee, Additional Registration Fee, Customs Duty, QP of COE, and road tax.
[10] The Excise Duty was 45% previously, but was reduced to 41% in April 1994 as a measure to offset the effect of the introduction of the Goods and Services Tax (GST).

temper the full effects of the market allocation system in the VQS, the concern that car ownership would be within reach of only the rich persisted.

The VQS Review Committee actively sought the views of various groups of stakeholders and interested parties, ranging from academics, the Feedback Unit's Transport Group and members of the public, and convened three dialogue sessions. The Committee submitted its report, and the government responded in Parliament in March 1999. Its recommendations were largely adopted.

The Committee concluded that the VQS was "an effective and necessary tool in the government's demand management strategy", that had achieved its objective of moderating vehicle growth to 3% annually, a rate that could be sustained by the road network without increasing traffic congestion. If Singapore were to discard VQS and rely solely on road pricing to manage traffic congestion, ERP charges would have to be very much higher (about $20 per day), with attendant adverse impact on businesses in the CBD. As they were charged daily, ERP charges were also less likely to be acceptable than a one-time high upfront ownership cost. Therefore, the VQS was retained as one of the key pillars of traffic management to complement usage restraint measures, in line with the government's two-pronged approach of balancing ownership and usage measures. The VQS would remain an important pillar of this traffic management strategy while the shift from a primarily ownership-based system to a more usage-based system would take place gradually.

As with the first Select Committee, the Review Committee also examined the issue of alternative methods of allocating COEs, and its conclusion was no different — that "bidding remains the most efficient and equitable means of allocation". Nonetheless, it recommended refining three areas of the VQS:

First, the Committee felt that the prevailing formula then (based on the number of de-registered vehicles in the previous quota year, plus 3% growth in the total vehicle population of the previous quota year) brought about a demand and supply mismatch, due to a one-year lag between de-registration and recycling of COEs. It suggested improving its responsiveness to demand from owners who de-registered their vehicles, by computing the annual COE supply based on the *projections* of vehicle de-registrations for that year, rather than *actual* de-registrations in the previous year. The projections would be based on past years' trends and the age profile of vehicles de-registered. Where the projections were over- or under-estimations of the actual numbers, an adjustment to the quota would be made in the following year. The proposed formula was:

$$COE_N = Projected\ Deregistrations_N + x\ Vehicle\ Population_{(N-1)} +$$
$$(Actual\ Deregistrations - Projected\ Deregistrations)_{(N-1)}$$

This new formula was implemented in the quota year beginning May 1999. It was used for the large part of 2000s, before it was changed again.

The VQS adopted the bidding system to attain allocation efficiency. In fact, from the land transport point of view, "there should only be one category of COEs for allocation efficiency" too, as noted by the Review Committee. This would mean merging all vehicle categories.

Again, it was recognised that there would always be social concerns to be addressed, in addition to efficiency considerations. For one, the Committee noted that commercial vehicles and motorcycles catered to different groups — commercial vehicles catered to businesses purposes, motorcycles, and the lower income groups. Hence, the recommendation was to retain the respective COE categories for commercial vehicles and motorcycles, to shield them from competition from car buyers when bidding for COEs.

As for the four car categories then, the Committee opined that "there was general consensus that merging them into a single category would provide greater liquidity and reduce price distortions." This was because having multiple car categories could lead to small quota sizes in one or more of these categories, which was considered an undesirable as small quotas could lead to anomalies and fluctuations in their Quota Premiums, and could also be manipulated more easily. However, public feedback revealed the preference for some segmentation of small and big cars to be retained for social equity reasons. Many were concerned that, if the car categories were merged, small car buyers would be squeezed out by luxury car buyers who had higher bidding power.

In view of these concerns, the Committee recommended that the number of car categories could be reduced from four to two or three. This would provide car buyers with greater range of car models within a category, while reducing some of the distortions associated with small quota sizes. The final number would hinge largely on the resultant quota sizes of the new categories. Based on this consideration, the Committee suggested merging Categories 1 and 2 into one category, and Categories 3 and 4 into another. For the first two categories, merging them would address the relatively small quota size and limited choices of car models in Category 1. The convergence of QPs in both categories over the past year or so also lent support to their merger. Likewise, the QPs of Categories 3 and 4 had been trending in similar fashion.

The Open Category had allowed the vehicle mix to evolve over the years to cater to changing preferences for different types of vehicles. Therefore, the Committee recommended retaining it.

With the government accepting the recommendations on re-categorisation of COEs, the four car categories were merged into two new categories, A and B. The commercial vehicle, motorcycle and Open categories were retained and renamed categories C, D and E respectively, as shown in Table 6.

Even as the Select Committee of 1990 had concluded that using the Lowest Successful Bid (LSB) for COE tenders was the most appropriate, many continued to call for it to be replaced with the Pay-As-You-Bid (PAYB) system. Such calls arose as the COE prices had increased substantially[11] since the VQS was implemented, and were based on the belief that PAYB would discourage unduly high bids. The Review Committee of 1999 revisited the issue, but came to the same conclusion that the PAYB system was not necessarily superior to the LSB system, which should be retained.

Similar arguments against PAYB were those raised in 1989 — confusion arising from differing values of cars of the same model and age; unhappiness among successful bidders as some had to pay more for the same COE; and greater reliance of vehicle buyers on dealers to bid on their behalf, because of the difficulty of buyers deciding on the right bid price.

Table 6.   Categories of Certificate of Entitlement (COE).

Vehicle type	Old COE Category	New COE Category
Cars with engine capacity below 1,000cc	1	A
Cars with engine capacity between 1,001cc and 1,600cc; and taxis	2	
Cars with engine capacity between 1,600cc and 2,000cc	3	B
Cars with engine capacity above 2,000cc	4	
Commercial Vehicles	5	C
Motorcycles	6	D
Open Category	7	E

Source: Land Transport Authority.

---

[11] COE prices for cars in Categories 1 to 4 were all less than $10,000 in the first year of implementation; they had increased to more than $30,000 by end of 1997, the year before the review.

Further examination of COE bidding activities, which were not available earlier in 1989, revealed converging bids towards the LSB, with successful bids generally clustered in a range of 20 to 30% higher than the LSB. There was no evidence found of irresponsible bidding in tenders under LSB system. The government therefore accepted the Committee's recommendation for status quo. However, many among the public remained unconvinced, and in the subsequent years, calls in support of PAYB continued unabated, especially during periods of high COE prices. This issue was once again revisited and reviewed in 2013.

Not only did the COE price increases led to public calls to relook at the bidding system, it triggered similarly strong views about the role of motor vehicle dealers. Then, as is now, they submitted the majority of COE bids, and almost all successful bids were from them. It was not difficult to see why many perceived the dealers to be manipulating the bids and driving up the COE prices.

The Review Committee considered whether to limit them from bidding for COEs on behalf of buyers. Motor vehicle dealers have long had an established role in the market. They commonly provided car buyers package deals that bundled both the cost of the car and the COE price. With every incentive to maintain their market share in the competitive motor vehicle market, the dealer would absorb as much as possible any increases in COE prices once the package deal is agreed upon with the buyer, thus providing him some degree of cushion from COE price fluctuations, especially increases.

The dealers also saved the buyer the hassle of deciding the amount to bid and actually doing the bidding. As part of the package deal, the dealer also paid the bid deposit (which was at least 50% of the bid amount), on the buyer's behalf, as well as arranged the financing for the full cost of the car, including the COE price.[12] Hence, many buyers preferred to rely on dealers for COE bidding for its convenience and lower cash outlay. Hence, the Review Committee decided that banning dealers from bidding for COEs could affect these buyers and this was not practical. It also felt that the role of the government was to ensure the VQS was simple, fair and transparent enough for individual buyers to submit their own bids, if they chose not to rely on dealers. It should not intervene in commercial decisions on how vehicles should be packaged, sold and at what price.

---

[12] If a buyer were to bid for his own COE, he might be able to secure financing only for the price of the car, as invoiced by the dealer, and not for the COE.

Nonetheless, the Committee opined that the public's concerns could be addressed with a better flow of bidding information, through sources such as media releases, the LTA websites and offices. An open bidding system, which was also another key recommendation from the Review Committee that was implemented, would help in this respect.

With the advent of IT technology since 1990, the opportunity to improve the bidding process arose. The Review Committee explored the feasibility of improving the then closed bidding system that had several shortcomings. First, it limited bidding information to the COE bidder, who did not know how the rest of his competitors were bidding. For inexperienced individual bidders, this was particularly problematic because they could be bidding without much knowledge of the prevailing demand for COEs, and this could lead to unrealistic bids and fluctuations in the COE prices.

The Review Committee explored the feasibility of an open bidding system that would be transparent with the bidding information and enable bidders to make more informed decisions. With both individual buyers and dealers having access to the same amount of information on prevailing bidding situation, individual bidders could be better equipped and encouraged to bid for their own COEs, thus reducing their reliance on dealers to bid on their behalf.

The Committee assessed that the prevailing technology allowed a move to electronic on line bidding system that would be sufficiently secure and accessible via ATM, the Internet or through telephone. In recognition of reservations among some quarters, it suggested the open bidding system be tried out on the Open Category (Category E) so as to assess the mechanics and efficacy of the system and the public's acceptance.

The government accepted the recommendation, and LTA proceeded with a trial in June 2001. At the start, only Category E (Open Category) was involved, and the open bidding exercises were held every two weeks, and alternated the closed bidding exercises. Bidders submitted their bids through ATMs, the Internet and phone banking by keying in their reserve price. The reserve price was the maximum amount that the bidder was willing to pay for the COE. The system would revise the bid upwards at an increment of $1 in real time, according to the current COE price, until his reserve price was reached. When the bid reached the reserve price, the bidder had to revise the bid upwards to remain in the running. The bidder could revise his/her price upwards to remaining in the running, and there was no limit to the number of times he could revise the reserve price during the exercise.

The trial allowed LTA to test the system, assess users' behavior and the possible impacts of the system. With feedback that the system was easy to use and useful to bidders, LTA replaced the closed bidding system fully with the open system from April 2002. Two tender exercises were held on the first and third Mondays of each month, each of which would end two days later on Wednesday.

The Boom Years (2000s)

The 1990s saw major policies on travel demand management been implemented and revised — the introduction of the Vehicle Quota System (VQS) in 1990, and its subsequent review in 1999, the introduction of Electronic Road Pricing (ERP) and vehicle tax rationalisation exercise in 1998. With these main pieces of the land transport policy in place, the next decade saw fewer major policy changes. There were however continual adjustments of these major schemes, to bring about a deliberate and measured shift towards lowering the statutory or fixed taxes to make ownership more accessible, and relying more on usage restraint measures to manage travel demand.

This rebalancing of ownership and usage measures was driven by the popular narrative of this period that Singaporeans could have more cars as long as they do not contribute to congestion. In fact, such a nexus between higher car ownership and more extensive usage control was amplified by the government's earlier announcement in 1997, prior to ERP's implementation, that between 15,000 and 20,000 additional COEs would be introduced, if ERP proved to be effective in managing congestion. 5,500 of these COEs were actually released that year, and another 5,000 were added into the quota in 2002.

Continual lowering of fixed statutory costs

In the same year (2002), the government again lowered the ARF rate for cars by 10 percentage points to 130% of OMV, and the Excise Duty, from 31% of OMV to 20%. The road tax was also reduced by an average of 20% in 2002. As the ARF rate had seen significant reduction since its peak of 175%, the PARF rebates, which were also pegged to the OMV, had become disproportionately high relative to the ARF. Therefore, the PARF values were reformulated in the same year to be tied to the ARF paid instead. This was so that as the ARF rate was reduced, the PARF rebate would also be reduced correspondingly.

The ARF for cars was later reduced two more times — to 110% of OMV in 2004, and then to 100% in 2008. The latter turned out to be its lowest level since the government started reducing it in 1990.

The road tax was reduced twice as well. In 2007, it was cut by 8% for cars as well as for motorcycles. A year later, a larger reduction of 15%, which applied to all vehicles, was made as part of the package of measures introduced together with the implementation of ERP to more locations.

These reductions in statutory costs lowered the outlay of owning a car considerably. The Off Peak Car (OPC) scheme also saw increased popularity. The additional OPC rebate further reduced the upfront statutory costs, already lowered by the ARF and ED revisions, and made owning a car even more affordable. To make it even more attractive, the scheme was further refined in January 2010. The revised OPC (ROPC) scheme provided the perk of unlimited usage on Saturdays and eves of five public holidays.[13] The removal of usage restrictions was accompanied with a reduction in the annual road tax discount from $800 to $500 (subject to a minimum payment of $70). Owners of normal cars could also opt into the ROPC scheme and enjoy a cash rebate of $1,100 every six months. Previously, this rebate was paid only at the point of de-registration of the OPC when it was still eligible for PARF benefits. The OPC population reached its zenith in 2010 (Fig. 9), when there were more than 50,000 OPC on the roads, an eight-fold increase from about 6,000 at the start of the decade.

Around the same period, the COE supply, particularly of cars, saw an upswing, and the combined yearly quota for cars (Categories A and B)

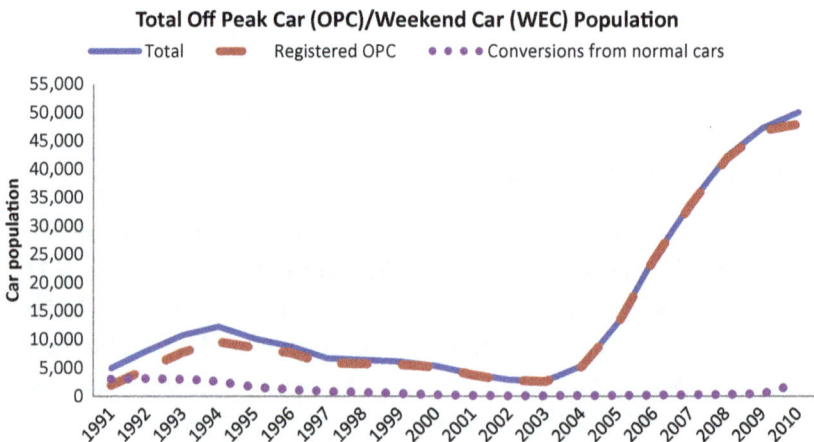

Fig. 9.   Trend of OPC population.

Note: * WEC scheme was introduced on 1 May 1991. OPC includes Revised OPC.

---

[13] New Year, Lunar New Year, Hari Raya Puasa, Deepavali and Christmas.

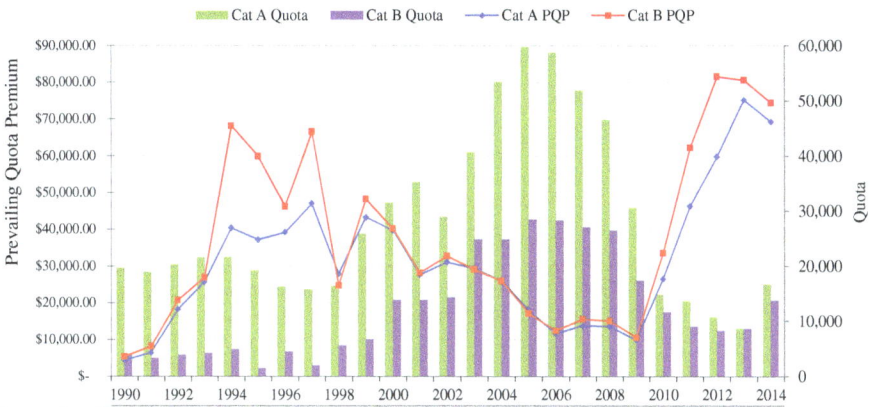

Source: Land Transport Authority.

Fig. 10.   Quota and Quota Premiums of Category A and B cars.

Note: In May 1999, Categories 1 and 2 were merged to form Category A; and Categories 3 and 4 were merged to form Category B. The prevailing quota premium (PQP) is a moving average of the quota premiums (QP) over the last 12 months (1990–1997), however with effect from November 1998, the PQP is based on three months moving average of QP. The PQP of Cat A and B prior to 1999 were calculated based on the PQPs of Categories 1, 2, 3 and 4 weighted by quota.

exceeded 70,000 between 2004 and 2008 (Fig. 10). The COE prices varied at a substantially lower level compared to the 1990s. Coupled with a relaxation of vehicle loan conditions,[14] car ownership became much more accessible. The car population grew from about 383,000 in the start of 2000 to 577,000 by the end of the decade (Fig. 11). The proportion of car-owning households increased from 38% in 2004 to 46% in 2012, even as Singapore's population grew and many more Singaporeans became first time owners.

Expansion of ERP

With the growth of the vehicle population, and more roads beginning to experience congestion, the ERP scheme was quickly extended beyond the RZ and expressways to a number of major arterial roads. In fact, this took place just one year after ERP was implemented. To manage traffic congestion along the radial

---

[14] In 2003, the Monetary Authority of Singapore lifted the financing restrictions on vehicle loans put in place in 1995, that limited the maximum financing to 70% of the car price and maximum loan period to seven years.

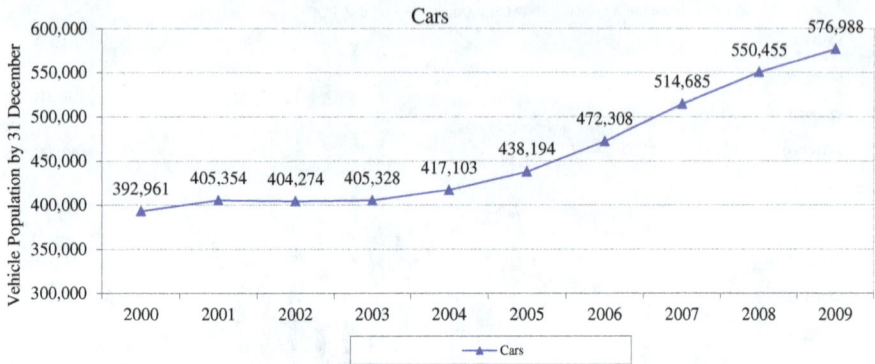

Source: Land Transport Authority.

Fig. 11.   Car Population 2000–2009.

expressways and arterial roads leading into the CBD, seven new gantries[15] were installed. Collectively, these pricing points formed part of an Outer Cordon. However, unlike those of the RZ cordon, ERP rates of these points of the Outer Cordon were determined individually based on the traffic speeds of the expressways or roads where they were located, and did not have uniform rates. Another three gantries were introduced in the next three years — one at the PIE slip road into CTE in 2000, and two along Dunearn Road in 2002, once traffic conditions warranted pricing.

While the ability of ERP to support different rates for different time windows and to enable road pricing to be responsive and targeted to varying traffic conditions was a significant advantage, it resulted in charges of adjacent charging windows differing significantly in some instances, which led to some rather unexpected and peculiar phenomena on the roads. Many motorists resorted to slowing down (and even stopping altogether) before the gantries in order to enjoy lower rates in the next window. In other cases, they would speed up to avoid paying higher charges in the next window. To deter such unsafe behaviour, graduated ERP rates were introduced in February 2003. Under this change, where the difference in charges of adjacent windows was at least $1 per PCU, the rate for the last five minutes of the preceding window and the first

---

[15] These seven gantries were at Ayer Rajah Expressway (towards city) before Alexandra Road, Pan Island Expressway (PIE) (towards Changi) after Adam Road, Mount Pleasant Road slip road into PIE (towards Changi), Central Expressway (towards city) before Braddell Road, Thomson Road, Bendemeer Road and Kallang Road.

five minutes of the succeeding one would be set at the average of the two windows' rate.

There was a hiatus in the expansion of ERP coverage for two years, before another set of new gantries was introduced in August 2005. This time, the ERP was not to address city-bound traffic in the morning peak periods, but to manage the increasing traffic along the Central Expressway (CTE) travelling away from the city after working hours. The introduction of evening ERP between 6pm and 8pm on a congested stretch of CTE[16] broke new grounds in that road pricing was now to tackle congestion from largely home-bound traffic. The rationale communicated to the public was that traffic congestion, at any time of the day, would lead to sub-optimal utilisation of the road network, and therefore had to be managed appropriately. ERP was therefore necessary to relieve congestion as alternative solutions, such as traffic engineering methods of road widening and traffic light timing adjustments, were either not quickly available or had been exhausted.

Two months later, in October 2005, another road pricing approach was adopted with the erection of two new gantries in the Orchard Road area. The introduction of these two gantries, together with the existing ones in the vicinity, created a new Orchard sub-cordon, within the original RZ cordon. The rationale for the sub-cordon was to recognise the significant differences in traffic characteristics between the Orchard Road area and the rest of the RZ, and allow the rates in the Orchard sub-cordon to be set differently. For example, as a shopping belt, Orchard Road had relatively light traffic in the morning, before the opening hours of the retail businesses, and could therefore attract zero or very low ERP rates, whereas the rest of RZ experienced much heavier commute traffic which warranted much higher rates. The division of the RZ cordon allowed road pricing to be further targeted and effective in tackling localised congestion within the city area. Motorists who travelled in less congested parts of the RZ or during less congested times could now pay less or no ERP charges. At the same time, such a sub-cordon helped to discourage through-traffic along Orchard Road, which made up as much as 30% of its total traffic. The initial operating hours of the gantries of this sub-cordon were 12 noon to 8pm on weekdays and Saturdays.

The expansion of ERP coverage remained unabated. In November 2007, four more gantries were installed. At two priced locations, the gantries served to address the growing outbound evening traffic from the city, just like the first CTE evening ERP gantry installed in 2005. The eastbound East

---

[16] The ERP gantry was installed on northbound CTE between PIE and Braddell Road.

Coast Parkway before the Rochor Road exit and Ophir Road slip road saw evening ERP from 6pm to 8pm. Similarly, the northbound CTE traffic before the PIE faced another gantry, which would start its operation at 5.30pm and end at a considerably late time of 10.30pm. To tackle growing morning traffic along the expressways, a new gantry was erected along the southbound Bukit Timah Expressway (BKE) before the PIE entry, and the operating hours of the existing southbound CTE gantry before Braddell Road were extended by more than two hours to between 7.am and 11.am, from between 7.30am and 9.30am before. The growing coverage was intended to encourage drivers to consciously plan their trips and to consider route or mode alternatives.

The year 2008 saw numerous major refinements and adjustments to the ERP implementation. In April, another four new gantries,[17] which operated in the morning hours, were added to the Outer Cordon to strengthen the effectiveness of ERP in managing traffic over a larger road network. Three months later in July 2008, more substantial changes were introduced as part of the Land Transport Masterplan 2008 to enhance ERP's effectiveness in addressing evolving traffic conditions.

A new Singapore River Line (SRL), consisting of five new gantries[18] operating between 5pm and 8pm, was introduced to manage congestion within the RZ better. This line of gantries was created in response to worsening evening congestion in the city area, and was designed to cut down through-traffic by discouraging outbound motorists from travelling through the city for their trips out. The SRL also resulted in the creation of two other sub-cordons within the RZ — the Bugis–Marina Centre and the Shenton Way–Chinatown sub-cordons. Together with the Orchard sub-cordon, the RZ had now three sub-cordons, which allowed ERP rates to be differentiated within the RZ for more targeted pricing within the city area. The evening ERP operating hours for the RZ were also extended to end at 8pm, instead of 7pm. With these ERP measures targeting the evening peak traffic in the CBD, the travel speeds within the RZ quickly saw marked improvements. Through-traffic across the SRL was reduced by about 30%, with corresponding speed improvement to between 22 km/h and 26 km/h, from below the minimum threshold speed of 20 km/h. For the entire RZ, travel speeds between 6pm to 8pm also went up by nine to 15%. So substantial were the improvements in traffic conditions that

---

[17] The four gantries were along Kallang Bahru and Geylang Bahru, Lorong 6 Toa Payoh, Upper Boon Keng Road, and Upper Bukit Timah Road.
[18] The gantries were at Eu Tong Sen Street after Tew Chew Street, New Bridge Road, South Bridge Road, Fullerton Road (eastbound), and Fullerton Road (westbound).

ERP rates along the SRL and the Shenton Way–Chinatown sub-cordon were reduced two months later in September 2008, and outside of the usual quarterly ERP rate review in November.

Motorists travelling to the city area on Saturdays were also not spared. Saturday ERP, which was originally implemented for the Orchard sub-cordon, was extended to the Bugis–Marina Centre sub-cordon. The Saturday ERP operation in the former area also started an hour earlier at 11am, instead of 12 noon.

The method of computing traffic speeds to determine rate changes was also changed. Then, the average speed was used, but it was assessed that the minimum average speed thresholds (45 km/h for expressways and 20 km/h for arterial roads and RZ) set in 1998 to trigger a rate increase were too low, and at such levels, traffic could easily deteriorate into the unstable 'stop-start' conditions. To avoid such situations from occurring frequently, the 85th percentile speed (which meant 85% of the vehicles were at or above this speed) was adopted instead to create a sufficient buffer from the 'unstable' zone. This measure was also assessed to be more reflective of actual traffic conditions, and in line with international traffic engineering practices.

The rate structure was also revised. Since ERP was implemented, the initial ERP rate was $1 per PCU, and each subsequent rate change was $0.50 per PCU. However, it was found that motorists had, over time, become less sensitive to such rate changes. To strengthen the effectiveness of the rate structure in influencing road usage behaviour, the initial rate and subsequent rate change were doubled to $2 and $1 per PCU respectively.[19] These changes were phased in progressively, starting with the CBD cordon, followed by the Outer Cordon in November 2008, and then to the remaining gantries in February 2009.

The changes brought about some improvements in the traffic conditions, both in the city area and along the roads and expressways beyond. As a result, only one additional ERP gantry, at westbound PIE near Jalan Euros, out of six locations[20] along the Outer Cordon identified for ERP implementation in November 2008 was actually put into operation.

---

[19] For ERP rates that were still in half-dollar values, e.g., $1.50, would be progressively simplified to multiples of $1 (by using $0.50 adjustments) during the quarterly rate revisions triggered by speed changes.

[20] Four of such locations were Commonwealth Avenue, Jalan Bukit Merah, Alexandra Road and westbound Ayer Rajah Expressway (AYE). The fifth, Serangoon Road, experienced congestion, but because ongoing improvement works at the Little India area and Woodsville Interchange which could affect the traffic conditions there, ERP was deferred pending further monitoring and review.

Another pair of major changes relating to ERP operating hours took place about three years later, in October 2011. While the evening ERP proved to be effective in keeping congestion at bay along northbound CTE, there had been repeated feedback questioning the need for the ERP hours to be extended late into the evening. Then, one of the two gantries on the northbound CTE operated between 5.30pm and 10.30pm (before PIE), and the other, between 6pm and 8pm (after PIE). After a review, LTA decided to shorten the operating hours of the first gantry to end at 8pm, to be in line with the end times of the second gantry, as well as that of the evening ERP gantry along eastbound ECP. The rationale for departing from the earlier stance that traffic congestion, regardless of the time of the day, needed to be managed appropriately, was that most of the trips on northbound CTE in the later part of the evening were home-bound, so motorists could accept and tolerate some level of congestion for such trips in the absence of ERP.

This change was also significant in that feedback from the travelling public was acknowledged and considered in the review. Nonetheless, evening ERP was not removed entirely to provide home-bound motorists flexibility and options in choosing their travel timings. For motorists who appreciated the option of reaching home earlier under improved traffic conditions could continue to do so by travelling before 8pm during the ERP operating hours. Those who did not wish to pay ERP could either use CTE after 8pm or travel along alternative arterial roads such as Thomson Road that were not priced in the evening.

Another change was also in consideration of the concern of another group of stakeholders affected by ERP. To help businesses in Chinatown, the operating hours of the southbound SRL gantries[21] were shortened to end at 7pm.[22] This was so that motorists from the Bugis–Marina Centre area, who would like to patronise businesses in Chinatown, could enter Shenton Way–Chinatown sub-cordon earlier without incurring ERP charges.

Following the announcement of the shortening of evening operating hours at the SRL that helped the Chinatown businesses, business associations in the adjacent sub-cordons of Orchard and Bugis–Marina Centre, as well as members of the public, also called for a review of Saturday ERP there. Many advocated for its total removal in the belief that more shoppers and taxis could be attracted into these areas.

---

[21] These gantries were at New Bridge Road, South Bridge Road, southbound Fullerton Road and southbound Bayfront Avenue, the last of which was included in the SRL as the Shenton Way–Chinatown sub-cordon was extended to include Marina Bay areas in February 2011.
[22] The SRL gantries covering northbound traffic continued to operate up to 8pm.

LTA's assessment was that ERP was still needed on Saturdays, given the prevailing traffic conditions and levels of congestion in the two sub-cordons. Nonetheless, applying a rationale similar to that for shortening the evening ERP operating hours at the CTE earlier, it was felt that it might be possible to strike a balance between motorists who were willing to face slightly more congestion and inconvenience to avoid ERP charges and those who prepared to pay for a better travel experience. This was because trips into these two sub-cordons on weekends were largely for leisure or shopping. Therefore, to provide greater flexibility and choice to motorists, Saturday ERP for the Orchard sub-cordon was adjusted to start at a later time of 12.30pm, instead of 11am, and the ERP rates for Bugis–Marina Centre sub-cordon reduced from $2.00 to $1.00 between 12.30pm and 2pm, and between 5.30pm and 6.30pm. These changes applied from June 2012.

LTA decided against doing away with ERP on Saturdays fully because doing so would lead to more congestion and less than optimal usage of the roads. Those who did not wish to pay ERP charges had alternatives in the extensive public transport services to the areas and the options to shop on Sundays when ERP did not operate.

After a break since November 2008,[23] the geographical coverage was once again extended. To address both morning and evening peak hour congestion building up along the Ayer Rajah Expressway (AYE) between Jurong Town Hall Road and North Buona Vista Road,[24] LTA decided to implement ERP there to manage the traffic conditions, which were expected to worsen with increasing traffic demand arising from future developments in the area, such as the Jurong Gateway area. The set of four gantries, three for eastbound AYE and its slip roads, and one for westbound AYE, began operations in August 2014.

The lean COE years — Late 2000s–2010s

Since its start in 1990, the allowable vehicle growth rate (VGR) under the VQS had been set at 3% per year. The Review Committee of 1999 had also examined the VGR as part of its work then, and recommended the government review it to ensure that it could be sustained by the transport network

---

[23] These excluded adjustments in the location of existing ERP gantries affected by road network changes and other developments over the years, and the extension of Shenton Way-Chinatown sub-cordon into the adjacent Marina Bay areas in 2011.
[24] Specifically, the congestion was along city-bound AYE between Jurong Town Hall and Clementi Road during both the morning and evening peak hours, and Tuas-bound AYE after North Buona Vista Road during the evening peak hours.

without congestion rising to an unacceptable level. While the government decided to maintain the 3% VGR rate then, it proceeded to conduct regular reviews in later years.

The first of such reviews was conducted in 2005. The MOT and LTA weighed factors such as the scope for expanding the road network capacity, the prevailing traffic conditions, the extent of road charging, developmental plans for public transport, and the perennial aspirations of Singaporeans to own cars. In view of the various considerations, it was decided that the VGR be maintained at 3% a year for the next three years. However, the stage for a slowdown in vehicle population growth was already set when Minister for Transport Mr Yeo Cheow Tong stated in 2006 that it would be impossible to maintain at that level indefinitely.

After 19 years, the vehicle growth rate allowed under the VQS was finally reduced in 2009. Setting the stage for this reduction, Minister for Transport Mr Raymond Lim announced in 2008 that the allowable VGR was to be reduced from 3% per year to 1.5%, starting from May 2009. Even as aspirations to own cars remained widespread and persistent, and transport policies of the past, including the use of ERP, had enabled more to attain such aspirations than would otherwise be possible, a VGR of 3% would be untenable in future. While the VQS had effectively constrained the growth of vehicle population to a compounded annual rate of about 2.8% per year from 1990 to 2008, well within the target of 3%, the vehicle population still reached more than 925,000 by 2009. The increase in the car population was particularly rapid, having increased by almost 40% in the preceding 10 years.

As a result, as Mr Lim noted, congestion levels had increased by 25% since 1999, after the ERP was introduced, and more roads faced congestion during peak hours. Road expansion, at a rate of about 1% per year over the past 15 years, could not catch up with the increase in trips that came with a bigger vehicle population. Even so, by 2008, about 12% of land in Singapore was already used for roads, close to 15% for housing. With the pace of road expansion for the next 15 years projected to be halved to about 0.5% each year, the VGR would have to be similarly reduced. Otherwise, the ERP coverage would have to be much more extensive, and its charges higher, which Mr Lim highlighted as the "the key trade-off we have to make, to maintain smooth flowing roads". Indeed, as could be seen in the numerous policy changes in the subsequent years, the future trade-offs continued to be become much sharper.

In addition to reining in the vehicle population growth by halving the allowable VGR to 1.5% per year, the VQS formula for determining the COE quota was also reviewed shortly after. This was due to the difficulty in keeping the vehicle population to the target VGR since the revised quota formula suggested by the Review Committee was adopted. The vehicle growth rate in the later part of the 2000s was higher than the 3% VGR target, at around 5% per year (Fig. 12). In contrast, the earlier part of the same decade saw only about a 1% annual growth.

Whilst the formula was intended to make the COE supply more responsive to demand from owners who de-registered their vehicles, projecting the number of vehicles de-registered to be close to the actual number proved to be a challenge. As a result, the number of vehicles that would be registered in the later part of the decade was over-estimated. While adjustments to the quota made in the following year would eventually account for the deviations of the estimated de-registration numbers from the actual, they did not prevent mismatches in demand and supply from year to year.

So, in 2009, a review of the VQS formula was conducted, and it was decided that from April 2010, the formula would be based on the actual number of vehicle de-registrations. Essentially, this was an acknowledgement of the difficulty in reversing the pre-1999 formula. The key difference was that the recycling of COEs from de-registered vehicles would be conducted every six months, instead of every year. There would now be two six-month recycling periods, with the COE quota from February to July of each year based on the actual vehicle de-registrations between July and December of

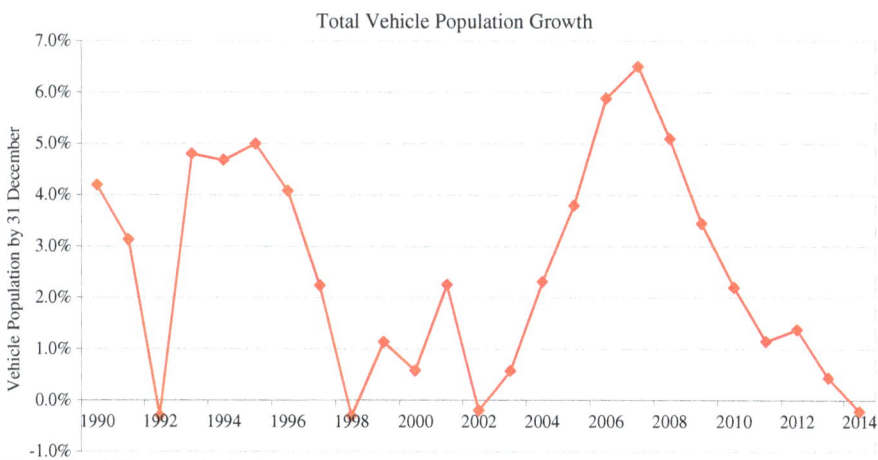

Source: Land Transport Authority.

Fig. 12.   Year-on-year growth of vehicle operation.

the previous year (in addition to the growth rate of vehicles and other adjustments), and quota from August to January, based on de-registrations between January and June.

While the revised formula did away with the need to project vehicle de-registrations in the year ahead, and its need to adjust the quota for any over- or under-projection, there were still about 17,500 COEs that needed to be deducted from the quota because of over-projections under the previous formula, which would have accounted for these excess COEs in subsequent quota periods. It was decided that these excess COEs would be 'clawed back' over 22 months from April 2010 to January 2012.

However, to mitigate the impact of these 'claw-backs, which exacerbated the reduction in quota sizes in the midst of a downward trend in de-registrations (and hence quota), especially in the car categories (Categories A and B), it was soon announced in early 2011 that the adjustment period be extended by another 24 months till January 2014, instead of ending in January 2012. By then, close to 8,000 COEs were deducted from the quota, with the remaining 9,600 adjustments to be spread over the next 36 months.

In the same year (2011), another review of the VGR was conducted. This was because the prevailing VGR of 1.5% would still outstrip both the annual road expansion of 1% in recent years, and of 0.5% expected for the next decade. To ensure that the vehicle population growth remained sustainable, the government announced in October 2011 a further reduction of the VGR to 0.5% from August 2012[25] till the end of 2014. This reduction came just three years from the last.

However, less than a year after the announcement of the reduction of VGR, and before it was implemented, a series of measures were introduced in 2012 to ease the transition. This was in view of the continuing reduction of de-registrations and consequentially, of the quota, which pushed the COE prices to record highs in the second quarter of that year.

The first measure was to defer the lowering of VGR to 0.5% to February 2013, from August 2012. Instead, the rate for August 2012 to January 2013 would be 1.0% in August 2012 and 0.5% from February 2013 to January 2015. The second was to defer the remaining adjustments to the over-projections of the COE quota for Quota Years 2008 and 2009 for one year from August 2012 to July 2013, such that the adjustments would complete in January 2015,

---

[25] It is actually a reduction from 1.5% to 1.0% in 2012, with the VGR remaining at 1.5% for the first half of the quota year from February to July 2012, and falling to 0.5% for the remaining half year from August to January 2013.

instead of January 2014. Both measures provided temporary relief to the tight COE supply situation by increasing the monthly quota — the deferment of VGR reduction made available about 10% more COEs from August 2012 to January 2013 than would be otherwise, and suspension of over-projection adjustments, 7% more from August 2012 to July 2013.

The third measure was to cut the contribution rate to the Open Category or Category E, i.e., to reduce the proportion of recycled COEs of de-registered vehicles allocated to Open Category. This would be done in two steps — from 25% to 20% from August 2012 and 15% from February 2013. This would mark the first time the contribution rate was reduced in the VQS, and the effect would be to allow each category, from A to D, to retain more recycled COEs from deregistered vehicles.

The rise of 'social equity' concerns (2010s)

Despite the three measures, concerns about whether aspirations to own cars could still be fulfilled and about fairness in COE allocation, continued to dominate the public discourse, as they did since the start of the VQS. Many felt that car ownership had become less accessible with greater competition from car buyers with more financial means for the ever-shrinking number of COEs. Such concerns were especially heightened since the early 2010s as the COE prices rose with tighter COE supply. Correspondingly, several policy interventions were made in 2012 and 2013.

In 2012, there was increasing feedback about the influence of taxi companies on COE prices, because they also participated in COE bidding in Category A.[26] Many believed that they were at a distinct disadvantage when competing with the taxi companies, who had the resources to put in multiple bids at high prices.[27] In response, the government announced in July 2012 two changes that would take effect from August 2012. First, taxis would be taken out of the COE bidding process. Taxi companies could only pay for COEs of new taxis based on the PQP of Category A, irrespective of whether these new taxis were to replace de-registered taxis or to add to the existing fleets. In other words, taxi

---

[26] Since the start of the VQS, taxis were grouped together with cars with engine capacity between 1,001cc and 1,600cc (Category 2), and then cars with engine capacity of 1,600 cc and below (Category A). This is even though most of them, especially with diesel engines, have engine capacities above 1,600cc. This is a concession given to taxis for their public transport role.
[27] Taxi companies then could register a taxi by bidding for a new COE or paying the PQP. The latter option was open only if the new taxi was to replace a deregistered one within six months.

companies would no longer participate in the COE bidding market, and could only be COE price takers.

Second, the COEs used to register taxis for fleet expansion will be extracted from the Open Category (Category E), instead of Category A. The rationale was that this reflected the wide variety of vehicles spanning both Categories A and B that were registered as taxis. As the quota of Category E was contributed by all vehicle types, extracting taxi COEs from this category meant that Category A would no longer be the only category from which COEs were drawn to register taxis; instead, this burden was to be shared more equitably among all categories. Thus, more COEs became available for Category A car buyers than would be otherwise.

Since the VQS was introduced in 1990, the Additional Registration Fee (ARF) rates for cars had been progressively reduced over the years. From the record high of 175% of the Open Market Value (OMV), it had been reduced to 100%. This downward trend was finally reversed in 2013 when Deputy Prime Minister and Minister for Finance Tharman Shanmugaratnam announced in his Budget Statement that the flat ARF rate for cars, taxis and goods-cum-passenger vehicles would be replaced with a three-tiered ARF structure.

The ARF for cars with OMVs up to $20,000 remained at 100% of OMV. Two more tiers for more expensive cars were introduced — higher rates of 140% and 180% would be applied to OMV values between $20,001 and $50,000, and above $50,000 respectively. This meant that cars with OMVs higher than $20,000 would pay a higher effective ARF rate. A more progressive vehicle tax structure to address social equity considerations was thus introduced.

Still, concerns remained. Many were now concerned about luxury cars, with significantly higher OMVs than mass market cars, encroaching into Category A, as more of them met the engine capacity criterion of small cars. Such a phenomenon was not unexpected as technological and regulatory developments in the major car markets had led to more cars with smaller but powerful engines. Nonetheless, many buyers, and even the motor dealers, of mass market cars in Category A saw themselves at a disadvantage competing for COEs with luxury car buyers. Others also felt that, with an increasingly limited number of COEs, individuals owning more than one car were depriving others of the opportunity to own a car.

Hence, Minister of Transport Mr Lui Tuck Yew announced in May 2013 that LTA would consult publicly for views on whether and how the VQS could be refined. Specifically, LTA solicited suggestions on two areas. First, the public was asked about ways to better delineate Category A from Category B so that buyers of mass-market cars do not compete directly with buyers of premium car models in the COE bidding. Second, it was about whether and how multiple car ownership could be addressed through the levying of surcharges, from the perspective of enhancing social equitability. As with past reviews, LTA invited the public to submit views and suggestions. A month-long online survey and a few focus group discussions with members of public, academics and motor dealers were conducted.

The review was completed in September 2013. There was general consensus that the COE system remained fundamentally sound and effective in managing the vehicle population. When controlling car ownership in particular, the market-based approach remained the most appropriate way to allocating the COE. This is because car ownership, as a non-basic good, could not be treated the same way as necessities such as housing, health or education, and it would be difficult for the government to decide who should own cars, and how many cars each should be allowed to own.

Nevertheless, LTA agreed to refine the COE categorisation criteria for cars. It acknowledged that over 80% of survey respondents felt that the current criteria should be changed, so that Category A could retain its purpose of catering to the mass market cars better. The existing Category A/B threshold of 1,600cc engine capacity was retained, but a new criterion of engine power was added. A Category A car should now meet both the criteria of having engine capacity not exceeding exceed 1,600cc and engine power not exceeding exceed 97 kilowatts (kW) (or about 130 brake horsepower (bhp)). This set of categorisation criteria was to better delineate mass market from premium cars.

The change was implemented on cars registered with COEs obtained from the February 2014 bidding exercise onwards. LTA also committed to review the criteria every few years so as to keep pace with market developments and technological improvements.

LTA had considered other alternatives. Many survey respondents were in favour of using the Open Market Value (OMV) as the categorisation criterion, and most preferred the threshold to be set at $20,000. While acknowledging that OMV was the best proxy for the value of a car, LTA found practical challenges if this criterion were to be adopted, because OMV could differ quite significantly among different batches of the same car model, due to exchange rates fluctuations and specifications changes. It was possible that the same model could be in either Category A or B at different times. The new twin criteria of engine capacity and engine power were a suitable alternative proxy for car value, and unlike OMV, these attributes were invariant over time, thus providing absolute certainty on the COE category of the car.

There were other suggestions, but they were either assessed as not likely to be effective, e.g., using carbon emissions (many mass market models had higher emissions than premium ones), or impractical to implement, e.g., using brands (difficult and controversial for the authorities to determine which brands are mass market or premium).

On multiple car ownership, the survey revealed that a levy or tax surcharge on those with more than one car was popular. However, there was very mixed views expressed in the focus group discussions on how this could be designed and implemented. The ease with which the surcharge could be circumvented, by registering the subsequent cars using another person's name for example, and the difficulty in enforcement, weighed against its adoption. Certain groups, such as those multi-generation households, would also be inadvertently penalised. Others were also concerned if such a measure was even fair fundamentally, and worried about the signal this could send against society's meritocratic system.

LTA therefore decided not to implement a multiple car ownership surcharge. Instead, it would continue to rely on measures outside the COE system to address social equity concerns, including the tiered Additional Registration Fee introduced shortly before and the road tax which had already a progressive structure.

In addition to the two issues on which LTA sought feedback, two perennial topics were inevitably raised by the survey and focus group discussion participants. These two issues were examined by the VQS Review Committee in 1999, and LTA came to the same conclusions with this review. The first was on the a Pay-As-You-Bid (PAYB) auction system to replace the current Lowest Successful Bid (LSB) system. Despite the periodic reviews over the years that concluded otherwise, many remained convinced that the PAYB system would induce more conservative bidding and therefore lower COE premiums. LTA consulted again experts in auction theory, who explained that the current LSB system encouraged truthful rather than aggressive bidding, i.e., buyers bid the true amount they were willing to pay, and did not resorte to strategising or gaming their bids. Data from actual COE bidding also suggested that bidders were generally cautious, and very few bids were substantially above the final COE price. In contrast, an open PAYB system was not likely to lead to a superior outcome. As COE prices were ultimately determined by demand, a change in auction system would not necessarily lead to lower COE prices, as many would believe.

The second popular suggestion was to ban motor dealers from bidding for COEs. As with the 1999 review, the latest review again found that many focus group participants recognised the role motor dealers played in handling all the paperwork for bidding, registration, trading in, financing and insurance. Some also noted that the futility of preventing buyers from getting dealers to bid for them by proxy. Even if an effective ban were to be in place, the COE price might not be lowered because it ultimately reflected the buyers' willingness to pay. To obtain an indication of the level of support for this suggestion, LTA had also conducted an earlier survey in February 2013, and found that a sizable minority of about 43% disagreed with the ban. In view of the mixed feedback, LTA decided against the suggestion.

After 19 years without a change in the allowable vehicle growth rate (VGR), the Vehicle Quota System (VQS) saw two revisions (2009 and 2012), and a sharp drop in VGR from 3% to 0.5% within a short period of three years. Even so, a few months after the VGR was cut to 0.5%, the Land Transport Master Plan of 2013 announced that the necessity of a further reduction would be reviewed, given the limited scope of road network expansion. That review came a year later in 2014. The outcome is yet another VGR cut — there would be a further reduction in to 0.25% per year, effective from February 2015 for a period of three years. The policy rationale was the familiar but no less compelling one: that a limited scope for road network expansion made it impossible to sustain the same pace of vehicle population growth. This was especially with the number of vehicles approaching the one million mark.

In the same review, LTA announced a second cut to the contribution rate to the Open Category (Category E). From the prevailing 15%, which was set in 2012, the rate was reduced to 10% from February 2015 onwards. With more COEs from de-registrations returned to their respective categories with a reduced contribution rate, the impact of the VGR cut on the quota of each category was mitigated. Thus, the move helped maintain a more stable supply of COEs as the VQS transited to an even lower VGR.

A new frontier — demand management in Public Transport

Over the past few decades, Singapore had invested much effort on curbing demand for owning and using private transport, particularly cars, to manage congestion on the roads. These ownership and usage measures were also intended to encourage more to use public transport, which was the only sustainable way in meeting increasing transport demand in the long term. As laid out in the Land Transport Master Plan 2013, the target was to have 75% of peak period travel to be on public transport by 2030.

However, as more took to public transport for the journeys to work and school, the public transport network could not increase its capacity fast enough. As public transport ridership built up significantly in the past five years, so did the congestion on the trains and buses. While the government quickly took steps to increase the public bus fleet through the Bus Service Enhancement Programme[28] from 2012, there were limitations on how fast the rail network capacity could be expanded. It typically took about a decade, from inception to opening, to deliver a new MRT line. Even improving the capacity of the current MRT lines could only be done progressively. Additional trains were added to these lines only from 2014, as it took several years to build them. The upgrade of their signalling system, which would enable more trains could arrive at closer intervals, commenced in 2011, but would take up to 2016 (North–South Line) and 2018 (East–West Line) to complete.

Hence, in the short-run, the challenge was to meet the peak-hour public transport demand, given the capacity constraints. The response was to introduce travel demand management measures to bring about a more optimal use of existing capacity, while new capacity was being added over time. Unlike the corresponding measures to tackle road congestion which sought to suppress demand primarily, the objective of travel demand management in public transport was not to discourage its use, but rather to shift some of the travel demand during peak periods to off-peak periods. This would help reduce the level of congestion and make the travel experience more pleasant during peak periods.

Fare Discounts for Early Travel Discount

Travel demand management on public transport started modestly with an initiative from SMRT on its MRT network. It offered a 10 cent fare discount for commuters exiting at nine city area stations[29] before 7.30am on Mondays to Saturdays (public holidays excluded). The discount only applied to commuters who entered stations on the North–South and East–West lines (NSEW lines), and Bukit Panjang LRT. This was to encourage city-bound commuters to complete their travel before the morning peak travel periods. In August 2008,

---

[28] BSEP added 550 buses by end 2014, and would see a total addition of 1,000 buses by 2017.
[29] The nine stations are Bugis, City Hall, Dhoby Ghaut, Lavender, Orchard, Outram Park, Raffles Place, Somerset and Tanjong Pagar.

it was packaged into its Early Travel Perks programme, which was launched together with discount offers on breakfast at various partnering food and beverage outlets near the nine city area stations.

In October 2011, SMRT tripled the fare discount to 30 cents under its Early Travel Discount scheme. It also extended the eligible period by 15 minutes from before 7.30am to 7.45am. There were some encouraging responses: it was observed that about 3% to 4% of commuters had shifted their travel time out of the morning peak period.

To provide more incentives to commuters to travel during the morning off-peak hours and ease crowding during the peak periods, the scheme was further enhanced in August 2012. The discount was raised further to 50 cents. The scheme was also expanded to more entry and exit stations. Commuters who entered via the Circle Line (CCL) stations were now eligible. For exit stations, the number was increased to 14 stations to include city stations along the CCL.[30] To focus on shifting peak period commuters during weekdays, the scheme no longer applied on Saturdays when travel demand during the morning period was much lower.

SMRT's efforts to encourage rail commuters to make changes to their travel schedule and travel earlier before the peak period was given a big boost from the government, less than a year it enhanced its Early Travel Discount scheme. In a bold attempt to spread the morning peak hour crowds to the pre-peak period, thus alleviating the peak hour crowding on city-bound trains, LTA launched a one-year trial from June 2013 to offer free travel on the rail network for commuters ending their journeys in the city area before the morning peak period.

The government-funded trial was a significant enhancement of SMRT's Early Travel Discount scheme, which offered a 50-cent discount for pre-peak travel on SMRT network to 14 stations. Not only did the Pre-Peak Free Travel scheme allowed completely free pre-peak travel for those exiting the city stations before 7.45am on weekdays, it provided a further fare discount of up to 50 cents for those exiting between 7.45am and 8.00am, to allow those who travelled slightly later some discounts as well. It also had a wider coverage — it was extended to trips commencing on SBST's North East Line (NEL), and Sengkang and Punggol LRTs and included two NEL stations, to bring the total number of eligible city stations to 16.[31] The trial progressed smoothly, and

---

[30] The additional stations were Bayfront, Bras Basah, Esplanade, Marina Bay and Promenade.
[31] With the opening of the Downtown Line Phase in December 2014, two other stations, Downtown and Telok Ayer were added subsequently.

initial results showed an encouraging decrease of about 9% in the number of commuters exiting the city stations between 8am and 9am. The reduction in ridership sustained over the succeeding months, with about 7% of commuters having shifted out of the morning peak period. A more even distribution of morning rail ridership was attained, with the ratio of commuters exiting at the city stations during morning peak (8am to 9am) to pre-peak (7am to 8am) fallen to 2.1 from 2.7. With such encouraging outcomes, the scheme was subsequently extended by one year to June 2015.

Rewarding Early Travel

In the same year the SMRT's Early Travel Discount scheme was enhanced, another initiative was launched. In January 2012, a group of researchers from Stanford University and the National University of Singapore, with support from LTA, started a trial called 'Incentives for Singapore's Commuters' (INSINC). It was a reward system in which participating adult commuters earned credits according to the distance travelled in the rail network from Mondays to Fridays. To encourage them to shift to off-peak periods to reduce peak-hour congestion, extra credits (three credits per kilometre) were given for making shoulder-peak travel, or 'decongesting' trips, which were to start between 6am and 7am, and between 8am and 9am. To further encourage travel during the decongesting periods, the reward mechanism favoured those who made decongesting trips regularly by giving them higher reward status (with higher prize quanta and chances). Hence, the more rail trips, especially decongesting trips, participants made, the greater the opportunities to be rewarded financially. There was also a social and fun element — participants were given bonus credits for inviting friends and successfully signing them up, and for attaining targets set in a weekly "magic box offer", e.g., learning about INSINC facts, or making more off-peak travel.

The credits were automatically calculated using details of all the trips tracked using the participants' transit cards. These credits could either be used to redeem for a small monetary prize at a fixed exchange rate, or accumulated to obtain higher chances in online games or in regular raffles to win higher amounts of cash, which could be up to $100.

The INSINC trial got off to a promising start. Within six months, close to 17,000 commuters participated, and the scheme managed to prod almost 10% of them to shift their morning peak period trips to off-peak periods. The encouraging start prompted LTA to announce extending the scheme for another 18 months so as to study its longer term effectiveness. In June 2012, in a bid to attract up to 60,000 participants, LTA announced more enhancements that took effect progressively from August 2012. More commuters could participate as the programme was extended to two more groups of commuters — adult commuters who used NETS FlashPay cards,[32] and concession pass holders who were senior citizens and students from universities, polytechnics and the Institutes of Technical Education (ITE). From November 2012 onwards, the maximum cash prize was doubled to $200, and up to six times more credits per kilometre were given to participants with higher status levels,[33] which are determined by the number of decongesting trips. Within a few months of expanding the scheme, the number of participants more than doubled, from around 17,000 in June 2012 to 41,000 in November 2012. By April 2013, LTA announced that the number stood at more than 85,000, far exceeding the target of 60,000 set.

In August 2014, INSINC was rebranded Travel Smart Rewards (TSR), with a few other adjustments to the programme. The one-hour decongesting periods were shifted 15 minutes earlier pre-peak (6.15am to 7.15am), and 15 minutes later post-peak (8.45am to 9.45am). A monthly lucky draw with a top prize of $1,500 was also introduced.

A separate reward structure, called the Corporate-Tier Travel Smart Rewards, was introduced for employees of companies that enrolled in LTA's Travel Smart Network programme. They would receive five to eight credits per kilometre, instead of three to six credits for the general public participants, as well as higher sign-up bonus credits.

Getting the Support of Employers

Initiatives such as Pre Peak Free Travel and Travel Smart Rewards would only be possible if working commuters were able to have flexible work arrangements that allowed them to arrive at their workplaces during off-peak periods. In a survey conducted in September 2013, LTA found that, among commuters who shifted their travel to take advantage the free travel or discounted fare, almost all (95%) would continue to do so. However, for those who did not shift, majority (two out of three) of them did not have flexible work arrangements.

---

[32] INSINC enrolled only commuters with EZ-Link cards initially.
[33] Previously, a flat three credits per kilometre were awarded for a decongesting trip. The new credit structure was: 3 credits for Member, 4 credits for Silver, 5 credits for Gold and 6 credits for Platinum.

So LTA doubled up its efforts to get more employers to implement flexi-work arrangements and other supporting measures for their employees who wanted to travel during off-peak periods and enjoy the benefits of such early travel initiatives. In October 2012, it launched Travel Smart, a two-year consultancy pilot that sought to work with employers to implement plans to facilitate their employees to engage in both off-peak travel and sustainable travel. An initial batch of 12 organisations joined the pilot, and with the help of the consultant, introduced workplace travel plans to put in such flexi-work arrangements as staggered work hours, IT facilities for telecommuting, and locker and shower facilities for their employees who chose to walk or cycle to work. The Public Service also supported the Pre Peak Free Travel Scheme with its work practices that offered public officers flexibility to report for work early. In a July 2013 survey, 8% of public officers working in more than 40 public agencies located in the city area shifted their work start time and travelled before the morning peak period.

Gaining from the experience of the pilot Travel Smart programme, LTA launched a new programme, Travel Smart Network, on August 2014 to support more employers to provide work policies and environments that would enable their staff to take up off-peak commute trips and sustainable modes of travel. The programme targeted employers with more than 200 employees located near MRT stations. These employers could benefit from the programme through three initiatives: Corporate-Tier Travel Smart Rewards (described earlier); a Travel Smart Consultancy Voucher of up to $30,000 for companies to defray expenses to engage consultants to study travel patterns of their employees and tailor travel demand management action plans for their own employees; and a Travel Smart Grant of up to $160,000 per year over three years for co-funding of the cost of company-led initiatives to support the adoption of flexi-travel arrangements.

Off-peak Monthly Travel Pass

The efforts to encourage commuters to travel outside of the peak periods continue unabated. In early 2015, the government announced a two-year trial, starting in July 2015, on an Off-peak Monthly Travel Pass that would offer a one-third discount off the prices of Adult Monthly Travel Pass and Monthly Concession Pass for persons with disabilities and senior citizens. This pass would allow unlimited travel on the bus and train services any time except for the two 2.5-hour peak periods of between 6.30am and 9am and between 5pm and 7.30pm. It would target commuters able and willing to adjust their schedules to travel outside of the peak periods. The government would support this trial by fully funding the expected cost of $10 million a year.

Preparing for the future

Over the past few decades, Singapore had adopted a proactive approach to managing travel demand, especially that for private transport, through a broad suite of vehicle ownership and usage measures. These measures were refined and adjusted over time to respond to private transport demand changes that was driven strongly by economic and population growth. These measures, while arguably unpopular and certainly controversial even to date, had proven to be effective.

Without VQS and the upfront vehicle taxes, the vehicle population would be very much higher than what the road network could support. If the car population were to be allowed to grow unrestrained in the early years, Singapore would have double the number of cars alone (Fig. 13). This would not be inconceivable with the rapid economic growth in the past few decades.

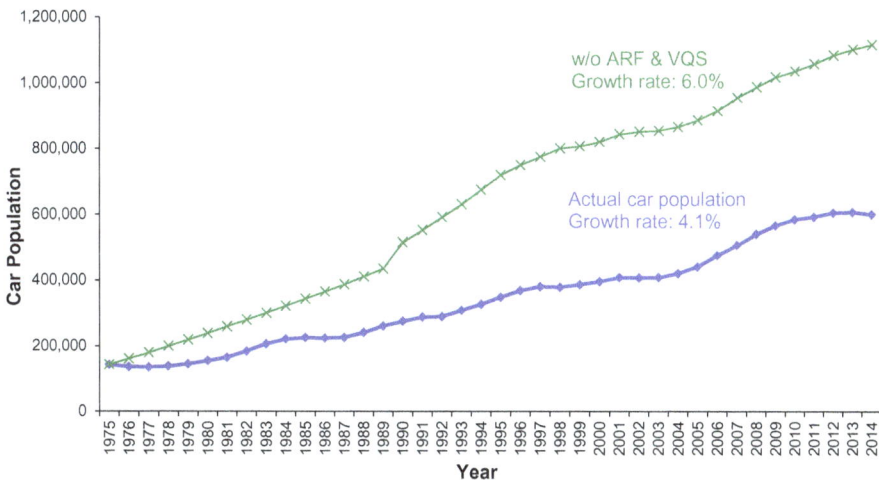

Fig. 13.　Estimating Car Population without Ownership Pricing.

Note: Assuming Singapore's car ownership per capita of 64 cars per 1000 people in 1975, it would have taken 20 years to reach the car ownership per capita of 200 cars per 1000 people, around that for New York/Tokyo. Subsequently, the car ownership per capita would remain constant, and the car population growth rate tracked that of the human population.

Similarly, usage measures had kept a firm lid on road congestion. In particular, the traffic volume entering the Central Business District (CBD) had been successfully reined in, first by the Area Licensing Scheme (ALS) from 1975, and then by the ERP in 1998. This was despite the rapid economic expansion and substantial growth in employment and developments within the CBD. Even as the vehicle population has more than tripled, the ALS and ERP systems had kept the morning peak traffic into the CBD to a similar level 40 years ago (Fig. 14).

Such measures are poised to be continued. With the ERP system approaching two decades in age, its equipment was becoming increasingly difficult and expensive to maintain or upgrade. The physical gantries also posed constraints to the expansion of the system to meet future road pricing needs due to their high costs and large land space needed.

LTA started preparations to develop the next generation ERP system. It identified the Global Navigation Satellite System (GNSS) technology as a feasible option for this system. An open tender was called for a System Evaluation Test (SET), which was a technological trial to identify the most suitable tech-

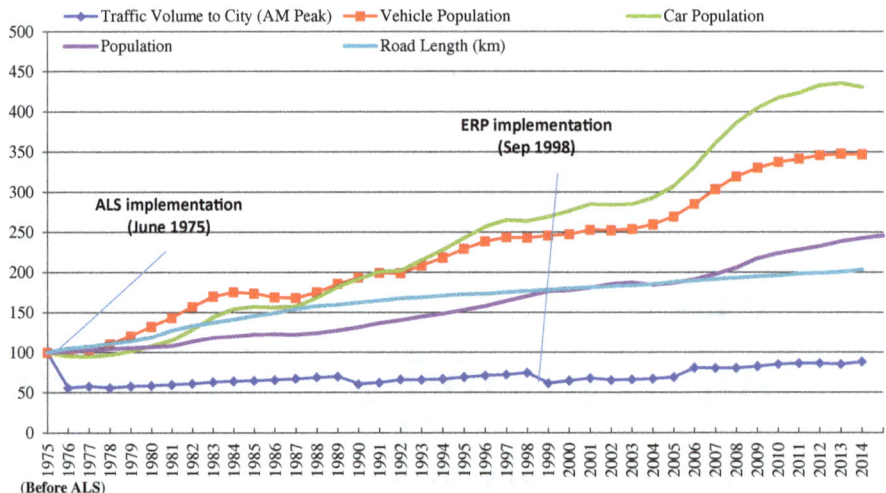

Source: Land Transport Authority.

Fig. 14.   Effect of Road pricing on traffic volumes into the City (AM peak hours).

nological solution, and it was awarded in May 2011. Four consortia that met the SET requirements were invited to participate and test the feasibility their proposed solutions.

The 18-month SET ended in December 2012, and the results showed that a GNSS-based road pricing system was technologically feasible in Singapore, and the appropriate solutions were available for a highly built up urban environment. So in October 2014, LTA called a tender to develop ERP2, the next generation road pricing system, and invited three consortia,[34] all of which were earlier involved in the SET, to participate. The tender closed in March 2015, and the award of the contract was expected in the second half of 2015, with the actual implementation of an ERP2 system to begin in 2020.

Apart from congestion pricing, the ERP2 was to be designed to provide several value-added services and convenience to motorists. Through an interactive intelligent device called the on-board unit (OBU), similar to the in-vehicle unit of the current ERP system, ERP2 was to deliver a range of services such as dynamic and customised traffic information and electronic payment of parking fees. Each OBU in turn would serve as a traffic-sensing device that would supply real time data that would be most invaluable for traffic management and transport planning.

On the ownership restraint measures, the control over vehicle population growth through the VQS is likely to be tightened ever further. In a Committee of Supply (COS) debate in 2015, Senior Minister of State Ms Josephine Teo noted that, while the vehicle population growth rate was lowered to 0.25% per annum (which took effect just a month earlier), it would likely need to be lowered to zero%. This set the stage for a zero-vehicle population growth for the future.

This represented the clearest indication that the intent to fulfil car ownership aspirations of as many Singaporeans as possible was no longer pursued. Many in the public could recall the government's announcement in 1997 that additional 15,000 to 20,000 COEs could be released, if ERP were to be effective in managing congestion. This nexus between imposing more restraint on vehicle usage to curb congestion and allowing higher vehicle ownership was clearly no longer possible nor sustainable.

In fact, the Land Transport Master Plan 2013 had earlier served notice of such a development. It questioned that "the conventional wisdom that we can

---

[34] The three consortia were NCS Pte Ltd & MHI Engine System Asia Pte Ltd, ST Electronics (Info-Comm Systems) Pte Ltd, and Watchdata Technologies Pte Ltd & Beijing Watchdata System Co Ltd.

have more cars as long as they do not contribute to congestion". It highlighted that the costs of higher car ownership on society were not just about greater congestion on the roads, but also the opportunity costs of setting aside more parking spaces at residences, workplaces, schools and destinations of interest. Less land would then be available for other types of development, and localised parking and traffic problems would also become more prevalent. The paradigm that one could own a car as long as it was not used frequently was no longer tenable.

With a possible zero-growth situation in the horizon, the demand for vehicles will continue to outstrip the supply. This will be particularly so for private cars, as both the population and the economy continue to grow. Under such a context, the government has acknowledged that car ownership will not be within easy reach for many. As Ms Teo noted in her COS 2015 speech, private car ownership is unlikely to be a low-cost option.

It is therefore fully conceivable that the perennial concerns about the accessibility of car ownership and the fairness of the COE system will persist. Calls to revise how the VQS allocates COEs are also expected to continue, if not intensify, in future. However, as shared by Ms Teo in the same COS 2015 speech, such efforts in "tinkering with the COE allocation framework" would likely to be futile. She noted that, with a fixed quota of COEs, any adjustment to benefit certain groups of buyers would disadvantage some other groups, ending in a "zero-sum game."

On the usage demand management front, while the current strategies of cordon pricing (for the RZ) and point charging (at expressways and arterial roads) have managed to contain congestion well for more than 15 years, future congestion management will have to account for changing travel demand brought about by further population and economic growth. New and more flexible charging schemes will be needed. One of those being explored is distance-based pricing, which is assessed to not only be more effective in managing traffic, but also more equitable than the existing point and cordon-based pricing schemes, as motorists will pay charges proportionate to the distance travelled on congested roads. With the new ERP2 system expected from around 2020, distance-based charging can become a real possibility then. The ERP2 system will also cater to Off-Peak Car (OPC) schemes that are more flexible and responsive to the current one. For example, OPC users may only need to pay for using their cars for shorter periods rather than the whole day, or for using them only on uncongested roads. The next decade will therefore likely see a significant shift in how usage demand is managed.

As both ownership and usage measures continue to be applied, and possibly tightened, in the foreseeable future, it would neither be as accessible nor low cost as it was in the past to own or use private vehicles. In land-scarce Singapore, the sustainable, and probably the only way to meet travel demand challenges is for Singaporeans to rely more on public transport and other allied modes of travel, such as walking and cycling. Travel demand management is an integral part of the strategy to move the society from a mobility model that is car-centric, to a more sustainable one based on high quality public-transport services. This plan was articulated by Prime Minister Lee Hsien Loong in November 2014 at the launch of the Singapore Sustainable Blueprint 2015:

> "We will aim for a 'Car-Lite Singapore by promoting and developing other modes of transport, making them convenient. We have to rely less on cars on the roads because we cannot keep on building roads — more roads for more cars. So we will provide more options for Singaporeans that are better than cars."

Chapter 4

Expressways and Traffic Management

A. P. Gopinath MENON

This chapter provides a brief history of traffic management and expressways in Singapore from the colonial to present-day period. It provides the rationale behind the technical and implementation decisions and dwells on the future of traffic management in Singapore.

1. Introduction

1.1. *Land transportation*

Highway engineering had dominated the field of land transportation engineering for many years. With the advent of the motor car, there were fundamental shifts in people's habits as a result of the freedom to travel. The car became a much-desired possession in developed and developing countries, and soon the spectre of traffic congestion began to surface. What was good for the individual did not seem so good for the motorists at large anymore. Many authorities believed they could build their way out of congestion. Massive road building programmes such as the autobahn systems in Germany, the inter-state highway systems in the United States began to take off. Intercity travel became convenient. While useful to improve accessibility, it was soon realised that the growth of traffic on the roads always outstripped the supply of road kilometers. Within the city, it was another matter altogether. Building new roads in heavily built-up areas was not easy, and neither did new roads improve traffic flow significantly.

The well-known planner Lewis Mumford wrote, "Cities are meant for the care and culture of men, not for the constant passage of cars." Yet, a city would find it difficult to be vibrant without streets and cars, because streets provide an agreeable sense of busyness and bustle while cars provide accessibility, without which a city's vitality would be greatly diminished. Therefore, in many major cities, it is not unusual to find an inordinately large amount of space, i.e., about 10 to 15% of the total land area, devoted to accommodating streets.

Therein lies the dilemma — it would be nice to have no streets in the cities because streets also bring about traffic congestion, environmental concerns on atmospheric pollution, traffic accidents, traffic noise as well as unattractive street furniture. However, if we restrict accessibility unduly, cities lose their attractiveness in drawing people.

Coping with the daily demands of traffic movements presents one of the most pressing challenges for cities throughout the world. When the number of vehicles using the road was low, there was little need for traffic management. The vehicles had the road to themselves and they could move as they pleased if they did not cause any accidents.

Highway engineering no longer dominates the transportation field nowadays. Land transportation is now looked at in total, involving cars, public transport, pedestrians and goods vehicles — the emphasis is on moving people and goods, and no more on just vehicles. Transportation is considered as a total system and the emphasis is on improving the system, with roads and traffic management being the two most important components.

Emergence of the high speed road

In the 1920s, there was a superhighway craze with double and triple deck streets being seen as the solution of the traffic congestion problem. It was assumed that these types of highways would be the rule rather than the exception fortunately, such scenarios did not materialise.

Germany was the first to construct a high speed road called the Autobahn with limited road accesses in the 1920s. The objective was to alleviate massive unemployment. It was a 20-kilometre stretch of rural highway from Bonn to Cologne. Design speed was set at 165 km/hr. The road had a divided carriageway with two lanes in each direction. Italy also started the high speed road, the Autostrade, around the same time. After the Second World War, most countries saw such high speed roads not only for better travel, but also as strategic roads for defence purposes.

Junctions of roads had always been on the same level. It was in Paris in 1906 that the idea of the multilevel clover leaf interchange (another name for junctions of high speed roads) took shape. When the cloverleaf and similar interchanges came common, they became large consumers of land. The more complex of them came to be known as spaghetti junctions.

In the United States, a National Highway System, called the Interstate Freeway System was sanctioned by the Federal Aid Highway Act of 1956, and it was completed 35 years later. In 1958, Britain started its motorway system which provided the convenience of long distance travel at high speeds. High speed roads have now become common in developed and developing countries. While the building and control of roads have mainly been the responsibility of the government, the private sector has also got into the act. The build-operate-transfer (BOT) model for new toll-charging high speed roads has meant that many countries are now able to afford such expensive infrastructure.

These roads go by various names in different countries — freeways, motorways, expressways.

Emergence of traffic management

For many years, reliance had been placed on building of new roads or widening old ones as the way of handling traffic problems. The earliest traffic management was in the way of road markings, traffic signs and traffic signals to ensure discipline of road users and, in the process, improve road safety. By 1927, the American Association of State Highway Officials (AASHO) was already producing manuals on traffic signs with the publication of the first Manual of Uniform Traffic Control Device (MUTCD). In 1933, the Road Research Laboratory was set up in Britain with some of its earlier work being on traffic management and safety.

By the late 1960s, traffic management had become well-entrenched as an essential component of any worthwhile transportation strategy for a city. Traffic management is cheap as opposed to massive road construction. The budget could easily be found. It takes a short time to implement a traffic management project on the site. The schemes are seldom as disruptive to residents and the community as road construction is and surprisingly, the initial results and improvements are good. What traffic management does, is to optimise traffic of an existing road network.

For traffic management to be successful, it requires obedience of traffic rules and regulations from road users. Such rules are part of the Highway Code which all motorists have to pass before obtaining a driving licence. But obedience to the rules is another matter altogether. In many cities, traffic management fails woefully due to rampant disobedience of traffic rules. Traffic enforcement by the police can help, but there is a limit to how much enforcement can be instituted. Therefore, many authorities resort to selective enforcement, i.e., enforcing randomly at certain locations and for certain violations. The best way to ensure obedience is by educating road users on the reasons and the need for rules so that they feel inclined to follow them. It must also be remembered that non-motorised users such as cyclists and pedestrians are not obliged to study the Highway Code.

In the late 1990s, advances in computer, optics, sensor, satellite, and mobile phone technologies gave rise to a whole new field of Intelligent Transport Systems (ITS). The ambitious goal then, was to provide effective and efficient communications between the three players in any traffic system, namely the road user, the vehicle and the road space, so as to enhance safety and improve traffic flow. ITS has given a new respectability to traffic management.

1.2. *Singapore Pre- independence period (pre-1965)*

Early road hierarchy in Singapore

Since the 1920s, the land use-transport planning of the colonial government was dictated by the General Improvement Plan, which was not a very comprehensive one. In 1951, the Singapore Improvement Trust which had been administering the General Improvement Plan was given the task to prepare a Master Plan that would guide land use and development for a 20-year period up to 1972. The Master Plan consisted of a series of maps supported by written statements to regulate private and public development through zoning of land uses and reservation of land needed for public purposes such as roads, parks and utility plants. The final report of the Master Plan was approved by the Governor in August 1958.

The working committee on the Master Plan proposed planning standards for residential, community space and business areas. For the first time, there was a road hierarchy, namely road classification into five types — arterial roads serving the island as a whole; major roads connecting centres of population; main thoroughfare which is a local road carrying bypass (through traffic); development roads providing frontage to building developments; and minor roads. The current classification of roads has also five types and took cue from this Master Plan. The optimum width of a traffic lane was to be at 10 feet (3.1, metres) and cyclists and pedestrians were to be segregated from vehicles.

There were no expressways in the road hierarchy because in 1958, the idea of the high speed roads was just catching on.

Early traffic management in Singapore

In 1896, Herman Katz of Katz Brothers, a general importer in Singapore, brought in a single-cylinder belt-driven 5hp Benz of 1894 manufacture for a Mr B. Frost of the Eastern Extension Telegraph Company. This was the first car in Singapore. At the turn of the century, as more people became convinced of the functionality of the motor car, the problem of errant drivers dashing around had the police force already worried. In 1907, the Straits Automobile Club was formed with 50 members, and was later renamed the Singapore Automobile Club. This became part of the Automobile Association of Malaya in 1932, and later became the Automobile Association of Singapore in 1952. So the motorists' lobby was a very powerful and influential group which started as soon as the first car arrived in Singapore. By 1913, Singapore already had about 550 cars with a population of about 330,000 persons, 4900 by 1927 and 11,000 by 1940.

The first road public transport started in 1905 with electric trams along five lines — linking the harbour, the town, Paya Lebar and Geylang. It was run by the Singapore Electric Tramway Company. Rising public dissatisfaction led to a commission to inquire into the state of the electric trams. The operation was taken over by the Shanghai Electric Construction Company to improve tram service. Since the tracks were in bad condition due to inadequate maintenance, it was decided to replace the trams with trolley buses which appeared around 1929. A new company, the Singapore Traction Company Ltd (STC), took over the operations. The STC was given a 30-year monopoly to operate road-based trolley and omnibuses within the town limits. Some private operators also provided transport services to the rural areas which had never been served well, using old 14 seater buses.

By the 1930s, there were complaints from motorists of the lack of sign-posts, poor street lighting on many roads and an increase in road deaths. These were attributed to a lack of coordination among the different departments in dealing with traffic matters. In 1937, the colonial government after consulta-tion with the Municipal Commissioners set up a traffic committee headed by George Trimmer, Chairman of the Singapore Harbour Board, with the task of studying the traffic conditions in the island and to make recommendations for their improvement. The members were from the Municipality, the Police, the Rural Board, the Automobile Association, the Straits Settlement Association and the Motor Traders Association.

The Study report in 1938 concluded that there were about 15,500 vehicles which were being used in the town area, and that most of the roads were planned and built when the number of vehicles were about half of this number. The main reasons for obstructions to traffic were pedestrians on the road, vehicles reversing on the road and fixed obstructions. The report suggested providing kerbs for roads, proper footpaths, better positioning of bus stops, separate tracks or lanes for bicycles, tricycles and rickshaws, prohibiting some vehicles from using certain roads at certain times and better road markings. Use of more one-way streets was also recommended.

The replacement of the manually operated traffic signals that were operated by the police by automated ones; and linking them up with a master controller were recommended. A speed limit of 30 miles per hour and the use of police cars to deter speeding were to be implemented. The report asked for more pedestrian crossings and convictions for any pedestrian who walked heedless of traffic.

Systematic education of road users was essential, the general public was informed through the cinema, press and broadcasting while the Traffic Police undertook road education programmes for children and storytelling for people who could not read and write.

To solve parking problems which were unrestricted and free in the business part of the town, short term (of up to one hour), long term (of up to two hours) along the roadside and garage (longer than two hours) were proposed. Garage parking was not to be allowed on public roads but only off the road area. It was also suggested that new business buildings should provide car and bicycle parking for a large percentage of its tenants. Legislation should be used to get the owners to comply.

Five footways, alleys and passages were being obstructed by indiscriminate bicycle parking. It was recommended to provide bicycle parking spaces and stands. This was considered urgent as the population of bicycles was growing rapidly.

The report also dealt with the complaints about poor street lighting, and that many of the accused in court cases involving road accidents had pleaded poor street lighting as contributory to the accidents. The report proposed better standards of street lighting, especially as many of the roads did not have sidewalks and had rickshaws using them.

The report noted that there had not been sympathetic cooperation between the Municipality and the Police, resulting in delays in enacting traffic legislation. It noted that previous Traffic Committees constituted in the 1920s had not been effective. It advised the formation of a Traffic Advisory Committee with members from the Police, the Municipality and public bodies involved in traffic matters.

The report recommended that a mobile squad be formed within the Police because no speed limits could be enforced without such a unit. The existing method of driving tests was antiquated and needed improvements. All horn sounding of stationary vehicles were to be prohibited. Furthermore, sounding of horns was also to be prohibited in residential areas.

This was one of the most comprehensive reports on traffic that was produced, and it set the tone for much of the traffic management schemes in the early half of the last century. Some recommendations were followed up immediately. However, the Second World War had intervened and put a stop to more of these improvements.

Post-war period

After the war, the new roads constructed in the 1950s attempted to separate motorised traffic from pedestrians and cyclists by providing separate tracks or combined paths for both pedestrians and cyclists. The first automatic traffic lights appeared at the junction of Serangoon Road and Bukit Timah Road in 1948. Zebra pedestrian crossing appeared, soon after they were successfully introduced in Britain in the early 1950s.

This was the period when roundabouts (or circuses as called locally) at road junctions became popular. The roundabout junction was tried out in Britain in 1909, but its popularity was more pronounced in France. In 1931, a proposal for a roundabout was put up for Newton Circus, after which more new junctions were built as roundabouts. The advantage of the roundabout junction is that there are fewer conflicts of moving vehicles as their movements do not cross each other, but rather, the movements merge or diverge, resulting in fewer accidents. At the height of its popularity, there were about 50 roundabouts on the island. Many of these roundabouts took large tracts of land that were landscaped, and some such as the ones at Farrer Road/Holland Road and Tanglin Road/Grange Road even had water fountains in the middle in order to beautify the landscape.

1.3. *Post-Independence — 1965 onwards*

The Singapore Concept Plan

The first Town Plan was prepared by Lieutenant Philip Jackson for Stamford Raffles in 1823, four years after the founding of modern Singapore. After that, there were various attempts with the next complete one appearing in 1958. After independence in 1965, the Singapore government made a formal request to the

United Nations for assistance in implementing a comprehensive long-range planning project to prepare a long-range plan to guide the future physical development of the island. The project was done over a period of five years from, 1967 to 1971 and they came up with the Singapore Concept Plan.

What has the Singapore Concept Plan achieved? It has set Singapore on a course of orderly development, following good land-use and transport planning principles, in which we are very fortunate when compared with other cities.

With a Concept Plan already in hand, the government took the bold step of coming up with an overall transport policy statement, which comprises of:

— Minimisation of travel demand by the careful arrangement of land use.
— Design and operation of the transport system to supply adequate service and to influence the choice of travel mode, both for private and public transport.
— Administrative and fiscal measures aimed at cutting down the demand for excessive usage of private vehicles, especially for the work trips.

Out of the policy came the transport strategy consisting of five main pillars.

— Integrated land use-transport planning.
— Construction of a modest road network.
— Traffic management to maximise utilisation of the road network.
— Improvements to public transport.
— Travel demand management.

This chapter is devoted to two of the pillars — the construction of a modest network, touching only on the development of the expressway network, and on traffic management.

1.4. *Expressways*

Expressway network

The Singapore Concept Plan had recommended a Ring Plan for the development of the island. The city is located in the south of the island. The island has the water catchment nature reserves in the middle of it. The Ring Plan had a belt of high density development to the west and east of the city towards Jurong and Changi, with a ring of heavy urbanised development in the form of New Towns around the other three sides of the water catchment. There was to be a transport network linking all the development areas with expressways and a mass transit system through the city area. The general form of a ring is an economical one for transport corridors and public utilities such as water and electricity.

For the road network, the plan proposed an expressway network around the central catchment area linking all the development areas, such as the city, Jurong, Changi and the New Towns. Since there was already a wide range of commercial, banking and social activities in the city area, it was expected to remain a desirable location for such activities. The road hierarchy of 1958 was reclassified into five types:

- Expressway (Category 1) forms the primary network where all long distance traffic movements should be directed.
- Major Arterial (Category 2) interconnects expressways and minor arterials as well as with other major arterial roads.
- Minor Arterial (Category 3) distributes traffic within the major residential and industrial areas.
- Primary Access (Category 4) forms the link between local accesses and arterial roads.
- Local Access (Category 5) gives direct access to buildings and other developments

Today, there are a total of 10 expressways with a total length of about 170 kilometres. The expressways are shown in Fig. 1. They are:

- **Ayer Rajah Expressway (AYE)** — The AYE links the west to the city (part of it in viaduct).
- **Bukit Timah Expressway (BKE)** — The BKE connects the north to the Pan Island Expressway.
- **Central Expressway (CTE)** — The CTE connects the north to the city (with two short sections in tunnels).
- **East Coast Parkway (ECP)** — The ECP connects the east to the city.
- **Kranji Expressway (KJE)** — The KJE connects the west to the north.
- **Pan Island Expressway (PIE)** — The PIE runs across the island from east to west.
- **Seletar Expressway (SLE)** — The SLE connects the north to the north east.
- **Tampines Expressway (TPE)** — The TPE connects the east to the northeast.
- **Kallang-Paya Lebar Expressway (KPE)** — The KPE connects the south to the northeast (almost completely in tunnels).

Source: Land Transport Authority.

Fig. 1.   Expressway network.

- **Marina Coastal Expressway (MCE)** — The MCE connects the southern end of the KPE, and the western and eastern end of the ECP and AYE respectively to the new Downtown (in undersea tunnels for part of it).

Characteristics of expressways

A typical expressway has a road reserve of 55 to 70 metres from drain to drain. There are road shoulders on both sides for broken-down vehicles to wait for assistance. There are usually three to four traffic lanes each of 3.3 to 3.7 metres width in each direction with a wide central wide median, where roadside trees are planted. There are vehicle impact guard rails next to the drains and along the central median. Shrubs are planted in front of the guard rails to soften the metallic appearance. The street lamp posts are usually on both sides or in some cases, along the centre median. There are white edge markings next to the center median and between the left lane and the road shoulder (Fig. 2).

The edge markings next to the shoulders are grooved to give a rough jolt when a vehicle goes over them. These vibra-lines alert the inattentive driver if he drifts out of the left lane onto the shoulder.

Fig. 2.   Typical expressway.

Some U-turns with lockable swing gates are provided at regular intervals which are to be used in the case of emergencies to divert traffic. This provision is not critical because interchanges (expressway junctions) on our network are located not too far away from each other.

Left slip roads which lead out of the expressway have special diverging lanes for vehicles to slow down before getting off. The gore area between the left slip road and the main expressways is a potential area for accidents, and hence it is usually protected with crash cushions.

Left slip roads from the side roads that enter the expressways are provided with acceleration lanes to allow entering vehicles to accelerate to the prevailing speeds along the expressway to merge with the expressway traffic. The speed limits on the expressway vary between 70 km/hr to 90 km/hr while the speed limit of the roads connecting to the expressway is only 50 km/hr. There are "start of expressway" and "end of expressway" information signs for vehicles entering and exiting the expressway.

Cyclists and pedestrians are not allowed and all pedestrian crossings are either pedestrian overhead bridges or underpasses. Vehicles are not allowed to park or wait on the expressway. If vehicles break down, they are expected to be moved to the road shoulders.

Bus stops are not allowed on the expressway, but exceptions had to be made to retain a few specially designed bus stops with bays on two expressways which were former major roads that were upgraded to expressways.

All junctions are grade separated (at different levels) unless the joining roads are with left-in, left-out arrangements, which are provided with diverging and merging lanes on the expressway. The grade separation is done by flyovers or vehicle underpasses. The right turns are through left turn loop structures (partial clover leaf arrangement) or by direct right turn structures. The other alternative is to bring the right turns to the ground system with traffic lights, thus not impeding the main straight flow at the flyover or underpass. Under most flyovers, rain shelters are provided as small bays off the expressway for motorcyclists to congregate and wait during rain, away from the left lane of the expressways where they can become very vulnerable to accidents.

Vehicles with posted speed limit of 40 km/hr (construction vehicles such as ready-mix concrete trucks and mobile cranes) are not allowed to use expressways unless with special permit. There are kilometre markers on each expressway from end to end. These are used for numbering exits. For example "Exit 8, Marine Vista" on the East Coast Parkway (ECP) means that it is located at 8 kilometres from the beginning of this expressway. The kilometre markers can also be used by vehicles in distress to give locations when calling for assistance.

There are no rest areas on the expressway network due to the short lengths. Motorists can easily get off the expressway as the interchanges are closely spaced.

Expressway tunnels

The expressway tunnels are cut and cover concrete boxes with three to four lanes in each direction, with a solid wall between the two directions (Fig. 3). Cross over doors are located at certain intervals between the two directions so that motorists can get from one part of the tunnel to the other in the event of emergency evacuation. Tunnels are fitted with mechanical and electrical equipment for lighting, ventilation and fire-fighting. There is a round-the-clock monitoring of traffic conditions, ambient conditions and status of traffic incidents from the control centres. Vehicles are advised to drive in the tunnels with their headlights switched on.

At the exit of the tunnels, motorcycles tend to congregate on the left lane when it rains, causing safety problems to themselves and others. Therefore, special protection laybys are provided in the tunnels to keep them away from the left lane, but still under shelter (Fig. 4).

Source: Land Transport Authority.

Fig. 3.   Expressway tunnels.

Source: Land Transport Authority.

Fig. 4.   Motorcycle protection layby at exit of tunnels.

Public buses have been discouraged from using the tunnels unless there are no alternatives, because if there is a bus breakdown, large numbers of people will be in the tunnels together with moving traffic. Vehicles carrying dangerous goods are banned from using the tunnels. Dangerous goods are inflammable materials, explosives, corrosive chemicals, bio-hazardous and poisonous materials, the movement of which are regulated by law.

Tunnels are under 24-hour surveillance from control centres. The most disastrous incident that can happen in a tunnel is a fire. A fire rapidly consumes oxygen in the tunnel, and this can be fatal to those who are trapped in the tunnel. Smoke can also affect the visibility of those abandoning their vehicles and trying to escape through the staircase that lead out into the open.

The tunnels are all built with safety devices to minimise the risk of a fire and to contain a fire. Also, regular emergency exercises are conducted among the Land Transport Authority (LTA), the Police and the Singapore Civil Defence Force (SCDF) to fine tune the operational procedures needed in the event of a fire.

Expressways upgraded from arterial roads

The network of expressways ran along some of the older roads like Jalan Toa Payoh, Paya Lebar Way and Ayer Rajah Road. Expressways were to be three lanes in each direction, at a speed limit of 70 km/hr, with no traffic signal junctions, no bus stops, no pedestrian footpaths and limited number of side roads. These requirements were necessary to ensure high fast speed travel without causing road safety problems. These restrictions caused unhappiness because the residents who lived in the vicinity of these former roads found it inconvenient to be denied of these conveniences. Many meetings had to be held with the residents and the grassroots organisations to explain and persuade the residents on the expressway programme. Generally, most agreed reluctantly, but the one issue that was unacceptable to them was the removal of bus stops. Hence, it was necessary to retain specially designed bus stops on some of these roads that were upgraded to expressways. Another sore point was the loss of direct access to some minor roads from these former roads, but if the expressways were to function well, these had to be cut off.

Roadworks on expressways

No works other than pruning of branches of trees, or routine or corrective maintenance work are allowed on the expressways and tunnels. When such work is done, workmen and moving traffic have to be protected by the use of truck-mounted attenuators, placed well in advance of the work location. The attenuators shield the workers, all of whom are required to wear reflective vests. If any approaching vehicle unwittingly crashes into the attenuator, the attenuator just collapses and absorbs the crash energy. The vehicle may be damaged, but the occupants are unlikely to suffer injuries.

Our expressways

The earliest expressways were upgraded major arterial roads. Others were on reclaimed land or in forested areas; so land acquisition was mainly confined to cemetery land and some farms. Hence, the expressway construction for the most part proceeded smoothly. The expressways were designed to have three to four lanes in each direction with a fairly wide central median. There were no footpaths because pedestrians are not allowed to walk along them. Instead, there is the "shoulder" meant for vehicles that break down to wait away from the main traffic lanes. There is a full white line between the left lane and the shoulder. The shoulder width is slightly less that one full lane's width so that vehicles do not use them for travelling. The first road shoulders were not made up to a full road standard, but used a gravel surface to distinguish them from the main lanes. However, the gravel had a tendency to end up in the side drains and block drainage flow. Subsequently, the shoulders were all made up of the same asphalt road surface as that of normal lanes.

The East Coast Parkway (ECP) which was built on reclaimed land connecting Changi Airport in the east to the city, was the first expressway to be opened in 1981. There was controversy over the bridge over Kallang River. There were many ship repair yards in Tanjong Rhu in the Kallang Basin. Since tall ships had to go in and out, a normal bridge would have been an obstruction on Kallang River. A high level bridge, or a bascule bridge (a bridge with an opening that could be lifted when ships passed by) were the two options. Finally, the decision to build a high level bridge, the Benjamin Sheares Bridge, was made because an opening bridge would need vehicles to be stopped and cause traffic congestion on the expressway. However, the moment work started on the building of the bridge, a decision was made that all the shipyards had to move out of Kallang Basin. A high bridge which is 10-storeys tall and restricts access to the area beneath it because of the long steep concrete ramps would not have been needed. These ramps from the side roads have very steep gradients, causing heavy goods vehicles to slow down when climbing them. Double decker buses are also not allowed on these steep gradients due to their high centre of gravity. Nevertheless, the bridge offers the best panoramic view of the city and the river mouth, and was featured in the old

$50 currency note. In the earlier days, the bridge was a favourite spot for couples to take wedding photographs, but this trend had to be discouraged due to safety considerations.

One day in the future however, the bridge may have to be taken down and replaced with a lower level bridge to improve accessibility to the Marina centre area.

The other expressway that caused controversy was the Central Expressway (CTE) cutting across the Orchard Road belt. One option was to construct it on a viaduct (a long flyover) and the other was to build it in tunnel. The Singapore Concept Plan had proposed the CTE as a combination of level roads and flyovers. The Public Works Department (the predecessor of the Land Transport Authority) examined the underground alternative in the early 1980s for environmental reasons, despite the much higher costs. The wise decision was made to minimise the severance effect of large overhead structures cutting across the Istana area and commercial Orchard Road belt, by building the two first road tunnels to undercross this area in 1991. Although it cost more, the city has been spared of the sight of large overhead structures carrying moving vehicles. Perhaps we took cue from Boston, where it was said that "a monstrous overhead highway cut ruthlessly through the centre of Boston and hopelessly destroyed human amenity". In recent years, Boston has demolished this overhead highway and brought it into tunnels.

Construction was difficult because the tunnels went under the Singapore River and over the Mass Rapid Transit (MRT) train tunnels.

This was not to say that viaducts (long double decker roads) could not be considered for roads. The Keppel viaduct, which runs over Keppel Road, is part of the Ayer Rajah Expressway. That part of Singapore abuts the port and the old, now-defunct railway station and hence, is not as critical as commercial, shopping or residential areas. Hence, the cheaper option of the viaduct was chosen.

Bukit Timah Expressway (BKE) ran across the nature reserve in the centre of the island, and there were some concerns from the Nature Society and National Parks Board on the negative environmental impacts. One of the issues was that the central catchment area should not be polluted when the vehicles started using it. Mitigation measures had to be taken to prevent pollution of the water catchment areas.

Tampines Expressway (TPE) had an interchange over Lorong Halus. Lorong Halus was designated as a dumping ground for rubbish in between 1970 and 1999. The rubbish that accumulated over the years contained organic and inorganic substances that leached chemicals into the surrounding area. Leaching is a process by which soluble substances are washed out of the soil. Some may be poisonous and corrosive and harmful to health. Extreme care had to be taken during the construction to avoid any unnecessary health risks to the workmen, and the corrosion of the piles that were to be used for the foundations of the flyover.

The original expressway network has been expanded with each of the Concept Plan reviews. There was reluctance in building the first road tunnels due to the advice from foreign experts on the difficulty in building and operating road tunnels. However, after the Central Expressways was built in tunnels over certain sections in 1991, tunnels became a possible option for new roads where there were land constraints. The Kallang–Paya Lebar Expressway (KPE) had to go in tunnels under the Paya Lebar Airport runway in 2008. Marina Coastal Expressway (MCE) had to go in tunnels under the sea in 2013. There are proposals to build a North–South Expressway which will also be partly in tunnels.

Expressway pavement and structures

Roads are usually built over a good earth foundation. Compacted mud tracks were the first roads in the early 20th century, but they deteriorated as the number of vehicles over them increased. Binder and block (gravel and large granite stones) were compacted over the flattened mud track to be the road surface. Binder and block construction required large blocks of granite laid over the compacted mud road by hand, called block pitching and the interstices filled with fine dust before compacting by using heavy road rollers.

As road technology progressed, a layer of stone-filled sheet asphalt (a mixture of bitumen, stone and sand) was laid to be the road surface. A thin layer of bitumen was spread over the binder and block before laying this sheet asphalt of 75-millimetre thick and rolling to compact it to become the final road surface. It was usual to leave roads without putting on the final surface for three to four months so that the vehicles using it would compact it naturally, as it was deemed that mere rolling by the rollers were not sufficient to compact the surface adequately. Roads can also be in concrete instead of asphalt. As early as 1925, both types of road surfaces were being used in the island. Concrete roads were used where the road foundation was in swampy areas.

Over the years, structural road design improved with better compaction techniques and better quality of asphalt. Expressways are almost exclusively built with asphalt surfaces, and concrete surfaces are confined to heavily used surface junctions and bus bays. All bridges, flyovers and tunnels are concrete structures. The concrete slab is sometimes overlaid with asphalt to become the final road surface.

The heavy tropical rainfall requires good drainage of the road surface to prevent aqua planing, which is a process by which the tyres of a vehicle are riding on a thin film of water instead of on the road surface. Water is led off the road surface by the normal cross-fall of the road to the sides, except at bends where drains are required in the central median due to the super elevation (banking) of the road. In some areas, open drainage asphalt porous mix for the road surface was used to drain the surface of the water quickly. Such a surface reduces the water spray from moving vehicles and improves the forward visibility for drivers.

Flooding of the network is uncommon and there have only been a couple of instances when this has happened, in conjunction with unusually heavy rainfall and high tides when flooding occurred on many parts of the island.

Bridges have been built across many of the rivers since the first bridge across Singapore River called the Presentment Bridge was built in 1823. The first few bridges were of wrought iron trusses so popular in Europe during the period. In the 20th century, concrete bridges became popular.

There were some railway bridges that went across roads, but the bridge over a road (i.e., flyover) started with the construction of expressways, the first one being a reinforced concrete bridge of the Pan-Island Expressway (PIE) over Thomson Road. The later flyovers were in prestressed concrete. Unlike the building of a bridge over the river, construction of flyovers poses greater risks to the public, especially during construction when it is required to place heavy concrete beams across the road when traffic is moving under it. So it was necessary to close the roads below to traffic over a weekend and clear the area before the launching of these beams to rest on the bridge piers.

A structure such as a flyover is supported on piles driven into the ground, so the structure will not settle. However, the road that joins the structure on an earth embankment is not on piles *and with repeated use and if on soft underlying soils will settle over a time.* This is caused by the consolidation of the underlying soil. This results in a bump at the meeting point between the road and the flyover after some time. One way to get over the problem is to top up the meeting point with new asphalt surface until the settlement stops in the future. A more permanent measure is to treat the underlying soil of the road to make it more stable before constructing the earth embankment, but at a greater cost. This is usually done by installing vertical sand drains in the soil to draw out the water and expedite any consolidation before the construction of the embankment. Since carrying out reconstruction work frequently on high speed roads is dangerous, the permanent measure has become a standard feature, when necessary (Fig. 5).

From 1990 onwards, all flyovers and underpasses have been given names, and these are shown by plaques installed in the vicinity. It eases identification for locations by motorists and for asset maintenance records.

Source: Land Transport Authority.

Fig. 5.   Flyover.

Method of procurement

Construction projects were always let out by competitive public tenders. Some were let out on a design and build basis, and others followed the construction based on the design of LTA. Project management and supervision were by the LTA engineers. Most projects were completed within the time schedule. Delays were caused not by technical matters, but on issues such as land acquisition (in a few cases) and resettlement of residents involved.

Unique issues on our expressways

Some of the unique issues and problems that Singapore had in building and operating the land networks were:

It was mainly an urban expressway network. Hence, it went through many heavily populated areas and there was necessity to keep out pedestrians and cyclists. There were requests to fence off the expressways to prevent pedestrians and even stray dogs from wandering onto the expressway, but this did not prove to be a major problem. Instead, vehicle impact guardrails were installed, next to the shoulders on both sides and also on both sides of the central median. Besides discouraging pedestrians from walking on the expressway, the guardrails also prevented vehicles from leaving the expressway. This has proved to be effective, as is evident from the many tell-tale signs of impact on the guard rails.

Most of the earlier expressways were completed in stages instead of being completed at one go. This meant that there was congestion when the heavy and fast traffic flow from the expressway met the first traffic signal junction, until an interchange was built at that junction. Many motorists did not understand this problem of staged construction and blamed the authorities for poor planning.

Without these interchanges right from the beginning, the presence of traffic signals meant that very high speed traffic had to come to a stop at them suddenly, and this had a tendency to result in some rear end types of accidents, if drivers were not alert. To improve the situation, many of these "interim" expressway approaches to the traffic signals were coated with calcined bauxite chips in an epoxy resin to improve the skid resistance.

Two of the expressways, the Bukit Timah Expressway (BKE) and the Ayer Rajah Expressway (AYE) lead into the Singapore checkpoints for vehicles proceeding towards Malaysia. There are rigorous checks on immigration, customs and security, which result in the service rate at the checkpoints being lower than the arrival rate of vehicles from the expressways. Hence, long queues can extend into the end sections of these expressways on some evenings.

Because it is a dense network, there are many closely spaced interchanges. Therefore, vehicles entering and vehicles exiting have to weave across each other, reducing capacity of the lanes. This is overcome in a few cases by having service roads parallel to the expressways or single lane flyovers over the expressway to connect two expressway junctions which are close to each other, without such traffic having to come on to the expressway.

Maximum speed limit is set only at 90 km/hr and there are some older stretches where speed limits are 70 km/hr. "Reduce Speed" signs are installed at the exits of the expressways into all other categories of roads. Interchanges are large consumers of land. Therefore, our limited land situation also means that the interchange loops have tight radii and hence, the speed limits are brought down to 40 km/hr on some of the loops.

With the exception of cases such as the junction of CTE/PIE, complicated spaghetti junctions have been avoided. The vacant land under the flyovers cannot be effectively used due to problems of accessibility from the main roads. One such location has been used for parking lots, but this needed special service roads to connect to the slip roads on the ground system of the flyover.

The expressways are lushly landscaped with beautiful roadside trees in the centre median. They have a tendency to make the street lighting less effective and at times, mask directional signs.

In an urban setting, expressways often run close to residential areas. As a result, some residents are affected by traffic noise, especially at night. There have been requests from some residents to close off the expressways at night, but this is not practical. Noise barriers are not fully effective as most of these adjacent residences are high rise blocks. Trees are planted to shade these blocks from the expressway, but their effect in noise abatement is minimal.

Future of expressways

Expressways play an important role in our land transport. It is estimated that 55% of all daily traffic flow is on the expressways. Since 12% of the total land area is already taken up for land transport, it is unlikely that there will be new expressways at surface level; instead, there may be some in tunnels and some in viaducts. One of the disappointments of expressways has been that they attracted more usage than they were designed to accommodate. This excess demand reduces the level of traffic service provided. We have to use electronic road pricing to curb this excess demand. The era of the large-scale expressway construction is probably over — the challenge will be on how to use traffic management to get more out of the existing ones.

In Europe for example, the road shoulders are being converted for use as traffic lanes during the peak periods, with proper and safe traffic management. Some other possibilities are operating expressways in a tidal manner, the current three lanes: three-lane arrangement being changed to four lanes: two-lane arrangement in the predominant direction during the peak periods. This can only be done for long stretches. If there are interchanges in-between the operational length of the tidal flow, it will pose problems.

In the United States, ramp metering is used to limit the number of vehicles entering the freeways during the period of heavy usage. Traffic signals are placed along the side roads (i.e., ramps) leading to the freeway. The joining vehicles are only able to enter the freeway when the green light comes on, which happens when the system spots a safe and acceptable traffic gap in the freeway stream. The ramps need to be long to hold the queue of joining vehicles, as otherwise, they will choke up the adjacent road network. Ramp metering increases the efficiency of the main freeway flow at the expense of the side road flow.

2. Traffic Management

2.1. *Providing information to drivers*

Traffic signs

Motorists need instructions on how to use the road network. This requires traffic signs, directional signs and lane markings at the appropriate locations to advise them on what they can and cannot do. A well-thought out and clear methodology of sign installation is a pre-requisite to good traffic management. These signs make driving less ambiguous, less tiring and less hazardous.

The earlier traffic signs and markings followed the British practice. In 1968, the United Nations published its first Protocol on Road signs and markings. The traffic signs were in four categories — white triangle with red border and a symbol or yellow diamond with a symbol as warning signs (e.g., Bend Ahead sign), blue circular sign with a symbol as mandatory signs (e.g., Turn Left sign) and white circular signs in a red border and a red diagonal over a symbol as prohibitory sign (e.g., No right Turn). A blue rectangular sign with a symbol indicated an information sign (e.g., Expressway Ahead sign) (Fig. 6). There were exceptions such as the hexagonal "Stop Sign", the inverted triangle "Give Way" sign and the circular red background "No Entry" sign. Legends (i.e., lettering) were avoided as far as possible.

Singapore adopted the recommendations of the protocol and our signs are international. The peculiar change is in the U-turn sign. Most cities use a prohibitory "No U-turn sign" at every centre opening on a road. The absence of this sign indicates that a driver can make a U-turn at the opening. This would have required a large number of such signs and they are often knocked down when they are erected on narrow central divider of roads. In Singapore, this sign is not used — instead a blue information U-turn sign is used where U-turns are allowed. When there are no such information signs, it is illegal to make a U-turn.

Indiscriminate use of traffic signs can result in clutter, misinformation and eventually, a disregard for them altogether. It is not advisable to fix more than three traffic signs on the same pole due to the burden placed on the driver to comprehend a lot of messages from the symbols. The golden rule is that signs should only be put up if they are relevant and will aid the driver. We have planned and installed a system of traffic signs after a methodical review. This has resulted in better-ordered traffic flow on the road network.

Source: Land Transport Authority.

Fig. 6. Types of traffic signs in use.

Directional signs

Directional signs were rarely used prior to the mid-1960s. They were used more to advertise trade fairs and exhibitions, and were therefore on a temporary basis. However, when new roads were being built and important destinations such as the airport was constructed, such signs became more important. There were two schools of thought on directional signs — that the regular drivers did not need the directional signs as they were familiar with the roads, and the other view that as the road network became more complex, it would be difficult to become familiar with the directions, especially at night. It was thus decided on a directional signing programme.

The first few directional signs were internally illuminated signs, using Perspex boxes in which there were fluorescent tubes , controlled to work by time switches. They were used sparingly. One problem was that Perspex sheets were damaged easily even by loose flying gravel from the road, and that the fluorescent lights, if not maintained regularly, tended to flicker, thus becoming irritants to drivers. One reason for using lighted signs was that prior to the 1970s, drivers drove with only their parking lights switched on unless the street lighting was dim.

It was made compulsory to drive at night with vehicle headlights on in 1970. This meant that retro-reflective sheeting could be used for directional signs (and for traffic signs). Such sheeting works on the principle that when a beam of light is shone on it, the beams are reflected back to the eyes of the driver, hence making the sign very prominent at night.

The expressway network was expanding and needed better directional signs. Directional signs are rectangular in shape with green or blue backgrounds and white or yellow legends. It was decided to use only the English alphabet for the legends, although there were some suggestions of also including the Chinese script. This would have made the signs much larger, and also in a country where there were four official languages, it was decided to stick to a neutral language, English. The legends on the directional signs should be large enough to be read well in advance, and there should not be more than four to five legends on a sign. Lower case lettering (capital and small letters) was adopted because it allowed for better recognition from a far distance. The legends can be arranged in a stacked form — right, left and straight or in map form. The directional sign manual specifies the sizes of letters to be used, the spacing between various letters, the colours to be used and the locations at which they were to be placed. For each destination, it is good to have an advance sign and a confirmation sign near the junction to reassure motorists. A minimum of two signs are used but in some cases, an additional advance sign is added (Fig. 7). Across wide roads, overhead gantry signs are erected.

It is not possible to expect the directional signs to be a street directory. Motorists are advised plan their journey in advance, especially if they are proceeding to an unfamiliar location.

Source: Land Transport Authority.

Fig. 7.   Directional sign.

In the beginning, it was decided to use yellow letterings for the destinations that could be reached by the expressways, but the contrast between the green background and the yellow lettering was poor, resulting in unsatisfactory visibility of the signs, especially at night. A change was thus made to use a blue box with white lettering on the green signs to indicate destinations reachable by expressways. The other new standards were *white background with black lettering* for facilities such as hospitals, and brown background with white lettering for tourist facilities such as the zoo.

The problems in erecting directional signs, which tend to be much larger than normal traffic signs, are the difficulty in finding a place in heavily built-up areas in the city, and the fact that on tree-lined major roads, the tree foliage grows fast and can mask the signs. This requires regular pruning of the branches by the National Parks Board.

Lane marking

In the earlier period, the demarcating of separate lanes was poor, and vehicles did not line up at junctions properly when they were stopped. This resulted in unruly and often dangerous manoeuvres resulting in unsafe situations. Lanes did not line up across junctions, some lanes disappeared halfway and some lanes narrowed down significantly after junctions. In 1974, a massive exercise was held to regularise the width of lanes, mark the lanes to line up across junctions, and convert any excess road not needed to footpaths. These raised howls of protests from motoring organisations such as the Automobile Association of Singapore (AAS) that road space was being taken away from motorists. What was stressed to them was that only excess space that vehicles did not use was taken away. Traffic lanes were to be between 3.0 to 3.7 metres with the left lane, which was used by larger vehicles, and shared with cyclists at 3.7 to even 4.0 metres in some cases. There were claims that drivers will still follow the old practices of not sticking to lanes or switching lanes. This happened at first, but over time, drivers realised they could all move better and in an orderly fashion if they followed the lane markings. This was a tremendous breakthrough in traffic management on the roads.

The initial road markings were painted on the road in chlorinated rubber based paint. They were usually marked out and painted manually at night to minimise traffic congestion. Strings and chalk were used for setting out the markings and were then painted over. Rubber-based paints faded quickly and had to be repainted at a three-month intervals, each time causing disruption to traffic. In the 1980s, thermoplastic paints using machines were introduced. A thin layer of thermoplastic paint that is heated is laid on the road with which it bonds. In some cases, glass beads are laid on top of the hot paint for better illumination at night, but the beads would wear away very fast. Thermoplastic paints can last up to two years and hence, repainting can be done at longer intervals. Thermoplastic paint has better anti-skid properties than a thin film of chlorinated rubber-based paint.

The skid resistance of paint in whatever form is lower than that of the road surface, whether asphalt or concrete. Lane markings do not pose problems because they are narrow, but when larger areas of paint are used for arrow markings, motorcycles which stop on them in rainy weather at junctions may face some skidding problems. The problem was quite critical when rubber-based paint was used, but conditions improved with the use of thermoplastic paint.

Another problem with lane marking is that they are difficult to spot at night in wet weather due to the reflection from the wet surface. There were some attempts to use raised road studs made of ceramic material at intervals along the lanes. This did improve visibility, but motorcyclists who unfortunately tend to ride in between lanes experienced discomfort and complained about skidding. Hence, their use has been limited.

Thermoplastic lane markings have also been used as edge lines on expressways to keep vehicles within the left lane. This is because expressways do not have raised road kerbs.

Besides just lanes and arrows, road markings have also been used for marking yellow boxes at junctions. Yellow boxes at junctions introduced in 1971 have been effective in preventing locking of junctions during traffic congestion. Locking happens at a junction when a long queue of vehicles on one approach stops across the junction blocking traffic from the cross road from moving. Vehicles are not allowed to enter and remain in the yellow box markings unless they are able to clear it before the cross traffic starts to move. This ensures that junctions are kept clear of traffic at all times. Along the junctions of minor roads with major roads, individual lanes (especially the left lane of the major road) were boxed, so as to allow vehicles from minor road to join the traffic on a major road. Box lanes are also used on the left lane at the exit of the bus bay, in order for the bus to get out if there is already a queue of vehicles extending to the bus bay. There is, of course, a fine if a driver is caught in the yellow box holding up the traffic from the side roads.

At some junctions, a common practice is to flare the approaches to provide extra lanes for vehicles to queue up when they come to a stop at the red light. As the vehicles move across the junction when the lights turn green, it is necessary to bring the number of lanes down to the original number. This is done by

Source: Land Transport Authority.

Fig. 8.   Merge arrows.

using merge arrows and merge signs. This system increases junction through-put marginally (Fig. 8).

At junctions, there is a traffic island which separates the left slip road from the straight through lanes. Painted chevron markings are used on the road to ensure that vehicles are led away from the traffic island. Pedestrians seek refuge on the island when crossing at the junctions. When this traffic island is small and less than 5 square metres in area, it is usual to paint the island on the road rather than construct a traffic island which could be too small to be noticed by the vehicles. One danger of using painted islands is that pedestrians also seek refuge on this painted island, which may be unsafe.

Legends (i.e., letters) have also been painted on the road to guide vehicles. This can be in the form of destination names of expressways, bus lane operation hours, 'give way to buses' and so on. As explained earlier, markings on the road may cause vehicles to skid and thus, should be used with care.

Lane markings can also double up for another purpose and be given legal meanings. For example, a single white line while demarcating opposite directions of flow can also indicate no parking on the roads in both directions. Double yellow lines on the side of the road indicates no parking for 24 hours, while single yellow lines indicates no parking from 7.00am–7.00pm. Zig-sag white lines on the side of the road on approaches to pedestrian crossings indicate no stopping or overtaking, double yellow zig-zag lines on the sides of the road indicate no stopping at any time, and single yellow zig-zag lines indicate no stopping from 7.00am to 7.00pm (Fig. 9).

Double white lines drawn on the centre of a road with no centre dividers indicate that the drivers are not permitted to cross over to the other side to overtake. Later on, double white lines have also been used between lanes in the same direction to keep vehicles within a particular lane.

One area of traffic signing that can be often neglected is the locations around worksites along the road; a diamond orange sign is used as the warning sign around worksites to indicate their temporary nature and to highlight these construction sites. These signs together with the barricades, traffic cones, and warning lights steer vehicles safely through worksite locations. These temporary signs are removed as soon as they are no longer applicable. There is a "Temporary signing manual" that has been prepared for this purpose. The manual gives recommendations on the types and placing of the signs.

Source: Land Transport Authority.

Fig. 9.   Double zig zag lines indicating no stopping or waiting at any time.

Junction improvements

There was a major programme to convert roundabouts to signal-controlled junctions. The roundabout works well on the 'give way to the right' rule. This means that a driver has to give way to other vehicles circulating in the roundabout. But once the vehicle is in, the driver has the right-of-way. This system works well if the vehicle numbers from each of the approaches is low or about the same. However, if the number of vehicles from one particular approach becomes high, the other vehicles from other approaches have few opportunities to enter the roundabout; whereas at a signal control junction, every approach knows that it will get some green time to clear the junction. Many of the roundabouts began to give problems when traffic flow increased and thus, had to be converted. This did not pose many problems if the number of approaches were four or less and if the roundabout did not have any beautification structures as water fountains. Two examples where water fountains had to be demolished for conversion to a signal-controlled junction were Farrer Circus (at the junction of Queensway/Farrer Road/Holland Road and Tanglin Circus (at the junction of Tanglin Road/Orchard Road/Grange Road). Newton Circus with eight legs constructed across a canal could not be converted; a flyover was constructed over it to take two main vehicle flows away, and in 1982, traffic signals were installed within the circus without conversion. Similar treatments of installing traffic signals at roundabouts were used at some other locations. Roundabouts added a sense of beauty to our junctions and were generally safer than signalised junctions. On hindsight, there may not have been a dire need to convert all of them; signalising the roundabout may have been adequate.

The use of left slip roads and right turn storage lanes at junctions started mainly from the conversion of roundabouts to signalised cross junctions. The left slip road takes vehicles away from the junction, thus bypassing the traffic signals. The right turn storage lane is a special lane formed to hold the right turn vehicles while they are waiting to get an opportunity to turn. The right storage lane ensured that straight through traffic is not blocked by these right turners. Furthermore, turning pockets are drawn within the junctions, advising the right-turning vehicles on where to wait without blocking the opposing straight flow (Fig. 10).

Traffic signals

Manually operated traffic signals were used by the Police in the 1920s. The first set of traffic lights operating automatically on a fixed time allocation of green times to each approach of a junction, was at the junction, of Bukit

Source: Land Transport Authority.

Fig. 10.   Left turn slip roads, right turn storage lanes, turning pockets and yellow box.

Timah Road/Serangoon Road in 1948. At the first few junctions, there was only one pole in the centre of the junction with four sets of red, amber and green signals facing each direction. The electric controller for operating the signals was on the side of the road. During rush hour, the Police used to be at these junctions to take over, switch off the signals and control the junction manually if necessary. There would be a mobile pedestal on the side of the road which could be placed at the centre of the junction for the policeman to stand on and direct traffic.

The traffic signals operated in red-amber-green and green-amber-red sequence. Green meant "proceed carefully", amber alone meant "prepare to stop" and red meant "stop". The red–amber would appear together before the green came on. This practice continued until the mid-1960s when the red-amber-green was changed to red–green, without the amber appearing. This was because many vehicles started moving when the red-amber appeared before the start of the green. In the beginning, arrow signals were not used. A full green meant that drivers could go straight or turn left. Drivers could also turn right if vehicles in the opposite direction were sufficiently far enough for them to make this turn safely. Otherwise, they had to remain within the junction and only turn right when the amber signal came on. The use of the right green arrow signal to permit this turning was only introduced in the late 1960s. Subsequently, the left green arrow and red and amber arrows signals were introduced to facilitate right turning movements when junctions became more complicated.

The single pole in the middle of the junction, which used to get knocked down by errant drivers, was replaced by traffic signal poles on the sides of the junctions. Each approach had at least one primary pole on its side near the stop line; and at least a secondary pole on the opposite side. Tertiary poles in the form of horizontal poles were introduced in the late 1960s emulating Japanese practice. Pedestrian "green man" and "red man" signals at the junctions were introduced only in 1976, although these pedestrian signals had been in use since the early 1970s at isolated pedestrian signal crossings.

In the city, there were closely spaced traffic signals along the main roads, which if operated randomly would cause unnecessary traffic congestion. It would be preferable to get the traffic signals to operate in some coordinated fashion (i.e., in sympathy) so that vehicles starting from green on the first junction would see the lights turning green as they approach the next junction and so on. This is called the "green wave". The first simple system was set up in 1966, when about 36 junction signals along the city roads of North Bridge, South Bridge, New Bridge, Hill Street, Bras Basah and Stamford Roads had a hard wire connection to a master electric signal controller at the junction of Stamford Road/Hill Street. This master would send out two pulses each minute. Each of the electric controllers at the junctions connected to it would receive these pulses and was required to turn to green for a specific approach at a fixed period of seconds after receiving one of the pulses. This fixed period varied from junction to junction and was set so that the "green wave" could be achieved for a certain road. This rudimentary system worked well for some years. As the city network grew and more traffic signals were installed, no attempt was made to extend the system. One of the shortcomings of the system was that the system operated similarly every hour and did not take into account variations in the traffic flow.

The manual control by Traffic Police had been a regular affair at many junctions even after traffic signals had been installed. The policemen would switch off the traffic signals, place a raised pedestal at the centre of the junction and control manually or use a switch to manually control the traffic signals from the electric controller. One of the problems of such manual control was that the policeman would only try to clear as many vehicles at his junction without regard to what was happening at the adjacent junctions.

Another problem was that the policeman has no "amber indication". A traffic signal would show the amber when the green time has expired irrespective of whether vehicles were still approaching. The policeman will look at the traffic and has to wait for a "thinning" of traffic before he puts up his hand to require vehicles to come to a stop. So when there was constant stream of heavy traffic from one approach, he had problems of deciding when to require them to stop. The amber period at a junction is the clearance period meant to clear the junction of all vehicles before a new movement starts. It is set at three seconds except at large junctions where it could be extended slightly. At many junctions, an appropriate all-red period is also added. There have been a lot of discussions on how long the amber period should be. When any amber period starts, vehicles that are close to the junction cannot stop due to the fear of being hit from behind. Those who are further away cannot clear unless they beat the red light. There is a zone in between called the dilemma zone, where the driver is not sure whether to beat the amber or to come to a stop. If the amber period is extended, the dilemma zone gets shorter, but the lost time at the junction when no green light is showing increases. It is all a compromise setting.

Green arrow signals both for right turn and left turn are common. Green arrows are terminated by flashing them, unless there are amber arrows which, together with red arrows, were introduced later to control complex junctions.

By the mid-1970s, traffic signal controllers were being converted from the electro-mechanical type to microprocessors with a lot of computing capability. Furthermore, area traffic control systems using a central computer connecting to all signal controllers at junctions by telephone copper wire cables and controlling them to provide "green waves" in the traffic system were becoming popular. A cycle length is the time that a traffic signal takes to go through one sequence of red, amber and green times and generally in the range of 60 seconds to 150 seconds. The phase splits are the green times given to different approaches. The offset is the time difference between the start of the green time of adjacent junctions so as to provide a green wave for moving traffic along a particular road. In 1981, the first area traffic control system for the pedestrian signals and traffic signals at 151 junctions in the city was installed with two central computers (one main and one on hot standby) in a control room at the Ministry of National Development (MND) building on Maxwell Road. A library of six plans for each junction based on the traffic count data was installed in the computer, which would be implemented by a time table. So there will be different green times and different times at which each junction turned green relative to the adjacent junction (i.e., different offsets) so as to maintain a "green wave".

The control centre was manned by an operator who did not manually intervene as the plans were all pre-set using an optimisation programme called TRANSYT. An advantage was that faults at the traffic signals were spotted immediately as soon as they occurred, which allowed for the standby maintenance crew to get them repaired within an hour.

The positive results were immediate, and traffic speeds in the city rose by about 20% due to the elimination of random stops at traffic signals. The drivers noticed and complemented the system. The better traffic flow was estimated to have saved about $35,000 daily for motorists due to reduced fuel consumption. The Central Fire Station is located at Hill Street. There were special pre-emption plans that the operator at the Fire station could call for that would override the system, and give special "green wave" for the fire engines at about six of the nearest junctions. The Fire Service mentioned that this helped them to save a significant three minutes in getting to a fire site.

Motorists generally agreed that the system was working well. The dual computers at the control centre ensured a full standby and hence, breakdowns were rare, and when the system did go down once in not coordinating the signal operations, many calls were received saying that something was wrong. This attested to the fact that drivers were able to notice when the system was not working well. It can be said that this was the single most effective traffic management scheme introduced in Singapore.

All manual police control stopped after the Area Traffic Control (ATC) implementation. The Traffic Police only took over junction control in cases of emergencies and during visits by foreign heads of state when large convoys of cars move in unison and may need additional green time so as to not be separated.

The pre-set plans had to be updated every two years as they would otherwise be out of date, especially when traffic volumes are ever increasing. Updating was a laborious process because it needed collection of traffic data from each junction to work out the new pre-set plans. The pre-set plans also assumed stable conditions and could not accommodate breakdowns, accidents, and roadworks. The pre-set plans were not traffic-responsive because not enough automatic data was being collected by the system.

Vehicle actuated systems had been in use in Singapore from the early 1970s. Instead of fixing the time for each green period, vehicle actuated signals used wire loop sensors placed under the road surface to detect vehicles at the approaches. This information was used to gauge demand and apportion the green time accurately. So if there was no demand from a particular approach, it would not get any green time.

The next step was to upgrade the "green wave" system from a fixed time system to a traffic responsive system. This required the placing of wire loop detectors under the road surface at approaches to all junctions to gauge demand and at the same time, provide "green waves" as necessary. The upgrading was done in 1988 and is called the Green Link Determining System (GLIDE) based on special software from Sydney, Australia. The system implements the best phasing (i.e., cycle times, phase splits and offsets) for the current traffic situations for individual junctions as well as for the whole network. Incremental adjustments are made to the signal timings based on data coming in from the junctions. This is based on the automatic plan selection from a library of plans in response to the traffic data derived from loop detectors. The system did not produce more green time. There is only one hour of total green time to be shared among the various approaches. It ensured that the green was shown to the vehicles that required it most; and at the same time, the system kept the "green wave" as far as possible. The improvement over the fixed time system was about 4% which drivers would hardly notice. But the system was much more flexible and could cater for unusual situations such as accidents and breakdowns. From the 1990s onwards, the system was extended to all junctions and pedestrian signals on the island (Figure 11). The control centre with the main computer is at the Intelligent Transport Control Centre on River Valley Road. There are 12 regional computers under the control of the central computer, located usually under the green space below flyovers of expressways controlling the 12 regions to cover the whole island. With this decentralised system, the telephone lines connecting to the signal controllers are shorter, thereby ensuring reduced losses in communications and reduced costs.

The first traffic lights used incandescent bulbs in a reflector. The life of the lamp was about nine months, and bulb failures were common. Lightning strikes common in Singapore often blacked out the traffic signals. Signal poles were damaged by accidents. In the 1960s during Singapore's turbulent years, traffic signal poles were favourite targets of rioters. The other problem with incandescent bulb reflectors was that when a person was travelling in the east-west or west-east direction and the sun was directly behind him, the traffic signals looked like they were all lighted up due to the "sun-phantom" effect.

In 1976, in conjunction with the Area Traffic Control for the city, the incandescent bulbs were replaced with halogen bulbs which saved energy and also were much brighter, thereby reducing the "sun-phantom" effect. A new preventive maintenance system was devised — the green and red signals were

Source: Land Transport Authority.

Fig. 11.   The green wave.

relamped every nine months, and the amber signals at 18 months (due to its lower usage) irrespective of whether they had failed. The maintenance contractor was paid for the re-lamping, but if any lamp failed at any time, the contractor had to attend to it without payment. The contractor was also to service the microprocessor controller and clean the signals on each of his scheduled visits.

In 2000, there was another upgrade of replacing the halogen lamps with Light Emitting Diode (LED) lamps. These are even more energy efficient, are much brighter, and lamp failure rate is very low. All installations are now also protected against lightning strikes.

At large junctions and junctions located close to each other; there is a possibility of the driver spotting a signal not meant for him. In such cases, long hoods and louvres are used to mask the signal from those who do not need to see it.

The red light always had only one meaning, that is to stop. Some cities use the Left Turn on Red (LTOR) system (or Right Turn on Red in countries that drive on the right side of the road). This means that a driver can turn left against a red light if there are no cars that will collide with him on the cross road, and if there are no pedestrians crossing. Owing to pressure from motorists who had seen such schemes overseas to try out such a scheme to improve traffic flow, 44 minor signalised junctions were permitted with LTOR.

LTOR is a privilege, but some drivers took it as a right and this had caused problems. In some cases, pedestrians were startled and upset to see the left turning vehicles cross against them when they were having the "green man" indication. There were some instances when the first vehicle did not want to turn left against a red light and the following vehicle sounded the horn. This prompted complaints from residents who were irritated by the frequent sounding of horns. Based on such feedback, many of the LTOR junctions have been converted to signals that have left green arrows, which does not give the left turning vehicle a blanket right to turn against a red light, only when the left green arrow appears. There are only about 16 LTOR junctions at minor road junctions. The LTOR has been a failure in the local context and on hindsight, we should not have bowed to pressure to introduce it.

2.2. *Pedestrian facilities*

Pedestrian footpaths

The earliest town plan for the island demarcated land for roads, schools and government buildings, and an accompanying ordinance stipulated that all houses constructed of brick or tiles have a common type of front, each having an arcade of a certain depth, open to all sides as a continuous and open passage on each side of the street. This led to the foundation for the five-footway, which is still prevalent in Singapore along the frontages of many shop houses. The five-footway was the fore runner of pedestrian footpaths. When there were no shop houses, there were no footpaths and pedestrians walked on the road. This was not a problem when the number of vehicles on the roads was low.

The newer roads built in the 1950s and later had footpaths because the Master Plan of 1951 had set up standards for the road reserve. In an effort to cut down on the space used for roads, new standards were set up in the 1960s where the roadside drains were covered with concrete slabs to serve as footpaths.

A concerted effort to improve pedestrian safety started in the 1970s. This involved surveying old roads and whether excess space was available to be converted into footpaths. Where no such space existed, existing open roadside drains were provided with concrete slabs to serve as a footpath. The five-footways of shop houses were upgraded and the abutting roadside drains were provide with concrete slabs to provide a wider footpath.

The height of the road kerb is about 150 milimetres above the road level. Therefore the footpaths were a step higher than the road. Not much attention

Fig. 12.   Kerb cut ramps at footpaths.

was paid to the needs of the pedestrians in the earlier days. As long as a foot-path segregated them from the vehicles, it was considered adequate. Senior citizens and the disabled had problems negotiating the footpaths. Singapore's first disability code appeared in 1990 for buildings, and by 1997, the code also required street furniture to be friendly. Kerb cut ramps were provided where the footpath met the road (Fig. 12). These ramps brought the footpath to the road level removing the obstruction of a high kerb. This has now become a standard requirement for all footpaths.

Pedestrian crossings

The first form of pedestrian crossings were just two parallel white lines across the road. Pedestrians crossed at their own risk. At times, police were on duty to help them to cross. In 1951, pedestrian zebra crossings were introduced in Britain, which worked on a first-come-first-served basis. They were introduced to Singapore roads in the mid-1950s, with it being a common sight at many roads where there were human activities. The zebra crossing uses wide zebra stripe markings across the road with two flashing yellow beacons. There are stop lines in both directions where the vehicle has to stop to give way to crossing pedestrians. An experiment was carried out in the late 1980s to replace the beacons with solar powered LED lights on a circular disc. Over time, it was found that the lights became weaker due to insufficient charging of the battery by solar power. It was then decided to revert to the use of the normal beacons, powered from the electric mains.

When there were many pedestrians who were crossing, the zebra crossings held up traffic on major roads like Paya Lebar Road and Yio Chu Kang Road. In order to alleviate the situation, the pedestrian signals which shared the green time between vehicles and pedestrians with marked pedestrian crossing lines (cross walk) were introduced in the late 1960s. These replaced some of the zebra crossings. These pedestrian signal crossings were termed as mid-block crossings at they were not at the junctions. There were campaigns to teach pedestrians how to use these crossings. A "red man" meant 'do not cross', a "green man" meant 'cross' and a "flashing green man" meant 'do not leave the kerb to start crossing, but to complete crossing if already on the crosswalk'. The "flashing green man" time was set so that a pedestrian who had started the crossing just as the flashing started would be able to get to the kerb on the opposite side before the red man appeared. Pedestrian pushbuttons were provided for pedestrians to call for the "green man" to appear. If there was no demand then the green would remain with the vehicles.

In an education programme, large footsteps in yellow were marked on the pedestrian crosswalk to tell pedestrians where to cross. These were made large to attract the attention of pedestrians. Pedestrians always had problems crossing at junctions even with the normal traffic lights as they had to do in stages and look out for gaps in the traffic stream to cross. This was tiresome and dangerous. In 1975, the authorities made a thorough review of pedestrian safety, out of which came more pedestrian footpaths, pedestrian malls and pedestrian crossing facilities. It was complemented by the anti-jaywalking legislation, which required pedestrians to use a designated pedestrian crossings such as zebra crossings, pedestrian signal crossings, pedestrian bridges and pedestrian underpasses if they were within 50 metres from where they were crossing.

But this also required that traffic signal junctions become pedestrian crossings. The Highway Code had a rule that turning traffic has to give way to pedestrians. This resulted in a scheme to incorporate the pedestrian "green man" and "red man" signals to normal traffic signals. The "green man" time lasted for part of the time that the green for the parallel vehicle movement appeared. Vehicles who were turning right or left had to give way to the pedestrians who were crossing (i.e., with the flow of traffic). In other words, the turning traffic shared part of the green time with the pedestrians. The straight traffic was not affected. Traffic could turn freely if there were no pedestrians crossing, but had to give way if there were pedestrians crossing. The "green man" for pedestrians only lasted for part of the green time with the "red man" appearing later so that vehicles could turn without being impeded. Whenever a green arrow appeared for any turn, the "red man" came on to stop pedestrians.

This scheme was first implemented in 1976 at the junction of High Street and North Bridge Road. Police were at hand and there were signs explaining the scheme for about three months, after which motorists and pedestrians became familiar with it. With this system, every junction with traffic signals became a safe pedestrian crossing. It was the single most effective pedestrian safety scheme introduced in Singapore. There were comments on the scheme being dangerous because pedestrians and turning traffic shared the green time. The old system of pedestrians looking for gaps to cross at junctions was even more dangerous because they did so without any guidance. Another option was to stop all vehicle movements and give the pedestrians their own phase, where they could cross in all directions including diagonally. This option called the "scramble" was tried out at one junction in the city, but did not find favour with pedestrians and was not proliferated.

The pedestrian signals have gone through many evolutions during the past 40 years. The Visually Handicapped Society asked for beeping sounds at the pedestrian signals so that their members could use them safely. The first such one was tried out near their headquarters at Thomson Rise. The pedestrian signals gave off a beeping sound for the visually handicapped to sense its presence. During the operations, there would be a different tone of the beeping sound when the "green man" flashed so that they will know they still had time to cross.

The next improvement was the fixing of countdown timers for pedestrians at these signals in the late 1990s. When the flashing green man appeared, the countdown from the number of seconds that the flashing would operate (e.g., 20 seconds) to 0 seconds started. This gave the assurance to the pedestrians that they had sufficient time to complete the crossing. It also helped to speed up straggling pedestrians (Fig. 13).

While the able-bodied pedestrian had no problems using the crossing, the senior citizens found they needed more time to cross. There is now a scheme at some signals whereby a senior citizen can use his public transport fare card (issued to all over 60 years old to get discounts when travelling on public transport) to tap the pushbutton to get an extra five to eight seconds. This is being extended to all pedestrian signals and is called "green man plus" (Fig. 14).

Fig. 13.   Countdown timers.

Fig. 14.   Green man plus.

With the introduction of pedestrian signals, zebra crossings were relegated to minor roads or left slip roads. All slip roads are provided with zebra crossings. The zebra crossings are either placed directly under street lights or flood lights to improve visibility at night. Zebra crossings are provided liberally across minor roads at areas with high pedestrian activity.

However, for other higher level of crossings such as mid-block pedestrian signal crossings, pedestrian overhead bridges and underpasses, there was a need for a warrant or justification. There are many ways of coming up with this warrant which is empirical in nature. Conflict points are a method of doing this. Conflict points are derived by multiplying the vehicle traffic volume per peak hour that the pedestrian has to cross in one movement (V) (hence the maximum one directional volume for roads with centre dividers and two directional volume for undivided roads) by the number of pedestrians crossing within that hour along a stretch of 50 metres (P) from either side of where the crossing is to be located. Usually for a normal 50 km/hr speed limit road, if PxV exceeds 300,000 (called a warrant W) a crossing is warranted. For a 70 km/hr road, the W is reduced by a ratio of 5/7 and so on. If the majority of crossing pedestrians are children or senior citizens, W is reduced by multiplying it by 0.75.

After justifying a crossing, then it is necessary to decide on the type to be provided, either at surface as a mid-block pedestrian signal or grade-separated, based on the road and traffic conditions.

Pedestrian crossings can be grade separated, that is at different levels than the road. The first pedestrian overhead bridge was a steel structure spanning Collyer Quay in 1965. It was later converted into a shopping bridge to connect the old Change Alley to Clifford Pier. The height of the bridge above road level has to be 5.4 metres to ensure that over-height vehicles do not hit the bridge when traveling. The highest vehicle on the road is the double decker bus at 4.3 metres. The height limit sign of 4.5 metres is fixed on the bridge. There are about 36 steps to climb to use the bridge. The earliest bridges, up to the end of 1970s were all steel bridges. In order to beautify their stark appearance, metal claddings were used, but were later removed in favour of planting creepers along the columns and the railings. Sometimes these attracted monitor lizards and led to complaints from users.

There were also instances of over-height vehicles hitting the bridges and damaging them or bringing them down. The usual culprits were overloaded trucks traveling during the late nights or mobile cranes whose drivers forgot to bring their boom down before setting off. The problem got so bad that the legislation had to be changed to deter such irresponsible behaviour. Currently, any driver who causes the vehicle to collide with any building or structure over the road could be liable for a fine not exceeding $5000 or an imprisonment for a term not exceeding two years, one of the stiffest penalties ever.

In the 1980s, prestressed concrete bridges were introduced. They were sturdy and presented a sleek appearance, and a lesser possibility of dislodging in the event of an impact. Purpose-made flower troughs were provided for planning bougainvillea flowers (Fig. 15). Later on, covers were provided to some of them and all were lighted to ensure security at night.

Constructing pedestrian bridges in a mainly urban environment posed many problems. They cannot always be located at exactly where they are needed due to the presence of utility service lines such as water pipes, and electricity cables on the sides of the roads, where the bridge foundation has to be. Another issue is that many residents object to the presence of a nearby bridge because it overlooks into some of their rooms, and to prevent this, special masking screens have been used.

As the population has aged since 1965, many of the senior citizens are unable to use the bridges because of the many steps to climb. In some instances, concrete ramps have been provided in addition to a staircases.

Fig. 15.   Pedestrian overhead bridge.

These ramps take up much more space than staircase and they are very long, prompting complaints of the long distance that pedestrians have to travel to cross a road.

Escalators were provided when the bridges connected to buildings had security personnel. When an escalator breaks down and stops, the steps are more difficult to climb. If security personnel are around, they are able to set it in motion again. On free standing isolated bridges across roads, it will take much longer to get the maintenance staff to arrive and put the escalator back into operation. Hence, escalators are not used for bridges outside buildings. Of late, a few bridges are trying out lifts (elevators) after having some experience with using them at MRT stations. There are more than 400 pedestrian bridges in Singapore and they do pose problems for senior citizens and the disabled. Perhaps the programme should have been reviewed at mid-stage.

The first pedestrian underpass across Fullerton Road connecting Empress Place to Queen Elizabeth Walk was also built in 1965. It was a walk-through underpass in that pedestrians did not have to descend. The later underpasses were across East Coast Parkway when it was built during the construction of the road. Two major underpasses in the city area built in the 1970s were the ones across Orchard Road at Plaza Singapura and across Scotts Road at Shaw House. The advantage of the underpass when compared to the overhead bridge was that its depth was governed by the depth of the utility service lines that it had to go under and not by height control of 5.4 metres. Hence, fewer steps will be required for pedestrian underpasses. Underpasses are popular in crowded areas but in lonely locations, many pedestrians shun them at night due to fears of security. However with the vast number of underground MRT stations being built, their usage is getting more popular. Underpasses are more expensive to build than overhead bridges.

The first underpass to have escalators was the one at Scotts Road in 1982. Drainage sumps will be required to collect, store and pump out rainwater to the nearest roadside drains. The entrance to any pedestrian underpass has to be significantly above the last known flood level, especially if the underpass connects to the basement of a building. Otherwise, flood waters will flow into the basement.

With the technology for underground space under roads and junctions being quite developed mainly as a result of underground train line construction, the concept of underground malls started with the city mall construction in the early 2000s. More and more of such malls are being proposed in the city area.

Source: Land Transport Authority.

Fig. 16.   Covered link ways.

Accessibility for pedestrians mean that they are able to get to their final destinations without hassle; and without being dangerously exposed to vehicles. Pedestrians need to get to buildings and transport terminals such as bus stops and train stations. These are all linked up with footpaths or walkways, and when crossings are required, safe crossings are provided. Covers are provided in many cases so that pedestrians can walk sheltered from the sun and the rain (Fig. 16).

Pedestrian safety and convenience have featured very high in our priorities. The pedestrian problem is a difficult one in a highly urbanised situation faced in many cities. Traffic engineers need to spend a large amount of time and effort to alleviate this problem so that a city does not become hostile to the very people who bring life to it.

2.3. *Making best use of roads*

Jackson's town plan of 1823 had laid out the streets in a gridiron fashion and this facilitated the conversion of many of them into one-way street pairs. About 60% of the road network in the city works as one-way pairs. One-ways have the advantage of simplified junctions, easier linking for closely spaced traffic signals to provide the "green wave" and the ability to make full use of roads with an odd number of lanes. Traffic flow is thus smoother on one-ways. However, the distance between the pair has to be less than 200 metres because one-way pairs inconvenience bus passengers who have to walk further to their bus stops.

Nicoll Highway with Merdeka Bridge had two lanes in each direction with a very wide centre median when it first opened in 1956. This was Singapore's first high speed road and it soon became the most popular route from the east to the city. In 1968, the wide median was converted into three lanes which allowed for the tidal flow to give priority to the main traffic flow. The centre three lanes were operated by a swing gate. From 12 midnight to 12 noon, the three lanes operated towards the city — there were five lanes to the city and two to the east. From 12 noon to 12 midnight, the three lanes operated away from the city — there were five lanes to the east and two to the city. When the East Coast Parkway (ECP) opened in 1981 connecting the east and the city, the prominence of Nicoll Highway diminished although the tidal flow system continued until 1994, when Merdeka Bridge underwent reconstruction and Nicoll Highway was converted to a dual three-lane road.

The next attempt at tidal flow was on the approach to Jurong Island from Jurong Pier Road in 2008. Jurong Island is a high security area with many petrochemical industries. there are many security checks at the entry check-point to Jurong Island. This resulted in long queues along the approach road (i.e., Jurong Pier Road) which extended into the nearby expressway. Jurong Pier Road has two lanes leading in and out at the surface level and a dual four lane flyover to it. During the morning peak period, the number of vehicles leaving Jurong Island is low. So under the tidal flow scheme, the four lanes on the flyover lead towards Jurong Island in the morning — the operation is six lanes to Jurong Island and two lanes leading out from 6.30am to 9.00am. This system reduces the queue length on Jurong Pier Road, because there is more queueing space, although it does not make clearance of vehicles any faster at the checkpoint (Fig. 17). There is no need for tidal operation in the evening as there are no rigorous controls for vehicles leaving Jurong Island.

2.4. *Handling roadside friction*

Road lanes are meant for the passage of vehicles, but there are many roadside activities because the road is a public space. This causes roadside friction. The first of this is kerbside parking and loading/unloading. This problem was identified as early as 1930s and it still persists. Road side parking has been

Source: Land Transport Authority.

Fig. 17.   Tidal flow towards Jurong Island.

slowly phased out on all but the less important roads. Their demise came when bus lanes were introduced on all major roads. There have been and still remain a few areas with shop houses with commercial activity, which requires the loading/unloading of goods at the roadside. If such activities are completely prohibited, the businesses will suffer. A compromise was thus reached to allow loading/unloading activities during the off-period hours at these places. In all new building complexes, the developers have been required to provide loading/ unloading bays within the building since the 1990s, so as not to exacerbate the problem (Fig. 18).

Right from the beginning there were large warehouses, godowns and wet markets along the heavily populated areas in the city. On the roadside outside these areas, there was chaos when vehicles double parked and blocked many lanes. By good land use planning, these were relocated to the outer areas with

Source: Land Transport Authority.

Fig. 18.   Off-street loading bays.

proper loading/unloading facilities. Another problem was the presence of tyre, battery and audio shops along the roads where vehicles have to wait to be serviced. Denying the shops of the use of the roads would have affected their livelihood. Over a period of 40 years, these types of activities have been relocated to the new towns along special areas where they do not cause traffic congestion to main road traffic.

Itinerant hawkers selling their wares along the roadside have been a tradition in Singapore right from the beginning. Their stalls, make-shift chairs and tables used to occupy the roads and footpaths. Likewise, they have all been relocated to food centres in hygienic conditions away from the roads in the past 40 years. Today, there is popular alfresco dining but the tables only occupy part of the footway and not the entire road.

A problem that has yet to be solved is the roadside friction near primary schools during the school starting and ending hours when parents drive to drop or pick up their children. Majority of the Primary school children either use school buses or walk (if the school is close by) or are driven to schools. In general, more Secondary school students use the public buses. Schools do not generally allow the cars of parents to enter the school compound unless it is raining. It is difficult and unpopular to take enforcement action against these parents for the parking and waiting problems they are causing. Fortunately, the six schools that were in the city have been relocated to outer areas over the past 40 years, mainly because land values in the city are very expensive and the schools added to the congestion in the city.

2.5. *Speed control*

Speed control is an essential part of traffic management. Speeds themselves are not dangerous unless it is reckless speeding. However, higher speeds in traffic accidents cause greater damage or more injuries.

There are certain locations such as near primary schools and minor residential roads where it is desirable to get vehicles to travel slower. One way is by imposing lower speed limits than the 50 km/hr speed limit on urban roads. Unless there is good obedience, speed limit signs may not be effective. The other ways are to install road humps or speed regulating strips across the road. These physical measures force vehicles to lower speeds.

Speed humps are short raised portions of the road. Short and high humps may cause problems to unwary standing passengers in a bus. Bus friendly humps are wider with gentler slopes. The other option is speed regulating strips which are short raised thermoplastic strips placed at short regular intervals across the road. Vehicles going over them get an uncomfortable feeling and thus slow down. However, they could be irritating to nearby residents who are disturbed by the rattling sound when vehicles speed over the strips.

2.6. *Bus priority measures*

Buses are one of the main modes of our public transport system. In the city, they account for about 50% of public transport trips, the others being train and taxi. The bus will continue to remain as a popular mode due to its flexibility and its convenience. Bus speeds are only about 55% of the speed of other vehicles as they have to call at bus stops. Buses travel in the mixed traffic stream and are subject to any delays caused in the stream. The other location where they get delayed is at bus stops, especially when many passengers board the bus.

Bus lanes

Traffic management schemes are necessary to speed up the movement of buses. One of the major projects is the designation of the kerbside lane as a bus lane on all major roads in the city, where bus volumes exceeded 40 buses per hour during the peak hour. Only scheduled buses and bicycles are allowed to use the bus lane. In most cases, bus lanes need only operate during the morning and evening peak hours. This is important as there will not be many buses on some roads in an exclusive bus lane during the off-peak hours, which will lead to resentment by other vehicles who see an underutilised bus lane being denied to them. The normal bus lanes operate from 7.30am to 9.30am and from 5.00pm to 8.00pm on weekdays, except on Sundays and Public Holidays. However, when bus volumes are consistently high, full day bus lanes operating from 7.30am to 8.00pm, from Monday to Saturday, except Public Holidays, are provided. Bus lanes are identified by road markings and traffic signs.

Bus lanes have speeded up bus movements by at least 15% as buses are able to travel from junction to junction or from junction to bus stop unimpeded. There is no doubt that other vehicles experience a deterioration in traffic conditions when one lane is taken away for the exclusive use of buses. The only justification for the scheme is that traffic management should be aimed at moving people, not vehicles. A bus can carry between 60 and 120 passengers whereas an average car in the city carries 1.2 to 1.4 passengers. Therefore, a bus lane carries more people than any other lane.

Bus lanes, which are drawn on the left lanes of roads, have other negative impacts. They cause problems for left turning vehicles at junctions. Usually a short break is provided in the bus lane just before the junction for left turning vehicles to enter and form up to turn left. If the bus lane is fully utilised, these left turning vehicles do not get an opportunity to enter the lane and miss out on turning. This has caused resentment from drivers of other vehicles. Therefore, when left turning volumes are high, the bus lane provides longer breaks for left turning vehicles to enter to turn left.

Bus lanes are cheap to implement and are very effective. They produce immediate results and the bus passengers see a very visible scheme that helps them to move faster. We started marking bus lanes as early as 1974. All major roads with high bus volumes (more than 40 buses per hour) have a bus lane if the number of lanes in each direction is at least three (Fig. 19).

Contraflow bus lanes which have bus lanes in the opposite (contra) direction to main flow were studied, but found to not be suitable for our traffic conditions. Bus rapid transit systems were not considered because mass transit railway was the preferred option.

Bus priority at junctions

Bus lanes give priority to buses between junctions but not at junctions. A way to give priority to buses is to allow buses to make movements that are prohibited for

Source: Land Transport Authority.

Fig. 19.   Bus lane.

other vehicles at junctions. For example, at a busy junction, all right turns may have been banned in an effort to speed up traffic flow. Buses could be exempt from such a restriction, permitting them to move faster than other vehicles.

Buses can also be given priority at traffic signals. The simplest system is to allow the bus to get the green light (or the green B-signal) before the other waiting vehicles. This six to eight second headstart permits buses waiting at the stop line to move out and form up in the correct lanes before others start moving. This is especially useful if buses have to turn right at the next junction. If the buses had moved together with all the traffic, they would have had problems weaving across the moving stream of traffic to turn right at the next junction. However, the usefulness of B-signals is limited if the first vehicle waiting at the stop line is not a bus.

Minimising Delays at Bus Stops

Proper location of bus stops is another traffic management measure that can be used to speed up buses. Our buses have been one-man operated since 1975. This meant that the driver (called the bus captain) has to collect or check the fare. The use of automatic fare cards has speeded up times for the boarding of buses.

Nevertheless, when too many buses call at one bus stop, bus queues form and the delay experienced by the first bus is added on to the delay experienced by other buses, which are in the queue.

These bus queues can be shortened by providing more kerbside space at bus stops. For example, bus stop locations where many passengers catch buses are staggered so that half of the services call at one stop and the other half at the next stop. This is popularly called "leap-frogging" operations. Alternatively, where space permits, multiple parallel bays can be provided at the same bus stop so that buses can stop to drop or pick up passengers simultaneously.

It was common to provide bus bays along major roads to require buses to stop away from the left lane. Nevertheless, buses also experience some delay when they pull out of the bus bay at bus stops because other vehicles refuse to give way to them. It is estimated that the delays caused can add up to 10% of bus journey time. All buses carry stickers at their back exhorting other vehicles to give way to buses emerging out of bus bays. This appeal has had a fair degree of success. Legislation has been enacted to require vehicles to give way to buses getting out of bus bays when the necessary markings and signs are installed (Fig. 20).

Bus shelters always come second best when compared with train stations. To make the passenger experience better, bus shelters have been upgraded with

Source: Land Transport Authority.

Fig. 20. Give way to buses exiting from bus bays.

better designs, comfortable seats, and lighting and information panels on bus services. Tastefully designed and lighted advertisement panels are positioned without impeding passengers' movements or view of approaching buses. The panels have been let out on a contract basis for a fixed number of years in return for the advertising agency's regular cleaning and maintenance.

Since 2008, bus arrival panels have been provided at popular bus shelters to provide real-time bus arrival information. The computation of the predicted timing for bus arrival is carried out by the Public Transport Operators (PTOs) while the Land Transport Authority (LTA) control centre provides the platform to integrate information and display them on the panels. The purpose of the information, (which is accurate to plus/minus two minutes) is to manage the anxiety of waiting passengers (Fig. 21).

2.7. *Goods vehicles*

The earliest goods vehicles were bullock carts. Later on, pick-ups, trucks and containers took on this role. Due to their poor maneuverability and larger sizes, these vehicles tend to move slower. They were always advised to keep to the slow lane which is the extreme left lane of the road. The problems of loading/unloading have been discussed. Another problem was the overnight parking (12 midnight to 6.00am) of these vehicles. Usually they are hired during the daytime and they have to find a place to park at night. The parking lots provided in the new towns are mainly for residents and visitors. Designated parking lots for goods vehicles are rare. There is also a practice of some companies to allow

Source: Land Transport Authority.

Fig. 21. Bus arrival information signs.

their drivers to take the goods vehicle home so that they can report to any work place on the next morning without having to go to the company to pick up the vehicle. This meant that the drivers used the goods vehicles as personal transport and parked near to their homes, usually on the narrow roads of private housing estates where there were no parking restrictions, causing nuisance. Some even dared to park on the shoulders of expressways overnight which resulted in a few accidents in the late 1980s.

In 1995, a vehicle parking certificate (VPC) system was introduced. Under this scheme all owners of goods vehicles with maximum laden weight greater than five tonnes had to furnish proof of having an overnight parking lot. These heavy vehicle parks which are located some distance away from residential areas are provided by the public and private organisations. The owners are able to rent a parking lot in these heavy vehicle parks to obtain a vehicle parking certificate, which confirms that the particular vehicle has an overnight parking lot available when he renews his annual road tax. He has also to produce the VPC when he buys a new goods vehicle. Heavy vehicle drivers who were used to having the convenience of parking their vehicles close to their homes are unhappy as they now have get to the heavy vehicle park early in the morning and get home after parking the vehicle in the special lot which could be some distance away from their homes (Fig. 22).

Source: Land Transport Authority.

Fig. 22.   Heavy vehicle parks.

Another problem is rat running, especially by goods vehicles. This is caused by vehicles taking short cuts through narrow residential roads in order to avoid major roads and traffic jams. Those who travel overseas will find that many of their taxis resort to rat running. This can be prevented by having a clear hierarchy of road types and by discouraging such rat runs by making theses short cut routes more difficult to negotiate. Some roads in these estates are made no-through roads or have access cut off from some main roads. The residents may face some inconvenience, but their environment is more peaceful. Goods vehicle over 2.5 tonnes are prohibited from using some minor residential roads by prohibitory signs.

2.8. *Motorcycles*

About 15% of the vehicle population consists of motorcycles, mainly owned by the less well-off members of the population. The motorcycle is an unstable vehicle and the rider and the pillion have little protection in an accident. Their main issue is road safety, especially when they are using high speed roads such as expressways. When the first two expressways were opened in 1981, there was a proposal to look into restricting motorcycles from using expressways in the interests of their own safety. But in those days, Changi Airport, which employed many people, was accessible only through the two expressways and banning motorcycles would have caused a lot of hardship.

Motorcyclists have a tendency to ride between lanes and keep close distances to vehicles in front when the red light appears. Motorcycle lanes were never considered as an option to stop this practice. Protection of motorcycles on expressways and tunnels by rain shelters have been discussed earlier in the chapter.

Another safety problem faced is when making right turns at signalised junctions. Their small size and their propensity to accelerate to turn right before the opposing traffic starts when the green appears, often recklessly, tend to involve them in accidents. Their smaller sizes also make them less noticeable by other vehicles.

2.9. *Bicycles*

Bicycles were popular in the 1950s. It is a popular and cheap means of transport. Parents used to send their primary school children to school by carrying them on the central bar. Secondary school students cycled to school and many schools had covered sheds for parking of bicycles. Some roads like Macpherson Road were built with separated cycle tracks. Owners had to register their bicycles with the Registry of Vehicles.

Unfortunately when the car population increased and public transportation improved, the usage of bicycles decreased and many of the tracks were converted into roads for cars. Recreational cycling still continued to be popular. A recent survey showed that one in two households own a bicycle. National Parks Board (NParks) and the Public Utilities Board (PUB) started the beautification of drainage reserves and converted them into park connectors for pedestrians and cyclists.

Cycling started to pick up again when the trains started operating. Passengers started cycling to the stations to catch a train, popularly known as bike-and-ride. Hence bicycle lots became common in train stations (Fig. 23).

Passengers on the trains and buses are allowed to bring in foldable bicycles during the off-peak periods of Mondays to Fridays, 9.30am–4.00pm and after 8.00pm, and on the whole day on weekends. The scheme has yet to catch on.

Most cyclists now consider that the road is not safe for them and thus cycle on footpaths, which brings them into conflict with pedestrians. Cycling on the footpath is illegal; however Tampines Town started a scheme

Source: Land Transport Authority.

Fig. 23.   Bicycle parks.

by widening the footpaths for sharing between pedestrians and cyclists, and changing the bylaws to make it legal for cyclists to use the footpaths. The initiative is working well.

Currently, there are plans to further encourage more cycling and many initiatives can be expected. These will be a mixture of separate bicycle tracks and widened shared footpaths.

2.10. *Roadside and off-street parking*

Parking by the roadside had always been a common practice. This gave rise to the start of "jaga kereta" boys. These were street urchins who stationed themselves at popular spots, gave directions to drivers to park their vehicles and agreed to guard their cars in return for a small tip. As unemployment was rife, every popular spot had "jaga kereta" boys . Giving a tip was voluntary, but there were complaints of cars being vandalised if the tip was not given.

The Parking Places Order was invoked in 1965 which allowed roadsides to be designated as parking places and for the government to collect a charge for parking of cars. Paid parking was to regulate parking to meet demands. It provided jobs for many as parking attendants, at a time when unemployment was high. It also reined in menaces such as "jaga kereta" boys and illegal itinerant hawkers who occupied roads.

The job of the parking attendant, usually young ladies clad in blue uniforms, was to issue advice notices when cars entered the lot, collect parking fees and issue a receipt when the car left the lot. They would position themselves in the middle of the area under their charge or at the entrance of the car park. The initial charge was set at 25 cents per hour.

This was later changed to coupon parking. Motorists had to buy and keep paper parking coupons and tear the tabs for the year, month, date and time of start of parking on individual coupons and display them on the dashboard of vehicles. Each coupon came in denominations of 50-cents and $1.00. A 50-cents coupon allowed parking for 30 minutes, for example. Today, roadside paid parking is only allowed on minor roads. All roadside parking is controlled, the parking rates are determined and the revenue collected by the Urban Redevelopment Authority (URA). On local access roads, parking is free. Other car park lots are in off-street car parks — surface, multi-storey or underground. There are very few stand-alone parking garages, most of them are part of building complexes. The Parking Planning provisions require all new building developments to provide their own parking lots according to certain minimum standards.

The Housing and Development Board (HDB) controls car parks in the new towns. Other off-street car parks are in private hands and the parking rates are determined by market forces, with no government involvement.

Most off-street car parks permit season (long-term) parking and hourly parking. Season parking is for monthly parking and is meant mainly for tenants of the buildings. Hourly parking is meant for visitors.

Many off-street car parks are now using electronic parking systems, using physical barriers for entry and exit, which is explained later under intelligent transport systems.

2.11. *Blocking of roads and footpaths*

Piped water had been available since the mid-1850s and the sewerage system started in 1912. These service lines were in underground pipes laid in road trenches, mainly relying on gravity for their flow. Electricity transmission started in 1927 and telephone services were available as early as 1879, and the normal method of transmission was by posts and overhead cables.

A policy decision was made in the 1960s to disallow overhead cables. Hence electric power and telecommunications had to be laid underground. Under the current laws, the utility companies have the authority to enter private property and lay services underground, but this is unpopular except when the service lines serve that property. Hence these lines are mainly laid within the road reserve, which is public land. Such works on the roads which block traffic lanes, have to be properly managed as they cause congestion, safety problems and may affect the structure of the road.

In 1966, a Road Opening Coordinating Committee was set up under the Public Works Department with representatives from all the service utility departments. In this way, other than for emergency repair works, the utility departments could work with each other and the roads department to ensure there was coordination to minimise unnecessary road trenching works. They have come up with a code of practice for works on public streets and a code for traffic control for temporary works. Some of the requirements are that there should be no road opening of a newly resurfaced road for one year, no road openings on expressways, and for large projects, a lane for lane replacements have to be provided to minimise traffic congestion.

At the planning stage of new towns and industrial estates , all the authorities get together to work out the layouts of service lines, roads , and underground train lines to ensure that the best of coordination is achieved so as to prevent abortive work.

Blocking of roads is not the only issue, sometimes footpaths are also blocked by obstructions placed by commercial establishments and restaurants. The Streetworks Act makes it an offence to block free passage, and the Land Transport Authority officers carry out regular patrols to ensure compliance.

2.12. *Intelligent Transport Systems*

Towards the 1990s, the cold war between the east and the west was coming to an end. Many defence firms had technologies that they had invented for defence purposes, which could now be declassified and used for other purposes. Intelligent Transport Systems (ITS) had started. ITS was brought about by the advances in new technologies in telemetering, advanced telecommunications, microcomputers and optics. ITS aims to use the communications among the three players — namely the driver/road user, the vehicle and the road network. It could be argued that the installation of the first traffic signals at a junction and the provision of control panels on the dashboard of vehicles could have been considered as the first use of intelligence in traffic control. However, any communication among the three players had been minimal — for example, a red light at the traffic signals merely asks the road user (pedestrian and motorist) to stop; the control panels of the vehicle give basic indications on the health of the vehicle. ITS aims for a much higher level of intelligence and sophistication of communications. Working together as a system, ITS can improve traffic safety and traffic flow.

ITS can be classified into three, mainly Traffic Management System (TMS), Traveller Information Systems (TIS) and Vehicle Control Systems (VCS). TMS and TIS are usually in the realm of the traffic authorities and transport providers and VCS in the realm of vehicle manufacturers.

ITS is exciting because it brings in new dimensions in traffic management, more than the mundane physical road improvements. However, ITS also appears daunting as it is assumed to be so complex with many interdependent players — namely the road user, the road network and the vehicle. There is some confusion on what roles the authorities and the vehicle manufacturers are responsible for. To many, the specter of wiring up all the roads for connecting ITS equipment looms large, although this is not true nowadays with wireless technologies. Furthermore, many decision makers who hold the purse strings do not understand what ITS is or have not been exposed to ITS. While they understand that a flyover built at a junction will improve traffic flow, it takes much longer to convince them that spotting and removing incidents quickly on an expressway have equal benefits.

The expressway network was shaping up well by 1990, and breakdowns of vehicles along them were causing some concern. Drivers were advised to move such vehicles to the breakdown shoulders. These vehicles had to be removed quickly to ensure that the incidents did not lead to secondary accidents. Therefore, pay phones were installed at strategic locations for drivers in distress to call their preferred workshops to attend to the broken down vehicles. The 1990s saw the widespread use of beepers to contact each other. In the age of beepers, many drivers stopped on the expressway shoulders to attend to messages they received from their beepers. This became a problem and in one single year, there were about nine serious accidents caused by these drivers.

To improve the situation, the payphones were disabled and made free so that drivers could only call the police, the fire service or the Automobile Association of Singapore line for assistance. These were called SOS phones. There were many complaints from the drivers that they were denied the opportunity to attend to urgent business due to the removal of normal telephones. However, since safety was the main consideration, the phones were not reverted to normal use.

When the Central Expressway tunnels were opened in 1991, a control centre was set up to monitor the traffic flow on a 24-hour basis inside the tunnels. This was done by means of using detector cameras at close intervals within the tunnel, which indicated slowdowns of traffic flow. Closed circuit television cameras were then used by the operators in the control centre to zoom into the particular area to spot the incident and send a tow-truck crew to remove the breakdown or call the traffic police in the event of an accident. There were also SOS phones located at frequent intervals along the tunnel that the driver in distress could use to call the CTE control centre for assistance. The tow-truck crew are stationed at the Land Transport Authority's Integrated Transport Systems Centre (ITSC) at River Valley Road or they patrol the tunnels. They have to respond within eight minutes after receiving the directions from the operators at the control centre to attend to an incident. The Operator at the control centre switches on variable message signs (VMS) in the tunnels and on the approaches to the tunnels to forewarn approaching motorists of the incident, so they could exercise more caution or decide not to use the tunnels, but divert to other roads. The tow crew's duty is to clear the broken down vehicle from the tunnel as soon as possible.

Incidents in the tunnels or expressways cause other vehicles to slow down to watch, and this could result in secondary accidents. The worst thing that can happen in a tunnel is a fire which could be set off unintentionally by accidents.

While it was considered necessary to provide such immediate monitoring and assistance in the tunnels, it was considered expensive to extend the system to open expressways. Along these expressways, incidents that occurred were not reported unless they were accidents. Sometimes, broken down vehicles could be left unattended on the expressways for long period of times, and approaching vehicles were not forewarned of any danger caused by the broken down vehicles.

After a major incident of a mobile crane hitting an overhead bridge brought the expressway network to a stand still for a few hours in 1995, the decision was made to extend a similar system as that for CTE tunnels to all existing and future expressways. Today, the system called Expressway Monitoring and Advisory System (EMAS) is operated from the ITSC for all expressways, and a similar system is operated from the control centre at Marina South for the Marina Coastal Expressway (MCE) and Kallang Paya Lebar Expressway (KPE). EMAS handles non-recurrent congestion on the expressways. The detection and collection of traffic data for EMAS is through the subsystem of video cameras with computerised processors. The subsystem has an intelligent incident detection algorithm that sends alerts whenever traffic speeds drop or vehicle occupancy increases suddenly. The equipment is able to pick up incidents and relay this information to the control centre. The operator will immediately use the CCTV cameras to verify the incident and call out the standby tow crew and the traffic marshals to the location and Traffic Police if it is an accident. The normal types of incidents are accidents, mechanical breakdowns, blown tyres, no fuel or litter/debris dropped on the expressway.

Upon arrival at the site, the crew tows the vehicle to the nearest car park outside the expressway, for the driver to call for his own repair service. If necessary, the traffic marshals do temporary traffic control at the location. They can also perform on-scene investigation for minor accidents. Serious or fatal accidents still require investigations by the Traffic Police. The detection time is the time required for the operator to detect the incident, which is very fast. The response time is the time for the crew to reach the location and this is specified in the contract with the tow service crew who are on standby at a few vantage locations. The recovery time is the total time at which the incident is cleared after detection. If it is a breakdown, this should not usually last more than 30 minutes. If it is an accident involving injury, then the recovery time is much longer as the Police investigation team has to take some measurements before clearing the vehicles away. Once an incident or an accident happens, the operator displays messages on the variable message signs on the approaches to the incident to forewarn approaching drivers, similar to what is done for the tunnel operations. It can be said that EMAS has been responsible for saving many lives and preventing secondary accidents along the expressway network (Fig. 24 and Fig. 25).

The Surveillance system was then extended by closed circuit television (CCTV) cameras at junctions called J–eyes to spot any unusual traffic conditions or obstructions that impeded traffic flow. Operators could then take appropriate measures such as adjusting the traffic signal timings. Variable message signs are those that can vary the messages shown to motorists. When there are no messages to be displayed, these signs are blanked out. These variable messages, much appreciated by motorists, are now extended to main roads in addition to the expressways. Messages inform motorists of roadworks, road diversions and traffic conditions.

The taxi companies started to use the global network navigational systems to locate the position of their taxis and despatch the nearest taxi to a caller who booked a taxi. The authorities took this opportunity to collect the information to estimate the cruising speed of taxis. This was combined with the vast amounts of data on speeds obtained from the detection cameras on expressways and major roads to work out the prevailing speeds on roads to display on

Source: Land Transport Authority.

Fig. 24.   ITSC control centre.

Source: Land Transport Authority.

Fig. 25.   Variable message signs.

the Internet as traveller information. Drivers can view photos of current traffic conditions from web cameras that are posted on the Internet and mobile applications.

The next step was to provide parking guidance systems from the control centre. This started with the car parks within Marina Centre and was later extended to HarbourFront and the Orchard Commercial belt. The system collects data at the various participating car parks. There is a central computer system for processing the received information on available parking spaces in each car park, and sending information to electronic panels positioned at key locations on the number of available parking spaces at the various locations. With real-time information, motorists are able to make an informed decision on which car park to use. This reduces the incidence of vehicles circulating around unnecessarily within the area looking for available parking spaces (Fig. 26).

Many possibilities are now available to give relevant and timely information to motorists and other road users. The LTA's i-transport Platform integrates the control and operation of various Intelligent Transport Systems (ITS), for better traffic control and incident management of our road network.

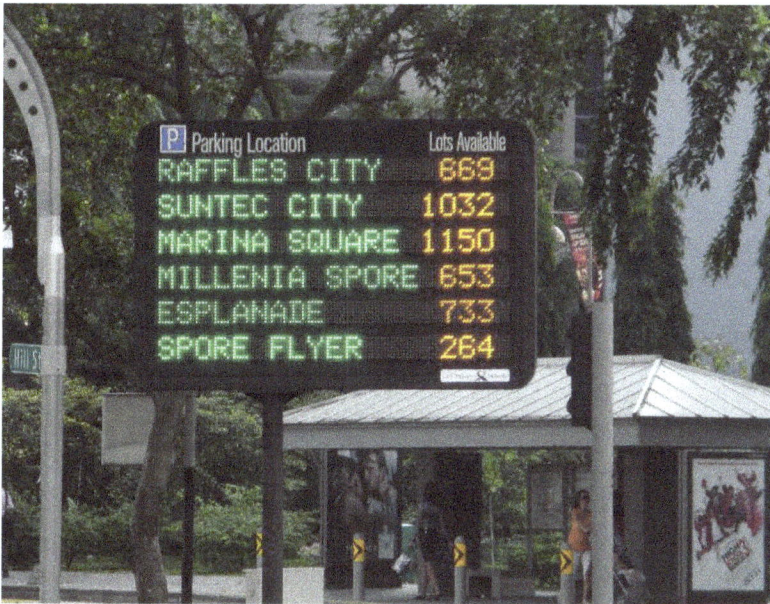

Source: Land Transport Authority.

Fig. 26.   Parking guidance system.

It presents a unified view of the traffic conditions on the road for the benefit of the road users. By using knowledge-based decision module which proposes appropriate traffic action plans, it eases the incident management performed by the operators. It also integrates the traffic data collected by each ITS and perform data fusion computation in order.

Traveller information

Real-time traffic information is disseminated to the public via radio broadcasting, the Internet, personal navigation devices and mobile applications. Traffic information is shared automatically to the media company and the radio broadcasting stations for airing. Real-time traffic information, traffic camera images, and carpark lots availability information are also available on the ONE. MOTORING website (www.onemotoring.com.sg) and MyTransport.SG (www.mytransport.sg) mobile application. These aid motorists and public transport passengers in planning their journeys on the go, to avoid congestion and reduce travel time, especially during peak hours.

In addition, the traffic information is also shared to mobile developers, industry players and research institutes via DataMall@MyTransport (http://

www.mytransport.sg/content/mytransport/home/dataMall.html) so that they could create more traffic-related applications to assist motorists to make informed decisions of their journey.

Motorists can also receive real-time traffic information via portable navigation device (PND) or in-vehicle devices. They can subscribe for real-time information via Radio Data System-Traffic Message Channel (RDS-TMC) or mobile data network from TMC services providers.

Another major ITS scheme was the electronic road pricing (ERP) system implemented in 1998. Since 1975, Singapore had a congestion pricing system whereby vehicles had to pay for entering a restricted zone in the city area. The objective is to charge vehicles for the use of roads at times and at places when and where they cause congestion. The initial manual scheme used paper area licences which vehicles had to display for recognition by the police officers stationed at the overhead gantry points at the entry to the zone.

In 1998, this was made automatic. All vehicles had to install an on-board unit (called the In-vehicle Unit or IU) which communicated with a set of radio antennae on the overhead gantry. By the use of direct short range communications, the system instructed the IU to deduct a charge from an i/c chip card called the CashCard, which was an electronic purse (Fig. 27). If the deduction was unsuccessful, a violation photo of the rear licence plate would be captured for the issuing of summons against the violator.

Source: Land Transport Authority.

Fig. 27.   In-vehicle unit and CashCard.

The IUs, which are colour-coded for different types of vehicle are permanently fixed to the vehicle — on the handle bar for motorcycles and on the windscreen for other vehicles. Different classes of vehicles are charged differently based on the passenger car equivalents e.g., if a car is charged $1, then the motorcycle which has a passenger car equivalent of 0.5 pays 50 cents. The rates can also change at half-hourly intervals. The system works on a multi-lane free flow without any barriers that are needed on toll road systems.

In the later stages, the ERP system was extended to the expressways and some major arterial roads as a travel demand management measure on these roads.

The ERP IU and the CashCard can also be used for payments in the Electronic Parking System introduced at many off-street car parks. A vehicle approaches the entry barrier with a CashCard in the IU. The IU/Cashcard and the time of entry are recorded by a central computer and the barrier opens for the vehicle to enter. When the vehicle approaches the exit barrier, the computer recognises the IU, the time spent in the car park and instructs the IU to deduct the appropriate parking charges before opening the barrier.

Maintenance practices

All road facilities and systems need regular maintenance to continue to serve the road users. Without proper maintenance, effectiveness of all measures aimed at improving traffic flow and safety can reduce considerably.

Prior to the 1980s, maintenance was mostly corrective due to the need to reduce costs. This meant that a facility or a system had to be on the verge of a breakdown or be damaged before repair works were done.

The concept of preventive maintenance took root in the 1980s. The concept is that it is cheaper in the long run to intervene and do preventive maintenance of a road facility at certain points in its life to extend its life rather than wait for it to breakdown and then reconstruct.

In the 1990s, the concept of predictive maintenance was introduced. This method used measurements collected from the road facility in terms of its appearance and structure, and use a computer programme to schedule maintenance priorities.

As for systems, it may be necessary to upgrade the hardware and software at certain points in their life so as not to become obsolete or run out of a supply of spare parts.

It can be said that there is a sophisticated regimen for road maintenance. This is important to ensure that the maintenance dollar is spent wisely so as to keep our assets in good condition.

Usefulness of speed-flow curves for setting targets

Rather than making arbitrary decisions on the acceptability of prevailing traffic conditions, speed-flow curves were derived for lanes for expressways and for city roads by the Centre of Transportation Studies of the Nanyang Technological University, an example of which is shown (Fig. 28).

Site measurements of speed and density were taken from which the speed–density of traffic flow from which the speed–flow curve is derived. The local curve figures are much more modest than those of the United States Highway Capacity Manual.

The levels of service proposed are:

A = Very Free flow
B = Reasonably free flow
C = Stable flow
D = Heavy but acceptable traffic flow
E = Flow nearing road capacity
F = Unstable flow with stop-go conditions

Good practice is to aim for level of service D to E during the peak periods, C to D during the normal off-peak periods and A to B in the quiet periods.

Speed-flow graph

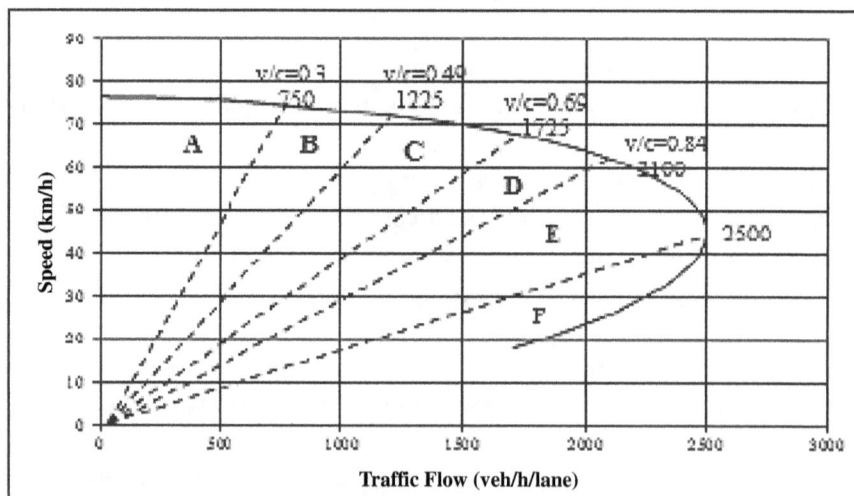

Source: Public Works Department Singapore

Fig. 28. Speed-flow graph for expressways.

Of course, there are instances when the levels of service get to F during the peak periods, then the aim is to keep that for only short durations.

When there are incidents or accidents, then the level of service can deteriorate to F and hence the need for handling non-recurrent congestion effectively.

The speed flow curve is also used to adjust the electronic road pricing charges at three-monthly intervals. The desired 85 precentile speeds on the city roads (a basket of typical roads) are in the range of 20–30 km/hr and on the individual expressways in the range of 45–65 km/hr. If the measured speeds fall below the range, the charges are increased by $1 for the corresponding three months, and if the speeds are above the range, the charges are reduced by the same amount.

Standard software in traffic management

There are many off-the-shelf computer software programmes such as SIDRA, TRANSYT, VISSIM, and VISSUM which are now in popular use to aid in analysing and recommending traffic management. They should be calibrated for local use before they are used in the analysis. SIDRA is used for individual junctions while the others are for use on a road network basis.

2.13. *Future of traffic management*

With issues of sustainable development and methods to mitigate the effects of global warming due to the use of fossil fuels, the emphasis for handling transport will shift to getting the best out of the road network for moving people. Hence, traffic management and innovative ways of using ITS still have a role to play. Lane capacity, which is the maximum number of vehicles a lane can carry, is determined today by the minimum distance a driver keeps from the vehicle in front due to fears of safety. With automatic driverless cars or co-operative driving, whereby vehicles lock into an intelligent highway and are moved in convoys, this minimum distance could be reduced. This could increase lane capacity and increase permitted speeds without endangering road safety.

On public transport, urban trains have their own right of way and do not suffer delay caused by other vehicles. Other means would have to be found to speed up buses which share the road with other vehicles. The car is inefficient when it has a low occupancy of one to two occupants; whereas a car that carries four occupants is an efficient mover of people. Hence, car sharing and carpooling are expected to become popular if the proper infrastructure and systems are put in place to facilitate such practices. So will non-motorised transport when more facilities are provided to make their movements safer and more comfortable.

One of the drawbacks of the current traveller information is that it is based on the historical data of the traffic situation that happened in the last few minutes — it is reactive. Traffic situation can change quickly and the information may become out-of-date. What is needed, and which are under experimentation, is to provide predictive information so that the system predicts what is going to happen in the next few minutes to the traffic flow based on what is happening now. Such information will help both private transport and public transport users.

Initially road information was given by static signs, now much of it is given by variable message signs. The driver only pays attention if such messages are relevant to him. The next step will be to give tailor-made information rather than general information to the driver at on-board units within the vehicle.

The current navigation systems are expected to be complemented by prevailing traffic conditions from the road network so that the driver is routed through the most efficient route to his destination.

References

1. M G Lay 1992, Ways of the World USA: Rutgers University Press.
2. Malaysian Branch of the Royal Asiatic Society 1973, Singapore 150 Years: 150 anniversary of the founding of Singapore.
3. Cheong Colin 1992, Framework and Foundations: A History of the Public Works Department Singapore: Times Editions.
4. The Straits Times Media Reports.
5. Koh, Tommy T.B. 2006 Singapore: The Encyclopedia, Singapore: Editions Didier Millet; National Heritage Board.
6. Various unpublished reports of the Public Works Department and Land Transport Authority.

Chapter 5

Road Safety

HO Seng Tim and CHIN Kian Keong

This chapter provides a review of road safety in the development of land transport in Singapore over the years. The focus is on engineering road safety, together with education and enforcement collaboration efforts with major stakeholders such as the Traffic Police and Singapore Road Safety Council. The chapter begins with a reflection of the past road accident statistics together with related factors like vehicle and population growth, comparing the road safety situation then and now. Next we look at how priorities have changed, with the focus moving from pedestrians to motorcyclists and recently, to cyclists. We elaborate on how Singapore's road safety has improved and also how the city state has become glob-ally benchmarked to be among the safest countries in the world. We recall some inspirational accounts of Singapore's success story in road safety. These accounts will flow with key milestones under the three pillars of road safety: engineering, enforcement and education. Next, we highlight some of the key road safety initia-tives that we have developed over the years, with examples of engineering road safety schemes that have been effectively used over the years. These will include major national road safety projects including School Zones and Silver Zones to make our roads safer for young school-children and the increasing number of seniors that comes with an ageing population. Finally, we conclude with some key learning points for the success of road safety in Singapore.

1. Introduction

Road safety in Singapore has improved much over the past 50 years. To understand the change in road safety over the years, Mr. Ho Seng Tim asks his active 76-year-old Singaporean mother about the changes in road safety over the past 50 years in

---

Singapore. His mother replied, "In the earlier days, there were not many vehicles on the roads, so there were less traffic conflicts. Nevertheless, it wasn't easy to cross the roads. Now, with significant road improvements, I feel much safer to cross roads." However, his mother is concerned with the behaviour of some road users, including speeding motorists who for their own convenience do not give way to other road users. This could explain why besides engineering, education and enforcement efforts are also needed to enhance road safety. Mr. Ho Seng Tim tells his mother that he has good news for her: Over the years, much has been done to make Singapore's roads safer. His mother smiled as she looks forward to more road safety initiatives such as the Silver Zones to enhance safety for seniors.

Fifty years ago on 9 August 1965, Singapore separated from Malaysia to become wholly independent as the Republic of Singapore. Much has changed since then, including road safety, which has evolved over time in tandem with the development of land transport in Singapore over the years. This chapter focuses largely on the engineering road safety efforts, together with education and enforcement collaboration with major stakeholders like the Traffic Police (TP) and Singapore Road Safety Council.

The past 50 years of transportation in Singapore has seen much road safety improvements. Singapore has achieved one of the lowest road fatality rates in the world. In 2014, the number of road fatalities per 10,000 registered vehicles was 1.6, whereas the number of road fatalities per 100 million vehicle kilometers travelled was 0.7. The fatality rate, normalised by 100,000 population, has decreased to 2.8.

Global recognition was accorded to Singapore's efforts on road safety. In December 2007, the Land Transport Authority (LTA) and the TP won the Prince Michael International Road Safety Award in the United Kingdom. Then in August 2011, the LTA was awarded by the Institute of Transportation Engineers (ITE) to receive the Edmund R. Ricker Transportation Safety Council Award in the United States. Singapore is honoured to be recognised for her high level of road safety excellence. In fact, officials from many countries often visit Singapore to learn and share experiences on improving road safety.

2. Road Safety Past and Present

2.1. *Road safety in the earlier years*[a]

Road safety in Singapore has come a long way. It is now a major part of every traffic policy and plan, being of utmost priority in the mind of every professional

---

[a] Extracted from face-to-face interview with A/P Gopinath Menon held in October 2014. A/P Menon was the Chief Transportation Engineer of Singapore with Public Works Department

in the related industries. Below highlights some of the key milestones in the earlier years of road safety management in Singapore.

2.1.1. *1960s*

There were two major road safety projects in the 1960s. In 1964, Singapore's first pedestrian bridge was constructed next to Clifford Pier at Collyer Quay, near Change Alley (*Remember Singapore, 2012*). Traffic was heavy along Collyer Quay, and pedestrians often had to risk their lives to cross the street in the city centre. The bridge proved to be very effective; starkly reducing the number of vehicle-pedestrian conflicts after its completion. At the same time, a walk-through pedestrian underpass was constructed at Empress Place.

After Singapore's independence in 1965, road safety became a huge concern. The National Safety First Council (NSFC) was then formed in 1966 to focus on improving road safety. It then gradually branched out to embrace safety in other aspects, including home safety, work safety and fire safety. The National Safety First Council later renamed itself as National Safety Council in the 1970s. The Singapore Road Safety Council (SRSC) was subsequently set up in 2009 by the Ministry of Home Affairs (MHA) to look into road safety specifically. Since its formation, the SRSC has rolled out numerous road safety campaigns, such as the road safety education programme, "Anti-Drink Driving" and "Safer Roads" campaigns (*SRSC, 2011*).

Bus bays to get buses to stop off the road became a mandatory feature where space permitted. They allowed buses to stop for passengers to safely board and alight without obstructing traffic. However, along busy stretches of roads, bus bays also made it challenging for buses to turn out of the bus bay. Today, we have the "Mandatory-Give-Way-To-Buses" schemes and bus lanes to facilitate bus movements on the roads.

2.1.2. *1970s*

Improvements in road safety took a major turn in 1970 when the Green/Red Man aspects were installed onto traffic lights in Singapore. In the beginning, few pedestrians used these new signalised crossings. In an effort to encourage higher usage, the Public Works Department (PWD) painted footprints on the road to direct pedestrians to these crossings. These were first installed at High Street, North Bridge Road and near the Capitol Theatre.

(1966–1995) and the Land Transport Authority (1995–2001). Currently, he is holding multiple appointments including Adjunct Associate Professor in Nanyang Technological University, Principal consultant in CPG Consultants and Vice-chairman of the Singapore Road Safety Council.

In 1973, the first pedestrian mall in Singapore at Orchard Road was started. In 1974, traffic accident statistics were generated and closely analysed. Associate Professor Gopinath Menon highlighted that pedestrians turned out to be the most vulnerable road user group. In an effort to enhance road safety for pedestrians, the authorities rolled out a series of implementations. These included setting the pedestrian crossing rules such as making jaywalking illegal.

The most successful scheme that was implemented was perhaps making it mandatory for turning traffic to give way to pedestrians at junctions. It was originally stipulated in the Highway Code, but was not enforced. This was an opportunity that the authorities quickly recognised. Together with the introduction of the Green/Red Man aspects, it became mandatory for left or right turning vehicles to give way to adjacent crossing pedestrians. There was significant opposition for this implementation when it was first rolled out, with many citing safety concerns. However, the authorities recognised the benefits of the scheme and spent three years (from 1974 to 1976) pushing hard for it. As we look back now, more than 30 years have passed and this scheme has been and is still very effective. With the implementation of this scheme, every signalised traffic junction became a crossing for pedestrians. Slip roads at traffic junctions were then installed with zebra crossings to facilitate pedestrian movements between the junctions and the sidewalks.

In 1974, in collaboration with architects, major walkway programmes were started. There were three large-scale projects at Boat Quay, Changi Village and Tanjong Katong Road. With these walkway programmes, a Walkway Unit was subsequently formed. It then gradually became a standard for footpaths to be constructed alongside every road in Singapore.

2.1.3. *1980s*

In the 1980s, Vehicular Impact Guardrails (VIGs) were not part of the design for expressways. Hence, motorists who veered off-course might tumble off the road, sustaining serious injuries. A 10-million-dollar programme was then started to install VIGs along all expressways. It started on the East Coast Parkway (ECP) and is now a standard feature on all expressways.

In the 1980s, telephone booths were installed along expressways to allow motorists to seek assistance if their vehicles broke down. However, many motorists stopped to make phone calls in those days when beepers/pagers were common. It was then decided to either remove these telephones or to convert them such that they can only be used for emergency calls to the police, fire station and the Automobile Association of Singapore (AA) for tow services.

Remote monitoring cameras Expressway Monitoring Advisory System (EMAS) were then installed subsequently to replace the emergency phones to assist motorists in the event of an emergency.

A project to provide rain shelters under structures along expressways for motorcyclists was also spearheaded. Before this programme, motorcyclists used to stop on the left lane of expressways during wet weather, which exposed them to oncoming traffic along these high speed roads. Now, you can find the unique Umbrella signs along expressways that symbolise these rain shelters.

In 1983, an island-wide programme on pedestrian overhead bridges was started to increase accessibility for pedestrians. In total, over 180 bridges were constructed. Today, we have 513 bridges (*Remember Singapore, 2012*). With it, there are now accessibility problems for these bridges which need to be addressed. Apart from bridges, ramps were provided at crossings. These ramps were made of a slight gradient from the side table to the road pavement. This made it easier for pedestrians to cross the road, especially for the mobility-impaired and seniors.

In 1983, Shell-grip-Calcined Bauxite was first applied at nine black spots. It increases friction and hence reduces the tendency of skidding, especially for motorcyclists. Today, Calcined Bauxite is still applied at black spots to increase skid resistance.

Truck-mounted attenuators to protect workers on high-speed roads were made mandatory in the mid-1990s, which then spurred the similar use of crash cushions at gore areas.

Despite all the positive implementations that enhanced road safety in Singapore, there were schemes that turned out ineffective as well. One such scheme was the Left-Turn-On-Red (LTOR). It was originally pushed for by policymakers because of its success overseas. However, it was not popular with local road users, and the authorities received an immense amount of feedback asking for it to be removed. Many motorists felt that it was very confusing to turn, or even to move off, when the red light was showing. Many of the locations implemented with the LTOR scheme have since been changed to the Green Left-Turn Arrow.

In the earlier years, the focus of road safety has been on the most vulnerable road users such as pedestrians and motorcyclists. Motorcyclists were often involved in numerous accidents on the expressways. The authorities had planned to ban motorcycles on expressways before, but the plan did not come to fruition as there was only one way via the ECP to Changi Airport, where many motorcyclists work at.

With the ever-expanding public transport network in Singapore, more people are turning to buses and MRT trains as a choice mode of transport. This has reduced the number of vehicles on the roads, taking into account the increasing population as well, which may have led to reduced traffic conflicts and hence, safer roads over the years.

2.2. Past road accident statistics (1965–2014)

Singapore is a highly urbanised country. Since 1965, the vehicle and resident populations have increased by 4.5 and 3 times respectively (Fig. 1). However, road fatalities have decreased by 46% from 288 in 1974 to 155 in 2014 (the TP has no record of road traffic accident data for the 1960s). Figure 2 depicts past accident fatalities as well as key transport initiatives such as the opening of the first two major expressways (Pan Island Expressway and the ECP, spanning between the east and west) in 1981. The fatality rate hovers up and down over the years but subsumes a general decreasing trend. In 1998, the Road Safety Engineering Unit (RSEU) of the LTA was formed. This was followed by the launch of a series of major road safety programmes such as the Road Safety Assessment Programme in 1999 and the Black Spot Programme in 2005.

In the 1970s, most of the traffic fatalities were pedestrians, contributing about 40%. This percentage dropped to about 30% with almost equivalent

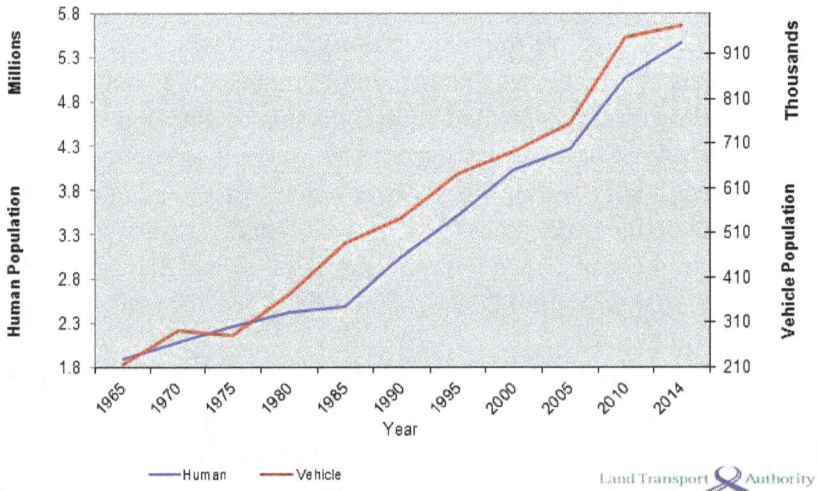

Fig. 1.   Human and vehicle population in Singapore (*LTA, 2014a*); (*Department of Statistics, 2015*).

No. of Fatalities

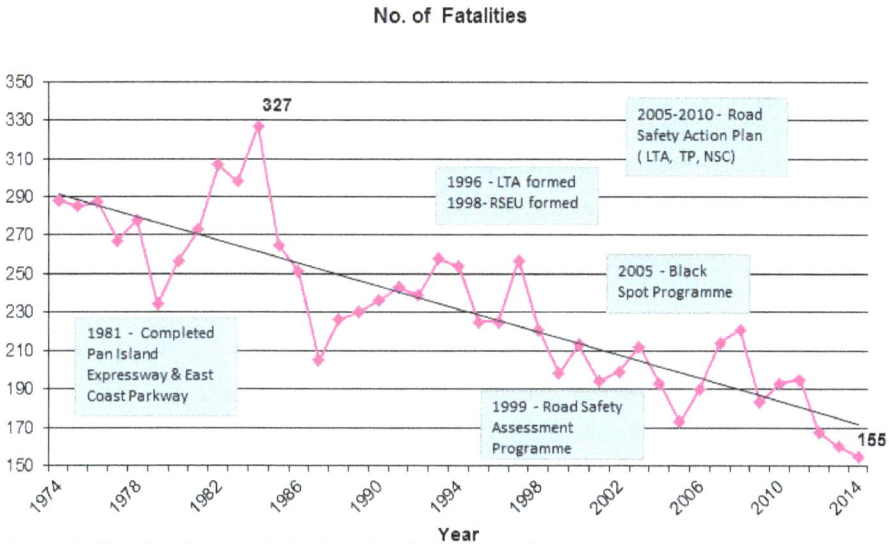

Source: *Traffic Police, Singapore Police Force. Land Transport Authority.*

Fig. 2.   Road fatalities and major transport initiatives in Singapore (*Traffic Police, 1974–2014*).

percentage contributed by the motorcyclists and pillion riders in the 1980s.[b] Pedestrian fatalities continued to decrease to about 29% till now while motorcyclist fatalities picked up to almost 50% since mid-1980s. Among the pedestrian fatalities, seniors consistently made up about half of the cases. In addition, with the growth of the ageing population, there should be more focus to enhance safety for this group of road users. Another area of concern is young children pedestrians, as they are shorter and therefore are less visible, and are also less experienced on the roads. Cyclists continued to form about 10% of the total road traffic fatalities throughout the years. Fatalities of accidents involving heavy goods vehicles remain a concern, due to their greater size and impact.

A closer look at some national statistics over the past 11 years from 2004 to 2014 shows possible challenges in road safety in the years ahead (*LTA, 2005; LTA, 2014a*). Singapore's resident population is high and has increased by 30% from 4.2 million to 5.47 million; while the vehicle population has increased by 43% from 0.7 million to 1 million. On the other hand Singapore's land area has only increased by 3% from 699 square kilometres to 718 square kilometres. About 12% of our total land area is taken up by roads.

---

[b] Statistics were computed from Traffic Police's traffic accident records.

The road network is highly urbanized, having increased by 10% from 3,188 km to 3,495 km of paved roads. Another 21 km (North-South Expressway) of road length will be added after 2020. The road infrastructure is well developed and well maintained with lighting on every public road. Compared with many developed countries, car ownership is low with usage demand management and other controls. The vehicle fleet is modern with motorcycles taking up about 15% of all motor vehicles.

The rail length has increased by 41% from 109 to 154 km. Rail network will be expanded to 360 kilometre, including 50-kilometre of Cross Island Line (CRL) rail line and 20-kilometre Jurong Region Line (JRL) by 2030. Besides the MRT, public bus services also play an important role in our daily lives. Both public buses and the MRT, with their large passenger capacities, are the most efficient means of moving large numbers of people. Public buses have made tremendous improvements over the years, in terms of quality and comfort. In the 1970s, the buses were old, relatively slow and non air-conditioned, but, all are now fully air-conditioned and are equipped with better features. The two main operators, SBS and SMRT, have a fleet of more than 4,200 buses as of 2013, an increase of about 20% as compared to 3,547 buses in 2004. We also see an increase in the taxi fleet from 20,407 to 27,695 over the past 10 years.

Road safety has improved over the years. Besides the decrease in accident fatality numbers, fatality rates (normalised by key national figures such as human and vehicle populations and total vehicle kilometres travelled) in general are also on decreasing trends (Fig. 3).

To get a better feel of the road safety performance, let us look at the road fatality rates over the recent past 11 years from 2004 to 2014. Fatalities per 100,000 population has decreased by 40% to 2.8 from 4.7. Similarly, fatalities per 10,000 registered vehicles also dropped by 16% to 1.6 from 1.9; and fatalities per 100 million VKT (vehicle kilometres travelled) reduced by 36% to 0.7 from 1.1.

2.3. *Road safety records regionally and internationally with other countries*

We are proud to share that the road network in Singapore is among the safest in the world with our road safety records improving over the years (*Koh et al., 2006*). Figs. 4 to 6 show three benchmarking graphs to compare Singapore's road safety performance with other countries: Australia, Canada, Hong Kong,

Fatalities Rate in Singapore

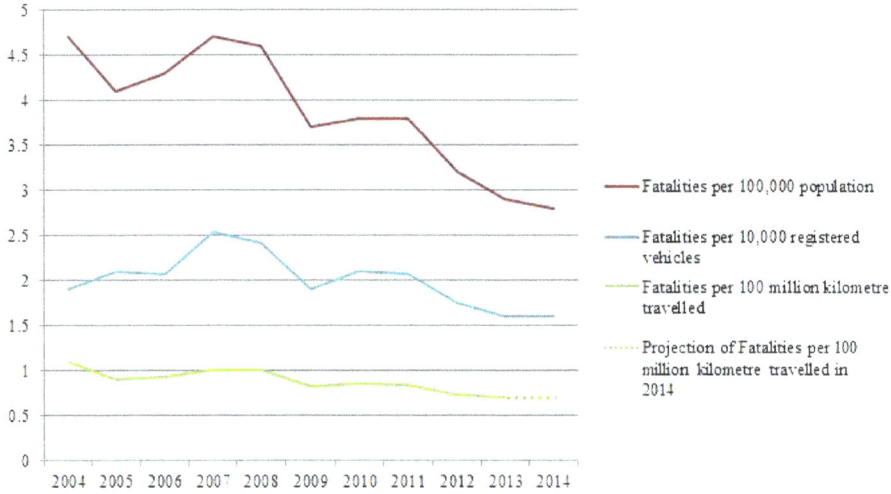

Fig. 3. Road fatalities rates 2004–2014 (*LTA, One.Motoring*); (*Traffic Police, 2004–2014*); (*Department of Statistics, 2015*).

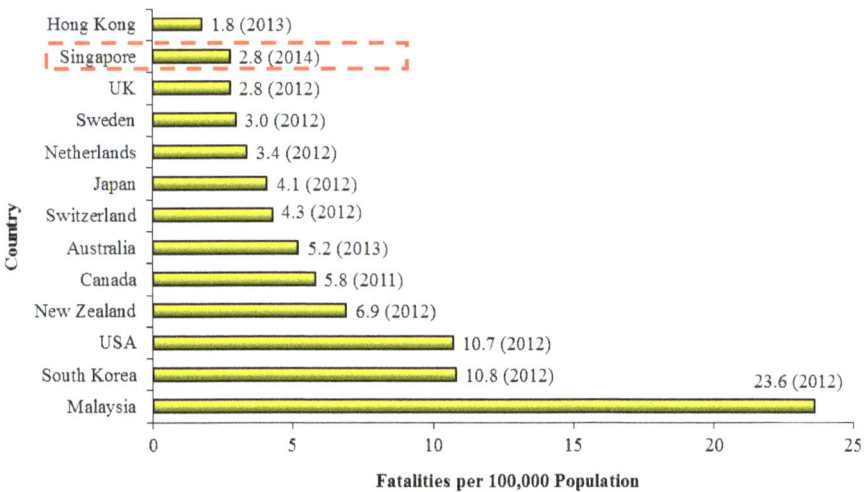

Fig. 4. Fatalities per 100,000 population among countries (*LTA, 2014a & 2014b*); (Traffic Police, 2014); (UK, Sweden, Netherlands, Japan, Switzerland, Australia, Canada, New Zealand, USA, South Korea, Malaysia: IRTAD Annual Report 2014); (Hong Kong Transport Department, 2014).

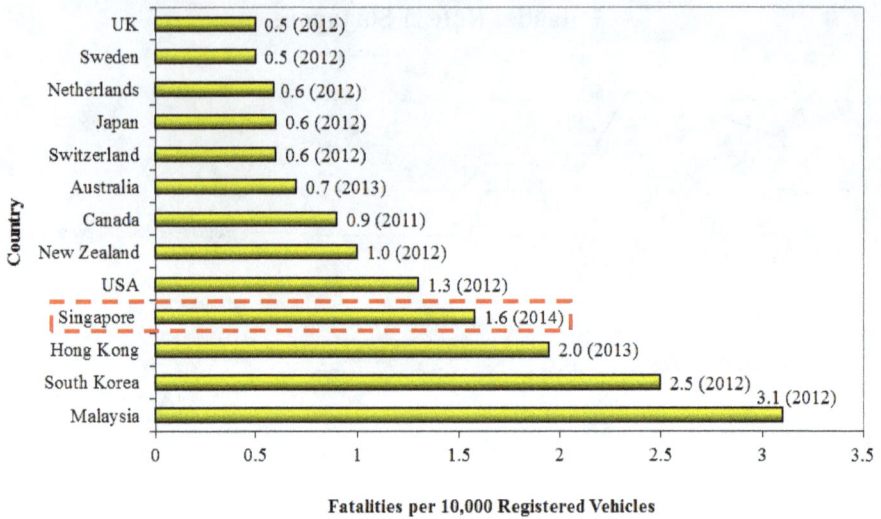

Fatalities per 10,000 Registered Vehicles

Fig. 5. Fatalities per 10,000 registered vehicles among countries (LTA, 2014a & 2014b); (Traffic Police, 2014); (UK, Sweden, Netherlands, Japan, Switzerland, Australia, Canada, New Zealand, USA, South Korea, Malaysia: IRTAD Annual Report 2014); (Hong Kong Transport Department, 2014).

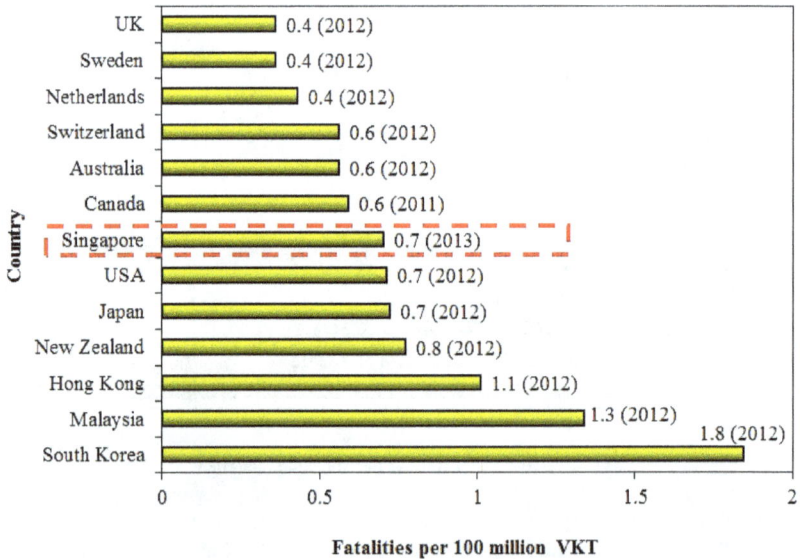

Fatalities per 100 million VKT

Fig. 6. Fatalities per 100 million VKT among countries (LTA, 2014a & 2014b); (Traffic Police, 2014); (UK, Sweden, Netherlands, Japan, Switzerland, Australia, Canada, New Zealand, USA, South Korea, Malaysia: IRTAD Annual Report 2014); (Hong Kong Transport Department, 2014).

Japan, Malaysia, Netherlands, New Zealand, South Korea, Sweden, Switzerland, United Kingdom and the United States of America. Some countries were primarily selected due to the well-known high standards of road safety performance displayed in each nation. Some countries located within the Southeast Asian region are included due to its close proximity to Singapore and similar culture or habits of the people. Hong Kong, on the other hand, is selected for comparison as it is a relatively small country with an extremely dense population, much like Singapore.

Our road safety records gradually show good improvement over the years. This is largely attributed to the continual efforts by our road safety practitioners and professionals in exploring ways in terms of the engineering, enforcement and education aspects to prevent or reduce road accidents. Globally, our roads are recognised to be among the safest. However, there is still room for improvement when we compare with other developed countries such as UK, USA and Japan.

2.4. *Vulnerable road users*

While taking care of road safety, the group of vulnerable road users is normally the focus of actions. The basic mode of transport is still by foot and some human-assisted wheeled transport, to transport goods and passengers. Bicycles surfaced as one form of affordable wheeled transport. However, as people became more affluent, motorised transport including motorcycles replaced bicycles and people could travel further. The use of motorised transport continued to rise and the focus of transportation shifted to motorised vehicles. This subsequently led to the realisation that motorised transportation is and would be the source of traffic congestion and pollution, and these resulted in many jurisdictions to relook at greener options such as walking and cycling.

In 2014, there were 155 deaths and 9,858 injuries on the roads, with 10 fatal accidents involving drink-driving. Out of the 155 fatalities, almost half or 48% involved motorcyclists/pillion riders and 29% involved pedestrians (Fig. 7). These are the two most vulnerable groups of road users. The groups of road users that are more susceptible to fatal accidents are young motorcyclists (20–24) and senior pedestrians (aged 60 and above). Speeding is a major cause of fatal and injury accidents. Based on the accident statistics in 2014, the TP has identified motorcyclists/pillion riders and pedestrians as two areas of concern (*Singapore Police Force, 2014*). While cyclists make up about 10% of the road traffic fatalities, the number of cyclists has been increasing over the past 10 years. Safety efforts may need to be enhanced to focus more on this group.

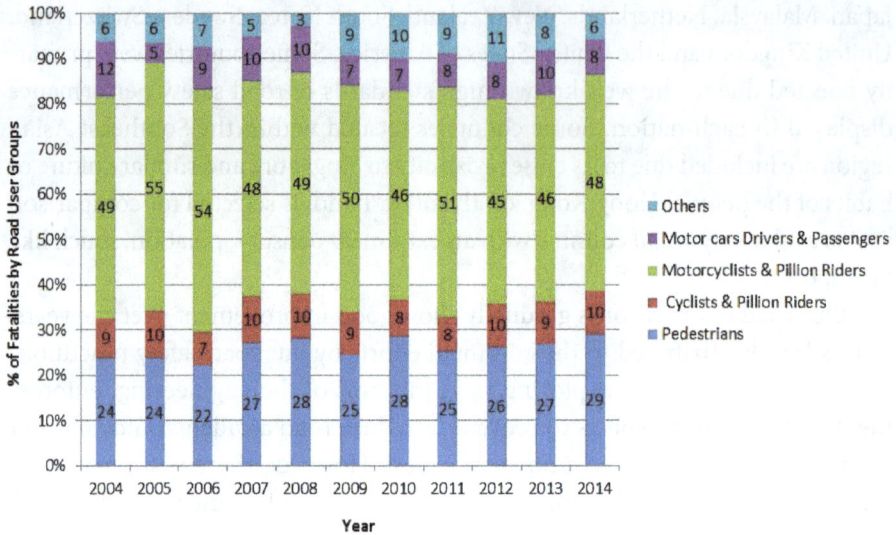

Fig. 7.   Proportion of fatalities by road user groups (*Traffic Police, 2004–2014*).

3. Singapore's Success Story in Road Safety

We recall some inspirational accounts of Singapore's success story in road safety in this section. These accounts flow with some key milestones under the three pillars of road safety in terms of engineering, enforcement and education.

3.1. *Option of roads & transportation division of the Public Works Department and the Traffic Police merged to oversee Road Safety in Singapore in the early years*

In the earlier years, there was a period when efforts were made to merge the then Public Works Department (Roads and Transportation Division) or PWD and the Traffic Police (TP). At that time, the PWD was responsible for the planning, design, implementation and management of road infrastructure engineering. The TP was responsible for road safety enforcement and education. The PWD and the TP often worked closely together to promote road safety with the other major stakeholders such as the then National Safety Council of Singapore.

For better coordination and collaboration efforts to enhance road safety, it was proposed that the two government agencies (the PWD and the TP) should merge under one Ministry. However, after much discussion and consideration, it was decided that it was better for PWD and the TP to continue functioning separately under the Ministry of National Development and the Ministry of Home Affairs respectively. Road safety and other policy matters could continue to be discussed at regular senior management meetings with the PWD, the TP and the other ministries.

In the 1970s, there was a roads and transportation unit in the PWD that was in charge of everything related to transportation — from erecting pedestrian overhead bridges and bus shelters, to an array of other road-building projects. That small unit has since grown as the work scope of the PWD expanded and then later became part of the LTA. The LTA is a government agency dedicated to developing a land transport system for all Singapore Citizens. The LTA was established on 1 September 1995, after the merger of four public sector entities: Registry of Vehicles (ROV), Mass Rapid Transit Corporation (MRTC), Roads & Transportation Division (PWD) and the Land Transportation Division of the Ministry of Communications (MCI).

3.2. *The setting up of Road Safety Engineering Unit, a dedicated unit in the LTA to focus on road safety engineering in 1998*

In August 1998, the LTA set up a dedicated Road Safety Engineering Unit (RSEU) with a mission to implement road safety engineering schemes and best practices to make our roads safer. This early decision was a wise move as later in 2004, it became recognised globally as a road safety best practice to a have a dedicated team to be responsible for road safety. The RSEU was set up with the planning, implementation and management of various road safety initiatives. Processes and techniques such as the Black Spot Programme, Crash Site Investigation, Road Safety Assessment and remedial treatment programmes are well established best practices that are being implemented by RSEU.

3.3. *Collaboration efforts with major stakeholders under the Singapore National Road Safety Action Plan 2005–2010*

The LTA works closely with other road safety advocates and major stakeholders. The World Health Organisation (WHO)'s Report on Road Traffic Injury and Prevention in 2004 recommended that "each country should prepare a national road safety strategy and plan of action". Following this, the first National Road Safety Action Plan (NRSAP) was officially launched at the National Road Safety Exhibition and Rally in 2005 (*MOT, 2003*). The Asian Development

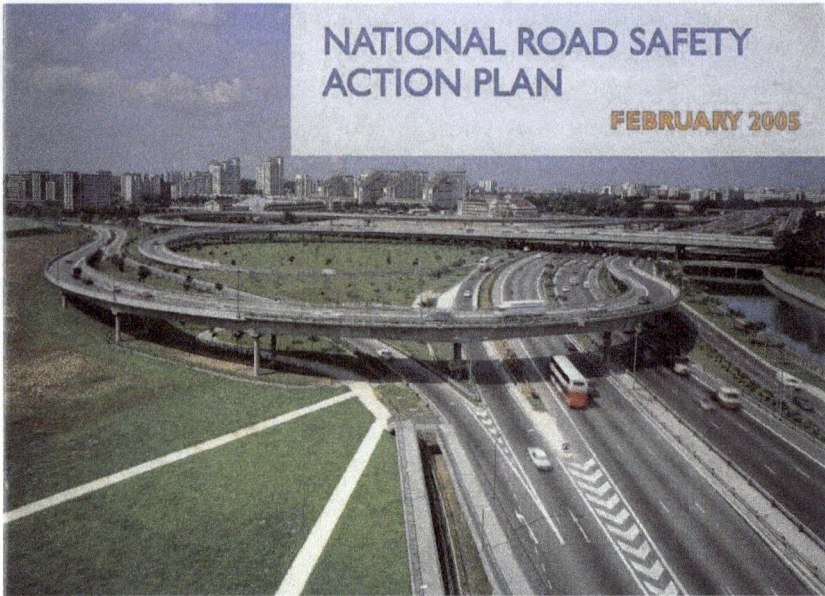

Fig. 8. National Road Safety Action Plan, February 2005.

Bank (ADB)-backed NRSAP is the result of a strategic collaboration between various government agencies, non-governmental organisations and the private sector, with the aim to set common targets on road safety strategies (Fig. 8). It is specifically tailored to the particular needs of Singapore. The NRSAP leads to greater emphasis and commitment on road safety issues and spurs more road safety initiatives.

The NRSAP was a timely effort by the various government agencies to come up with concrete initiatives to improve road safety. In particular, there is a need for a concerted effort by both road users and the relevant government agencies to reduce the number of road accidents.

Under the NRSAP, there are 14 sectors of action plans, being categorised under five Es. They are:

Encouragement

(1) Co ordination and Management of Road Safety — An effective road safety executive structure and a multi-disciplinary workgroup with adequate resources to make a strategic change in enhancing the level of road safety.

(2) Road Safety Funding and Role of Insurance Industry — An effective and sustainable funding mechanism by identifying and encouraging all

stakeholders with vested interests to provide financial support for the development and implementation of coordinated and effective road safety programmes.

Education

(3) Road Safety Publicity Campaign — Implement effective road safety publicity campaigns and educational initiatives based on comprehensive analysis of traffic accident data. The TP and the LTA, with the support from private sector partners, have a long history of implementing road safety publicity campaigns.

(4) Road Safety Education of Children — Inculcate road safety in children in an effective manner so as to ensure a strong foundation for lifelong road safety attitude and behaviour.

(5) Driver Training and Testing — Formulate an effective driver training and testing system, with emphasis on drivers and riders between the age of 18 to 28, to ensure they are adequately trained and tested before being allowed to drive or ride on the road.

Enforcement

(6) Enforcement — Bring about effective traffic law enforcement that encourages safer road usage and promotes smooth traffic flow.

(7) Legislation — Implement a framework of laws and regulations that can be easily and effectively enforced, and be periodically updated as required.

Emergency Preparedness

(8) Emergency Assistance to Accident Victims — Provide the relevant medical intervention at the accident site and the transportation of victims to the most appropriate hospital for treatment through better co ordination among the various parties involved.

Engineering

(9) Road Accident Data System — Set up an accurate, up-to-date, comprehensive and accessible accident database that will allow a clearer understanding of the scale, nature and characteristics of the road safety issue for the formulation of effective road safety measures and countermeasures.

(10) Safe Planning and Design of Roads — Incorporate safety practices in the safe planning and design of roads that are "safe to build" and "safe to use" across the various government agencies.

(11) Vehicle Safety Standards — Improve road safety by enhancing vehicle safety standards.

(12) Improvement of Hazardous Locations — Identify and prioritise remedial treatments for accident-prone locations, through accident trend analysis.

(13) Road Safety Research — Identify and quantify factors contributing to road accidents by building on a framework of knowledge to enable sharing, discussion and application.

(14) Road Accident Costing — Identify methods to obtain accurate road accident costing as there is currently no organization or government body that collates data for road accident costing.

3.4. *Memorable win on global stage — the Prince Michael International Road Safety Award in 2007*

In 2007, Singapore was bestowed a prestigious Prince Michael International Road Safety Award (UK) for Road Safety Management for its efforts to improve safety (see Fig. 9).

These annual awards — well known among road safety professionals around the globe — were first given out 20 years ago. Outside of the United Kingdom, only two other countries received the international accolade in 2007; the United States and Russia.

According to Mr Adrian Walsh, secretary and director of Roadsafe, "We receive a large number of international nominations and our international judging panel seriously considers about 12 each year. We only consider nominations which fit the criteria and will make a large difference to road safety."

What made Singapore's nomination stand out was how the LTA and the TP work in tandem to tackle road safety issues. The award panel considered this a "best practice" which could be adopted in other countries. Emphasising the need to work together for road safety, the LTA's then chief executive Yam Ah Mee said, "We would like to share this award with our public, because we all work together to make our roads safe."

3.5. *The LTA wins international award for improving road safety — the Institute of Transportation Engineers, Edmund R Ricker Transportation Safety Award in 2011*

Singapore's road network is rated one of the best in the world, in both design and safety. It received the recognition at the annual Institute of Transportation Engineers (ITE) awards ceremony held in St. Louis, Missouri (USA), on August 2011.

Fig. 9.   At the award ceremony held at The Savoy in London on 11 December 2007 are (from left to right) the LTA's Dr. Chuai Chip Tiong, Director of Traffic & ITS Operations; Mr. Ho Seng Tim, Head of Road Safety Engineering; His Royal Highness, Prince Michael of Kent; and Mr. Alvin Lee, Ex-Head of Research, Planning and Organisational Development, and Deputy Superintendent of the Traffic Police.

The LTA was presented with the Edmund R. Ricker Transportation Safety Council Award for its efforts in improving safety for road users in Singapore. Only one award is given out in the individual and organisation categories each year (see Fig. 10).

Past winners of this prestigious international award include the World Bank, and city administrations in America and Australia for their efforts in promoting road safety.

3.6. Formation of the Singapore Road Safety Council

In December 2009, Singapore Road Safety Council (SRSC) was set up by the Ministry of Home Affairs (MHA), supported by the Ministry of Transport (MOT), the LTA and other road safety-related stakeholders. The council aims to work together with governmental and non-governmental agencies to conduct research and campaigns to look into road safety issues such as accident prevention and education of road users. They also serve as the official body for

Fig. 10.   At the award ceremony held at St. Louis, Missouri, USA on 16 August 2011 are (from left to right) Ex-ITE President, Mr Robert C. Wunderlich; the LTA's Mr. Ho Seng Tim, Head of Road Safety Engineering; and Saifulbahri Bin Rasno, Deputy Director of Infrastructure Enhancement.

Singapore to engage with other international road safety councils/programmes in contribution to global road safety outcomes and also a forum for the exchange of ideas on matters pertaining to road safety.

The council has been launching numerous road safety outreach programmes focusing on the two most vulnerable groups of road users — senior pedestrians and motorcyclists. The programme aimed to spread road safety awareness through public education campaigns in the heartlands where the seniors congregate.

In support of the United Nations Decade of Action for Road Safety, the Singapore Road Safety Council, together with the Automobile Association of Singapore, People's Association and the TP, jointly organised a nation-wide campaign called Safe Roads Singapore on 6 March 2011. As a result of the strong support from donors, partners and the community, 500,000 road safety pledges were collected during the campaign's pledge gathering exercise. This means that half a million road users in Singapore had pledged to be more vigilant on the roads.

3.7. *High-profile intervention with Pedestrian and Cyclist Safety Committee (PCSC) and Safer Roads Industry Taskforce (SRIT) committee*

The Committee on Pedestrian and Cyclist Safety (*MOT, 2013*) was also set up in February 2013 to re-examine ways to make Singapore's roads safer for pedestrians and cyclists (Fig. 11). This Committee was chaired by then Parliamentary Secretary for Transport Associate Professor Muhammad Faishal Ibrahim, and comprised of members from the LTA, the TP, Singapore Road Safety Council and the Ministry of Education (MOE). The Committee, which prioritised the safety of school children and seniors, introduced many new initiatives. First on the committee's to-do list was enhancing safety within school zones. Besides tougher enforcement by the TP, speed limits around schools were selectively reduced and provided with additional measures like speed humps, flashing LED lights, more signs, and installing railings at both sides of the road in school zones, etc.

Other notable initiatives that were spearheaded by this committee include the inauguration of the Silver Zones and cycling towns. The School Zone scheme, to improve safety for school children, has seen success since its implementation. With an ageing population in Singapore, the LTA has now introduced Silver Zones to further enhance the safety of seniors. Silver Zones aim to enhance the safety for seniors by altering the character of the street to slow vehicles down.

The Safer Roads Industry Taskforce (SRIT) was conceived by the Ministry of Home Affairs and the TP, to serve as a key plank for their engagement strat-

Fig. 11. Pedestrian and Cyclist Safety Committee.

egy under the Safer Roads Singapore Plan. It was chaired by then Senior Minister of State for Home Affairs Mr Masagos Zulkifli, with the then Parliamentary Secretary for Transport Associate Professor Muhammad Faishal Ibrahim as the vice-chairman (*MHA, 2013*). Its inaugural meeting was held in October 2013.

The Safer Roads Industry Taskforce's objectives serve as a platform for the government and industry to share and co-develop initiatives to promote safer driving for vocational drivers. More specifically, the Safer Roads Industry Taskforce aimed to achieve the following:

(1) Review current road safety situation and existing programmes for vocational drivers;
(2) Propose road safety initiatives and programmes for vocational drivers; and
(3) Create platforms for vocational drivers and employers to continue with the development and sharing of road safety initiatives.

The Safer Roads Industry Taskforce's members included various representatives of stakeholder groups from the industry, unions, government and other key stakeholders. Over its one-and-a-half-year tenure, the committee took reference from best practices internationally by consulting widely with these stakeholders, in its attempt to apply them locally.

4. Key Road Safety Initiatives

In Singapore, the LTA and the TP are the two government agencies responsible for managing road safety. The LTA provides a safe physical road network for road users, while the TP enforces traffic regulations, complementing this with public education programmes. In addition to the efforts made by the government, the LTA and the TP partner with other non-government organisations such as the Singapore Road Safety Council and the private sector to enhance and promote road safety (*Ding and Ho, 2014*).

The LTA's partnership with the TP has been well-established over the years. Now with the new player, Singapore Road Safety Council, which came onboard in 2009, road safety is able to be dealt with in a more holistic manner. One recent outcome of this successful partnership among the LTA, the TP and Singapore Road Safety Council is the Singapore Road Safety Month (aligned with United Nations (UN) Decade of Action for Road Safety 2011–2020 launched in over 100 countries) held in May 2013 and April 2014. The LTA also collaborates with the private sectors, institutions and road users to achieve major improvements in road safety.

Road accidents result in loss of lives and injuries, and lead to congestion and delays for road users. Thus, it is important for the LTA to continue in its efforts to enhance road safety, especially for the vulnerable groups including the seniors.

Road safety professionals can use engineering measures to prevent and/or reduce road accidents. However, there is also a need to champion greater road safety awareness among all road users. With more vehicular traffic and a growing population each year, the LTA has implemented various road safety measures to ensure road fatalities in Singapore are kept low.

Road safety strategies

While safety is of paramount importance, we need to have a balance between road safety and traffic efficiency. The LTA implements the world's best practices for road safety management and delivers safe roads to the public, through the planning, implementing and managing of various road safety initiatives. Processes and techniques such as Road Safety Assessment, Black Spot Programme, Crash Site Investigation and remedial treatment programmes are well established best practices being used.

The LTA's role in enhancing road safety focuses on three key road safety strategies as illustrated in Fig. 12. The first strategy is the Accident Prevention

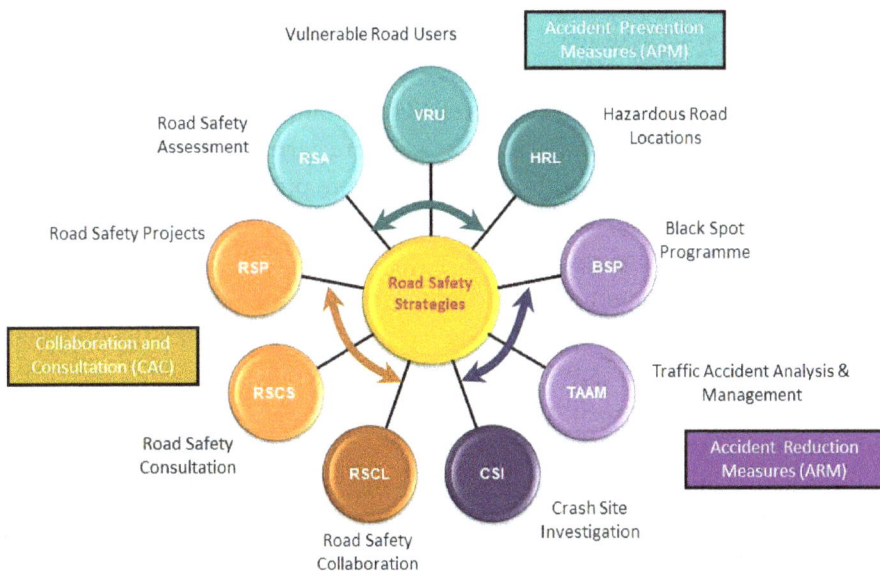

Fig. 12.   Framework of road safety initiatives in Singapore.

Measures (APM) to treat locations before they become black spots, while the second strategy is the Accident Reduction Measures (ARM) to treat locations where accidents have occurred. Lastly, the Collaboration and Consultation (C&C) strategy focuses on managing sustainable collaboration and consultation efforts with major stakeholders in road safety.

In short, the three key strategies have enabled the development of a sustainable framework to improve road safety in Singapore.

4.1. *Accident Prevention Measures (APM)*

Accident Prevention Measures (APM) aim to treat potential hazards/deficiencies before numerous injury accidents occur. Under this strategy, three programmes are carried out, namely Road Safety Assessment (RSA), Hazardous Road Locations (HRL) and Vulnerable Road users (VRU).

4.1.1. *Road Safety Assessment (RSA)*

Singapore's road network consists 164 km of expressways and 662 km of arterial roads (*LTA, 2014b*). It is thus necessary to carry out a formal and independent process to audit and treat potential road hazards to reduce the likelihood of accident occurrences or the severity if an accident does occur. Austroads Road Safety Audit (*Austroads, 2009*) procedure is adopted for RSA programme in Singapore. RSA programme is categorised under two categories — operation safety of existing roads and safety of those roads undergoing road developments or new road projects at design, construction and post construction stages.

The first RSA on existing roads was carried out on Pioneer Road in 1999. This was then followed by the expressways spanning from 1999 to 2000. Subsequently, assessments of arterial roads were carried out from year 2002 onwards. The roads identified for assessment were prioritised based on speed limit and accident occurrences. Arterial roads with higher speed limit and accident occurrences are assessed first.

During RSA, the auditors walk or drive through the roads as pedestrians and motorists in both directions during day and night to identify the potential hazards. Both general and site-specific observations are reported with measures recommended. These measures are then implemented or addressed with alternate measures if they are unable to be implemented due to site constraints.

The before and after photos in Figs. 13 and 14 show examples of hazards picked up during the audit. The section along Havelock Road (Fig. 13) is a sharp bend without delineation. As such, Curve Alignment Markers (CAMs) were installed to delineate the bend for better visibility so that the motorists can maneuver the bend better. Figure 14 shows a section of a damaged footpath

Fig. 13 (i)

Fig. 13 (ii)

Fig. 13.   (i) Before — No delineation, (ii) After — Bend delineated with CAMs.

along Lorong Ah Soo which could be a tripping hazard to pedestrians. The damaged footpath was then rectified by the maintenance team.

Assessment on all road projects that involves physical change to the road system, and to all projects related to the management and control of traffic on

Fig. 14 (i)

Fig. 14 (ii)

Fig. 14.   (i) Before — Damaged footpath, (ii) After — Footpath rectified.

the road system, are carried out under Project Safety Review (PSR). It was introduced in 2000 as part of the LTA's initiatives to enhance the safety of road users in new road projects. The purpose of the process is to eliminate safety issues before project construction (*Loh, 2008*). Under this process, potential safety hazards of proposed schemes, detailed design of projects (e.g. details of road infrastructure) and roads affected by construction works for duration of at least nine months were identified and mitigated or designed out. Assessment is also carried out after construction to ensure that projects are constructed as designed.

An example of the benefits of PSR is illustrated below (Fig. 15). An audit was carried out during the detailed design stage for a proposed pedestrian overhead bridge and the associated covered linkway connected to an existing bus stop. The steps at the platform of the bus stop were identified to be a potential tripping hazard to bus commuters. As such, the steps were removed as part of the design, thus improving safety for bus commuters (*Loh, 2008*).

Arising from issues that were frequently raised during the assessment, reviews are sometimes carried out on certain existing practices or designed to improve safety for road users. For example, open drains on centre medians could potentially be safety hazards. Errant vehicles that encroach onto the median, especially at tight bends, could overturn as a result. As such, open drains on medians and side tables with widths greater than 450 milimetres are recommended to be covered with concrete slabs or gratings (Fig. 16).

As of June 2015, 86% of the existing arterial roads had been audited, with the majority from high speed roads i.e., with speed limits of 60 km/h and 70 km/h.

4.1.2. *Hazardous Road Locations (HRL)*

Hazardous Road Locations (HRL) is a proactive measure that assists to prevent a potentially hazardous location from becoming a black spot. The locations are identified through infrastructure hits or public feedbacks. These infrastructure hits include hits on crash cushions, vehicle guardrails and traffic light poles.

Although these locations have not developed into black spots, the frequent hits indicate that there are potential safety deficiencies at these locations which need to be mitigated. As such, databases with infrastructure hits are reviewed on a regular basis so that such locations can be identified and treated; for example, arising from the frequent hits on crash cushions. This could also be an indication that certain expressway exits, where the crash cushions were

Fig. 15 (i)

Fig. 15 (ii)

Fig. 15. (i) Before — Steps at bus stop (*Loh, 2008*), (ii) After — Steps removed from bus stop (*Loh, 2008*).

installed at the gore, were not conspicuous enough to the motorists, leading to failed last-minute maneuvers by inattentive motorists. At these areas, a row of flexible posts are installed at a distance in front of the crash cushions to alert the motorists as a mitigation measure (Fig. 17).

Fig. 16 (i)

Fig. 16 (ii)

Fig. 16.   (i) Before — Open drain, (ii). After — Open drain covered with grating.

Fig. 17.   Flexible posts installed in front of the crash cushion.

4.1.3. *Vulnerable Road Users (VRU)*

Vulnerable Road Users (VRU), an accident prevention measure strategy, places emphasis on enhancing the safety of the relatively more vulnerable road users such as pedestrians, cyclists and motorcyclists. As compared to the general motor vehicle population, these VRUs do not have protection in the form of a metal chassis; hence their involvement in any road traffic accidents would possibly result in severe injury or worse, fatality.

Pedestrians are viewed the most vulnerable because they are not protected by any device nor gear; especially the young and seniors. Due to their age, majority of the young lack the ability to evaluate traffic conditions accurately and safely; while seniors are less agile, with poorer decision making skills resulting from slower perception and response time. Hence, more care is taken to provide specially demarcated areas, such as School Zones and Silver Zones to further enhance the safety of these pedestrian groups, respectively.

Interaction and conflict between cyclists and motorists are reduced through the introduction of off-road cycling tracks. These dedicated cycling tracks run parallel to main roads, as well as through housing estates and parks; helping cyclists get from one location to another. Other forms of safety enhancements for cyclists include a signalised bicycle crossing at mid block crossings, where there is clear demarcation to separate cyclists from pedestrians at busy traffic lights.

Motorcyclists, though the most maneuverable, however, are the least visible when traveling on the road. Especially during self-skidding accidents, motorcyclists face significant risks of crippling injury or even death, should they strike the support posts of vehicular guardrails during such accidents.

As such, possible areas of improvement for the different vulnerable road users are identified for further road safety enhancement. This is when the strategy of Road Safety Projects, under Collaboration and Consultation, complements the VRU strategy. Road safety initiatives that help to address specific road safety issues are explored under this collaboration of strategies.

4.2. *Accident Reduction Measures*

Accident Reduction Measures (ARM) aim to treat locations where accidents have occurred, and remedy existing problems that have contributed to accidents. This is to reduce the number and severity of future accidents. Under this strategy, three programmes are formulated, namely Traffic Accident Analysis and Management (TAAM), Crash Site Investigation (CSI) and the Black Spot Programme (BSP).

4.2.1. *Traffic Accident Analysis and Management (TAAM)*

In the past, accident points were manually recorded on a big map in the TP headquarters. Nowadays, the systems are all computerised. Accident information is logged into the database much more efficiently, which also allows the authorities to monitor accident rates much more effectively, taking pre-emptive measures swiftly.

With continued efforts in engaging stakeholders, educating the general public, infrastructure improvements, and deterrence with enforcement, Singapore strives to work towards even safer roads for the benefit of all road users.

Traffic Accident Analysis and Management (TAAM) is a customised in-house developed Geographical Information System (GIS) application that was commissioned in 2004. In 2011, it was enhanced to run on a web-based environment for some of the functions (Fig. 18). The system is designed with four principal functions: data management, data validation and amendment, data query and analysis, graphs and report generation (*Hau et al., 2007*).

Fig. 18.   TAAM homepage.

Concurrently, there is another independent system named Accident Location Registration System (ALRS) which is stationed at the TP headquarters. This system enables a TP investigation officer to mark the accident points on a GIS map. At the same time, the investigation officer will also key in the accident data in the TP's in-house system called Traffic Incident Management System (TIMS). Every month, both textual and spatial records are uploaded into TAAM for an LTA officer to analyse the accidents.

Users can access the system to make multiple queries to obtain accident statistics for specific locations and time periods, to query the database by driver, road, vehicles characteristics, and to generate reports (Figs. 19–22). The graphical interface of TAAM allows the viewer to see onscreen maps of accident locations and to select specific nodes and links of interest at a click of the mouse. The results of TAAM queries can be displayed in tabular form and in maps. Most importantly, TAAM helps the road safety engineers to identify black spots easily through the use of spatial tools. As such, they are able to focus on problematic sites and target the treatment in a cost-effective manner. It also serves as a database to store textual and spatial accidents in an organised, structured and integrated manner (*Hau et al., 2007*).

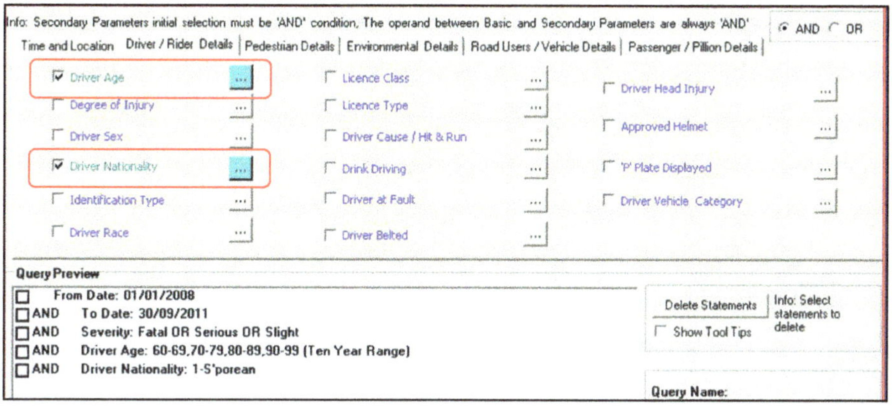

Fig. 19.   Database query by driver.

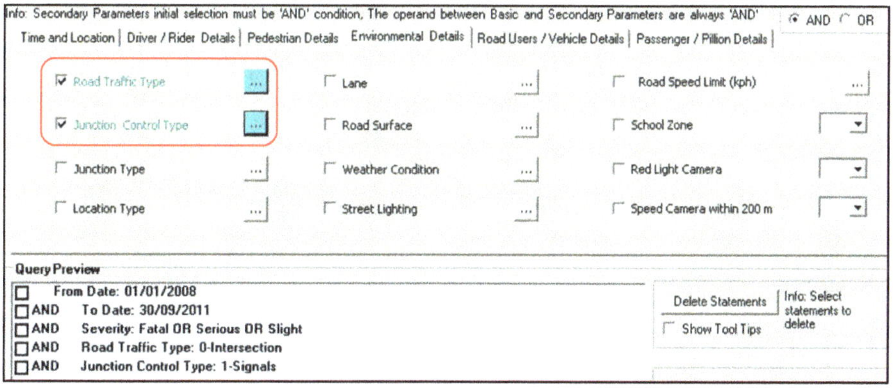

Fig. 20.   Database query by road.

TAAM has enhanced and improved the efficiency of the LTA in tackling road safety problems. It has helped to solve the problems faced by Singapore, pertaining to data collection, storage and analysis. With the support of TAAM, the LTA started the Black Spot Programme to identify problematic sites and carried out appropriate treatment at targeted sites. Textual and spatial accident information is also retrieved effectively and efficiently as and when required. The system analyses accident data from different perspectives to identify problematic sites and to apply more appropriate and cost-effective solutions to improve the situation for the safety of motorists and pedestrians (*Hau et al., 2007*). Figure 23 shows a typical display of accidents generated from TAAM.

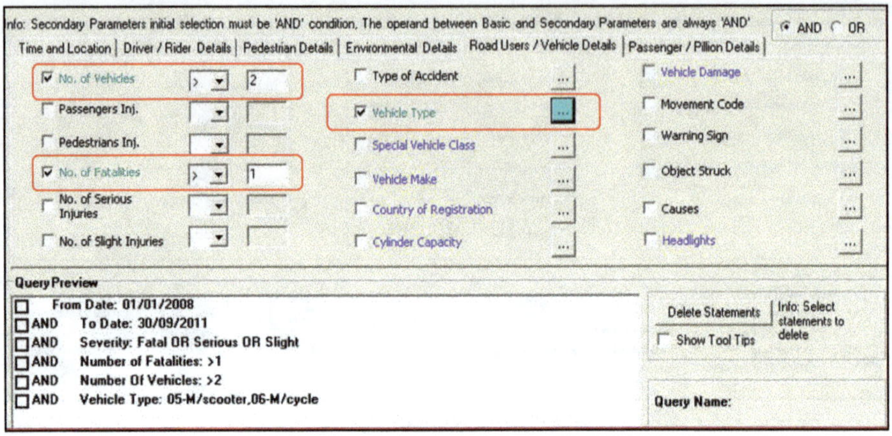

Fig. 21.   Database query by vehicle characteristics.

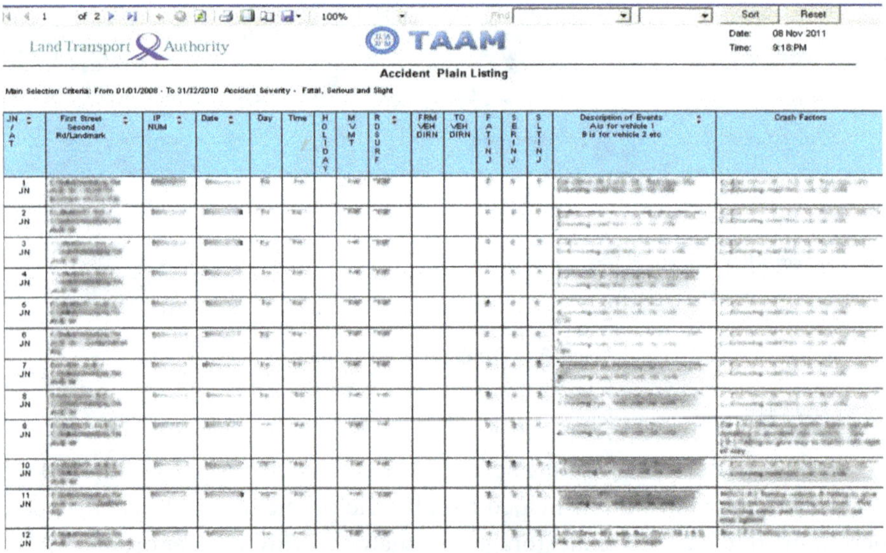

Fig. 22.   Database query to generate report.

4.2.2 Crash Site Investigation (CSI)

CSI is one of the core areas in Accident Reduction Measures. It was set up in 1998 to focus on fatal accidents that are not caused by drink/drug driving, vehicle malfunction and health-related issues. The TP will send out a weekly

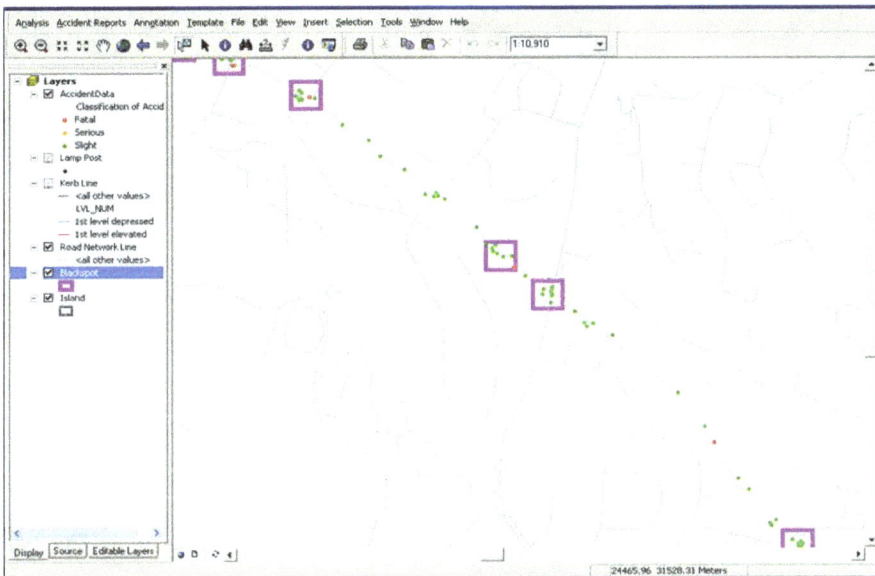

Fig. 23.   Accident clusters identified through TAAM indicating black spot.

fatal accident report to the CSI administrator. After which, a team of two will be assigned by the administrator to investigate the fatal location and record their findings on any engineering deficiencies. Thereafter, the team will propose safety counter-measures in a CSI investigation report. The CSI report and recommendations, together with site photos, will be submitted to CSI administrator. The team will then proceed to prepare and implement the recommendations.

Before 2005, when the team was assigned a fatal case to investigate, the fatal accident would be labelled and marked manually on a map showing Singapore's road network. This helps to provide a macro view of the location and frequency of fatal accidents on our road network. As more fatal accidents are marked, the team will be able to divert more attention and time at such accident prone locations for further mitigating measures to prevent future accidents and improve safety levels. This process has since stopped after the computerised-based TAAM was commissioned in 2004. Accident data is now stored electronically in a database, and it includes serious and slight injury accidents as well. Historical accident data for past three years can be generated at the accident location and used by the CSI team as part of their investigation and assessment. Some examples of CSI cases that have been investigated and treated with engineering improvements are shown in Figs. 24 and 25.

Fig. 24. Bukit Batok East Avenue 3 — Sealing up the large gap between existing railings to prevent jaywalking.

Fig. 25. Whampoa Road/Whampoa Drive — Painting of turning pockets, Dashed Pedestrian Crossing Line and Count-down timer.

4.2.3. *Black Spot Programme (BSP)*

Black Spot Programme (BSP) started in 2005 (*Hau and Ho, 2010; Koh and Ho, 2006*). This is another core area in Accident Reduction Measures. It is an important road safety tool to identify locations with high occurrence of traffic accidents so that appropriate engineering measures can be implemented to treat these black spots and make our roads even safer.

Black spots are identified from the past accident data through TAAM. Historical data from the past three or five years is retrieved for analysis. This time period is considered adequate to provide a reliable and accurate historical trend and pattern of accident occurrences. The accident results generated are then displayed in tables and map format. After the analysis is carried out, site

Fig. 26. Bedok Reservoir Road/Tampines Avenue 1/Tampines Avenue 4.

investigation is conducted to relate the accident results from past years with the road condition, the surroundings and characteristics of road users to ascertain the potential improvements or measures for the location. The findings collected from the site will be helpful to determine the appropriate treatments for the black spot location.

Many black spot locations have benefited from BSP since its launch ten years ago. On average, the LTA has successfully treated and removed five to 10 locations each year from its list of black spots. Locations like Bedok Reservoir Road/Tampines Avenue 1/Tampines Avenue 4, and Seletar Expressway/Bukit Timah Expressway have witnessed a big drop of over 70% in accidents after counter-measures were introduced.

Figure 26 shows the before and after treatment at the junction of Bedok Reservoir Road/Tampines Avenue 1/Tampines Avenue 4. Before treatment, the junction allows permissible right-turn movements between traffic gap and right-turn phase. After treatment, full controlled right-turns are implemented (Red–Amber–Green) arrows to reduce potential conflicts between ahead and right-turning traffic.

Figure 27 shows the before and after treatment along the slip road at Seletar Expressway/Bukit Timah Expressway. The treatment included lengthening existing speed regulating strips, reducing lane width with chevron markings and resurfacing the road to increase its skid resistance.

BSP is also a recommended best practice in many countries such as the United Kingdom, Australia and New Zealand. The LTA will continue to work towards reducing the number of black spots on our roads for the health and safety of all our road users.

Fig. 27.   Seletar Expressway/Bukit Timah Expressway.

4.3. *Collaboration and Consultation*

The earlier two strategies (APM and ARM) adopt a systematic approach by applying road safety engineering principles and essential tools. In the Collaboration and Consultation (C&C) strategy, it focuses on managing sustainable collaboration and consultation efforts with major stakeholders of road safety and implementing road safety projects.

4.3.1. *Road Safety Collaboration (RSCL)*

As aforementioned, the LTA and the TP are close working partners. Besides sharing accident data, collaboration on various road safety events, the TP and the LTA also work hand-in-hand on joint road safety works such as the review of speed limits in Singapore. The most recent comprehensive speed limit review on arterial roads was carried out in 2005 and 2006 on a region-by-region basis. This review led to the speed limit revision of 247 sections of roads over six phases. 78% of the reviewed roads have their speed limits revised from 50 km/h to 60 km/h and 18% from 50 km/h to 70 km/h (*Hau and Ho, 2014*). The purpose was to have consistent speed limits that match the driving environment for better compliance in the posted speed limit. A consultancy study was carried out in 2010 to evaluate the impact of this revision. Based on available data, the study found that the speed limit revisions did not result in any distinctive change in safety level or substantive change in traffic operating speeds (*Hau and Ho, 2014*).

The most recent speed limit revision on expressway was carried on Kallang–Paya Lebar Expressway (KPE) tunnel. The speed limit in the tunnel was increased from 70 km/h to 80 km/h on 29 December 2013, except for a short section, which has its speed limit retained at 70 km/h due to the tight geometry.

Apart from various collaboration efforts with major stakeholders locally, road safety collaboration (with international partners) provides an excellent platform for the LTA to promote its road safety engineering experience and explore business/commercial opportunities with international organisations advocating road safety best practices. This helps raise the standing of our local expertise in the international arena. Through the working relationship established over the years with various international bodies such as Global Road Safety Partnership (GRSP), Road Engineering Association of Asia and Australasia (REAAA), Asia–Pacific Economic Cooperation (APEC) as well as United Nations Economic and Social Commission for Asia and the Pacific (UNESCAP), the LTA has been contributing to the international community in the area of road safety expertise and exchanging best practices.

Besides collaborating with organisations, the LTA works with the local communities to enhance road safety by fostering sustainable partnerships with community leaders, and supporting the community's involvement in transport issues including road safety. From experience, the LTA has learnt that road safety features would not be as effective if the road users were not educated to use them correctly. Thus, the LTA reaches out to the communities in hopes of educating them, as well as to listen to their concerns. Brochures, leaflets, posters and interactive games for motorists and pedestrians have been developed to educate the various road users on the safe use of the road facilities.

4.3.2. *Road Safety Consultation (RSCS)*

The LTA provides consultancy in the areas of road safety engineering such as Road Safety Assessment, Road Safety Programmes for vulnerable road users and Black Spot Programme. The LTA also conducts road safety engineering training courses to share their experiences in road safety engineering and management with many countries. Road safety activities, programs and training carried out by the LTA include providing practical expertise to encourage capacity building and knowledge transfer. So far, the LTA has trained delegates from more than 35 countries, some of which include Gambia, Tunisia, Egypt, Yemen, Uganda, Brunei, Cameroon, Cambodia, Vietnam, Laos, Turkey, Abu Dhabi, Mauritius, Seychelles and Maldives.

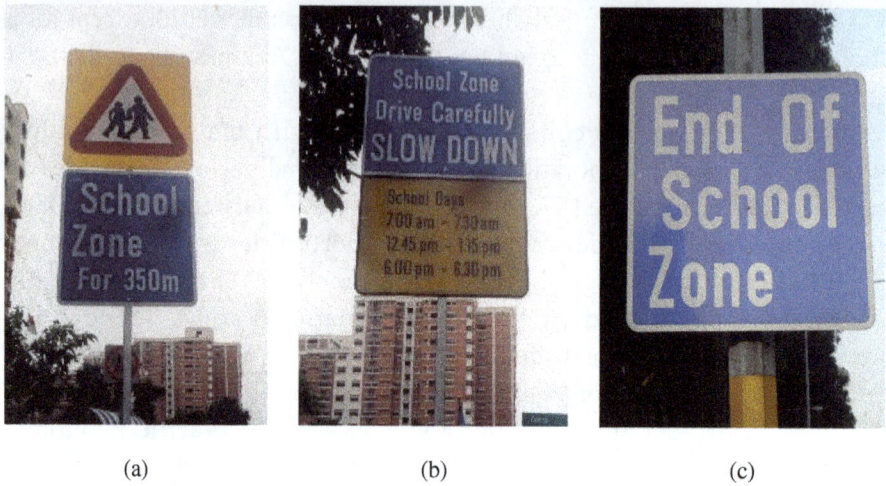

(a)	(b)	(c)

Fig. 28.   Left to Right — (a) Start of school zone, (b) Operation timing and (c) End of school zone signs.

4.3.3. *Road Safety Projects (RSP)*

Next, we highlight some of the key road safety projects (RSP) that the LTA has developed over the years with examples of engineering road safety schemes that have been effectively used.

RSPs are implemented to enhance road safety for various road users, especially vulnerable ones such as motorcyclists, pedestrians and bus commuters. These initiatives are based on international best practices and adapted to suit our local environment. The LTA has been continually exploring, evaluating and adapting new road safety projects that further enhance safety (*Koh et al., 2009*). Since its establishment, the LTA has introduced, implemented and developed criteria for more than 20 road safety initiatives.

To enhance pedestrian safety:

(1) School Zones

In the late 1990s, there was concern regarding the safety of young school-going children, especially those in primary education. This spurred the discussion to enhance road safety along primary school frontages. In 2000, the LTA introduced the concept of School Zone. The intent was to slow down motorists on school days during nominated times where there were activities along the school frontage.

(a)　　　　　　　　　　　　　　　　　　(b)

Fig. 29.　Left to Right — (a) Jumping Amber Lights and (b) Red pavements.

During the start and end times of primary school sessions, there is usually heavy traffic in the vicinity of schools. As school children are vulnerable road users with limited experience in using the roads, it is of paramount concern to keep such areas safe. The School Zone toolkit then, consisted of a set of traffic signs (see Fig. 28) to:

(a) Demarcate the start of a School Zone and its road length
(b) Highlight peak periods with relatively higher student pedestrian volume
(c) Mark the end of a School Zone

To complement the School Zone, a Safe Drive Zone (SDZ) was also introduced with additional involvement from the community (e.g. parents or residents, who volunteered as school traffic wardens, helped to look out for students' safety during peak hours). A noticeable component of the SDZ was the introduction of the Jumping Amber Lights (see Fig. 29). This feature comprised of a pair of amber lights that read "SCH" and "ZONE" which flash alternately. The flashing effect was meant to easily capture the attention of motorists to warn them to slow down and drive carefully as they are entering a School Zone.

The year 2004 saw the introduction of the Enhanced School Zone. This came along with two patches of red asphalt pavements and "SLOW DOWN"

Fig. 30. "40 km/h When Lights Flash" sign.

markings on the road for better visibility. The School Zone signs have been modified to make it easier for motorists to read while maintaining the purpose, which is to slow them down. The red asphalt pavement texture is meant to be highly conspicuous to alert motorists through visual and vibratory effects that they are travelling within a School Zone, thus prompting them to slow down. Similarly, the "SLOW DOWN" markings were also implemented at the start of a School Zone to achieve the same goal.

The latest enhancement in 2014 introduced the "40 km/h When Lights Flash" sign (Fig. 30). It displays a pair of amber aspects that flash alternately during school peak hours. During this time, drivers are to keep within the reduced speed limit of 40 Kilometres per hour as there will potentially be a heavier student pedestrian volume. The LTA has begun with a pilot phase which includes 10 School Zones, and will progressively expand implementation island-wide.

(2) Barrier-Free Accessibility

Since 2006, the LTA has been putting efforts into enhancing accessibility to major transport nodes (*LTA, 2011*).[31] Apart from the young, the seniors are another group of users to focus on. With ageing, the susceptibility to injury becomes greater. Seniors are more prone to serious injury or death as compared to young adults. They also suffer from common ailments such as muscle atrophy, arthritis and cataracts. These health conditions can reduce their ability to cross roads safely.

Under the LTA's vision to create a people-centred land transport system, the LTA aims to meet the diverse needs of the people. This includes providing a safer and more conducive walking environment for senior pedestrians as they require more time to cross roads, while managing the impact on traffic efficiency to avoid unnecessary traffic congestions.

To make our roads more pedestrian-friendly, the first step that the LTA took was to provide pedestrians with barrier-free accessibility (BFA) by ramping down selected parts of the sidewalks and kerbs that had relatively higher pedestrian volume. Apart from creating a more comfortable environment for seniors to get off the roads and onto footpaths easily and safely by eliminating the need for seniors to step up and down kerbs, wheelchair-bound pedestrians and other wheeled pedestrians (with prams or trolleys) could also benefit from this initiative. Senior-friendly features such as at-grade crossings are provided in areas where the seniors frequent, if traffic conditions allow. Apart from that, where bollards are installed, the clearance between them is also made sufficient for wheelchair-bound pedestrians to manoeuvre easily. In addition to that, railings are also selectively installed along centre medians to deter pedestrians from jaywalking and channel them to safe crossing points.

(3) Green Man+

Generally, the pedestrian green time is set based on the width of the road in accordance with a certain walking pace. However, as senior pedestrians require more time to cross the road, Green Man Plus (Green Man+) is introduced at areas frequented by seniors.

The Green Man+ is an initiative introduced by the LTA in October 2009 to address the needs of senior pedestrians who may require more time to cross the road (*LTA, 2014c*). This is one way in which the LTA leverages on technology to better meet the needs of our ageing population.

The Green Man+ uses Radio Frequency Identification (RFID) technology to detect senior pedestrians crossing the road. The RFID readers are mounted on the traffic light poles to detect the RFID cards, in the form of a Senior Citizen concession public transport card, held by senior pedestrians who wish to cross the roads.

Fig. 31.   Green Man+.

Senior pedestrians can activate the Green Man+ function simply by tapping their Senior Citizen concession cards on the reader mounted at the traffic light pole (Fig. 31). Once detected, the system will extend the 'green man' timing to allow them more time to make their way across the roads. The additional green man time will be extended ranging from three to 12 seconds depending on the width of the crossing.

With these changes, now extended to pedestrians with disabilities, they can enjoy more green man time at more pedestrian crossings. Instead of the Senior Citizen concession card, pedestrians with disabilities can similarly activate the Green Man+ function with the Green Man+ card issued to them.

Pedestrian crossings with the Green Man+ feature will be prominently marked. Besides the lighted LED indicators, when pedestrians with disabilities tap the Green Man+ Card, the new card readers will also produce a sound and vibration alert to notify the user that the green man time will be extended for the next cycle.

By 2014, a total of 500 pedestrian crossings have been implemented with the Green Man+ function.

(4) Silver Zone

The Silver Zone initiative is a national project to enhance road safety for seniors. This initiative was announced by Associate Professor Muhammad Faishal Ibrahim, then Parliamentary Secretary at the Ministry of Health and the Ministry of Transport, during the Committee of Supply (COS) in March 2014. The Silver Zone initiative consists of the implementation of a series of road safety engineering enhancements, as well as a 40km/h speed limit where feasible. Motorists entering Silver Zones will experience the gateway treatment which includes Silver Zone signs, three rumble strips (painted yellow) and speed limit road marking (painted white) (Figs. 32–34). Various road safety engineering measures will be applied to alter the character of the street, making the road a much safer environment for senior pedestrians. Examples include the use of pinch points with ramp-downs to create a road-narrowing effect and to allow pedestrians to cross the road with ease at a reduced crossing distance. The use of chicanes creates S-shaped curves along straight roads, which will induce a traffic-calming effect, slowing motorists down. Another innovative safety measure is "Eye-land", which is an expanded centre divider to facilitate a two-stage crossing (Figs. 35 and 36). A by-product of these measures is that they will also improve safety for other pedestrians, and not just seniors.

In the early stages when the idea of the Silver Zone was being conceived, its road safety toolkit and implementation criteria were contrived. In order to

Fig. 32.   Typical entrance before Silver Zone gateway treatment (Before).

Fig. 33.   Typical Silver Zone gateway treatment (After).

create distinctive Silver Zones, the LTA took a bold move to change the character of the roads within Silver Zones with fresh and innovative road safety engineering measures. It is important to plan ahead for the infrastructural needs of the seniors. Moving ahead, with an ageing population and increasing traffic volume on the roads, it is critical to provide a safer and more conducive walking environment for senior pedestrians while managing the impact on traffic efficiency. The Silver Zone will become an important initiative in balancing the needs of senior pedestrians and motorists.

The Silver Zone creates a safe and conducive walking environment for seniors to visit community centres, markets, dialysis centres, food centres, hospitals, welfare homes and polyclinics. This is achieved by drastically changing the character of the road with the implementation of the new road safety engineering measures. Since the announcement of Silver Zones in the COS 2014, there have been requests from Members of Parliament (MPs), grassroots and the public for the provision of Silver Zones in their estates.

Although the Silver Zones will drastically enhance road safety for seniors, some of the measures used may result in a decrease in traffic efficiency. To address this, the LTA engages the community to seek their understanding and also makes efforts to address concerns raised on the ground. This allows Silver Zones to be inclusive and garner greater public acceptance.

Fig. 34.   Chicanes creates S-shaped curves along straight roads which slow motorists down at Bukit Merah View.

Fig. 35.   Before the construction of "Eye-land".

Fig. 36.   "Eye-land" (an expanded centre divider) to facilitate a two-stage crossing at Bukit Merah View.

Since the announcement of Silver Zones during the COS in 2014, the initiative has become widely popular, receiving requests from Members of Parliament (MPs), grassroots and the general public for the provision of these zones in their estates.

At the zebra crossing

(5) Zebra crossings

The zebra crossing, as its name implies, symbolizes the striped pattern of a zebra with alternating black and white stripes that straddle across the road. They are provided to give pedestrians the priority in crossing the road.

The zebra crossings in Singapore come with safety features such as the blue 'Pedestrian Crossing' sign, which is mounted onto the individual flashing beacons painted in black and yellow; and zigzag lines to indicate to a motorist he is approaching a zebra crossing, as well as to prohibit parking near the zebra crossing. Road studs are placed before the stopline of the zebra crossing to delineate it to the motorists. The blue pedestrian crossing signs attached to the flashing beacon poles (Fig. 37) indicate to pedestrians and motorists the

Fig. 37.   Pedestrian Crossing sign mounted on flashing beacon poles.

position of a designated crossing. This further enhances the visibility and alerts motorists of the approaching zebra crossing.

Previously, the flashing beacons at zebra crossings were lighted up with halogen bulbs. Nowadays, these beacons have been replaced with LEDs for better reliability, and lower maintenance costs. Where lighting at a zebra crossing is insufficient, additional lighting known as the floodlight can also be installed to provide a greater contrast to the zebra crossings compared to the surroundings, allowing motorists to better look out for pedestrians.

(6) Raised Zebra Crossing

Raised zebra crossings are elevated by 100mm from the existing road level with its slopes painted in a black-and-yellow checkered design (see Fig. 38). This elevated design not only gives a vertical deflection to passing motorists which slows them down but also allows pedestrians to be more visible from afar to motorists, hence improving the safety of the pedestrian.

One of the locations where these raised zebra crossings are commonly found is near schools where children activities are higher. With this elevated crossing, shorter and smaller pedestrians will be more easily seen, and the speed of the moving vehicles is reduced, hence creating a safer environment for children to cross the road.

Fig. 38.   Raised Zebra Crossing.

(7) "LOOK" markings

The LTA has been constantly looking into pedestrian safety as pedestrians are the most vulnerable users on the road. "LOOK" marking was one of the initiatives arising from brainstorming sessions that was found to be feasible for exploration. From literature review, it was found that in other countries, the format of using the wordings "LOOK LEFT/RIGHT" was most common.

To cater to Singapore's multi-racial society, especially with the ageing population, the LTA adopted a more diagrammatic and intuitive design. As such, Singapore's version of "LOOK" markings was created. The 'O's in the "LOOK" markings were designed to look like a pair of eye balls, looking at the predominant traffic direction to remind pedestrians to look out for vehicles from that direction (Fig. 39). There are also two arrows at each side of the "LOOK" to remind pedestrians to look out for traffic from both directions, but with the bigger arrow pointing towards the predominant direction as a further reminder. The markings are painted on the road pavement, between the ramp of the footpath and first white strip of the zebra crossing, so that pedestrians could see and be reminded before they step onto the carriageway to cross the road.

"LOOK" markings were well received by the public. As such, it was expanded to zebra crossings in residential areas with a high senior population, as well as those in the school zones of primary and secondary schools.

Fig. 39. "LOOK" markings at zebra crossing along Prince Charles Crescent.

At the signalised pedestrian crossing

(8) Dashed Pedestrian Crossing Lines (DPCL)

The Dashed Pedestrian Crossing Lines *(DPCL)* provide a clearer differentiation from the continuous stop line and encourages more motorists to comply with the stop line (see Fig. 40). The programme to replace the previous continuous pedestrian crossing lines started in February 2009 after a successful pilot trial at Victoria Street. The programme was well received by members of the public as they felt that motorists became more compliant to stop before the vehicular stop line. Even the Singapore Association for the Visually Handicapped (SAVH) welcomed the new initiative as the new DPCL helped visually-impaired pedestrians to cross the roads safely by using their walking sticks to guide them within the pedestrian crossing.

(9) Green-Man Countdown Timer

At a signalised pedestrian crossing and faced with the flashing green man signal, would you still cross the road, or would you speed up your walking pace?

Green-Man countdown timers help to manage pedestrians' expectations, as well as make informed decisions, on whether to cross the road when faced with the above scenario. It serves as part of the overall efforts to enhance the "walking" environment. Countdown timers indicate the amount of time left in the pedestrian crossing phase (Green Man) before the Red Man appears.

Many existing signalised pedestrian crossings only have the flashing Green Man to warn pedestrians to increase their walking speed when the green phase

Fig. 40.   Dashed Pedestrian Crossing Lines.

for pedestrians is almost over. Since the remaining available green phase for pedestrians is unknown to them, they have to judge for themselves whether to cross if they are at the beginning of the crossing, or to what extent to increase their walking pace in order to cross the road safely in time. Accidents could result if wrong judgments are made.

The original intent for the provision of the Green-Man countdown timer was primarily to increase the pedestrian walking speed across wide junctions and at intersections where the volume of left/right turning traffic is high. However, the Green-Man countdown timer has gained popularity over the years as pedestrians find it useful with the indication of the amount of time left before the flashing Green Man terminates.

Information on the crossing time available not only encourages pedestrians to walk faster but also discourages pedestrians from starting to cross if they were uncertain about their ability to cross within the remaining flashing green time. This is especially useful for the senior population. On the other hand, motorists also reap the benefits of not being unduly delayed due to the straggling pedestrians who cross the road against the Red Man.

The Green-Man countdown timer was developed in the 1990s to improve pedestrians' safety at traffic junctions (*Koh et al., 2014*). It was first introduced

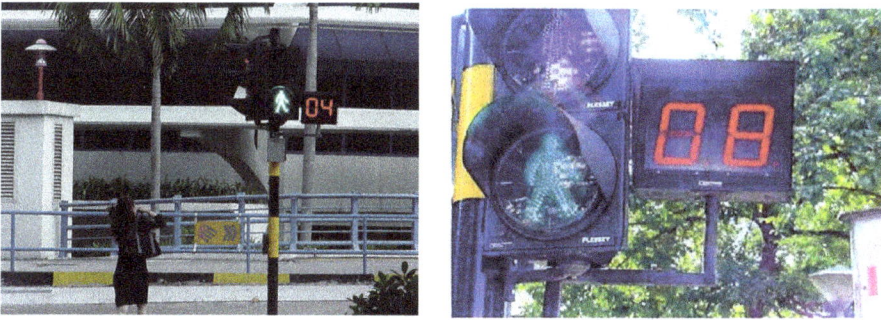

Fig. 41.   Old design of a Green-Man countdown timer.

in the form of a separate signal housing attached to the pedestrian signals, as shown in Fig. 41.

Subsequently in 2006, the design of the Green-Man countdown timer developed into the Integrated Pedestrian Countdown (IPC) (Fig. 42) that we see at many of our crossings now.

The IPC is first installed at crossings where there is a high volume of pedestrians or at locations where there is a high percentage of vulnerable pedestrians like children and seniors who use the crossing, such as in the vicinity of schools, markets and shopping centres. Subsequently, it will be implemented at all pedestrian crossings.

(10) Right turn vehicular conflict with pedestrians

Singapore adopts permissive right turns (left-hand driving) at traffic junctions; right turning motorists can look out for gaps in the opposing straight traffic and make a right turn, provided they give way to adjacent crossing pedestrians or cyclists. Typically, pedestrians or cyclists would be in the motorist's view by the time he/she is able to find a gap from the cleared straight-through traffic. Nevertheless, there would still be some occasions (during off-peak periods) where right-turn motorists may come into conflict with pedestrians and cyclists.

The most effective and straight forward safety measure is the fully controlled right-turn (i.e., Red-Amber-Green arrows — RAG) where right turning vehicles are given an exclusive phase to turn without coming into conflict with opposite straight-through traffic, pedestrians and cyclists. This measure, though highly effective in minimising conflicts, can have some impact on the public as they can experience traffic congestion. Typically, mitigating measures such as adjusting signal timings and allowing repeated right-turns are

Fig. 42.   Integrated Pedestrian Countdown (IPC).

implemented to complement this measure. The LTA can recall receiving lots of feedback from the public when RAG was first implemented. But gradually, the feedback died down, suggesting that the public required time to adjust to the new measure. In conclusion, there must be a good balance between safety and traffic efficiency for measures to work effectively and be well-received by public.

(11) Intelligent Road Studs (IRS)

In earlier years, right turning pockets (Fig. 43) were painted at signalised junctions with the intent for vehicles to stop and wait in them until it was safe to complete the right-turn manoeuvre. Over the past years, other additional forms of road safety measures were introduced in a bid to further enhance the safety

Fig. 43. Right turning pockets at a signalised junction.

of crossing pedestrians at the junction. One such measure was the Intelligent Road Studs (IRS).

IRS was introduced (Fig. 44) as another measure to enhance pedestrian safety. IRS are amber light-emitting diode studs. It is used as a safety device to provide additional warning to motorists to watch out for pedestrians crossing alongside when they make right turns in-between the opposing straight-through traffic. These studs are installed along the pedestrian crossing line, flushed with the road surface and blink during the green man phase. IRS is brightly illuminated during the day for maximum visibility and its luminance is automatically adjusted at night, so as not to blind motorists by the glare.

In 2005, the LTA piloted the IRS at two signalised junctions in the Toa Payoh estate as part of a three-month trial. The trial findings showed that 80% of the motorists were observed to give way to pedestrians after the installation of IRS, compared to 73% as before (*Koh et al., 2007*). Following the success of the pilot trial, the LTA has extended IRS to over 20 more locations over the past few years.

Fig. 44.   Intelligent Road Studs (IRS) at the junction of Toa Payoh Lorong 1/ Toa Payoh Lorong 2/Toa Payoh Lorong 6.

(12) "Give Way to Pedestrians" Sign

Apart from Intelligent Road Studs (IRS) that has been installed at selected junctions since 2005, a static advisory "Give Way to Pedestrians" (GWTP) sign was recently introduced in 2013 as an alternative cost-effective safety measure. The aim of the sign is to enhance pedestrian safety at junctions with permissive right-turning movement, reminding right turning motorists of pedestrians crossing within the adjacent pedestrian crossing, and to give way to them.

The GWTP sign (Fig. 45) comprises of a 'GIVE WAY' symbol, a right turn arrow representing turning traffic and a pedestrian crossing within the demarcated signalised pedestrian crossing. The combined use of fluorescent yellow and white reflective material, with a black border is expected to help increase the sign's conspicuousness. This sign is mounted on top of the traffic light pole diagonally opposite right-turning motorists to remind them to look out for and to give way to pedestrians as they make the turn.

The GWTP sign was first piloted at three signalised junctions.[c] Since the implementation of this sign at the pilot locations, more right-turning motorists

---

[c]Junctions of Rivervale Lane/ Rivervale Drive, Hougang Avenue 4/ Upper Serangoon Road and Punggol Road/ Punggol Central

Fig. 45.   "Give Way to Pedestrians" sign at the junction of Rivervale Lane/Rivervale Drive.

have been observed to give way to crossing pedestrians (*Chow, 2014*). This sign is one of the various measures in the road safety engineering toolkit for enhancing pedestrian safety. It can be implemented at more locations where necessary.

(13) Bollards at traffic island

Due to site constraints, there are some locations where it is not feasible to implement the full length of a raised kerb. In such cases, some retrofitting is required. To complement this, concrete bollards are installed to serve as a delineator to highlight to motorists the edge of the traffic island (Fig. 46).

(14) Cyclists

In the past decade, we have seen more and more cyclists plying the footpaths and roads around Singapore. To serve the growing needs of this group and to enhance safety between cyclists, pedestrians and motorists, a National Cycling Plan was launched in 2010 to provide intra-town off -road cycling path networks within public housing estates. The plan started with the LTA providing cycling tracks in five selected public housing towns (LTA, 2010), but the programme has since extended to cover all other HDB towns as a long-term goal.

Under the National Cycling Plan, besides providing infrastructure, engineering safety measures such as warning signs and markings are provided at critical points where pedestrian and cyclists could potentially come into conflict (LTA, 2013a). In addition, the LTA looks into the need to widen specific

Fig. 46. `Bollard with raised kerb at traffic island.

crossing points (Fig. 47) or provides segregated bicycle crossing at traffic signals (see Fig. 48). To encourage safe cycling, the LTA has also produced a safe cycling guide book, "Your Guide to Intra-Town Cycling", in collaboration with other agencies and partners (Fig. 49). This guide serves to inculcate safe cycling habits among cyclists.

Together with LTA, the Singapore Road Safety Council is also looking into establishing a National Cyclist Education Programme which will be launched in 2016. This programme serves to inculcate safe cycling skills and practices, enhance knowledge and understanding of cycling related infrastructure and educate users on the rules and code of conduct of cycling in Singapore.

(15) Motorcyclists

About 50% of road fatalities in Singapore involve motorcyclists or their pillion riders (*Hau et al., 2010*). While education and enforcement continue to be undertaken to enhance road safety, engineering measures are also intensified. Road surfaces at accident-prone locations are treated with high skid-resistance material to give better control for all road users, particularly motorcyclists. Exclusive right turns are used to regulate the right-turning movements at some traffic light junctions. Due to the inconspicuousness of the motorcyclists on the road, this measure helps to reduce the chances of motorcyclists involved in accidents due to misjudgment from either right-turning or on-coming vehicles.

Fig. 47.   Cycling path behind bus stop.

Fig. 48.   Segregated bicycle crossing at traffic signals.

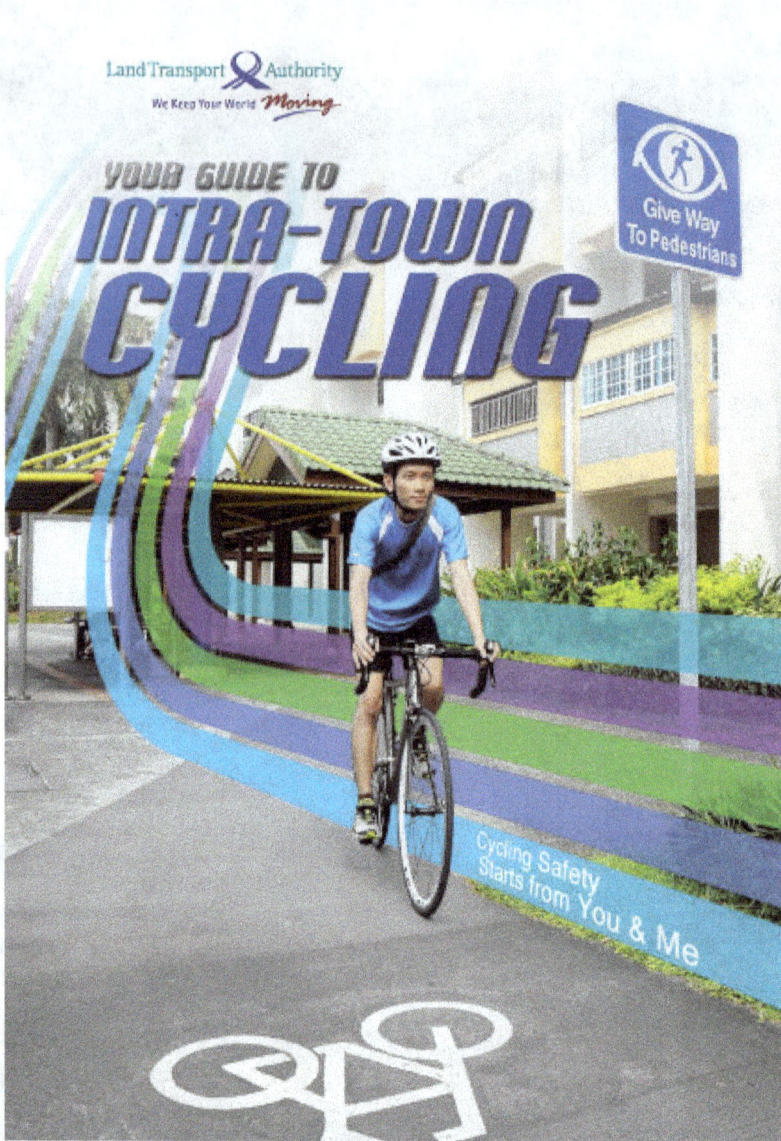

Fig. 49.   Safe cycling guide book — "Your Guide to Intra-Town Cycling".

Though they may not be the most vulnerable road user, motorcyclists run a much higher risk of injury compared to other motorists, due to the very nature of their vehicle. Especially in a country like Singapore, where land is limited and population increases rapidly, the number of vehicles (and hence, motorcycles) on the roads is sky-rocketing. Singapore also sees a substantial number of motorcyclists that travel into the island from Malaysia on a daily basis to work, thus contributing to the motorcycle count.

In the early 2000s, motorcyclist fatalities constituted about 50% of road traffic fatalities. To address this issue, studies were conducted and it was observed that the majority of fatal motorcyclist accidents were caused by speeding and self-skidding, especially on expressways. The LTA then began trialing a new type of material with higher skid-resistance, in a bid to reduce accidents caused by self-skidding. This material, Calcined Bauxite, is extremely resistant to wear and provides excellent grip especially when vehicles negotiate bends.

Over the years, Calcined Bauxite was implemented along stretches of expressways that have relatively higher counts of motorcycles self-skidding. Studies conducted after implementation have shown that such material has proven very effective, reducing the number of motorcyclist accidents by up to 77%.

Though motorcycles are most flexible in terms of manoeuvrability, they are also the least visible on the roads. Thus, a regulation was passed that it be made mandatory for all motorcyclists travelling along roads in Singapore to have their headlights switched on, regardless of the time of day. Motorcycles that are three years or older are also subjected to inspections by accredited boards, to ensure that the vehicle is safe to be ridden on the roads (*VICOM, 2012*).

Initiatives to enhance motorist safety include:

To enhance motorist's safety, several initiatives are implemented targeting to reduce their speed. In other cases, proper delineation and speed attenuators are installed.

Traffic calming measures

(16) Advance Warning Lights

Road junctions are signalised to regulate vehicular traffic flow and maximise traffic efficiency. However, a signalised junction or pedestrian crossing may be located where it is necessary but may lack adequate sight distance to motorists due to the presence of road bends or a cluster of trees along the road side. As a result, motorists approaching a signalised junction or pedestrian crossing may not have sufficient time to react and run the red light at the junction.

In 1999, Advance Warning Lights (AWL) was introduced at a signalised junction, located after a crest. The AWL was first implemented along Rochor Road before the signalised traffic junction that intersects with Beach Road. This location was chosen due to the crest that blocks the motorist's view of the traffic lights ahead. The response received from motorists was positive.

AWL is an active device, incorporating two flashing amber aspects mounted on a pole. Positioned in between the pair of flashing amber aspects is a "Prepare to Stop" information sign. On top of the amber aspects is a "Traffic Light Ahead" sign (Fig. 50).

AWL is generally installed some distance away before an intersection that is controlled by traffic signals. This pair of flashing amber signals is positioned along the approach to a signalised junction that may not be very visible due to the road geometry, such as a bend or crest. They will start flashing a

Fig. 50.   Advance Warning Lights.

few seconds before the downstream signal turns amber to warn motorists of the impending red light so that they are prepared to stop before the light turns red.

When AWL is flashing, motorists should get ready to stop at the traffic junction or pedestrian crossing downstream. It is a useful device to advise motorists of the change in signal phasing at the traffic lights downstream.

(17) Traffic Calming Markings (TrCM)

The Land Transport Masterplan (*LTA, 2013b*) outlined the LTA's intention to introduce new traffic calming measures to alert motorists to moderate their speeds, thereby enhancing safety. One of the new traffic calming initiatives is Traffic Calming Markings (TrCM).

Common traffic calming measures are physically based, such as road humps and speed regulating strips. These are effective in reducing motorists' speeds but may be unpopular with residents or motorists as these vertical deflections generate increased noise (to residents) and discomfort (to passengers) when vehicles travel over them. On roads with high traffic volume, physical calming measures can have an impact on the road capacity, leading to congestions. Hence, the introduction of TrCM, a non-physical traffic calming measure, is intended to overcome some or all of these posed difficulties, while still retaining the capability of reducing vehicle speeds.

It is also known as a form of perceptual countermeasure against speeding as it aims to manipulate the appearance of the traffic lane to a driver, by reducing the perceived lateral clearance. This will help to influence his/her behaviour and travelling speed to suit the road conditions.

This reduction of lane width is achieved by painting two parallel sets of white triangular markings on the side of road to create a visually narrower lane (Fig. 51). TrCM aims to trigger motorists' awareness of a proceeding hazard and the need to reduce their speeds. It has the advantage over traditional speed regulating strips as there is no additional noise produced when travelling across it. It is thus suitable to incorporate TrCM at places such as private residential, hospital, nursing homes and schools.

The LTA observed that overall speeds decreased by 10% at TrCM locations. About 64% of motorists surveyed found these markings effective in influencing them to slow down. Following the positive trial results, the LTA has extended the use of TrCM to more locations since 2009. The TrCM will be extended to other locations where appropriate or to complement existing physical calming measures when there is a further need to manage motorists' speeds.

Fig. 51.   Traffic Calming Markings (TrCM) at Eng Neo Avenue exit from PIE (Tuas).

(18) Broader Alignment Lane Markings (BALM)

BALM is another form of traffic calming measure that is applied along expressways and major arterial roads. It comprises of broader lane markings spaced at closer intervals from one another compared to the standard lane markings. It guides motorists to be aligned to their lanes better, and serves as a visual cue to perceptually narrow the travel lanes to encourage motorists to travel at lower speeds.

BALM was first implemented in March 2012 at two locations (near Mount Pleasant and Chantek Flyovers) along the Pan Island Expressway (PIE) (Fig. 52). Following these successful trials where speeds were reduced by five to 10% (Sim, 2013), the LTA extended these markings to other locations, including the Kallang–Paya Lebar Expressway (KPE) and Bukit Timah Expressway (BKE), and will be rolling this out at more locations in the near future.

(19) "Your Speed" Signs (YSS)

Using radar technology to detect vehicles, speeds, the electronic "Your Speed" Sign (YSS) serves to raise awareness of motorists' travelling speeds and also encourage them to comply with the speed limit to enhance safety of the roads (*Koh et al., 2010*). YSS displays the real time numeric speeds of passing vehicles (Fig. 53).

Fig. 52.   Left to Right — Before and After BALM at PIE(Changi) near Mount Pleasant Flyover.

Fig. 53.   Solar powered YSS at PIE(Changi) Exit to Tampines Avenue 5.

During the pilot trial in 2009, 72% of the surveyed motorists were satis-fied with YSS because they felt that it is effective in helping them to moderate their driving speeds at these locations. YSS is typically found at expressway exits where there are cases of speed-related accidents. However, it may not be suitable for application on multi-lane (more than two lanes) roads as motorists may get confused about whose speed is being displayed.

In the past, YSS was powered through tapping its power supply from the nearest street lighting's meter box, which would have required extensive trenching to lay the power cables. In order to save on the high costs from the trenching, from December 2011 onwards, the LTA came up with the use of an alternative power supply — the greener option of solar energy to power the YSS.

As of October 2014, there are five YSS installed in Singapore; of which, two are operating via solar power.

Delineators

(20) Curve Alignment Markers (CAMs)/Waveline

A device, known as "Sharp Deviation Sign" (SHS), was used to delineate road bends. SHS has two to three white arrows on a black background sign plate. They were installed along the kerb to help motorists negotiate the bend safely (Fig. 54).

Fig. 54.   The black and white sharp deviation sign.

Fig. 55. Curve Alignment Markers placed along road bend.

Since 1999, SHS has been replaced by Curve Alignment Markers (CAMs), which uses high retro-reflective sheeting.

CAMs are highly visible with a single black arrow on a retro-reflective fluorescent yellow background which helps motorists negotiate a road bend safely (Fig. 55).

CAMs are ideally used along bends with sufficient space behind the kerbs. However, there are bends which have space constraints such as a concrete wall behind the kerb, where lateral clearance is insufficient for CAM installation. For such a case, corrugated reflective sheeting (waveline) is more appropriate for use.

Waveline is the fluorescent yellow-green reflective sheets with black arrows, that is implemented at locations where CAMs are not effective or suitable due to site constraints or limited visibility (Fig. 56).

It has a unique shape of repeated raised strips that reflect light from vehicle head lamps across a wide range of angles. Waveline is also designed to give high visibility during both day and night time.

Waveline is often used at bends where the parapet wall at the centre median is narrow; or along exits and entrances of tunnels, where there is inadequate space to tilt the CAMs; or the walls at the outer edge of the bend have

Fig. 56.   Waveline used before a tunnel.

many noticeable scratch marks. Also, when there are repeated feedbacks from the public, waveline can be used as an additional measure to warn motorists of the bends at the locations.

According to the evaluation findings gathered after implementation of waveline, there was a decrease of 47% of motorists driving close to the wall as motorists were more aware of the wall edge. In addition, heavy goods vehicles also showed a huge decrease in the percentage of vehicles driving close to the wall.

(21) Enhanced Chevron Zone (ECZ)

A best practice in countries such as the United Kingdom (UK) is to use red pavements as a form of warning, where traffic should either not pass or pass with caution. Locally, red pavements have been implemented on sections of the carriageway in Enhanced School Zones (ESZs) along primary school frontages to warn motorists to keep a lookout for school children, and it has been well-received by the public.

As the usage of red pavement was well-received, it was also implemented at the Enhanced Chevron Zone (ECZ). ECZ is a red bus stop separator area implemented at bus stops that are located along an expressway (Fig. 57). In view of the high travelling speed of vehicles along the expressway, the zone is meant to raise motorists' awareness that they are approaching a bus bay, and thus to look out for exiting buses as well as not to encroach into the area. The bus stop separator area is covered with red pavement and vibralines. The red surface increases visibility and discourages encroachment into the chevron area, which will help to reduce conflict between mainstream vehicles and buses leaving the bus bay.

The first trial was carried out at bus stop (B08) along Ayer Rajah Expressway (AYE) outside the National University of Singapore (NUS) in early March 2008. Following the success of the trial, the LTA implemented ECZ at four more locations along AYE in April 2010.

(22) Crash Cushion (crash attenuator)

Before 1999, gore areas of expressways were treated with a short raised concrete platform before meeting the solid concrete parapet wall downstream. The chevron markings serve as a bifurcation to channelise the vehicular traffic to the slip road and the main expressway. Hence, they were drawn from the nosing of the raised platform and extended backward till the split of the traffic.

In 1998, a fatal accident took place at Tampines Expressway, TPE (PIE-bound) exit to Lorong Halus where a saloon car crashed into the bare concrete

Fig. 57.   Left to Right — Before and After ECZ at bus stop B20 (AYE towards Tuas, before exit to Jurong Town Hall Road).

parapet wall after mounting the raised platform. The LTA then began to look into adopting safety devices which were practised overseas for gore areas. Two trial sites at TPE (PIE) exit to Lorong Halus and AYE (Tuas) exit to Clementi Avenue 2 were selected for their frequent vehicular hits. After the successful trial at these two locations, crash cushions have since been widely used at all expressways in Singapore.

Crash cushions (Fig. 58) are able to absorb the impact from a vehicle up to the speed of 100 km/h (*TRB, 1993*). When a vehicle hits the crash cushion, the cartridges placed between the metal frames behind the nosing absorb the kinetic energy, and the metal frames are compressed forward against the back-up board, bringing the vehicle to a halt without causing serious injury to the vehicle occupants.

(23) Safety bollards at bus stops

In May 1998, a fatal accident took place at a bus stop along Bendemeer Road where three waiting bus commuters were killed. The fatal accident was caused by a sports car that lost control and crashed into the bus stop. This prompted

Fig. 58.   Crash cushion at Rochor Road.

the LTA to review safety measures at bus stops. Safety bollards were designed and tried out in front of bus shelters to eliminate direct impact on waiting bus commuters from the vehicles of errant drivers.

The safety bollard is an effective safety device for protecting bus commuters waiting at bus stops. Since its first implementation in 1999, the LTA has been receiving requests from members of public to install bollards at all bus stops. With the presence of the safety bollards, visually-impaired commuters can also use these bollards as a guide to remind them not to step forward into the bus bay.

Safety bollards were put on trial at five bus stops in 1999. In this trial, the LTA invited Nanyang Technological University (NTU) to conduct an opinion survey amongst bus commuters. The study showed that a majority of the public (86%) welcomed the idea and found it necessary to have safety bollards at bus shelters for safety reasons (*Ding et al., 2010*).

Out of the 27 accidents (1993–2014) occurring at bus stops, 16 cases involved vehicles crashing into the bus stops with bollards but there was no fatalities involving waiting commuters. As for the remaining 11 accidents, they all occurred at bus stops that were not installed with bollards at the time of the accidents. Three out of the 11 cases involved five waiting commuters being killed. There was one case in 2007 where a drink-driving driver careened head-on into the safety bollards at a bus stop. Due to the great impact, the first safety bollard was knocked down while the second managed to halt the errant vehicle from ploughing further into the bus stop. At the time of the accident, there were no commuters waiting at the bus stop and the driver was not seriously injured.

Safety bollards were first installed at bus stops that are generally located along high-speed roads or sharp bends, and at bus stops without bays. Subsequently, they are also installed along low speed roads and at bus stops with bays, in view that accidents can take place at random and in all site conditions. Ultimately, all bus stops will be installed with safety bollards. As of end October 2015, 88% of the 4,700 bus stops islandwide have been provided with safety bollards.

Besides the reflective sheeting that is pasted around the top portion of the bollard to enhance visibility and conspicuousness at night, other safety features such as a yellow band and raised kerb are installed in the proximity of the bus stop (Figs. 59 and 60).

Fig. 59. Safety bollards exceptionally "shiny" at night when shone on by vehicle headlamps.

(24) Spring-loaded posts (SLPs)

Prior to 1999, road marker posts (Fig. 61) were used at centre dividers and bus stop separators to highlight the existence of the raised concrete kerb. It is made of fibre glass in white and black, with reflectors on the post facing the traffic. These road marker posts are rigid at the base. As such, they have to be replaced whenever they are knocked by vehicles, as they are unable to bounce back to their original upright position.

Since 1999, road marker posts have been replaced by spring-loaded posts (SLPs), which is effective in alerting motorists approaching a centre divider. It is highly recommended to use a SLP to replace the conventional road marker posts for all narrow dividers.

SLP is a 1-metre high flexible post with a spring mechanism at the base which allows the post to bounce back to its original position. It is also able to withstand some degree of impact from a moving vehicle. Black and yellow retro-reflective sheetings are pasted on the entire post to enhance its conspicuousness (Fig. 62).

Fig. 60.   A car stopped by a safety bollard.

(25) Flexible Posts (FPs)

Flexible Posts (FPs) with yellow and black stripes are implemented on the road pavement to help improve the conspicuousness of gore areas with crash cushions. The bright color contrast also warns motorists of the road feature on the road (Fig. 63). Made of polyurethane material, FPs, compared to SLPs, are more capable of withstanding repeated impacts and wheel-over of vehicles without failure.

Fig. 61.   Road marker post on a centre divider.

Fig. 62.   Spring loaded posts on a centre divider.

After receiving numerous feedback on damaged SLPs installed before the crash cushion at a gore area, FPs were introduced in their place. It served as a means to reduce the number of hits on the crash cushions and the high cost of repairs.

Fig. 63.   Flexible posts installed on road pavement.

Based on the LTA's evaluation findings, the proportion of swerving motorists near to the gore area has decreased significantly after FPs were installed in front of the crash cushion. It was also observed that the FPs did not show signs of getting hit and remained in good condition after one year of installation.

5. Key Learning Pointers for the Success of Road Safety in Singapore

The Global status report on road safety 2013 presents information on road safety from 182 countries, which accounts for almost 99% of the world's population. The report shows that worldwide, the total number of road traffic deaths remains unacceptably high at 1.24 million per year (*WHO, 2013*). Between 20 and 50 million sustain non-fatal injuries. Road traffic injuries cause considerable economic losses to victims, their families, and to the country as a whole. Fortunately, Singapore roads are among the safest roads in the world. Over the past 50 years, Singapore has been successful in making her roads safer. Both road fatalities and road fatality rates are on the decreasing trend (Figs. 2 and 3).

International best practices such as the World Report on road traffic injury prevention (*WHO, 2004*) recommended one lead government agency for road safety efforts. In Singapore, the Singapore Road Safety Council (SRSC) can be viewed as the lead agency for road safety efforts. SRSC has the Land Transport Authority (LTA) and the Traffic Police (TP) as key members from the government. The LTA has a dedicated Road Safety Engineering Unit, focusing on engineering; and the TP has a road safety team to focus on education and enforcement aspects of Singapore's road safety efforts. In addition, SRSC collaborates closely with other major stakeholders like the Automobile Association of Singapore to promote education, enforcement and engineering efforts in road safety.

Singapore's road safety efforts are supported and encouraged by the senior management in both the LTA and the TP. In addition, high-level road safety intervention by the respective governmental ministries, non-governmental agencies and private sector has been encouraging. This has contributed to the success of road safety in Singapore. For example, the Pedestrian and Cyclist Safety Committee and Safer Roads Industry Taskforce Committee are led by political leaders. These leaders are appointed to champion the cause of road safety and engage the community. This has helped to establish and sustain national road safety efforts.

Singapore has a national road safety strategy and plan of action produced by key stakeholders including the LTA and the TP. For example, the national road safety action plan was launched in February 2005. This action plan aims to make the roads safer for all road users through 5Es: Encouragement, Education, Enforcement, Engineering and Emergency Preparedness. This plan is modeled after successful national and local plans that are in operation in many other countries. It is produced as part of the Asian Development Bank (ADB), ASEAN Regional Road Safety Project. It follows the Road Safety Action Plan Guidelines published by the United Nations, ADB and the World Bank. It has been specifically tailored to the particular needs of Singapore, and has also been endorsed by the Ministry of Transport.

Singapore has established many road design standards and data systems. These are important tools for the road safety professionals and policy makers to make decisions, assess, implement, evaluate, and monitor road safety performance. Many new road safety measures are implemented in line with best practices after careful before-after studies. These include dashed pedestrian crossing lines and crash cushions which are now adopted as standards.

Singapore regularly networks with international bodies involved in road safety to learn and share road safety practices and experience. These global bodies include the United Nations Economic and Social Commission for Asia and the Pacific (UNESCAP), Asia–Pacific Economic Commission (APEC), and the Association of Southeast Asian Nations (ASEAN). Singapore actively supports and encourages road safety participation in professional institutions such as the Road Engineering Association of Asia and Australasia (REAAA) and the World Road Association (PIARC).

Over the years, Singapore has managed to secure sufficient financial resources and developed a pool of human resources for road safety with capacity training. However, the challenge is to sustain and retain qualified road safety professionals. Road safety is a highly specialised field that needs a special calling and passion to serve the society by way of saving lives and reducing road injuries. One way to develop this passion is by encouraging road safety professionals to contribute technical papers and share their knowledge and experiences at conferences and seminars. For example, the LTA road safety professionals conduct road safety courses at the LTA Academy.

Careful studies must be conducted before implementing any road safety schemes. An example is when speed limits are adjusted. An increase in the average speed is directly related to both the likelihood of a crash occurring and to the severity of the consequences of the crash. It is considerably easy to raise the speed limits but on the contrary, lowering them will likely face strong resistance from the motoring community.

Smarter road safety schemes have also been designed to address public concerns. Some road safety improvement schemes may affect the traffic flow. For example, a simple traffic calming measure like a road hump will slow down traffic but at the same time, cause some noise on the road. Nearby residents and some road users may not welcome such road safety countermeasures. Hence, there is a need to explore smarter non-physical engineering measures such as traffic calming markings used in several residential areas. These markings create a visual effect of road narrowing and hence motorists will be more alert and slow down.

The LTA has a community partnership group that engages the community before implementing any major road safety improvement schemes. Road users have become more vocal in recent years and have high expectations of the authority to make the roads safer. Hence, the community partnership team responses to feedbacks and engages the grassroots organisations and the local community advisors and leaders to seek their understanding and support.

Sometimes, when emotions of affected road users are involved, the team needs to show empathy in dealing with individuals whose loved ones were killed or injured. Other examples include explaining to the community new road safety initiatives such as the Silver Zone to slow down motorists for the safety of the seniors in crossing the road.

In conclusion, this chapter has provided a review of road safety in the land transport development of Singapore over the years. Nevertheless, the need to balance road safety and traffic efficiency remains a challenge for the authorities like the LTA. Besides road safety engineering, education and enforcement efforts are also needed to enhance road safety. Ultimately, it is important for all road users to cooperate and recognise that road safety is a shared responsibility.

Acknowledgement

This chapter was written with contributions from the Road Safety Engineering Unit (RSEU, Transportation and Road Operations, Land Transport Authority, Singapore), and inputs from Associate Professor Gopinath Menon from Singapore Road Safety Council.

References

Austroads (2009). Guide to Road Safety Part 6: Road Safety Audit.

Chow, J. (2014). Safety boost at 240 zebra crossings. *My Paper*, 15 October 2014.

Department of Statistics Singapore (2015). Yearbook of Statistics Singapore, 2015.

Department of Statistics Singapore. Statistics, browse_by_theme, population and population structure, time series. Accessed on 13 November 2014.

Ding, W.K., Wah, T.H. and Ho, S.T. (2010). Improving Road Safety Of Commuters At Bus Stops. 7th Asia Pacific Conference on Transportation and the Environment, 3–5 June, Indonesia, Semarang.

Ding, W.K. and Ho, S.T. (2014). Road Safety Framework And Policies In Singapore. PIARC Routes/Roads, July 2014 edition, pp. 58–65.

Hau, L.P., Ho, S.T. and Chin, K.K. (2007). Using GIS Technology to Enhance Road Safety in Singapore. 14th Road Safety in 4 Continents Conference, 14–16 November, Bangkok, Thailand.

Hau, L.P., Koh, P.P. and Ho, S.T. (2008). Using ITS To Enhance Road Safety In Singapore. ITS AP2008 Conference, 14–16 July, Singapore.

Hau, L.P. and Ho, S.T. (2010). The Development of an Accident Black Spot Program in Singapore. REAAA Journal, Vol. 16, Nos. 1&2, pp. 36–41.

Hau, L.P., Ho, S.T. and Chandrasekar, P. (2010). Accident Characteristics By Road Types On Singapore Roads. 15th Road Safety in 4 Continents Conference, 28–30 March, Abu Dhabi.

Hau, L.P.and Ho, S.T. (2014). Impact of Changes in Road Speed Limits on Arterial Roads. 26th ARRB Conference — Research Driving Efficiency, 19–22 October, Sydney, Australia.

Hong Kong Transport Department (2014) — Summary of Key Statistics.

Koh, P.P. and Ho, S.T. (2006). Black Spot Management In Singapore. 12th REAAA Conference, 20–24 November, Philippines.

Koh, P.P., Ho, S.T. and Chin, K.K. (2006). Managing Traffic in Singapore: A Road Safety Perspective. 1st World Roads Conference, 27–30 September, Singapore.

Koh, P.P., Ho, S.T. and Chin, K.K. (2007). Use Of Intelligent Road Studs To Reduce Vehicle-pedestrian Conflicts At Signalised Junctions. 14th Road Safety in 4 Continents, Bangkok, 14–16 November, Bangkok, Thailand.

Koh, P.P., Ho, S.T. and Chandrasekar, P. (2009). Keeping Our Roads Safe Through New Road Safety Initiatives. 2nd World Roads Conference, 26–28 October, Singapore.

Koh, P.P., Ho, S.T. and Chandrasekar, P. (2010). Knowing Your Speed Can Save Your Life. 15th Road Safety 4 Continents Conference, 28–30 March, Abu Dhabi.

Koh, P.P., Wong, Y.D. and Menon, A.P.G. (2011). Acceptability of audible pedestrian signal noise. Transportation Research Part D, Vol. 17, No.2, pp. 179–183.

Koh, P.P., Wong, Y.D., Menon, A.P.G. and Oh, K.W., (2014). Technology-aided systems for safe pedestrian crossing in Singapore, REAAA Journal, October Issue, Vol. 18, pp.1–8.

Loh, K.F. (2008). Enhancing Safety of Road Users in New Road Projects.Safety News. Issue 12. October. pp. 04. http://www.lta.gov.sg/content/dam/ltaweb/corp/PublicationsResearch/files/ReportNewsletter/SafetyNews/Issue_12_LTA_News_Letter_Final08.pdf

LTA (2005). Land Transport Statistics in Brief. Land Transport Authority, Singapore.

LTA (2010). Changi–Simei, Bedok And Marina Bay To Have Cycling Infrastructure By 2014. News release, Land Transport Authority, 15 July 2010.

LTA (2011). Barrier Free Access Enhancements. News release, Land Transport Authority, 23 October 2011.

LTA (2013a). LTA Completes First Dedicated Cycling Path Network under the National Cycling Plan. News Release, Land Transport Authority, 14 June 2013.

LTA (2013b). Land Transport Master Plan, Land Transport Authority.

LTA (2014a). One. Motoring Portal, Information and Guidelines, Facts and Figures, Land Transport Authority, Singapore.

LTA (2014b). Land Transport Statistics in Brief. Land Transport Authority, Singapore.

LTA (2014c). One. Motoring Portal, Traffic Management, Green Man+ Brochure, Land Transport Authority, Singapore.

MHA (2013). Safer Roads Industry Taskforce. Media Factsheet, Ministry of Home Affairs, 6 December 2013.

MOT (2003). Opening Speech By Dr Balaji Sadasivan At The ADB-ASEAN Road Safety Programme. Ministry of Transport, Singapore, 2 October 2003. http://www.mot.gov.sg/News-Centre/News/2003/Opening-Speech-By-Dr-Balaji-Sadasivan-At-The-ADB-ASEAN-Road-Safety-Programme/

MOT (2013). COS 2013 — Speech by Parliamentary Secretary for Transport Associate Professor Muhammad Faishal Ibrahim on Meeting Diverse Needs, 13 March 2013. http://www.mot.gov.sg/News-Centre/News/2013/COS-2013—Speech-by-Parliamentary-Secretary-for-Transport-Associate-Professor-Muhammad-Faishal-Ibrahim-on-Meeting-Diverse-Needs/

Remember Singapore (2012). Lorong Chuan Overhead Bridge. Accessed 25 October 2014. http://remembersingapore.wordpress.com/old%20overhead%20bridge/.

Sim, R. (2013) Wider lane markings cut traffic speed. The Straits Times, 25 May 2013.

The International Road Traffic Accident Database (IRTAD (2014). Road Safety Annual Report 2014.

TRB (1993). National Cooperative Highway Research Program Report 350, Recommended Procedures for the Safety Performance Evaluation of Highway Features, Transportation Research Board, Washington, D.C.,

VICOM (2012). Inspection Price And Schedule List. Retrieved 20 October 2014 from http://www.vicom.com.sg/inspectionprice.htm.

WHO (2004). World Report on road traffic injury prevention. http://whqlibdoc.who.int/publications/2004/9241562609.pdf

WHO (2013). The Global status report on road safety 2013. http://www.who.int/violence_injury_prevention/road_safety_status/2013/report/en/

Chapter 6

Singapore: The Future of Logistic Hubs

Giulia PEDRIELLI, CHEW Ek Peng, LEE Loo Hay and
Chelsea WHITE

The logistics and supply chain industry has experienced fundamental changes in the past 50 years. Initially a support industry providing basic transportation and storage services for other industries, it has become a major value-added industry, playing a fundamental and strategic role for the growth of global trade. During this evolution, Singapore has played and continues to play a key role in sustaining and developing the growth of this industry, in large part due to long-term programmes of the Singapore government, and international and global agreements that have shaped Singapore's economic development.

Since the beginning of its industrialization, Singapore has heavily invested in both the development of robust and flexible key infrastructures such as the seaport and airport, becoming a de facto transportation hub in South East Asia. Long-term planning has guided Singapore to promote the development of knowledge-based industries, which has stimulated a high level of computerization and automation throughout Singapore's industrial base. The creation of new services and skills stimulated by the logistics industry has continuously added value to the economy, making the logistics sector one of the most important assets of the country. As a result, Singapore has increasingly attracted key firms in the logistics industry, such as third party logistics companies, many of which have established their Southeast Asian headquarters in the country.

All these have contributed in turning the island into being one of the main logistics hubs of the world today.

Nevertheless, several fundamental changes for the logistics industry are on the horizon, indicating the challenges that lie ahead for Singapore and the entire Southeast Asian region. To anticipate and respond to these challenges in a timely manner and to retain and strengthen its global leadership position in logistics, Singapore must creatively and strategically exploit new data technologies, transport solutions, and advanced automation technologies in order to provide the innovations and infrastructures needed for the future.

Starting from the growth and barriers that characterize the logistic industry, this chapter will delve into the historical development of Singapore as a key logistics hub focusing on some of the key driving factors of the country's growth. Finally, we will discuss the main challenges and how the next generation supply chain may transform to face them.

1. Introduction

Despite the severe global economic crisis faced in the past decade, global trade has experienced remarkable growth. In 2011, the percentage of world GDP due to international trade was 59%, an increase from 39% in 1990. Today's value of global trade is above US$20 trillion. Two key factors contributing to the growth of global trade are:

- The increase in political agreements between nations that have reduced regulatory trade barriers and lowered tariffs;
- Logistics innovations that have enabled the fast, reliable, and efficient shipment of goods over long distances and across national boundaries.

As a result, goods are increasingly no longer produced entirely in a single country. Often, supply chains across several national boundaries and value is added in each of the associated countries, taking advantage of what industry in each country does best. Indeed, in order to produce goods of higher quality, less expensively, and (often) faster, i.e., more competitive products, it is increasingly necessary to assign different stages of production to companies in different countries.

The growth of the global trade volumes, and eventually GDP, can be fostered by: (1) government policies that reduce (and perhaps eliminate) trade barriers, and (2) collaboration between the public and private sectors to understand where investment in infrastructure is most needed in order to enable an efficient global (and domestic) logistics industry. Otherwise, the cost of moving goods increases, increasing the total cost of the goods with no value added.

The work of the World Economic Forum (WEF) in January 2013 held in Davos-Klosters, Switzerland, resulted in a reference document "Enabling Trade: Valuing Growth Opportunities", which was written in collaboration with the World Bank and consulting firm Bain & Company. The report stated that the global community of nations must exert a strong and well-coordinated effort in order to abate supply chain barriers. It was estimated that the global GDP would show an increase of 2.6% if all countries could make their logistics performance improve by decreasing the barriers in supply chain to 50% with respect to the best regional performer.

Focusing on border management and infrastructures for transport, the report also observed that if countries could improve to the point of attaining 50% of the global (instead of the regional) best practice level (i.e., Singapore), then a jump of 4.7% could be observed in the GDP, a result that is six times larger than what could be obtained through eliminating import tariffs. Such a growth in global GDP would reduce unemployment by increasing trade and hence increasing the size of the job market. The following was concluded in the document:

> "(…) rather than persisting with trade agreements that revolve around deals to lower tariffs, governments should prioritize action that aims at lowering supply chain costs for operators. The reason is these will eliminate resource waste, whereas abolishing tariffs mainly reallocates resources. (…)" (*Enabling Trade: Valuing Growth Opportunities*", World Economic Forum. 2013).

Nevertheless, lowering supply chain costs for operators will require intense negotiation in several policy areas including the management of the operations at the border (e.g., customs clearance-related policies or "trade facilitation"), product standards, and transport barriers.

In summary, the logistics industry is an enabler of the movement of goods in supply chains and increasingly is adding value throughout the supply chain. As such, the logistics industry enables the growth of trade and hence the growth of GDP, thereby contributing to economic and societal growth and well-being. Logistics clusters are an apparent example of how supply chains and logistics are evolving. These clusters are attracting the talent required to create solutions to complex and global logistics challenges. Singapore is one of these clusters, setting the pace for Southeast Asia and the global logistics industry.

The remainder of the chapter will be structured as follows: Section 2 will quantify the growth of the logistics market in the period 1990s and 2000s, and will present how the public and private sectors have responded to trade growth in order to sustain and foster it. Section 3 will focus on the historical development of Singapore's logistics industry and the challenges that is has faced. Section 4 will present the main challenges for the future of Singapore as a logistics hub, while Section 5 will outline our perspective on next generation supply chains. Section 6 will present the conclusions.

2. Logistics Industry: Growth and Barriers

The last two decades have seen continuous growth in global trade volumes. Figure 1a presents the trade-to-GDP ratio from 1999 through 2010 in Germany, USA, and China. In Fig. 1a, we note the rapid rise of this ratio for China from 2002 through 2007, due to the economic reforms in the 1990s. We further observe the impact of the global economic crisis on this ratio for all three countries.

If we compare Fig. 1(a) and Fig. 1(b), we can notice that Singapore and Hong Kong have trade-to-GDP ratios that are considerably higher than those of Germany, USA, and China. Throughout the APAC region, both import and

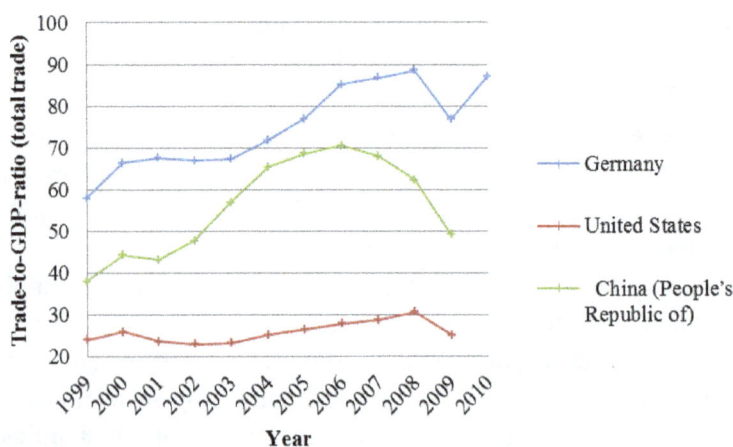

Source: World Trade Organization Database.

Fig. 1(a). Trade to GDP in Germany, USA and China in the period 1999–2010.

Trade (%GDP)

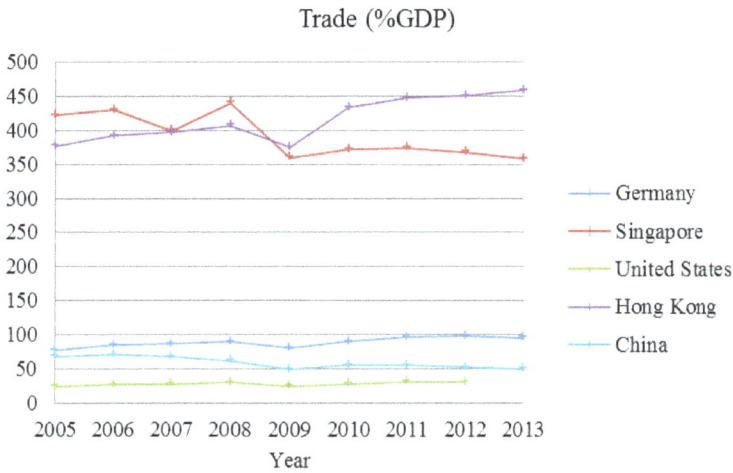

Fig. 1(b). Trade to GDP in Germany, Singapore, USA, Hong Kong and China in the period 2005–2013.

Marine Freight growth rate and forecast

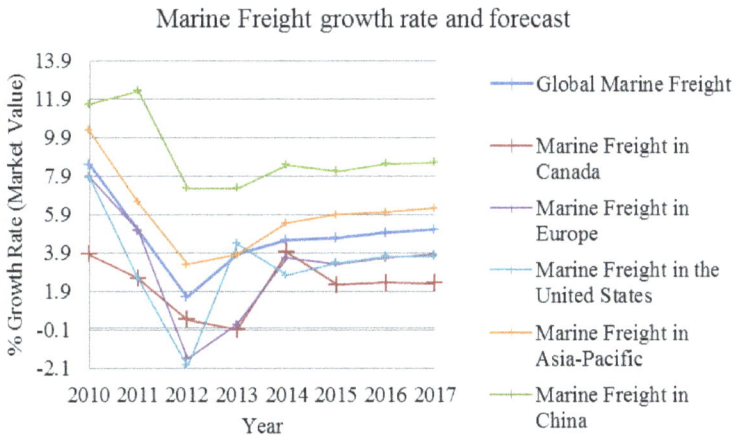

Source: Oriana Business Database.

Fig. 2. Marine freight growth rate.

export trade have experienced substantial growth with particularly high trade-to-GDP ratios in Singapore, Malaysia, Thailand, Indonesia, Australia and New Zealand (Bhatnagar et al., 1999).

The growth in trade volumes is closely correlated with the growth of the logistic industry. Growth rates and forecasts of marine freight and transportation services in various regions worldwide are presented in Fig. 2 and Fig. 3.

Transportation Services growth rate and forecast

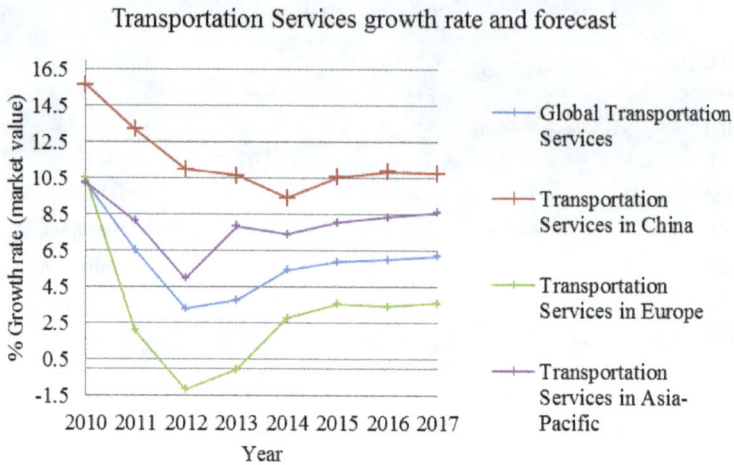

Source: Oriana Business Database.

Fig. 3.   Transportation services growth rate.

Projected growth rates of the logistics industry in North America and Europe are approximately 11% and 8% respectively (*18th Annual Survey of Third-Party Logistics Providers, Penske Logistics. Oct. 2013*). We remark that Fig. 2 and Fig. 3 indicate the close correlation between the growth of marine freight, an proxy of the growth of international trade, and the growth of transportation services.

There are several challenges associated with the need to sustain and foster growth in world trade and the consequent demand for logistics services. The WEF, in its annual *Global Enabling Trade Report* (GETR) (Hanouz *et al.*, 2014), identifies four potential barriers affecting supply chain efficiency: (1) market access, i.e., how well a country's policies facilitate the import of international products and the export of national goods to foreign markets; (2) border administration, i.e., how efficient is a country's border administration with respect to import and export trade; (3) telecommunication and transportation infrastructure quality, i.e., how these infrastructures facilitate the flow of goods in and out of the country; and (4) business conditions that contribute to global trade, e.g., safety, political stability. We remark that these four categories are identical to those used to determine the Enabling Trade Index (ETI), an indicator developed by the WEF to characterize countries worldwide in terms of their ability to sustain and foster domestic and global trade (Section 2.3). A summary of the main barriers within each of the four categories is reported in Table 1 (WEF, 2013).

Table 1.   Supply Chain Barrier to Trade.

Market Access (Domestic and Foreign market access)	Border Administration	Telecom and transport infrastructure	Business Environment
• Quotas • Import fees • Local content requirement • Rules of origin • Import/Export licenses	Efficiency of customs administration Efficiency in import/export procedures Transparency of border administration	Availability and quality of transport infrastructure Availability and quality of transport services Availability and use of information and communication technologies	Regulatory Environment (Investment Policy, hiring foreign workers, etc.) Physical Security

Source: Enabling Trade Valuing Growth Opportunity. World Economic Forum, 2013.

We interpret market access as access to both domestic and foreign markets, taking into account the tariff and non-tariff barriers that a foreign firm might face. Examples of these barriers include technical standards, safety and sanitary requirements, requirements for local content, and more generally political measures that hinder the transport of goods into and out of a country. Such barriers tend to favour domestic firms over foreign firms which, as a result, have to bare extra costs. As highlighted in the WEF Enabling Trade Report, examples of market access barriers are: (1) restrictions due to strict regulations on the origin of the goods (e.g., Indonesian markets and the Middle East countries); (2) referring to pharmaceuticals importers, clinical trials are required in Vietnam even for drugs already approved by the European Union's European Medicines Agency (EMA) or by the USA Food and Drug Administration (FDA); (3) and sea transport of goods within USA ports is restricted to USA-crewed, USA-built, and USA-owned ships. We remark that the third example is due to the Jones Merchant Marine Act of 1920, which apparently weakened non-USA companies and contributed to a reduction in environmental impact protection due to transport.

Concerning border administration, we can distinguish the efficiency of (1) customs, (2) procedures for importing and exporting goods, and (3) the accountability of the administration of the borders. The speed and the ease with which goods are imported and exported and the number and quality of services provided by national customs authorities generally define customs efficiency. Root causes of customs inefficiency include failing in making use of best practices in customs procedures, insufficient allocation of resources to customs agencies, and the lack of compliance with established procedures for importing-exporting. Results of these inefficiencies are frequent inspections, long waiting times, border delays, and burdensome requirements. The impact of such inefficiencies is particularly severe in some sectors such as the petrochemical industry, which is subject to multiple regulations from different agencies. Lack of standardized practices and of coordinating agreements among all agencies involved can cause a substantial drop in firm performance. Barriers associated with corruption are related to the transparency of border administration. Transparent border administration can help to identify if a firm is experiencing unusual delays that might be caused by refusing to pay a bribe.

The availability and quality of the transport and information technology (IT) infrastructures and their related services represent a critical element that contributes to logistics efficiency. A global supply chain might require the transfer of goods at different stages of their production cycle, crossing boundaries in a large geographical area requiring inland and coastal transportation. Such a supply chain is only efficient if the transport and IT infrastructures are appropriately developed. The number of airports, the number of lane-miles, road and bridge quality, and the level of congestion experienced at seaports and airports, on roadways, and at transport facilities are all indicators of the quality of the physical infrastructure and, to some extent, the quality of the IT infrastructure when this is adopted as a means to mitigate congestion. A high bandwidth and reliable IT infrastructure is required to track cargoes, monitor production rates and transport vehicle performance, and digitally (almost error-free, relative to paper-based) process transaction documentation.

A country's regulatory environment, safety, and security are key factors in defining its business environment. A country's business environment is a major factor considered by companies when deciding to expand geographically, relocate some of its functions, or expand some of its currently existing facilities.

We remark that the barriers listed in Table 1 are subject to interpretation. For example, a rule enforced at customs can be interpreted as a barrier to market accessibility and a negative with respect to customs administration.

The barriers listed in Table 1 are also not necessarily independent and one can cause or amplify the severity of another. An example of this is customs agents bribing, which is more likely to occur when regulations are particularly complicated. The impact due to barriers can be significant. As an example, the negative effect of custom delays is exacerbated when shipment tracking is difficult due to a poor IT infrastructure.

For the past 20 years, there has been a steady growth of trade agreements (between countries and regions etc.) and industrial clusters, including logistics industry clusters. Trade agreements increase market access and improve the business environment, fostering economic cooperation among countries and hence requiring the further development of logistics services to sustain such a growth.

Several authors (Krishnan, 2012, Sheffi, 2013, Notteboom *et al.*, 2014) have studied the positive impact the clustering phenomenon can have on trade efficiency. Industrial clustering leads to substantial improvements in operational efficiency due to the concentration of business knowledge and economies of scale, e.g., the consequent development of advanced infrastructure due to the possibility of sharing related fixed costs.

It is apparent how non-logistics industry hub development is correlated with logistics industry hub development. As non-logistics industry hubs develop, they attract logistics hubs, which in turn attract more non-logistics industry hubs, resulting in an economically virtuous cycle. However, if clustering can contribute to reducing infrastructure barriers, it also creates substantial challenges for the logistic services which must align with the industrial clusters in order to be able to serve the industry needs and face increasing competition.

Section 2.1 explores the trade agreement development, while the clustering is the subject of Section 2.2. Finally, Section 2.3 will focus specifically on logistics cluster and the way we can measure their performance.

2.1. *Trade agreements: improving market access and business environment*

Developing agreements to better integrate the economies of different countries to favour trade are a complex problem and they are generally defined at different levels, based on the integration resulting from the agreement itself. We identify seven main types of integration. The main differences between these types of agreement stem from the level of relationship implied between the economic policies (Balassa, 1997) 10: (1) Preferential Trading Area, (2) Free Trade Area, (3) Customs Union, (4) Common Market, (5) Economic Union, (6) Economic and Monetary Union, (7) Complete Economic Integration.

A Preferential Trading Area (PTA) is a trading bloc providing a preferential access for the countries exporting to the area for a specific category of products. PTA implies a reduction of the tariffs but the agreement does not lead to their complete abolishment. A PTA is generally established with a trade agreement and it typically represents the first step towards the *economic integration.*

Under a Free Trade Area (FTA), the participating states partially or fully abolish custom tariffs on their inner border. The zero-tariff will not be exploited at the regional level in case the goods originating from the territory of a member state of an FTA are subject to the rule of certificate of origin. Free Trade Area (FTA) agreements have minimal difference to PTA. In fact, almost any PTA has the potential of turning into a FTA according to the General Agreement on Tariffs and Trade (GATT). Adding free movement of services, capital and labour to the FTA leads to a Common Market.

If unified tariffs are introduced on the borders of participating countries, this leads to the creation of Customs Union (CU), which can be interpreted as a FTA and a common external tariff (CET). CUs are typically promoted to increase the economic efficiency as well as establishing stronger linkages between the countries both at a political as well as a cultural level. When shared currency is introduced, we have a Monetary Union (MU) that can have particularly important result for e-commerce companies serving different countries in the union, enabling firms to easily provide cash payment services (Chan *et al.*, 2002).

When a Fiscal Union is introduced (through a shared fiscal and budgetary policy) together with a Custom Union, then it then results in an Economic Union. An Economic Union is usually appropriate to increase the success probability and the effect of an economic integration, with examples of integration in economic policy concerning tax, social welfare benefits, and reductions in the rest of the trade barriers. The introduction of supranational bodies eventually leads to the Political Union.

There are several forms of multilateral agreement which apply to the different countries worldwide: common market trade, customs union, customs and monetary union (currency and customs unions) and monetary and economic unions (economic differs from monetary since the last one does not involve the concept of common market).

It is since the early 1990s that Regional Trade Agreements (RTAs), reciprocal trade agreements between two or more partners, have become widespread (Wilson *et al.* 2003). Indeed, around 585 notifications of RTAs were received by the GATT and World Trade Organization (WTO) up to 15 June 2014, and 379 of these were already operative by that day.

The demand for deeper integration between countries has certainly fostered the proliferation of RTAs (Baldwin, 2013, Capling *et al.*, 2011, Lejerraga, 2014). Specifically, there has been interest in the process of "*multilateralising* regionalism" as a mechanism to spread measures incorporated in RTAs across regional negotiations, and lead to convergence at the multilateral level. In fact, major efforts of liberalization in the past have highly eroded margins of preference, despite the same margins having been considered impossible to erode by most of the economy experts (Lejerraga, 2014). As a result, the most recent RTA provisions look for multilateral commitments characterized by increased depth (WTO-plus) as well as extension (WTO-beyond).

The Transatlantic Trade and Investment Partnership (TTIP) and the Trans-Pacific Partnership Agreement (TPPA) negotiation processes have motivated an important contribution in the research concerning agreements. As an example, the Centre for Economic Policy Research gave an estimation of the potential economic benefits resulting in the establishment of TTIP under different scenarios (Joseph, 2013). In particular, one scenario is if the TTIP agreement is implemented in its most ambitious version, i.e., with the full elimination of tariffs. In such a case, the center foresees a decrease of non-tariff barriers (NTB) by 25% on both goods and services and a reduction of NTB on procurement amounting to the 50%. Under such a scenario, the agreement would have an impact on the global GDP of about US$95 billion in the case of the US that become US$119 billion for the European Union. The 80% of the economic gains are actually to be brought back uniquely to the reduction of NTBs. In their work, Petri et al. (2011) gave an estimation of the increase of the global GDP amounting to US$104 billion, thanks to the implementation of the TTIP.

2.2. The emergence and advantages of Logistic Hubs

Other than trade agreements, a visible phenomenon we have increasingly observed in the industrial environment is the growth of *hubs*. Examples of this phenomenon are the Arab states and the US for the petrochemical industry, steel centers in Europe, fashion in Italy and ICT in the Silicon Valley (USA).

A considerable stream of research has focused on understanding the reasons behind the emergence of hubs and its deriving benefits (Amin *et al.*, 1992 Fujita *et al.*, 2001, Porter 1998, Krugman 1991, Long and Zhang, 2011, Hausmann and Klinger, 2006, Ciccone and Hall, 1996, Ciccone, 2002, Zhang and Tan, 2007, Hsieh and Klenow, 2009). In particular, several models have been proposed to explain agglomeration of Multinational Enterprises (MNEs) choosing their Host Country Head Quarters (HCHQ) (Xufei *et al.*, 2013). In such models, the internal structure of the company as well as the administrative and cultural distance are both interpreted as key factors in determining the choice of a firm for the country of its HCHQ.

There are three main positive effects generated by the cluster paradigm that can be identified through the literature review (Marshall, 1920):

- Improved access to the suppliers and the market;
- Pooling of the labor force;
- Technology and, more in general, knowledge (tacit as well as explicit) is easy to share.

In particular, Glaeser and Gottlieb (2009) stressed on the effect that phenomena of agglomeration have on the process of innovation, focusing on the easiness of idea exchange (i.e., tacit knowledge). Porter (1998, 1990, 2000) argues that firms can achieve their competitive advantage thanks to clustering due to the aforementioned positive externalities. Hausmann and Klinger (2006), focusing on the western countries, empirically shows that clustering displays a positive correlation at the local level with productivity. Ciccone (2002) analyses industrial investments and observes how the minimum requirements of capital for business are lower in a clustered industrial environment, and also observes the prevalent use of trade credit among firms in a clustered region with the consequently reduced external provision of working capital. Moreover, the degree of local industrial clustering has been associated with the emergence of non-state-owned (NSO) firms in the reference country. More specifically, studies have revealed how the NSO firms have larger export volumes as well as higher levels of total factor productivity (TFP). This generally does not related to state-owned firms, which are in fact less related to the presence of clusters.

The effect of globalization has contributed to reshaping the definition of clusters in terms of the relevance of the colocation replaced by the concept of (global) connectivity. As argued in Evers et al. (2010), a result of knowledge-based production and globalization might be the cooperation among firms with consequent outsourcing of parts of their administrative or productive units. As a consequence, the authors highlight how the creation of clusters is strongly related to the importance of the "tacit knowledge", as opposed to explicit knowledge which is stored and is typically available through public/private sources such as data bases. The authors state that:

> "The more important tacit knowledge is for production, the more localized production is likely to be (knowledge transfer hypothesis)" (*Evers et al., 2010*).

This interpretation is in line with the increasing spread of manufacturing networks due to the increasing standardization of knowledge, as opposed to the growth of knowledge hubs (e.g., Silicon Valley in the US or Biopolis in Singapore). Hence, hubs are characterized by a certain concentration of companies competing in sectors with competencies and skills related to the hub as well as common technologies or inputs.

Focusing on the "reach" of the considered hub, we can distinguish the case in which the hub has a regional aim (serves as a hinterland with national borders), as opposed to the case of global hubs, which extend beyond the national boundaries (Yue *et al.*, 2003).

Independently from the reach, hubs can emerge and decline in response to market forces (Yue *et al.*, 2003).

It can also be argued that, to some extent, the growth of clusters is related to the abatement of trade barriers. On the one hand, the exploitation of the economies of scale is indeed due to the concentration and the simplification of the supply network that can bring considerable efficiency in transportation and communication, while the establishment of certain industrial concentrations can favor government collaboration towards barriers abatement and efficiency of boarder procedures. It is also clear how the growth of industrial hubs and clusters has posed severe challenges to the logistics industry which has experienced the network changing their needs and always getting more global. Under the objective to answer to specific logistics needs and lower the costs, the logistic industry itself has reshaped into a cluster industry (Sheffi, 2013).

As a result of the spread of logistics hubs, such a phenomenon has received important attention and has been deeply analyzed in the last decade. In fact, the concentration of the logistics functions by logistics service providers (LSPs) as well as multi-national companies in a specific Global Logistics Hub (GLH) is one of the most apparent changes in the reshaping of the global logistics industry. Policymakers are competing to design and apply the correct strategies in order to attract multi-national companies, hence developing into a successful GLH (Tao and Park, 2004, Sheu, 2004, Lee *et al.*, 2009, Lu, 2003). In particular, the country has to provide the necessary background for supporting logistics activities in supplying several services, from transportation, consolidation and storage to more value-added activities such as inspection, assembly, packing, labeling up to documentation, and R&D services (Sheu, 2004, Chou *et al.*, 2003). Nevertheless, each logistics cluster can develop its own specificity, and different clusters can be known for different predominant activities. In this direction, Lee *et al.* (2009) propose to classify Global Logistics Hubs into two main categories based on the related performance measure used in each

Table 2. The functions of various types of GLH (*from Lee et al.,* (2009)).

Functions	Stage	Providing Firms	Types
Transportation	Transportation	Carriers	Transshipment
Warehousing		Forwarder	
Consolidation		CFS	
Distribution		Custom Brokers	
Assembly	Reprocessing	Manufacturing Firms	Re-export
Labelling		DC firms	
Packing			

category (Table 2): (1) Transshipment, and (2) Re-export hubs. In particular, the authors define the internal efficiency for the first category in terms of waiting time as well as the terminal rates. Differently, in re-export hubs, the key point is represented by the reprocessing efficiency due to reprocessing and customs costs, as well as manpower quality. Singapore, with more than 80% of the cargo being transshipped, falls in the first category (Notteboom *et al.,* 2014).

Another important contribution in the classification of Logistic Hubs is presented in Sheffi (2013), where different clusters are classified according to the following taxonomy: (1) modal-based classification (e.g., air logistic parks, port logistic parks), (2) scope-based classification (e.g., International, regional center); (3) functional classification (e.g., customs and taxation-advantaged places, single commodity logistic parks). However, none of the considered hubs are "purely" modal or functional. In relation to this, Yu *et al.* (2005) observed it is not possible to clearly identify the prevailing transportation mode for seaport and airpark since they have other modes of transportation and the multimodal parks do not reveal any dominant mode. Specifically, Yu *et al.* (2005) proposes to classify the logistics parks in China, taking into consideration that it has some of the biggest logistics hubs worldwide. What is revealed from these studies is that industrial clusters can only be effective if the infrastructure dedicated to the supply of the raw materials up to the final distribution are advanced and efficient (Sheffi, 2013), i.e., if an efficient logistic hub sustains the industrial cluster. In this regard, Sheffi (2013) recognized that in almost all logistics hubs, the following firms concentrate: (i) *3PLs*, and more generally, firms that supply the basic services such as logistics, transportation and warehousing; (ii) the firms departments providing the logistics operations, i.e., distribution to retailers, as well as manufacturers of after-market parts; (iii) the entire operations of companies with logistics as one of their main asset. Moreover, in logistics clusters we will usually find firms that provide services to logistics companies, such as those providing truck maintenance, law firms that provide insurance, software providers, and financial services providers.

Industrial hubs and logistics hubs are then strongly related, and countries are competing to be logistics hubs. However, there are several characteristics that a country should display to be a key logistic node. Specifically, in the last few years, several indicators have been proposed to evaluate the performance of the major logistics hubs nowadays as Section 2.3 will explain.

2.3. Characterizing the Performance of Logistics Hubs

Several indicators have been proposed for characterizing the performance of countries with respect to their logistics capability. In this chapter, we will mainly refer to Logistic Performance Index, Enabling Trade Index and the Ease of Doing Business Index.

The Enabling Trade index (ETI) is an indicator whose aim is to synthetize the ability of a specific country to benefit from trade. The ETI looks at the different countries and the related economies and tries to evaluate where and to which extent the infrastructures, the regulations and the institutional services provided by the different governments have the direct effect of addressing and eliminating trade barriers. The overall index is organized over four main sub-areas: (1) market access, (2) border administration, (3) infra-structure, and (4) operating environment (WEF, 2013, Section 2), for a total of 56 indicators used to evaluate the index. According to the ETI values for countries at different development values, Hanooz *et al.*, (2014) observed that developing countries are subject to a far larger variance of the indicator with respect to advanced economies. This is due to the fact that, in such locations, the political and industrial competencies can vary widely between neighboring countries.

Similarly to the ETI, the Logistic Performance Index (LPI) separates the international and the domestic trade. However, this is not done only with respect to one category, but instead, two separate LPI indexes are defined. The International LPI (iLPI) ranking is based on six dimensions of trade which are: (1) efficiency of customs and border management clearance, (2) quality of trade and transport infrastructure, (3) ease of arrange of competitively priced shipments, (4) competence and quality of logistics services (trucking, forward-ing and customs brokerage), (5) the ability to track and trace consignments, and (6) the frequency with which shipments reach consignees with in the scheduled or expected delivery times. The Domestic LPI (dLPI) looks in detail at the logistics environments in 116 countries. This measure was designed to provide more details concerning the local institutions, as well as the logistics supply chain models and the infrastructures as well as costs and time supply chain performance. The logistics constraints are not reduced at the gateways analysis such as ports of airports and they are investigated through four major determinants: (1) Infrastructures, (2) Services, (3) Border procedures and time at the border, and (4) Supply chain reliability.

The Ease of Doing Business Index (EDBI) was developed by the World Bank. The index focuses on two main aspects: the intellectual property and the complexity of business procedures. As the index scores high, this usually implies simpler and more effective regulations as well as a more effective protection of the intellectual property. The difference between EDBI and the two previous indicators (ETI and LPI, both domestic and international) stands in the fact that EDBI was developed to estimate the effect of the set of dispositions with a direct effect on business. In contrast, the index does not account for more general factors such as the proximity of a country to large markets, the level of infrastructures the safety of the nation or the inflation. Specifically, the following 10 determinants are considered: (1) procedures, time, cost and minimum capital to start a new business, (2) procedures, time and costs to build a new warehouse, (3) procedures time and cost to obtain permanent electricity, (4) procedures time and cost to register commercial real estate, (5) strength of legal rights index, depth of credit information index (related to the transparency of the procedures, which are also considered in the ETI index), (6) investors protection measured through the index of the extent of disclosure, the extent of director liability and the ease of shareholder suits (7) number of taxes paid, hours per year spent preparing the tax returns and total tax payable as share of gross profit, (8) number of documents, cost and time necessary to export and import (considered also by the previous indexes), (9) procedures time and costs to enforce a debt contract, (10) time cost and percentage of recovery rate under bankruptcy proceeding (WEF, 2013).

Figure 4 shows how Singapore is prevailing in terms of trade capabilities. This is reasonable compared to the statistics related to trade presented at the beginning of the section. Indeed, the country (as Hong Kong) has a ratio of trade and internal GDP which is not comparable to other countries (e.g., China, Germany, the US). As a result, it is particularly advanced in terms of market access and ease of starting a business. Germany is the best country in the world based on the LPI. Nonetheless, as with most of the European countries, it suffers from a tougher regulatory system that seldom restricts the possibility of making the business easier. More specifically, foreign direct investment and hiring of foreign labor are both limited by particularly strict

(a) International LPI score 2014

(b) EDBI score 2014

(c) ETI index

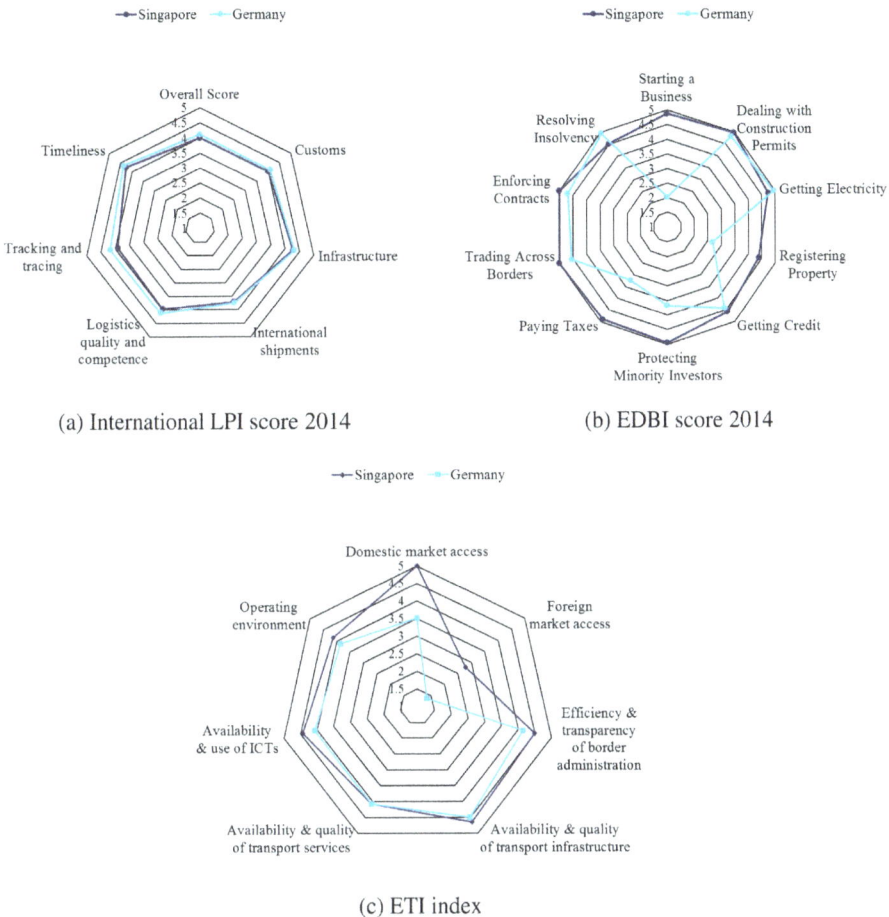

Fig. 4.   Comparison of Singapore and Germany on the main indexes.

legal policies. If not addressed, these factors will hinder Germany's future performance, especially considering the relative importance on trade in the movement of human and financial capital. Moreover, as with other EU economies, Germany suffers from a substantially complex structure of the tariff, having a particularly negative effect on agricultural products (WEF, 2013).

3. Singapore as a Global Logistics Hub: Historical development and driving factors

Singapore has experienced impressive economic growth since it became an independent republic in 1965. In the first decade after independence, Singapore was a 3rd world country with ambitions to become a part of the 1st world. The creation of jobs, the realization of needed infrastructure, and foreign investment were the means to foster and support economic development. Economic development in the first decade after independence was heavily capital intensive and the main industry was manufacturing. Since the 1980s, economic growth has largely been due to technology-intensive manufacturing performed by multi-national companies contributing an always increasing share of total value added with their high-technology products. By 1990, Singapore had established itself as a 1st world industrial power with a record growth rate. GDP growth per year has been approximately 7% during the period in 1960–2006. The knowledge-based economy is the paradigm that, starting from 1990, has brought Singapore to 1st world status in research and industrial entrepreneurship.

Singapore is currently a global leader for many industry sectors and particularly so for industries with high technological content such as electronics, pharmaceuticals, aerospace, and petro-chemicals. These industries have contributed significantly to the spectacular economic growth experienced by the country after the first industrialization phase (Table 3). The development in these sectors has also posed fundamental challenges to the development of the logistics industry in Singapore.

As for electronics, the effective management of the reverse supply chain is fundamental for cost reduction, improved efficiency, increased revenue, minimum waste, and the ability to recover value from returned goods and materials. Reverse logistics is also an important aspect in the increasingly

Table 3. Singapore established industries data.

Industry	Industry Data		Logistic Highlights
	%GDP	Employees	
Electronics	5.4	79,000	Reverse Logistics Green supply chain
Pharmaceutical	0.5	4,800	Regulations New tracking technologies
Aerospace	5.4	58,000	Rapid delivery of mission critical parts Integrated material management of spare parts
Oil & Gas	5.0	—	Tank storage Value added services

Source: EDB website.

predominant green supply chain paradigm. Indeed, reverse logistics can offer the opportunity to satisfactorily provide a response with respect to the increasing expectations for the firms' policies regarding environment protection and the conservation of resources.

In the pharmaceutical sector, many improvements can be traced back to innovations in the supply chain. As an example, a fundamental step was made by the National Healthcare Group (NHG) of hospitals in Singapore in 2008, when they decided to engage with local providers of real-time systems TCM-RFiD in order to achieve complete tracking of pharmaceutical products throughout the supply chain. More specifically, TCM-RFiD makes use of the Intelligent Medical Dispensing System (i-MDS), which provides the patient with the prescribed drug according to the prescribed dosage (Business Monitor International, 2014). These and other innovations that eliminate paper documents and hence reduce medical errors, significantly improve hospital workflow management.

The global aerospace and aviation industries are undoubtedly going through a period of impressive growth. Besides the growing demand for this transportation means, the planning capability of Singapore is surely fundamental to such a development. These industries require time to be developed. It is therefore a clear sign that the Singapore planners are foreseeing a future steady growth for air transportation. In fact, in the next two decades, the size of the global fleet is expected to double. Moreover, according to projections of the main aircraft manufacturers Airbus and Boeing, more than 30% of such overproduction will be delivered to Asia. As a result, Asia Pacific's fleet should triple its volume to 13,500 aircrafts (www.edb.gov.sg).

Singapore is also one of the world's leading energy and chemical industry production hubs. Jurong Island is an integrated complex for a large number of the world's leading energy and chemical firms (e.g., BASF, ExxonMobil, Lanxess, Mitsui Chemicals, Shell and Sumitomo Chemicals), and hence is a symbol of Singapore's leadership and growth in the petrochemical industry. Particularly relevant during the past two decades is the development of Singapore as a leader in research and innovation for this industry, a position of leadership that Singapore plans to retain in the future. The petro-chemical sector contributed S$38 billion to Singapore's manufacturing revenue in 2010, while this contribution was approximately S$28 billion in 2009.

As hubs of various industries have developed in Singapore, the demand for world-class logistics companies has also developed. As a result, Singapore has become a global hub for third-party logistics firms (3PLs), providing large scale services ranging from air freight, customs brokerage, ocean freight, local transportation, warehousing and distribution to customized supply chain solutions or other value-added services for products manufactured in or transshipped through Singapore. In addition, Singapore is a logistics hub for leading manufacturers across industries, including Diageo, Dell, Hewlett Packard, Infineon, LVMH, Novartis, ON Semiconductor, Panasonic, and Siemens Medical Instruments (Zhu *et al.*, 2002).

Singapore can now be considered a Knowledge Cluster and Global Logistic Hub. Section 3.1 will present the historical development of the country and the main milestones achieved, as it became a global hub. The main driving factors for this development will be the subject of Section 3.2.

3.1. *Singapore Historical development as a Global hub*

Reducing trade barriers and fostering foreign investments and trade has been the basis of Singapore's economic development since its independence from Malaysia in 1965. As a result of such a policy, more than 5,000 international firms are now located in Singapore, and a remarkable number of multinationals have established either their selling or operations Head Quarter on the island. In the early 1970s, the country achieved full employment with an average annual growth rate of 6.5% for the GNP. In 2000, the country population accounted for 4 million inhabitants, 25% of which were foreign. According to the World Bank, Singapore in 2013 was ranked 7th in the list according to the GNI per capita with an estimated value of US$54,040. According to these values, the small island results are comparable to countries such as Norway, Switzerland, Qatar, Denmark and Sweden.

To make this impressive growth rate possible, for two decades, the Singapore government has highly invested in the "immobile factors" such as the infrastructure, public services, land, property, political as well as social culture, and labor, always making sure that the growth of the country was appropriately supported. In fact, it is also apparent how the continued economic growth of Singapore relies upon the mobile factors such as capital, information and the talent which need to be attracted by the country (Phang, 2000).

Figure 5 represents the different phases in the development of Singapore, which are briefly explored in the next part of the section. In Section 3.2, we will analyze the driving factors supporting and contributing in establishing Singapore as a global logistics hub.

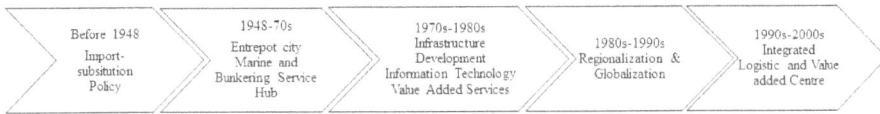

| Before 1948
Import-
subsitution
Policy | 1948-70s
Entrepot city
Marine and
Bunkering Service
Hub | 1970s-1980s
Infrastructure
Development
Information Technology
Value Added Services | 1980s-1990s
Regionalization &
Globalization | 1990s-2000s
Integrated
Logistic and Value
added Centre |

Fig. 5.   Phases in Singapore's growth.

Before 1948

Before the end of the Pacific War and the Japanese occupation, Singapore was mainly in charge of the provision of the goods to the inland which included Malaysia. This role remained active also in the first years after the end of the Japanese occupation, when Singapore still constituted the main supplier of goods for Malaysia, and this supply was also the most relevant business for the island. Nonetheless, the wish of the island to enlarge its market was clear when the port started its reconstruction plans after the Pacific War destruction.

1948–1970s

The internationalization of Singapore was apparent as it became one of the main bunkering centers in Asia in the 1960s, and a free trade policy island, making it a convenient stopping point for several shipping lines crossing the Asian sea. It constituted the natural entrepôt port for the region. The independence of the island from Malaysia on 9 August 1965 made the effort to internationalization a key for the success of Singapore, which started an important investment process to build the infrastructure required to become a global hub. Already in the late 1970s, Singapore understood the importance of value-added services as a key to attract the foreign business and investment. Warehousing services in terms of distribution centers were created, and chemical as well as refinery industries had their peak growth period.

In fact, the entrepôt function of Singapore did not decline due to the independence from Malaysia. Instead, with the shift in the manufacturing exports in the early 1980s, Singapore played an increasingly relevant role as intermediary for Southeast Asian trade in machinery and equipment and the procurement of electronics parts and components (Yue *et al.*, 2003).

1980–1990s

It is in the 1980s and 1990s that the logistic value-added services became a key factor for the country's growth. Distriparks such as the Alexandra and Keppel Distriparks represent the evolution of the distribution centers providing enhanced services with respect to consolidation and storage. Expertize in refrigerated cargoes as well as dangerous goods became a distinguishing feature for the country.

The opening of the Chinese economy constituted an important part in this decade. Back in 1978, the economic reforms began with the introduction of market principles. The process went through two different stages. At the first stage (in the late 1970s–early 1980s) agriculture was decollectivized, foreign investments were allowed to flow into the country and entrepreneurs were eventually allowed to start business. During the second stage of reform (in the late 1980s–1990s) much of the state-owned industry were privatized or con-tracted out. Price controls, protectionist policies and regulations were lifted, although sectors such as banking and petroleum remained state monopolies. It has been studied and assessed that a strong bidirectional relationship between logistics growth and economic development characterized and still character-izes China (Liu 2009) and it is apparent how this growth fostered the develop-ment of Asia and the APAC regions to exploit such growing market and remain competitive.

In this decade, the country started to promote the informatization of the industry. A clear sign of such a development was the fact that, while R&D was at the level of a 3rd world country up till the 1980s (Gross Expenditure on R&D (GERD) to GDP ratio was 0.86% in 1987), R&D investment intensity in Singapore started to quickly rise to the point that GERD increased by 13 times in the period from 1987–2006, like no other country ever experienced.

Understanding the possible limitations due to the size of the country, Singapore worked hard to increase its international presence. An example of this effort was joining Indonesia and Malaysia in the development of manu-facturing companies which would have then been followed by the establish-ment of high-tech/high-knowledge industries in the mainland. Under the agreed development plan, Malaysia and Indonesia were providing land and labor, which Singapore was lacking; Singapore, on the other hand, was providing infrastructure and administrative skills.

Turning Singapore into a hub for services at regional level was appar-ently the government target. In fact, such a role would have met the growing needs of growing markets such as China and India. To sustain this vision, the country worked hard in improving the quality of the linkages between the island and the rest of the world by investing in air and sea infrastructure. In 1997, as a proof of this effort, the Open Skies Agreement with the US was signed by Singapore, the first Asian country taking part into the agreement.

The Changi Airport was already ahead in the process of construction and it started its operations in 1981. Already by 1988, it was recognized as the best airport worldwide by the British Business Traveler. Considering the traffic of international passengers and the cargo handling, Changi is nowadays the seventh busiest worldwide (www.edb.gov.sg). The port of Singapore made similar advances through heavy investments in technological innovations.

Thanks to these strategies, companies established in Singapore were able to bring their business also outside the island.

During the 1990s, the competitiveness in Southeast Asia increased remarkably, impacting the growth rate of the country that had started to decline due to external shocks such as the Asian Financial Crisis in 1997. The government, as a response to such economic situation, launched the 1991 Strategic Economic Plan and identified 13 clusters, in manufacturing and services, for sustaining the country growth. The advantages and gaps of the country in each of the clusters were identified and strategies as well as initiatives to fill the gaps were examined and eventually executed.

An example of the plans rising from this analysis is Industry 21 (1999), a ten-year plan to make Singapore a solid and established global hub for knowledge industries in manufacturing and traded services, giving new emphasis to knowledge-based activities as the frontier of competitiveness.

2000s

The ASEAN–China Free Trade Area was signed in Cambodia (4 November 2002), as a free trade agreement among 11 nations that should have been involved by 2010. ASEAN–China represented the largest FTA in terms of population and third largest in terms of nominal GDP. The FTA came into effect on 1 January 2010. For ASEAN, an FTA with China was offering the Southeast Asian states a way to overcome the disadvantage of reduced size by pooling resources and combining markets. It is foreseen that, under the China–ASEAN FTA, an economic region of 1.7 billion consumers will enjoy a regional gross domestic product (GDP) of UD$2 trillion, while the total trade is estimated at $1.23 trillion (Wong and Chan 2003). Several agreements were also signed: (1) the Framework Agreement on Comprehensive Economic Cooperation (FACEC), (2) the Trade in Goods (TIG), and (3) the Investment Agreement (IA).

In 2002, ASEAN and China signed the TACEC between ASEAN and China under the two main objectives of (1) strengthening and enhancing economic trade and investment cooperation between the parties; and (2) progressively liberalize and promote trade in goods and services as well as creating a clear, liberal and facilitative investment regime. Moreover, the agreement between the parties was made to investigate new areas as well as to develop the best measures fostering a stronger economic cooperation. This had the remarkably positive effect of making the integration of the state members newer to ASEAN smoother, and it also favored the bridging of the existing development gap among the countries.

The TIG was signed in 2004 and then put into effect on 1 July 2005 by the ASEAN countries and on 20 July 2005 by China. According to the agreement, the six original ASEAN members and China had to proceed to the elimination of the tariffs to 90% by 2010. An exception was made for Cambodia, Lao PDR, Myanmar and Vietnam, whose deadline was extended to 2015. In particular, the Trade in Services Agreement (TSA) entered into force in July 2007 as a region/sector agreement (i.e., sector and region based agreement). Due to TSA, improved market access and national treatment was guaranteed to services suppliers/providers participating in the agreement.

On 15 February 2010, the (IA) was implemented to help the participating countries in creating a more transparent and facilitative environment, and give companies from ASEAN a competitive edge to tap on thriving opportunities in China.

Otherthan the agreements, the country was also involved in several fundamental development plans mainly in manufacturing, logistics as well as R&D. Manufacturing 2000 is one example which grouped the success strategies for manufacturing. The focus of the plan was on the need to retain a manufacturing base in Singapore in view of the manufacturing-services nexus, instead of imitating Hong Kong on the path of de-industrialization.

Nevertheless, the planning activities did not only focus on labor intensive industries. Indeed, strategies for services were grouped under the International Hub 2000 programme, which were aimed at developing Singapore as a global city and a hub for business and finance, logistics and distribution and communications and information.

In 2001, a review committee was nominated by the government to formulate a restructuring of the economy in order to maintain the leadership of the country. Within the results achieved, thanks to the strong policy to develop Singapore as a strategic value-added location for the logistics industry, the most influential companies in the 3PL as well as 4PL sectors found the country to be the ideal place to setup their Asian headquarters. As a result, the number of 3PL and 4PL increased and, thanks to the availability of efficient and high quality logistics services, the e-commerce industry grew as well.

In order to foster the concentration of companies, the creation of research centres and advanced education were necessary. In 2008, they reached the figure of 2.77% of the Gross Domestic Product (GDP). With such a growth, Singapore was ready to achieve the 3% representing the target result by 2010 (www.newsxinhua.net, 2009). As a response, the private sector also increased its R&D by 20.9% per year, from SGD4.23 billion in 2007 to SGD20.3 billion in 2008. It is apparent how the R&D activity was the main element pushing forward the development of the private sector (e.g., biochemical and electronics).

The strategy of Singapore for becoming a main hub was justified by the government intuition that all the major economic activities in the period such as telecommunication, shipping, air transport, finance, and information technology displayed the clear tendency to concentrate in a few nodes worldwide (Section 2.2). These nodes needed to have an extended hinterland to service and the infrastructure to link such a hinterland to the rest of the world. Singapore was a first mover, thanks to its stability which guarantees the country with the possibility to plan far ahead for developing world class physical infrastructures (such as airport, seaport, financial and industrial facilities), for creating policies for human resources, providing diplomatic lasting relationships with strategic countries, important investment and partnerships to guarantee its projects, to receive both the local as well as the global maximum competencies.

To establish herself as regional hub, Singapore based her strategy on providing added value to its economic hinterland by means of a variety of world class products as well as services and its reputation for reliability, quality and excellence. This story has been, and is, possible due to the development of strong infrastructures and communications as well as information technologies and policies facilitating business.

It is apparent how investments as well as the agreements were all in the direction of promoting the hub role of the country when faced with the barriers analyzed in Section 2. The next section will provide more details on the main drivers and strengths that Singapore could develop thanks to the political agreements as well as the massive investments performed.

3.2. Driving Factors for the development

At the basis of the development described in Section 3.1, Singapore has constantly improved several driving factors, guaranteeing steady continuous growth as a logistics hub: (1) transport infrastructures; (2) information and communication technology; (3) development of 3PL industries; and (4) government policies to support trade and lower logistics costs. In the following sections, these three driving factors will be separately considered, highlighting the main historical milestones related to the development of each factor.

3.2.1. *The Transport Infrastructure*

Singapore has been, a global example for the quality and efficiency of its transportation infrastructure in the past decade. In order to set the benchmark for the transportation industry, the development board has invested strongly in sustainability, promoting major public investments to favor the private industry growth as well as pushing for international competitiveness. At the local level, the strategic marginal social cost for pricing motor vehicle ownership and road usage are only examples of the government strategies to maximize the road efficiency (Willoughby, 2001, Phang, 2003).

What is unique to Singapore is that because of its reduced size and due to its sovereign status, urban issues are national issues, while some traditional urban problems (such as rural urban migration and city versus nation) do not exist. As a result of the ease of control and political stability, the infrastructure projects could always be planned and realized far ahead of demand. In fact, the ability of Singapore planners is a key element to explain how the island could quickly become a leading hub.

Undoubtedly, if the global transportation network is considered, the country is a focal sea and air hub node.

Changi Airport is constantly recognized through numerous international awards for its achievements in airport, retail and cargo facilities and provided services, and the same can be said for the Port of Singapore. The study of the impact of the logistics changes onto the ports' development has recently attracted an increasing attention. The seaport must be developed as part of an integrated logistics hub. As a result, not only the basic port-functions (airport), but also value-added logistics services must be provided. As an example, Robinson (2002) suggests to interpret seaports as an integrated element within value-driven chain. Heaver *et al.* (2000) and subsequently, Notteboom and Winkelmans (2001) discuss the fundamental topic of the new role played by the Port Authorities in a new logistics-based endeavor. Martin and Thomas (2001) specifically address the issue of the changes occurring at the level of the container ports specifically on a structural perspective (i.e., how the terminal should be changing in its design and management in order to increase its performance as a node in the supply chain). All these reasoning can be applied to air transportation.

The success of the Singapore seaports and airports has been the result of the concerted organization, efficient operations as well as updated equipment and competent labor, which are continuously and dynamically improved in order to maintain the leadership as one of the busiest hubs in the world.

At the present moment, Singapore's port activities are concentrated around connecting the oil and cargo markets from the West (Europe, Africa, Middle East, and South Asia) with that of East Asia. This is due to the large share of the country's GDP derived from oil, port and shipping activities. The competitive background makes it harder to always keep its dominance. In fact, through new services, solutions and performance, Singapore is expected to continuously raise the benchmark. However, the surrounding states such as Malaysia are also willing to increase their market relevance, and with China building pipelines to circumvent the Straits, are both scenarios to be carefully considered by the country to keep up its competitiveness.

The Singapore port currently accounts for 75 berths for a total quay length of more than 20 km. The two main operators are the Port of Singapore Authority (PSA) and Jurong Town Corporation (JTC).

Come 2027, the Singapore ports will all be consolidated in Tuas. The first phase of Tuas should come into operations in 2022, before the actual leases for the city lands will expire in Tanjong Pagar, Keppel and Pulau Brani. Before Tuas, the Pasir Panjang expansion at an estimated amount of US$3.5 billion is foreseen to increase PSA Singapore's maximum draft from 16 million to 18 million, resulting in the capacity to serve the last generation containerships of 18,000+ TEUs. The expansion will cover 2,500,000m² and add 15 new berths with a 6,000m quay. Singapore's present capacity of 35 million TEUs is

Table 4.   Total market throughput and main cargo segments (2010).

Segment	Mill. tons (TEU)	(%) of total
Container	289.7 (28.4)	57%
General cargo	24.0	5%
Bulk	12.6	3%
Oil	177.1	35%
Total	503.3	100%

approaching its saturation point and the Tuas development will add the vital new capacity.

> "[…] Tuas provides a suitable location because of its sheltered deep waters and proximity to both our major industrial areas and international shipping routes […] Consolidation at Tuas will eliminate this need for inter-terminal haulage […]" [*Singapore's Minister of Transport Lui Tuck Yew said when the project was announced last year*].

The Singapore Prime Minister Lee Hsien Loong said:

> "[…] The port has been very successful. It is growing; it is reaching its limits. So we are building a new port in Tuas, bigger, more efficient, almost double the present capacity. And when this is done, we can move from Tanjong Pagar to Tuas […]".

The reclaimed area counts for 10,000,000 m² covering the country from Shenton Way to Pasir Panjang. The Tuas project could make Singapore the world's biggest port in terms of capacity, according to Drewry Shipping Consultants. Sixteen new berths will be added to the container terminal of Pasir Panjang. Given the unprecedented size of the project, MPA put on top of the list the minimization of environmental impacts due to port developments started through a comprehensive programme which included the relocation of corals in the project vicinity and the placement of an Environment Monitoring and Management Plan (EMMP).

The air traffic was not less important to the eyes of Singapore's planners who had to tackle the hurdle of a lack of hinterland demand, hindering the development of the country as a Southeast Asian hub airport. The Changi Airport is a major air transport hub and it represents *the whole country air policy* whose main components are: strong airline company, liberal air policy, airport infrastructure investment decisions. This strategy was effective and brought Singapore to become among the most popular bases for international airline companies. The recognition of the air winning strategy came from the entire community: Changi was frequently awarded as best airport (one of 20 best airport awards in 2001), winning more than 430 awards since 1981, including 30 'Best' awards by 2012.

Despite the country's small domestic market, the Changi terminals began its operations in 1981. In 2001, the airport was connecting 159 cities of 50 countries through 59 airlines operating more than 3,250 services each week. This resulted in a total of more than 1.5 million tonnes of air cargoes and 28.2 million passengers. The airport was handling a number of passengers larger than 53.1 million in 2013, reaching an increase of 5% with respect to the result in 2012. Due to this performance, Changi was nominated the busiest airport in the world by passenger traffic and the second busiest in Asia (*http://www.aci.aero/*).

The role played by Changi in establishing Singapore as a logistics hub is apparent: the airport is one of the busiest cargo airports in the world, handling 1.85 million tonnes of cargo in 2013. This ability of answering demand peaks and support demand growth is thanks to the continuous upgrading investments that Singapore has done to upgrade the airport structures.

Urban planning was fundamental as well to make Changi integrated within the city: currently two expressways the East Coast Parkway and the Pan Island Expressway link the airport to the rest of the country, while the underground extension to the airport was completed in 2002.

1970–1981

The 1970s represented an important decade for transport development in Singapore, as the Tanjong Pagar seaport and the Changi Airport both started construction. The development of these new important transportation hubs will increase traffic in the island in terms of goods and people. The urban development perfectly reflects such a vision for change and, in the early 1970s, the government has its first discussion on the Mass Rapid Transport (MRT) development. Given its relevance, the development planning for the first stage of the MRT took up the largest part of the decade. Nevertheless, the planning was largely justified by the importance of the MRT construction as a means to foster the growth of both the financial as well as the business sectors.

Undoubtedly, containerization represented one of the main driver for the development of the Tanjung Pagar seaport. In fact, PSA envisioned a spectacular growth in the container business and its deriving globalization effect (i.e., the possibility to extend trade and increase the amount of transported goods outside Asia). As a response to this forecast growth, the East Lagoon (Berth 48–49, Tanjong Pagar) became operational in 1971, before even a single containership had made a call to the Port.

The wish of the country to become a hub for international trade required the improvement of the overall transportation infrastructure. The realization of the Changi Airport represents the most spectacular effort of the island in this direction. Led by PSA chairman Howe Yoon Chong, land-reclamation works involving over 52,000,000m^2, of landfill and sea fill, began in June 1975, when the airport at Paya Lebar was still being expanded. About 2 km^2 of mud

Fig. 6.   Changi control tower.

land were cleared and filled with 12,000,000 square metres of earth from the nearby hills, while another 40,000,000 m^2 of sand from the seabed were used to reclaim land. The first phase opened on 1 July 1981 with an investment of about S$1.3 billion and 8.1 million passengers had flown through the airport by the end of the year, whereas airfreight were at 193,000 tonnes, with a total of 63,100 aircraft movements.

When the Changi Airport started, the Changi Airfreight Centre (CAC) was established to provide cargo services. The 47-hectare CAC was a 24-hours Free Trade Zone (FTZ), where transshipment cargoes could be broken down and reconsolidated with minimum custom formalities (Changi Airport Group, 2013).

1982–1990

In 1982, the MRT project was finally confirmed for a total of S$5 billion (Rimmer, 1988; May, 2004). The statutory board for the MRT, i.e., the MRT Corporation (MRTC) was established in 1983, bringing the country into a new phase of development of its infrastructures. The roads developed by 1960 amounted to 800 km of roads, its length doubled in only 10 years with 3000 km roads developed in 1990. The country was growing together with its port, airport and industries.

In order to keep high standards in terms of service level and avoiding congestion problems (i.e., major service disruptions in airports worldwide),

Singapore always planned for airport developments far ahead of demand, contributing to the growth of the business in a proactive way. A clear example of this is, despite Changi initially being designed as a two-terminal airport, additional expansion plans were soon included, extending the initial design, and plans for a third terminal of the same size and functions as Terminal 2 was put into place.

Key investments were also made in the port sector, both in equipment and facilities. Other than the relevant investments in the modernization of the terminal equipments, the construction of maxi-warehouses detached from the port area was started towards the end of the 1980s, when Keppel Distripark was approved of representing one of the pioneering automated warehouse on the island. The complex was designed to occupy about 470,000m² for a cost of S$399 million. In 1988, the expansion of the port with the construction of latest technologies container berths in Pulau Brani was approved with a cost of S$870 Million.

Towards the end of the 1980s, Jurong Island was a clear illustration of the reactiveness of the Economic Development Board of Singapore (EDB). The Bayan LogisPark, Singapore's first integrated chemical logistics park, started operations on Jurong Island in 1973, serving as a regional transshipment, break-bulk and distribution center for the international chemicals industry. The park provided several services, from the basic handling of both solid as well as liquid bulk to the management of dangerous chemicals.

Despite the Western Country crisis and the oil crisis characterizing the period, EDB began its intensive promotion efforts with top players in the international chemical industries, while land reclamation was still in progress. Again, this shows the ability of city planners to embark on crucial investments far ahead of the realization for its demand.

1995–2000

Despite the general crisis in the economic sector, the Singapore government initiated a strategic logistic plan in 1997, and during the years of 1997/1998, the transport sector signalled a growth of 5.5%.

In 1995, the Land Transport Authority (LTA), controlled by the now Ministry of Transport, merged different political bodies having competencies in all the fields of land transport, from planning to implementation and management.

In January 1996, the white paper with the main policy guidelines for the new body was submitted to the Parliament and it was agreed with the complete development plan for the city MRT.

The Maritime Port Authority (MPA) was formed in 1996, which led to the corporatization of PSA in 1997. The start of the works for the forthcoming Pasir Panjang port constitutes a main fact in this period. The location of Tanjong Pagar in the proximity of the city center and the Central Business District (CBD) prevented further port expansions. In addition, the planning for a thorough relocation of the port and reclamation of land for business and residential purposes started to be more important in the planners' agendas.

A fundamental effort in the automation and informatization of the port gates was performed during the 1990s. In order to cater for the increase in container traffic in Tanjong Pagar, a new gate was developed. It had 14 lanes installed with high-tech furniture such as transponder readers, closed circuit televisions and a weight pad system to quickly clear the vehicular traffic. Expansion words of the storage areas was also on top of the priority list for PSA.

In 1996, there were four Distriparks: (1) Alexandra accounting for 212,000m^2, (2) Pasir Panjang with 196,000m^2, (3) Keppel with 112,000m^2, and (4) Tanjong Pagar for 75,000m^2. The services available at the Distriparks are more differentiated and customized to the respective distributions centers. The presence and growth of the Distriparks providing services such as warehousing, stevedoring, forwarding, customs, product insurance, agency, dedicated ICT solutions and barge operation, consolidation, distribution, road transport as well as value-added services (mixing, blending, repacking), boosted the increase of the concentration of 3PL and logistics company in general.

2000s

At the present moment, more than 3000 international and local logistics companies have their headquarter established in Singapore (EDB, 1998). In 2003, the Airport Logistic Park of Singapore (ALPS) was opened (Changi Airport Group, 2013). Situated within the Changi Airport FTZ, its advantage is that products can be brought in, reworked and flown out without leaving the FTZ. Since its launch, many global logistics players such as Bax Global, DB Schenker, DHL, Nippon Express and UPS have located themselves there. It has its own customs check point for exclusive use by tenants, thus allowing for very efficient turnaround of high-tech, high-value products. In 2010, CoolPort@Changi was set up within the FTZ as a dedicated facility for temperature-controlled, time sensitive critical cargo. Such a facility allows companies for unbroken cold chains and to handle a wide range of perishable commodities and pharmaceutical products. It has a handling capacity of at least 250,000 tonnes a year.

To meet the increasing needs of the aerospace industry, Seletar Aerospace Park was launched in 2007 as a dedicated park for the aerospace maintenance, repair and overhaul as well as the manufacturing of aircraft components.

Pasir Panjang is among the busiest and most advanced ports. Terminals 1 and 2 were completed in 2000. All five terminals (Tanjong Pagar, Keppel, Pulau Brani, Pasir Panjang I and II) had a built-in capacity to handle 35 million TEUs.

In 2012, PSA terminals handled 31.26 million TEUs of containers (PSA Annual Report, 2013).

3.2.2. *Information and Communication Technology*

It is clear how information and communication technology (ICT) services are fundamental for the development of logistics services: the shipper requires the complete computerized management of the order, from the purchase to the deliver, with full traceability of the product, while carriers and logistic companies that wish to automatize and optimize the internal operations as well as synchronize with other actors in the supply chain can do so by making full use of the available information.

The level of IT development of a country is a fundamental driving factor for its establishment as a logistics hub. The importance of the ICT services industry including software, content and other ICT services, came to be recognized with the rapid computerization during the 1990s. Undoubtedly, the Port of Singapore, under the main guidance of PSA (until 1997), and PSA and MPA (since 1997), was a pioneer in the computerization and use of ICT for improvements of port activities and port management. In the period between 1982–1984, PSA invested $6 millions to provide on-line terminals to the operational and engineering departments. The systems CITOS (Computer Integrated Terminal Operations System), CIMOS (Computer Integrated Management Operations), PortNet and TradeNet were all developed and started running in 1980–1990. In particular, TradeNet represented one of the first nationwide electronic data interchange (EDI) platforms for trade administration worldwide.

The success of the introduction and intense use of ICT in improving port operations and supporting ease of trade, probably fostered the development of the ICT industry (Bilbao-Osorio *et al.*, 2014). According to the Networked Readiness Index (NRI), Singapore is a global leader in ICT (WEF-GITR, 2014). NRI provides a useful conceptual framework to evaluate the impact of ICTs at a global level and to benchmark the ICT readiness and usage of different economies (Kirkman *et al.*, 2002). This indicator accounts for economic and social impacts of ICT in a country economy, as well as country environment (e.g., market conditions, regulatory framework, innovation-prone conditions), access and usage of ICT (Bilbao-Osorio *et al.*, 2014). In this section, we will try to understand how Singapore guaranteed its leadership in the sector and how the logistics industry was facilitated by this development.

1980–1985

In 1981, the government promoted its plans for the national information technology as part of the more general National Computerization Plan (NCP), with three main objectives: (1) computerization of the civil service, (2) training of software professionals, and (3) pushing for the growth of the local IT industry in order to acquire the capability to respond to the software and IT service market. In particular, the Civil Service Computerization Program (CSCP) started in 1982, proving the national computerization effort and giving further incentive to the private sector in the application of ICT. CSCP applications improved productivity within the government sector and helped in coordinating the different agencies and government ministries. The pool of ICT professionals created by the CSCP as well as technology transfers by foreign multinational companies contributed to the spread of ICT in the private sector. However, domestic R&D in ICT remained weak.

The company National Computer Systems PTE Ltd (NCS) was the main domestic enterprise in ICT (the counterpart of PSA for the Singapore Port). In fact, it was created in 1981 under the control of the National Computer Board (NCB) as its IT sub-section. NCS played a key role in driving the Singapore government in the process of computerizing the public services as well as the computerization of the civil service. As a result of such a supporting role, NCS was acquired by SingTel in 1997, thus becoming corporatized (exactly the same year when PSA was corporatized).

1985–1992

In 1985, the National IT Plan (NITP) was the natural continuation of the NCP plan in the past decade. The plan outlined seven pillars to construct the country IT strategy: (1) developing IT professionals and experts, (2) improving the ICT infrastructure, (3) promoting the ICT industry, (4) coordinating and collaborating between various ICT-promoting organizations, (5) establish a culture welcoming ICT, (6) encourage creativity and entrepreneurship, and (7) increase ICT applications in workplaces (Tan *et al.*, 1985).

This period was fundamental to establish the role of the ICT in Singapore. Indeed, the government and private companies collaborated in order to: expand computerization in the private sector, spread the use of ICT in key industries, promote alliances with international software firms to develop local engineering skills and IT segments. Concerning hardware, the installation of fibre optics and integrated services digital network (ISDN) was a key development of the period. The introduction of the Electronic Data Interchange (EDI) system was the fundamental enabler to improve port performance, leading Singapore to be among the fastest ports in the world. In order to improve domestic R&D in IT, there was a substantial increase in R&D expense in ICT, hence pushing the development of the still weak local industry in ICT.

1992–2004

In 1992, the NCB released the IT2000 plan promoting the vision of Singapore as an "Intelligent Island", based on a National Information Infrastructure (NII), with the role of creating a virtual network between Singaporean homes as well as workplaces all over the island as seen in Fig. 7 (Choo, 1997). The main target of this decade plan was undoubtedly the development of local skills in the IT market. The availability of skilled and trained manpower would have worked as a competitive advantage both within and beyond Singapore, leading to the establishment of the country as a regional ICT hub.

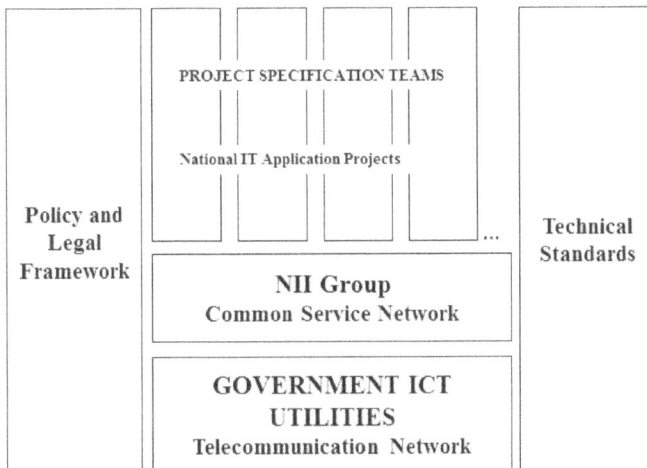

Source: National Computer Board, IT2000 Report.

Fig. 7. Framework for Singapore's National Information Infrastructure (NII).

Government ministries and agencies, both part of the group, were organized as shown in Fig. 7 under the NII framework. A high level steering committee coordinate policies across the different ministries and agencies.

Following the general policy of the country, according to which the government has the main role in the planning and executing strategic projects, the major IT and telecommunication projects were the responsibility of NCB and the Telecommunications Authority of Singapore (TAS) respectively.

NII led to the establishment of several EDI applications including TradeNet and PortNet for the port of Singapore. GIRO, the service for cashless transactions, was created, giving the possibility of transfers to bank accounts of merchants as well as government bodies. In the same period, Network for Electronic Transfer System (NETS) became operative with credit and debit card transactions at the point of sale. It was in June 1999 that Singapore-ONE was launched: the island-wide broadband high speed infrastructure of high capacity network and switches for multimedia applications such as e-commerce services, news, government transactions and fast-internet.

In order to maximize efficiency and coordination among different offices, the NCB was merged with TAS, and the Infocommunications Development Authority (IDA) was established.

The country was ready for the Technopreneurship 21 programme, launched in 1999 to foster innovation in ICT. Specifically, the Public Research Institutes/Centers were positioned to bridge basic research by universities with applied research in the private sector. The increasing investments in R&D are proof of such efforts (Table 5).

Given the infrastructure and improved available skills, with the Infocomm21 Masterplan, Singapore made a concerted effort in 2000 to transform the country into a dynamic and vibrant global ICT hub with a

Table 5.   R&D Expenditures by sector.

Sector	1993	1998	1999	2000	2001	2002	2003
Total	998.2	2492.3	2656.4	3009.5	3232.7	3404.7	3424.5
Private	618.9	1536.1	1670.9	1866	2045	2091.3	2081.2
Higher Education	157.3	305.8	310	338.3	367	430	457.5
Government	106.5	299.8	304.9	423.8	425.1	449.1	435.8
Public Research Institute	115.5	350.5	370.6	381.4	395.6	434.3	450

Source: Yearbook of Statistics Singapore, 2005.

competitive e-economy, an infocomm-savvy e-society, excellence in e-government and e-learning hubs.

2005–2015

A main planning decision on the infocomm sector's long-term growth was taken in May 2005, when a high-level steering committee established a 10-year masterplan where the role of infocomm technologies was clarified as being (1) the main enabler to push the competitiveness in industrial sectors of fundamental relevance for the country, as well as (2) having the fundamental set of technologies to connect the society at different levels of business.

In 2011, the IDA launched TradeXchange, an online trade platform to facilitate both the trading and logistics communities in the activity of exchanging relevant and sensible information. In particular, the platform was supporting the creation of trade documents between trading partners in a cost-effective and efficient way. Moreover, the customer could enjoy the possibility of having only one single platform for both trade transactions as well as enquiries to all the main government bodies (e.g., controlling agencies, maritime authorities, seaports and airports authorities and customs).

The reference programme nowadays is undoubtedly the living blueprint in iN2015. The programme was developed as a result of the collaboration of the private companies, with the main aim to support the transition of the country into a city based on information and communication technologies. iN2015's main objective is to support companies as well as talents, by fostering the design of new technological solutions prototyped by individuals as well as those developed by enterprises. The programme aims at connecting individuals and the communities with the different businesses by providing them resources both in terms of financial as well as human capital for both national as well as global. iN2015 will be the channel to for Singapore to access the resources from the world as well as to export the produced services, products and ideas to the entire world's marketplace.

3.2.3. *Third Party Logistics Company Development*

With the globalization of the trade, supply chains have become increasingly complex. Production, assembly, general value-added activity, as well as transportation and delivery are all distributed along the supply chain.

Companies cannot achieve logistics economies of scale under such an extended and complex setting. As a result of this impossibility, with the aim of reducing inventory costs as well as maximizing customer performance, companies have started to rely on logistics service providers to reduce the impact of supply complexity. The market of third party logistics providers has grown spectacularly in the past two decades as a fundamental condition to aiding efficient and effective global trade. This growth has been experienced also by Singapore as described in the following.

1960–1970s

Until the late 1960s, the main role played by Singapore freight forwarders was limited to local jobs such as clearing cargo from the port, transporting them to the warehouses and providing lighterage to ships anchored along the roads near the harbor. In the 1960s and 1970s, the industrialization and oil boom brought new opportunities to the domestic freight forwarding scene. Singapore-made products such as garments, electrical products and electronic sub-assemblies had to be shipped out by air and sea. Foreign freight forwarders were also attracted to Singapore as a source of cargo as well as a hub for the Asia–Pacific region. Companies from Germany (e.g., DHL, DB Schenker), Japan (e.g., Hitachi Express and Kamigumi) and the United States (e.g., Air Express, Airborne Freight) set up operations during this decade. On 6 September 1973, the Singapore Logistics Association was established as Singapore Freight Forwarders Association, with a membership of 70 companies and Mr Ho Shio Kwang (Sing Kwang Kee Pte Ltd) as the first chairman. The link with the Port of Singapore and PSA is clear and, in 1977, SLA participated in the PSA Quarterly meeting to promote better understanding and cooperation between the haulage industry and the port (sla.org.sg).

1980–1990s

Containerization brought about opportunities for freight forwarders to work directly with shipping lines to consolidate cargoes. As a result, the role of these companies changed from agents to principals in the chain, keeping under their control the entire transport of goods under single contracts. In fact, business strategies were mainly influenced by globalization and companies were now struggling to produce and design goods for a global market, modifying their supply chain in order to have global sourcing and supply (Cooper, 1993). As a consequence of the increased global trade, most of the current largest logistics service providers in Singapore were already established in the late 1980s (Table 6).

In the 1980s, the role of logistics and SCM as a key sector in the economy became more apparent due to the growth of the electronic components and high-tech sector. As a demonstration of this increased importance, the SLA

Table 6.   Net revenues of major logistics provider in Singapore in 1988 and 2009.

	Net revenues in 1988 (m$)	Net revenues in 2009 (m$)
Danzas AEI	490	3,624
CH Robinson	237	419
DSC	215	213
Exel	441	1,728
Expeditors	303	548
FedEx	104	545
Fritz	558	619
JB Hunt Dedicated	290	479
Menlo	248	445
Penske Logisitics	600	1,060
UPS logistics	307	1,021

was admitted as a member of the Zurich-based International Federation of Freight Forwarders Associations (FIATA) in 1983.

The Economic Development Board (EDB) actively pursued global third party logistics providers to set up base in Singapore and, at the same time, pushed for local logistic companies to upgrade. These logistic companies were supporting both the global distribution of products made in Singapore as well as the needs of the shippers in the region, supporting the role of Singapore as a regional *and* global hub.

An example of such local companies and their growth process is well represented by YCH Global Logistics.

"[…] Started as a transportation company in 1955 in Singapore by Yap Chwee Hock. In 1977 it changed from passenger to cargo transportation under the leadership of Robert Yap, Yap Chwee Hock's son, and became a leading cargo transportation carrier for the Port of Singapore. In the early 1980s it added warehouse leasing, warehouse management and freight forwarding services and later integrated and added services to become a full service 3PL. In 1992, it opened YCH "DistriPark," as part of the logistics cluster on the intersection of the Kayang Paya Lebar expressway and the Pan Island Expressway in Singapore. In the 1990s the company built a network of distribution centers in logistics clusters around Asia and, at the same time, developed a suite of supply chain management software applications focusing on manufacturing logistics, return management and consumer goods distribution. Today the company offers both logistics execution services and supply chain management consulting/solutions services with offices in 12 Asian countries. The sophistication of the software drove YCH in the 2000s to set its IT function as a standalone subsidiary, Y3 Technologies, developing and supplying IT applications to the logistics industry.[…]" (http://www.ych.com).

The 1990s represented a very important decade for the development of the 3PL sector in the Singapore economy. EDB estimated that the Singapore logistics sector was investing a record amount of S$1.1billion in fixed assets in 1996 which is even more surprising considering the steep increase of 70% with respect to the previous year. Together with investments, the logistics sector total business spending experienced a 40% growth in 1996, corresponding to more than S$280 million (Bhatnagar et al., 1999).

The commitment and the regional growth of global companies were also remarkable: in 1998, S$29 million were invested to set up the multifunctional Schenker Logistics Centre (SLC). It was the first 3PL in Singapore to be 'Good Distribution Practice' (GDP) certified. Bhatnagar et al., (1999) found that, through an industry survey, among, the 126 companies interviewed, 60.3% were using services provided by 3PLs, 73.7% of the firms who used to outsource those service also stated that they were in contact with more than one service provider at a time. The services mostly required from 3PL were: shipment consolidation, order fulfillment, carrier selection and freight payment.

2000s

The Singapore Logistics Association was established from the Singapore Freight Forwarders Association on 30 August 1999. In 2001, the Diploma/Certificate in Distribution Center Management was launched.

The automation in logistics became the keyword by the end of the 1990s until 2004, when the SLA's Automated Manifest System (AMS) was launched as a logistics facilitation portal.

More than 3000 international and local logistics companies were operating in Singapore in the early 2000 (EDB, 1998).

Global companies in new growth sectors have established Singapore as their key global and regional manufacturing base in Asia. These companies come from very disparate sectors including aerospace, biomedical sciences, chemical, energy, oil and gas equipment, technology, consumer goods, perishables and high value collectibles. This has required a fast and intense evolution for the 3PL sector, which had to be capable of offering increasingly complex services. This led to the establishment of new companies as well as the growth of providers with history of long presence. Table 6 shows the revenues of the major logistics industry in Singapore at the end of 1980s and the 2000s. The average growth of the logistics market is apparent. Most companies that were not present in the 1970s also became main players in the late 1990.

In 2010, 21 of the world's top 25 third-party logistics providers were based in Singapore or had set up regional or global headquarter functions. Currently, the key players are DB Schenker, DHL, FedEx, Gati, Kintetsu World Express, Toll Group, UPS, YCH Group and Yusen Logistics (www.edb.gov.sg). In 2011, a four-episode documentary was commissioned to Channel News Asia the topic of Singapore's logistics capability. The logistics business in 2011 was between S$22 billion and S$28 billion, with a particular attention to four industries closely related with logistics: 1) construction industry, fundamental for the growth of businesses, involving the logistics service from the port (particularly Jurong Port) to the construction site, 2) chemical industry requiring an expert supply chain in the transportation of dangerous goods, 3) art industry, a particularly important emerging market in Singapore, where it is clear how the value of the transported good determines the strategic function played by the logistic service; 4) pharmaceutical industry, in this case the perishability of the products represents the major challenge requiring advanced logistics services. In fact, the spectacular growth of the city in the past eight years has been facilitated and enabled by the logistics capacity of the city.

In 2012, DB Schenker initiated the first global Competence Center outside Europe with the objective of concentrating the development of different installations of Production Vendor Managed Inventory (PVMI) with a particular focus on manufacturing and electronics companies.

Currently, Singapore offers six logistics clusters: Changi North (APC Distribution Centre, Pan Asia Logistics Center), Loyang (Air Market Logistics Centre), Airport Logistics Park (ALPS) (Schenker mega-hub, Hi-Speed Logistics Centre), Changi South (C&P Changi Districenter), Pandan (CWT Commodity hub, CWT cold hub, Kim Heng Warehouse and Pandan Logistics Hub) and Gul Way (Precise Two).

3.2.4. *The Government Policies for Logistics Facilitation*

The Singapore government played a fundamental role in defining and facilitating the trade between countries. Through several bodies in different sectors, government presence has always been fundamental in the management of the country economy. In this section, we will review the actions of the Singapore government with a particular focus onto those oriented to the logistics sector.

1980–1997 — The Growth and Establishment of the Singapore Logistics

In 1986, the SLA designed a set of Standard Trade Conditions (STCs) to clarify the commercial relationship between freight forwarders and shippers after three years since the admission at the FIATA. At the same time, the SLA constitution was revised in order to promote Singapore's reputation for *reliability, integrity and high standards* in freight forwarding practice and management.

The scope and role of logistics changed dramatically during the 1990s: from its traditional role as a support to functional areas such as marketing and production with a scope limited to transportation and warehousing, it became more prominent as a critical factor for competitive advantage (Sum *et al.*, 2001). Market globalization, due to the recent advances in communication technology and the abatement of trade barriers, contributed to this important growth and the Singapore government played a key role in fostering the logistics market in the country. The Logistic Enhancement and Application Program (LEAP) was launched in 1997 with the following aims: (1) developing technical know-how; (2) developing man power skills; (3) developing infra-structures; and (4) enhancing the business processes (Sum *et al.*, 2001). The relevance of the programme resided in its target to develop Singapore as a logistics hub.

Under LEAP, a total of 16 projects were run. In particular, beginning in 1997, the following were started: (1) the Logistics Benchmarking, (2) the Balance Scorecard, and the (3) Cost of Quality Program. In 1999, the pallet standardization, the Electronic Container Seal and Promoting IT in Logistics were all launched by the SLA.

Thanks to these policies, despite the economic crisis caused by the floating of the Thai currency (Baht) during 1997–1998, Singapore's transport and com-munications sector grew by 5.5%. In this decade, the relevance of the logistics industry was recognized at the point when Singapore policymakers established logistics as one of the four pillar industries guiding the growth of the country for the coming millennium. In this endeavor, the Logistics 2000 plan was initi-ated (Sum *et al.*, 1999).

1998–2005s — Education and e-commerce, the e-logistics era.

Aside from industrial developments, the Singapore government had well understood how education was central for the development of local skills needed to sustain its growth. As a result, policymakers dedicated important investments to research as well as developing the education system. As an example, the two main public universities of the country, National University of Singapore (NUS) and Nanyang Technological University (NTU) – were further developed, with its own leading departments in several disciplines.

Part of this effort for education was directly related to logistics and supply chain systems. The Logistics Institute Asia–Pacific, involving the National University of Singapore and the Georgia Institute of Technology, and the MIT Singapore transportation initiative, which is part of the Singapore–MIT Alliance for Research and Technology, are some examples of higher education partnerships related to logistics established in the decade.

In line with these policies, the Certified Logistics Professional programs, the Strategic Manpower Conversion Program (SMCP) as well as the National Competency Roadmap were all initiated and partnered by International Enterprise (IE) and Ministry Of Manpower (MOM) in 2001 under the LEAP programme. The objective of the programme was to convert more non-logistics professionals to the logistics industry through conversion courses providing specific skills. Course fees were subsidized for up to 50% or for S$4000 and the participants could go for full time training while being offered monthly allowance up to S$800; on-the-job training allowance was also provided for three months at 50% of the salary, capped at S$800 if the participant used to work in the logistics industry for a minimum of six months after conversion.

The Certified Logistics Professional programs were launched by IE Singapore, IDA and Chartered institute of Logistics and Transport (CILT). Financial planning and investment industries, two sectors in which certifica-tions were already provided were taken as a reference to design these new programmes where professionals who were certified were recognised for achieving the minimum criteria in work experience and knowledge.

The 2000s were surely marked by the increase in the internet use in Southeast Asia, paving the way to e-commerce companies (Chan and Al-Hawamdeh 2002). Prompted by advances of the Internet, Singapore entered the information age. The Singapore Government recognized the importance of knowledge assets in the new economy and the need to invest to stay competitive. Hence, the *e-commerce Master Plan* was launched in 1998 under the main objectives of (1) providing guidance for the continuously growing electronic commerce in the country, and (2) supporting Singapore in establishing herself as a global e-commerce hub. Two ambitious targets to reach by 2003 quantified the aforementioned objectives: (1) a growth of S$4 billion in the worth of electronic transactions of product and services through the country and (2) a ratio of 50% of firms using e-commerce (IDA, 1997).[80] This programme was followed by the Local Enterprise e-Commerce Program in November 1998 to build up e-commerce applications and encourage local companies and support them in the implementation of e-commerce in their business operations. A total budget of S$9 million was set aside to assist those companies.

The partnerships are an important component of the LEAP framework, and from 1 July 2001, Singapore entered a Mutual Recognition Agreement with Australia. Such MRAs were developed in order to decrease or eliminate non-tariff barriers such as those coming from technical and regulatory requirements to products during the conformity assessment (testing, inspection and certification). According to the agreement, such conformity assessments had to be undertaken in the country of export avoiding double conformity checks. In order to make such a simplification possible, the involved political bodies in both countries agreed upon the testing procedures of the other countries involved in the agreement. As a result, the certificate issued by the Conformity Assessment Bodies (CABs) are recognized by all countries involved in the agreement.

2005–2014 — Singapore the Global Logistic Hub

On 25 May 2007, the Secure Trade Partnership Programme (STPP) was launched by the Singapore Customs as an optional programme providing certification. The main target of the programme was to lead companies in the adoption of robust security practices during trading operations, in light of the positive effect such behaviors would have over international security. The STPP provided a supply chain security framework to the companies guiding the entire process for improving SC security from the development of security measures and best practices, to the implementation and monitoring of the results. The reference of the STP program was the World Customs Organization (WCO) SAFE standard adopted in June 2005, developed for supporting and enhancing secure global trade.

In line with the need of compliance with global standards and in order to enhance the supply chain security, Singapore implemented the Regulated Air Cargo Agent Regime (RCAR) on 1 April 2008, a security measure to enhance air cargo security on commercial passenger aircraft, a requirement by the International Civil Aviation Organization (ICAO). This measure was necessary to ensure the safety and security of commercial passenger flights and it was proposed as a result of global terrorist threats. The principle underlying RCAR was that the operation of aircraft should be performed under maximum security guaranteeing massive control checks for all types of cargo being shipped through passenger transportation means. The safety of any cargo also has to be maintained in the different phases of the shipment as well as the different transshipment activities, by protecting the cargo from any unsecure interference and contamination.

RCAR has been a result of the relentless collaboration of the Singapore Police Force and the Civil Aviation Authority of Singapore.

In 2010, the Aviation Partnership Programme (APP) was launched to drive the adoption of industry-level standards and processes among local companies, organizations and research centers. Under this programme, several initiatives have been intiated. In particular, e-freight@Singapore is one of the most relevant to the logistics industry. This initiative, supported by the Civil Aviation Authority of Singapore (CAAS), gives support to companies ready to adopt paperless airfreight documentation and make use of ICT to give shippers and forwarders a wider set of solutions.

Also in this decade, the use of MRA was strategic: Singapore, South Korea and Canada signed their MRA on 1 January 2011. According to the MRA, the customs of Korea had to classify as "lower risk products" all the Singapore's Secure Trade Partnership-Plus (STP-Plus) exported goods. At the same time, Singapore agreed to classify as "lower risk" South Korea's Authorized Economic Operator (AEO) companies exported products. As a result, the product exported by AEO and STP-Plus firms enjoy smoother import practices in Singapore and in South Korea, respectively.

Singapore Customs General Director Mr Fong Yong Kian signed the MRA with Mr Atsuo Shibota, General Director of Japan's Customs and Tariff Bureau, at the 117th/118th WCO Council Sessions held from 23 to 25 June 2011 in Brussels. Japan was Singapore's sixth largest trading partner with a trade volume that already exceeded S$55.5 billion in 2010. Top exports to Japan included electronic chips and optical media, whereas the imports from Japan were mainly electronic integrated circuits and semi-manufactured gold. On 15 March 2013, the MRA signed with the Customs of the People's Republic of China came into effect.

Besides MRAs, the Zero GST scheme was also launched. Under this scheme, the ZG (Zero GST) warehouses could bypass the Goods & Services Tax (GST) process. The tax is instead paid on imports when they were released from the ZG warehouse for local consumption.

A strong effort was made also in the direction of business alignment with global standards. As an example, the Customs in Singapore, required to submit declarations for products (including those non-controlled and non-subject to duties) before their actual export (by sea or air) was performed from 1 April 2013.

These efforts are strongly connected and harmonized with the STPP as they are in place to help Singapore establish herself as a recognized hub for security and safety, both important in light of the growing paradigm of sustainable global supply chains. With this in mind, Singapore customs understand the key role played by the availability of advanced information for all products, both for import and export. To enhance Singapore's supply chain security, Singapore Customs requires all goods entering and leaving the country to have such information, hence enabling the body for measures such as timely export risk assessment.

As a result of effective management of advance data and collaborations with overseas Customs Administrations through MRAs, the Singapore Customs can effectively facilitate, legitimate, and secure trade.

4. The future of Logistics for Singapore in Southeast Asia

The logistics industry is facing several significant challenges that are arising from a changing environment, particularly in Southeast Asia. In order to continue its leadership position in this industry, Singapore must successfully address these challenges.

Urbanization and cross border trade are two key issues that are affecting and will continue to affect logistics in Southeast Asia. The growth of city populations and the increase in the number of mega-cities are the results of urbanization, and this will require advanced strategies for services such as efficient first mile pickup and last mile delivery. The increasing average life expectancy of the Asian population, the related growth of a relatively affluent middle class, and the increasing use of new channels of purchase (e.g., e-commerce, particularly across national borders) all contribute to the logistics challenges faced in Southeast Asia.

Southeast Asia is also economically diverse. According to the 2012 ETI, Singapore and Hong Kong rank first and second globally, and China and India are ranked 91st and 132nd, respectively, according to the EDBI. Some Southeast Asian countries apply the Value-added Tax (VAT), while elsewhere (as with Singapore), the Goods and Service Tax (GST) applies. Even within Southeast Asian countries using the same taxing structure, different rates are in effect in the different countries.

Southeast Asia is different from Europe and North America in a variety of ways. Both Europe with the European Union, and North America with NAFTA, have trade enabling agreements, both have large landmasses, and both are comprised of countries having relatively similar and generally stable political climates. Centralized management, vertical specialization, and the just-in-time movement of goods are effective corporate and supply chain strategies in Europe and North America that may not be effective in Southeast Asia. Two observations can be seen: (1) best practices in North America and Europe may not be good practices in Southeast Asia (they might even not be applicable); (2) what constitutes best practices for logistics in one Southeast Asian country may not apply at the regional level.

Overall, it is possible to specify the following main challenges characterizing the future of logistics in general and for Southeast Asia in particular, in the context of the demographic shifts described above: cross-border logistics, innovation, infrastructure, risk mitigation, e-commerce, big-data and sustainability. We will briefly present them individually in the remainder of the section.

Cross-border logistics

The Asian region is particularly affected by problems related to cross-border transportation. A country's trade-related regulations may not be clearly described or understood and it can change unpredictably. As a result, supply chain operations and design are often reactive, and there is a need for vigilance regarding the FTA and related general agreements and regulations. In addition, the regulations in one country may not match the regulations of the company's country of origin (e.g. US logistics companies, which are governed by FCPA). All these coordination issues among different regions lead, to the need for increased supply chain flexibility in order to efficiently cope with any disruption, and it requires deep knowledge of local laws and their interpretation. It is important to highlight the importance of local partners who can facilitate business and have good knowledge of the local situation; e.g., the tax systems of the Southeast Asian countries that are part of the supply chain (Wong and Chan, 2003).

Lack of Infrastructures

Lack of logistics infrastructure hinders economic development in many Southeast Asian nations. The lack of infrastructure slows the rate of supply chain efficiency improvement and can significantly reduce the usefulness of many efficiency enhancing transport innovations.

Innovation

The need of both breakthrough innovations and continuous improvement is a major determinant of future supply chain development. Innovations that have the sare improved the performance of European and/or North American supply chains may not have the same impact in Southeast Asia. Different supply chain strategies and innovations may be needed in order to meet Southeast Asian logistics challenges, and different transport technologies might be required to successfully deal with the variety of logistics infrastructures found in Southeast Asia.

Risk Mitigation

The risk in supply management is a fundamental component. There are several sides to the risk: risk related to the cargoes (nowadays shippers tend to leave the major responsibility of the whole transport chain in the hands of the logistic provider), and risk related to the infrastructure. If the insurance problem is common to the US and European supply chains, the disruptions due to modifications in the regulation system, and modification in the infrastructures, in the FTA and regions agreements as well as the political instability are all key aspects that require resilient supply chains and risk hedging methodologies which might be completely different from those adopted in other regions in the world.

e-commerce

E-commerce is having significant impact on, and represents a critical challenge for, the logistics industry. The diversity of delivery channels and the increasing demand for rapid last mile fulfillment that e-commerce generates are of fundamental concern. These challenges are amplified by a population that is increasingly urban, older, more affluent, and are in greater need of healthcare. Crowd-sourcing and warehouse consolidation may represent at least partial solutions.

Big-data

The volume, velocity, and variety of data received by logistics companies are ever increasing. Understanding the value in these data and extracting it for operational advantage is a challenge fundamental to all industries. Analytics has become a key component to competitive advantage. Leveraging the value in data can improve a supply chain's productivity, reduce risk, enhance stability, increase security, and help to enable autonomous logistics. Improved supply chain productivity can arise from various potential innovations, e.g., real time and congestion sensitive route and stop sequence optimization, predictive analytics, anticipatory routing, predictive network and capacity planning, crowd-sourced pickup and delivery.

Sustainability

Environmental sustainability is a fundamental issue of societal importance and the logistics industry must play its part in achieving national and global sustainability goals. Regulations and innovation will shape the success of achieving these goals.

5. The Next Generation Logistics Supply Chain

The supply chain and logistics industry is critical for Singapore's economy, accounting for 9.4% of the country's GDP. The pace of change in the global supply chain and logistics industry is increasing and creating opportunities for a substantial impact of supply chain and logistics research and innovation. All the challenges presented in previous sections and their associated complexities raise enormous issues and increase the need for new products, processes, and services (i.e., new innovations and their commercialization) that can take advantage of the opportunities created by these changes.

In fact, a data revolution is occurring in the logistics industry. The new technologies available to sellers and customers, the embrace of social media, the rapidly changing and highly customized demands, all require a change to traditional supply chain management strategies as well as supply chain infrastructures. Hence, future supply chains must be intelligent in their ability to exploit available data and technologies, adaptable in its capacity to answer to demand fluctuation as well as system disruptions, and efficient, deploying the most recent solutions in terms of automation.

It is apparent how the challenges in Section 4 and the opportunities in new technologies can lead to a vast number of scenarios, where new firms can find opportunities to grow, currently existing firms can adapt and benefit from significant opportunities, and some firms who fail to adjust to an ever-changing competitive environment will exit the market. We have identified five key drivers for logistics leadership: (1) data driven control, (2) automation, (3) future port city development, (4) regulatory system monitoring and analysis, and (5) risk mitigation, which we now discuss.

Data driven control

Access to large amounts of data arriving in real-time that contain high-value information and being able to quickly (in real time or near real time) transform these data into supply chain decisions is increasingly becoming a reality. Computer, sensor, and communication network technologies are enabling this reality. While the promise of such technology is great, concerns arise in data corruption, transmittal latency, cyber security issues, and system reliability. The main benefits available to future supply chains, thanks to the development of IT, will be: (1) increased electronic data sources generated from intelligent communication systems and vehicles as well as inventory information available, as an example, from RFID systems, (2) availability of efficient technologies to manage data, such as new distributed data technologies.

Nevertheless, in order for a supply chain to exploit such benefits, there is a substantial need for the development of a new generation of systems and software components that can explicitly extract in real-time the information value contained in large amounts of incoming data while adaptively updating distributions associated with any supply chain uncertainty.

Physical components with increasing intelligence (i.e., the internet of things) have to be thought of as part of larger and more complex system, and these systems themselves can be thought of as part of larger meta-systems that interact with each other and with their sub-systems and components. The greatest efficiency gains will often be generated when these meta-systems are optimized in part via data-driven decision support.

Automation

With the desire for high-speed, low-latency supply chains, automation is not just a key enabling technology but a technology paradigm whose implications will shape the design and management of modern supply chains for decades to come. Automation in supply chains goes beyond basic robotics in the form of, for example, automatic storage and retrieval systems and it will arise as a seamless and intelligent integration of automation hardware (e.g., advanced robotics) with automation software (e.g., optimization) into automation technology. Investing in automation will increase (1) availability of supply chains, (2) supply chain reliability, and (3) supply chain flexibility.

Indeed, standardizing processes reduces lead times along supply chains and lowers latencies, thus improving the availability of supply chains. Automation software will enable upstream integration of information, which will reduce myopic decision-making and hence reduce the likelihood of potentially dangerous misaligned processes. An example of such an upstream integration is a cloud-based navigation system incorporating both traffic information and simultaneously communicating its routing decisions back to the navigation system. Automation hardware will improve the efficiency of logistics operations. For example, driverless trucks will remove the need for driver rest and increase average driving speed and equipment utilization.

Automation will increase reliability by replacing parts of processes prone to human-error (in particular under high-speed, low-latency requirements) by automation technology, which typically leads to a significant reduction of variance and volatility. More importantly, it reduces the severity of low probability and high consequence events, which can lead to a dramatic reduction in the total cost of ownership.

Finally, last-generation automation can guarantee flexibility. Intelligent automation solutions (e.g., in warehouses, sea ports, airports) can be more easily adjusted to varying requirements, reducing overall setup and adjustment costs. Prime examples include the support of picking by machine vision and identifying and correcting picking errors while maintaining consistent inventory levels. Moreover, switches from high-throughput configurations to low-latency configurations can be more easily realized.

Leveraging automation technologies will require the close integration of automation hardware and automation software and a reliable communication and information infrastructure, enabling autonomous automated entities to communicate and interact with each other. This will ensure scalability and sustainability beyond the point of more traditional master-slave approaches with a centralized decision management.

Future Port City Development

Seaports and airports are important nodes in the global freight transportation network and for supply chains using them. Seaports enable national economic growth by facilitating global trade and providing opportunities to create value through production and services.

Maritime activities in Singapore contribute 7% of Singapore's economic growth and hence are a key pillar for the Singapore economy. To further enhance the competitiveness of Singapore's maritime industry, Singapore needs to improve its space and manpower productivity, enhance its efficiency, safety, and reliability, and seamlessly integrate a variety of different systems that support the port and maritime sectors. To achieve this, the Singapore government has started an initiative to develop a maritime hub in Tuas where it aims to consolidate all container operations in one terminal by 2027. This will help to provide greater economies of scale and increased efficiency by the adoption of new technologies and processes, such as automated container port systems, optimisation techniques and technologies, and green port technologies.

Regulatory system monitoring and analysis

The diversity of the regulatory environment in Southeast Asia represents a particular challenge to companies engaged in business in the region, particularly cross-border e-commerce. Intra-country regulation can be affected by regulatory environment uncertainty. A possible regional leadership role for governmental agencies, in partnership with university researchers, is to (1) help the industry anticipate regulatory changes in the region, (2) analyze the economic and societal impact of regulatory changes under consideration in order to better inform discourse surrounding regulatory policy formation, (3) play an active role regionally to ensure that any regulatory changes are as business friendly and societally beneficial as possible, (4) and help formulate standards (particularly data standards) to facilitate fast and efficient compliance with regulatory processes. The need for regulatory monitoring prompted UPS to offer a regulatory consulting service (termed as 'trade management services'), and Ernst & Young is offering a similar service in the Southeast Asian area.

Risk Mitigation

A major inhibitor for further improving the efficiency of supply chains and logistics operations is the risk of disruption. The more tightly coupled and integrated value-creation processes are, the higher the risk of disruptions and their amplification, often leading to propagation across the value chain. These risks, if not properly mitigated and managed, can lead to severe financial losses often offsetting the potential gain in efficiency. One of the key points will be the research and development of risk mitigation strategies in the context of supply chains and logistics both for endogenous disruptions (e.g., inventory misalignments) as well as exogenous disruptions (e.g., natural disasters), and the solutions have be provided in terms of risk management strategies and help policymakers in implementing emergency management strategies that incorporate best practices.

6. Conclusions

The logistics and supply chain industry is an industry sector that has experienced significant changes in the past 50 years. Indeed, starting as a service industry by providing basic transportation and storage services, it has become a major value-added industry playing a fundamental and strategic role in the growth of global trade. During this evolution, Singapore has played and continues to play a key role in sustaining and furthering the growth of this industry, due to long-term programmes of the Singapore government, and international and global agreements that have shaped Singapore's economic development.

In this chapter, we have summarized the historical development of the logistics industry in Singapore, observing how the country has generated industrial growth, developed its infrastructure, and put into practice the policies for trade and the use of IT to become the main logistics hub in Southeast Asia and one of the key logistics nodes worldwide. In so doing, Singapore has attracted multinational companies to establish their regional and global headquarters in the country, including the major logistics service providers worldwide. Nevertheless, the future is full of daunting challenges for the logistics industry, particularly in Southeast Asia. Supply chains will have to change in order to sustain trade growth by exploiting new technologies and responding to changing demographics. The ability of Singapore to exploit data to improve supply chain control, to effectively use automation, to develop robust solutions for supply chains of the future, and to develop an integrated hub with the airport and seaport as key components will be essential to maintaining Singapore's role as a leader of the supply chain and logistics industry.

References

Amin, A., Thrift, N. (1992). Neo-Marshallian nodes in global networks. *International Journal of Urban and Regional Research*, 16, pp. 571–578.

Balassa, B. (1967). Trade Creation and Trade Diversion in the European Common Market. *The Economic Journal*, 77, pp. 1–21.

Baldwin, R. (2013). *Multilateralising 21st Century Regionalism*. (HEID manuscript, Heidelberg, Germany).

Bhatnagar, R., Sohal, A. S., and Millen, R. (1999). Third party logistics services: a Singapore perspective. *International Journal of Physical Distribution & Logistics Management*, 29(9), pp. 569–587.

Bilbao-Osorio, B., Dutta, S. and Lanvin, B. 2014. *The Global Information Technology Report*. (The World Economic Forum, Geneva).

Capling, A., and Ravenhill, J. (2011). Multilateralising regionalism: what role for the Trans-Pacific Partnership Agreement?. *The Pacific Review*, 24(5), pp. 553–575.

Chan, B., and Al-Hawamdeh, S. (2002). The development of e-commerce in Singapore: The impact of government initiatives. *Business process management journal*, 8(3), pp. 278–288.

Choo, C. W. (1997). *IT2000: Singapore's vision of an intelligent island*. (Amsterdam, The Netherlands: Elsevier).

Ciccone, A., Hall, R.E., (1996). Productivity and the density of economic activity. *The American Economic Review*, 86 (1), pp. 54–70.

Ciccone, A. (2002). Agglomeration effects in Europe. *European Economic Review*, 46, pp. 213–227.

Chou, C. H., Ching-Wu, C., and Gin-Shuh, L. (2003). Competitiveness analysis of major ports in Eastern Asia. *Journal of the Eastern Asia Society for transportation studies*, 5, pp. 1–16.

Cooper, J.C. (1993). Logistics strategies for global businesses. *International Journal of Physical Distribution and Logistics Management*, 23(4), pp. 12–23.

Evers, H. D., Gerke, S., and Menkhoff, T. (2010). Knowledge clusters and knowledge hubs: designing epistemic landscapes for development. *Journal of knowledge management*, 14(5), pp. 678–689.

Fujita, M., Krugman, P. R., and Venables, A. J. (2001). *The spatial economy: Cities, regions, and international trade.* (MIT press. Boston, USA)

Glaeser, E.L., Gottlieb, J.D., 2009. The wealth of cities: agglomeration economies and spatial equilibrium in the United States. *Journal of Economic Literature*, 47, pp. 983–1028.

Hanouz, M. D., Geiger T. and Doherty, S. (2014). *Global Enabling Trade Report. Insight Report.* (World Economic Forum, Geneva).

Hausmann, R., and Klinger, B. (2006). *Structural transformation and patterns of comparative advantage in the product space.* (Center for International Development at Harvard University, Boston, USA).

Heaver, T., Meersman, H., Moglia, F., & Van de Voorde, E. (2000). Do mergers and alliances influence European shipping and port competition?. *Maritime Policy & Management*, 27(4), pp. 363–373.

Hsieh, C.T., Klenow, P., 2009. Misallocation and manufacturing TFP in China and India. *Quarterly Journal of Economics*, 124 (4), pp. 1403–1448.

Joseph, F. (2013). *Reducing Transatlantic Barriers to Trade and Investment. An Economic Assessment.* Final Project Report. Centre for Economic Policy Research, London.

Kirkman, G. S., Osorio, C. A., and Sachs, J. D. (2002). *The networked readiness index: Measuring the preparedness of nations for the networked world.* (Korea, 4, 20).

Krishnan, R. A. D. (2012). *A Case Study of Singapore as a Logistics Cluster.* (IGI Global).

Krugman, P. (1991) *Geography and Trade.* (Cambridge, MA: MIT Press).

Long, C., X. Zhang. (2011). Cluster-based industrialization in China: Financing and performance. *Journal of International Economics*, 84, pp. 112–123.

Lee, K. L., Huang, W. C., and Teng, J. Y. (2009). Locating the competitive relation of global logistics hub using quantitative SWOT analytical method. *Quality & Quantity*, 43(1), pp. 87–107.

Lejárraga, I. (2014), *Deep Provisions in Regional Trade Agreements: How Multilateral-friendly?: An Overview of OECD Findings.* (OECD Trade Policy Papers, No. 168, OECD Publishing).

Liu, S. (2009). A Research on the Relationship of Logistics Industry Development and Economic Growth of China. *International business research (Toronto)*, 2(3), pp. 197–200.

Lu, Chin-Shan. (2003). Market segment evaluation and international distribution centers. *Transportation Research Part E: Logistics and Transportation Review*, 39(1), pp. 49–60.

Marshall, A., 1920. *Principles of Economics*, 8th ed. (Macmillan, London).

Martin, J., and Thomas, B. J. (2001). The container terminal community. *Maritime Policy & Management*, 28(3), pp. 279–292.

May, A. D. (2004). Singapore: The development of a world class transport system. *Transport Reviews*, 24(1), pp. 79–101.

Notteboom, T. E., and Winkelmans, W. (2001). Structural changes in logistics: how will port authorities face the challenge?. *Maritime Policy & Management*, 28(1), pp. 71–89.

Notteboom, T., Parola, F., Satta, G. (2014). *State of the European Port System — market trends and structure update*. (Deliverable 1.1: PORTOPIA project (European Union)).

Penske Logistics. (2013). *18th Annula Survey of Third Party Logistic Providers*. Report

Petri, P. A., M. G. Plummer, and F. Zhai. (2011). *The Trans-Pacific Partnership and Asia-Pacific Integration: A Quantitative Assessment*. (East-West Center Working Papers — Economic Series, No. 119. Honolulu: East-West Center).

Phang, S. Y. (2000). *How Singapore regulates urban transportation and land use. Local Dynamics in An Era of Globalization*, (ed. Shahid Yusuf, Weiping Wu, and Simon Evenett).

Phang, S. Y. (2003). Strategic development of airport and rail infrastructure: the case of Singapore. *Transport Policy*, 10(1), pp. 27–33.

Porter, E.M., 1998. Clusters and the new economics of competition. *Harvard Business Review*, 76 (6), pp. 77–90.

Porter, M. E. (1990). *The Competitive Advantage of Nations*. (Harvard business review, Boston).

Porter, M. E. (2000). Location, competition, and economic development: Local clusters in a global economy. *Economic development quarterly*, 14(1), pp. 15–34.

Rimmer, P. J. (1988). *Rikisha to rapid transit: urban public transport systems and policy in Southeast Asia*. (Elsevier, New York).

Robinson, R. (2002). Ports as elements in value-driven chain systems: the new paradigm. *Maritime Policy & Management*, 29(3), pp. 241–255.

Sheffi, Y. (2013). Logistics-intensive clusters: global competitiveness and regional growth. (Handbook of Global Logistics. Springer New York).

Sheu, J. B. (2004). A hybrid fuzzy-based approach for identifying global logistics strategies. *Transportation Research Part E: Logistics and Transportation Review*, 40(1), pp. 39–61.

Sum, C. C., and Teo, C. B. (1999). Strategic posture of logistics service providers in Singapore. *International Journal of Physical Distribution & Logistics Management*, 29(9), pp. 588–605.

Sum, C. C., Teo, C. B., and Ng, K. K. (2001). Strategic logistics management in Singapore. *International Journal of Operations & Production Management*, 21(9), pp. 1239–1260.

Tan, C.N., S.K. Goh, S.T. Chua, J. Montiwalla, and S.M., Sung. (1985). *National IT Plan: a strategic framework*. (Working Committee, Singapore).

Tao, O. H., and Park, J. H. (2004). Multinational firms' location preference for regional distribution centers: focus on the Northeast Asian region. *Transportation Research Part E: Logistics and Transportation Review*, 40(2), pp. 101–121.

Willoughby, C. (2001). Singapore's motorization policies 1960–2000. *Transport Policy*, 8(2), pp. 125–139.

Wilson, J. S., Mann, C. L., and Otsuki, T. (2003). Trade facilitation and economic development: a new approach to quantifying the impact. *The World Bank Economic Review*, 17(3), pp. 367–389.

Wong, J., and Chan, S. (2003). China-ASEAN free trade agreement: shaping future economic relations. *Asian Survey*, 43(3), pp. 507–526.

WORLD ECONOMIC FORUM (WEF). (2013). *Enabling Trade: Valuing Growth Opportunities*. (World Economic Forum, Geneva).

WORLD ECONOMIC FORUM (WEF). (2014). *Global Information Technology Reports (GITR) Ranking*. (World Economic Forum, Geneva).

Xufei M., A. Delios and Lau, C-M. 2013. Beijing or Shanghai? The strategic location choice of large MNEs' host-country headquarters in China. *Journal of International Business Studies*, 44, pp. 953–961.

Yu, W., and Ding, W. L. (2005). The planning, building and developing of logistics parks in China: review of theory and practice. *China USA Bus Rev*, 4(3), pp. 73–78.

Yue, C. S., & Lim, J. J. (2003). *Singapore: a regional hub in ICT. Towards a Knowledge-based Economy: East Asia's Changing Industrial Geography*. (Institute of East Asia Studies, Singapore).

Zhang, X., Tan, K.-Y., 2007. Incremental reform and distortions in China's product and factor markets. *World Bank Economic Review*, 21 (2), pp. 279–299.

Zhu, J., Lean, H. S., and Ying, S. K. (2002). The third-party logistics services and globalization of manufacturing. *International Planning Studies*, 7(1), pp. 89–104.

Chapter 7

Development of the Port of Singapore: A Historical Review

Giulia PEDRIELLI, LEE Loo Hay, CHEW Ek Peng and
TAN Kok Choon

The Port of Singapore is, and has been for more than two centuries, one of the most important ports in Asia and, ultimately, of the world. Many factors concurred and are still at the basis of the Singapore leadership. Under each sharp market changes such as the arrival of containers, the oil crisis, and the establishment of hub and spoke model, the Port of Singapore could have been obsolete in a short period of time given the dynamics of its market position, as well as enter a new cycle of high growth.

The review will show how clever and far-sighted planning, such as equipment innovation, port expansion, and information technology, have all contributed to distinguish the port of Singapore to always be ahead, ready for new market challenges, showing great flexibility in its capacity to manage increasing volumes and substantial changes in the market demand.

The future plans for the port promises to bring Singapore to the next future highly capable to cope with diverse and critical challenges. The appropriate adoption and innovative integration of automation in container terminal operations will be one of the key factors in the next generation container port, which might be the first fully automated.

1. Introduction — Marine Trade Establishment and Growth in Asia

The port is Singapore's raison d'etre, its very reason to be.

(Minister Mentor Lee Kuan Yew, 2007).

This statement made by Singapore's first Prime Minister, Mr Lee Kuan Yew, in his inaugural Singapore Maritime Lecture at the Singapore Maritime Week 2007 (Singapore Nautilus, Issue 2, Q1, 2008), can easily be extended to several cities in Asia. In fact, the maritime tradition in this region is nothing new. About two thousand years ago, in China, the first compass, the si–nan, was invented: it was shaped like as a spoon, its handle points south when dropped into the water. The rudder was a Chinese invention as well, enabling long distance trips which were previously absolutely impossible to manage before the rudder was introduced in ship design.

Technology was not the only driving factors for the development and establishment of the maritime trade tradition in Asia. Nonetheless, it was an enabler to create connections within a region (China, Indonesia, India, Malaysia, Singapore, Thailand, and Vietnam) that constituted a natural centre of interest for western countries. Indeed, back in the 12th century, the extraordinary variety of cultures, populations and goods used to attract explorers from all over the globe such as Ibn Battuta, Faxian and Marco Polo.

"[…] Many languages are heard, all manner of habits are seen, and numerous varieties of food are consumed, truly, a more varies moving population it would be hard to find.[…]"

(From "A short history of the Port of Singapore"
March, 1922, Raffles Library).

1.1. Early settlement of marine trade in Asia

In the 1400s, the Maritime Silk Route which connected these different countries probably represented the oldest and largest port rotation in maritime history (Ma and Feng, 1997). Already at that time, Singapore constituted one of the major maritime centres, although routinely subject to pirates attacks. The mitigation of the phenomenon, attributable to the British intervention, took several years and nonetheless resulted in the flourishing of the Asian maritime trade and of the Singapore strait as one of the key passages till today.

Several countries in Southeast Asia along the Maritime Silk Route such as Malaysia, Vietnam and Indonesia were multicultural places where people used to speak Malay Arabic, Burmese, Cambodian, Chinese, Persian, Siamese and Tamil. The Maritime Silk Route connecting Quanzhou in South China to Malindi in Kenya, Africa, represented the most extensive exchange of goods and *costumes* all over the world (Fig. 1).

The route was more than 7,000 kilometers long, requiring several months of travel. The weather was a major determinant for choosing the voyage period in each direction of the route. Piracy also posed a serious threat for the sea trade and was one of the major concerns for the Asian governments, including Singapore. In fact, fighting piracy was one of the main tasks for the established British forces. Captain James Brook and Captain Henry Keppel (Keppel Harbour was named after him) represented two important personalities in this war during the seventeenth century.

Ensuring of safety in the strait fostered the development of the Asian maritime trade in the 19th century, at that time comprising three main routes: (1) the Chinese network, which linked Southeast Asia with the southern Chinese ports of Fujian and Guangdong; (2) the Southeast Asian network, which linked the islands of the Indonesian archipelago; and (3) the European and Indian Ocean network, which linked Singapore to markets in Europe and the Indian Ocean littoral. These networks were complementary and all involved Singapore as one of the main stops.

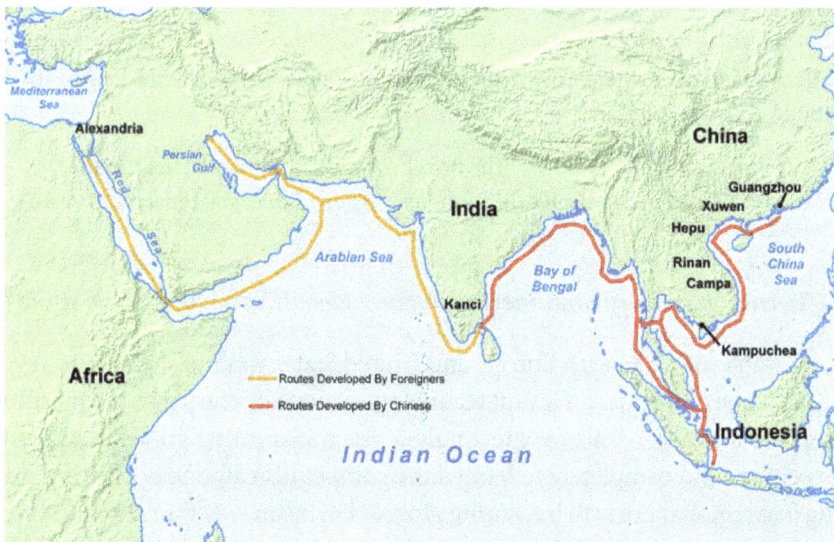

Fig. 1.   The Maritime Silk Route.

From the 19th century, the Asian trade has largely grown. In fact, the step from the 19th to the 20th century was marked by a major trend in the globalisation of trade which increased greatly.

However, the maritime operations were lacking enhanced technologies and good labour practices. The industry was highly inefficient due to labour-intensive activities such as loading and unloading of non-standardized (and largely heavy) cargoes. It is well-acknowledged that containerization in the 20th century was one of the major driving factors for the spectacular growth of global maritime trade.

> "[…] Containerization made shipping cheap and, by doing so, changed the shape of the world economy. […]"
>
> (From: Levinson, M. 2010. "The box: how the shipping container made the world smaller and the world economy bigger". *Princeton University Press*).

Section 1.2 will give a closer look at containerization and how it determined the maritime endeavour that we see today. Indeed, cities that had been centres of maritime commerce for centuries, such as New York and Liverpool, saw their waterfronts decline, unsuited for the container trade or were simply unneeded, and the manufacturing companies that endured high production costs in order to be close to their suppliers and their customers moved away. Shipping lines were crushed by the costs of adapting to container shipping.

> "[…] Even as it helped destroying the old economy, the container helped to build a new one. […]"
>
> (*From: Levinson, M. 2010. "The box: how the shipping container made the world smaller and the world economy bigger". *Princeton University Press*).

1.2. *Twentieth century containerization and globalization of marine trade*

The transatlantic trade with Europe and United States was not significant in volume until one of the most relevant technologies made its entry into the maritime sector: the container. This new technology for transporting goods marked the 20th century and completely redefined the competition among port cities, from being just regional ports to becoming global (Levinson, 2011; Li, 2011). Several ports failed to understand the magnitude of the containerization phenomenon while others, created ex-novo, became among the most competitive. As we will

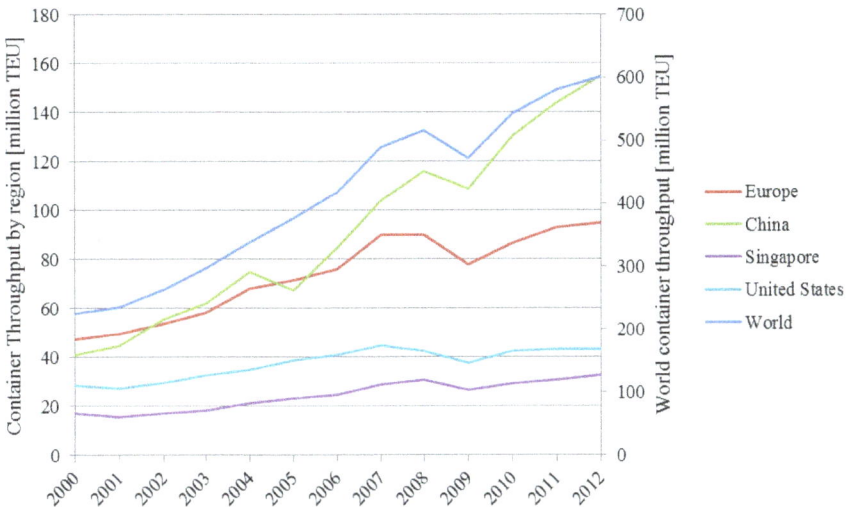

Source: Containerization International Yearbook, from several volumes.

Fig. 2.   Container throughput for 2000–2012.

see later in Section 2.4, Singapore was particularly far-sighted in appreciating and taking advantage of the phenomenon and succeeding in maintaining its competitiveness (Merk, 2013).

The impact of containerization can be well approximated by the growth of the world loading capacity that has risen from 1.5 million metric tonnes in 1970 to 8.748 million metric tonnes in 2011 (UNCTAD, 2011). Figure 2 shows the growth in the container market in recent years (2000–2012). Container ships carry an estimated 52% of global seaborne trade in terms of value (World Shipping Council, from *http://worldshipping.org*). Their share of the world fleet has grown almost eight-fold since 1980, as goods are increasingly container-ized for international transport.

In Fig. 3, we can further observe evidence of the impact of container trans-portation on sea trade growth: (1) the container throughput growth is strongly correlated to the sea trade growth; (2) the rate of growth of containerized car-goes is larger than the rate at which the trade grows, suggesting that increas-ingly more cargoes are transported in containers.

The growth of the containerized trade contributed substantially to the development of larger vessels in order to exploit the economies of scale, thus challenging the ports that, to remain competitive, needed to provide their ter-minals with increasingly advanced equipment in order to efficiently operate the increased vessel volume and size. For example, the first "Triple E" container

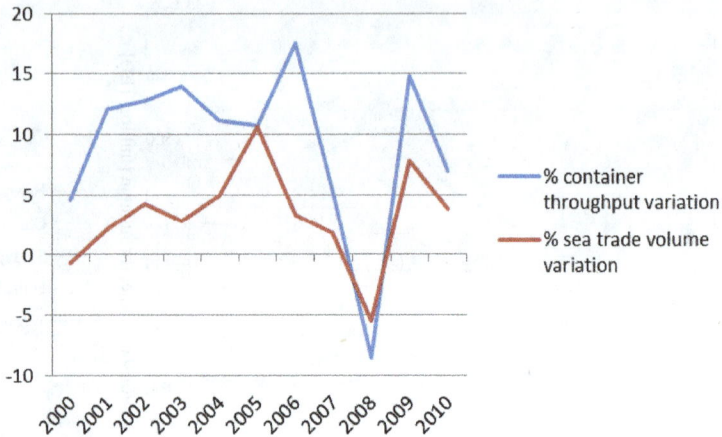

Fig. 3.   Sea trade growth and container transportation growth.

ships was delivered by Daewoo in South Korea to Maersk Line in Denmark in 2013, with a declared container-carrying capacity of 18,000 TEU. The same year, CSCL from China placed orders for even larger container ships, also from shipyards in South Korea that are scheduled to carry 18,400 TEU to be delivered in 2014. According to the Review of Maritime Transport (UNCTAD, 2013) The combined slot capacity of the world's containership fleet passed from 4,297,874 TEU in 1999 to 16,058,233 TEU in 2013. The average vessel size in 2013 was 2,755 TEU considering all shipping line companies, and 4,519 TEU considering the top 20 shipping line companies.

1.3. *Main ports today*

As a result of the ever increasing trade volume and the consequent changes in vessel specifications, ports were subject to serious challenges in the past two decades while experiencing a sensible change in the competitive environment. New ports in Asia and in China, have been developed and existing ports have been restructured or extended in order to respond to new trade needs. A possible way to measure the reaction of the present main port cities to the increase in the marine trade volume and, more specifically, of the containerized trade, is by looking at the land occupation, the storage space, the number of berths, and the percentage of dedicated container terminals with respect to general container terminals, as seen in Fig. 4.

First, the remarkable growth of the Chinese ports is apparent. In general, Asia is displaying much more growth than western countries. Presently, nine

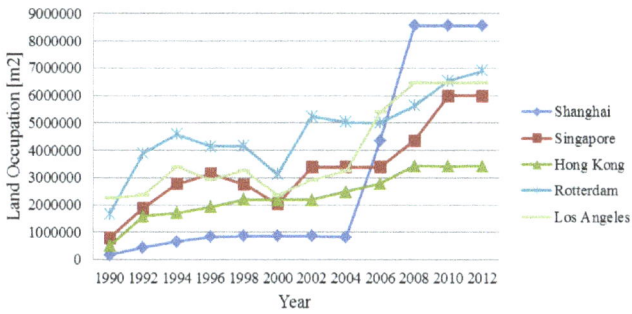

(a) Land Occupation

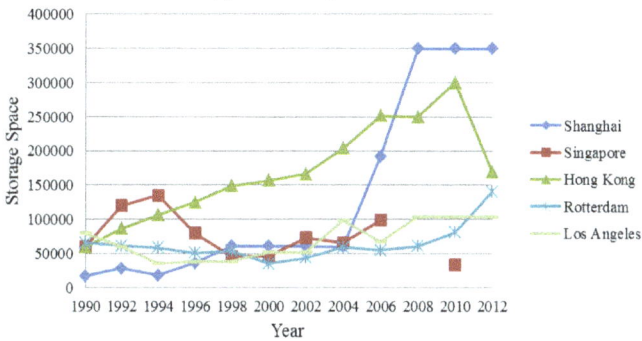

(b) Storage space (TEU)

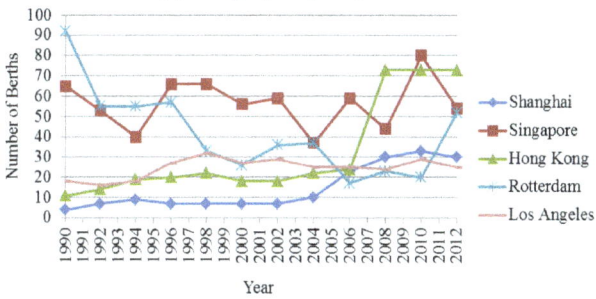

(c) Number of Berths

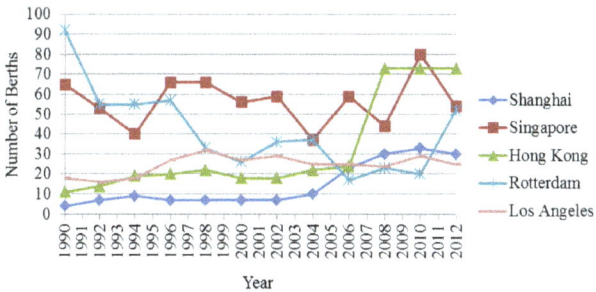

(d) Percentage of dedicated container terminals

Source: Containerization International Yearbook.

Fig. 4.   Port development measures for 1990–2012.

Table 1.   Main container terminals — years of development (data from year 2013).

Port	First container ship [Year]	Actual land occupation [m²]	Total storage capacity [TEU]	Total TEU shipped/ landed	Main port operators
Shanghai	1978	8,569,837	350,084	23,768,100	SCT
Singapore	1972	2,873,270	300,000	27,680,000	PSA
Hong Kong	1972	3,438,100	309,238	18,203,000	HIT, MTL
Busan	1987	4,277,778	325,155	14,194,334	HKC, KECT
Rotterdam	1978	6,139,600	1,253,000	11,145,804	APM,ECT
Hamburg	1968	7,900,000	5,932,550	132,275	HHLA
Antwerp	1989	7,384,073	2,786,917	8,648,475	PSA, MSC
LALB	1959	10,694,878	227,410	8,186,687	TraPac, Evergreen
NYNJ	1960	4,277,778	48,323	5,292,020	APM

among the top ten container ports in terms of container throughput are located in Asia (World Shipping Council, from *http://worldshipping.org*).

Following the classification by World Shipping Council Table 1 characterizes the major ports according to container throughput in Asia — Shanghai (China), Singapore, Hong Kong (China) and Busan (South Korea), Europe (Rotterdam (Netherlands), Hamburg (Germany) and Antwerp (Belgium)) and the United States (Los Angeles-Long Beach, LALB, (California) and New York–New Jersey, NYNJ).

As seen in Table 1, despite USA being the country that first adopted the containers for maritime transportation, the evolution of the container trade fostered an incredible growth at Asian ports, making them the busiest, with a noticeable difference in terms of container throughput handled. This fact is largely due to the globalization phenomenon undoubtedly facilitated by the spread of the container technology. Enabling global trade and marine transportation of goods keeping costs extremely low, containerization favoured the development of shipping routes connecting Asian regions and Asia, as a major export location, to both European and American markets.

According to the Maritime Transport Review (2013), while Asia continues to lead the global demand for container port services, sea trade growth is slowing down. However, compared to shipping, which is affected by an oversupply of vessel capacity and declining freight rates, the container port business is growing. Financing new port development projects is capital intensive. A recent study of the scale of future infrastructure demand examined nine economies (Brazil, China, France, Germany, India, Japan, Mexico, the United Kingdom, and the United States), collectively accounting for 60% of world GDP, and found that their annual spending on long-term investment totaled US$11.7 trillion in 2010. Extrapolating a range of growth forecasts and investment projections from external sources, the study estimated that developing countries will need an annual investment of US$18.8 trillion in real terms by 2020 to achieve even moderate levels of economic growth (*Group of 30, 2013*).

In Asia, port development projects are largely spurred by the importation of raw materials and increased industrial output. China currently leads the world in terms of port throughput and efficiency, and increasingly as a provider of expertise in port construction and management. As Chinese labour costs increase, some of the production processes are moving to neighboring countries, and Chinese companies are able to take advantage of this movement of trade through the provision of other higher value services such as expertise in port construction.

1.4. *Singapore's success as international maritime centre*

Singapore marine development is largely driven by the Port of Singapore Authority (PSA), one of its main stakeholder, fostering continuously its development as a Port City. In fact, PSA Corporation from Singapore has had an extensive corporate experience and expertise in port operations, development, construction and management through the many years of sustained growth and development in the Singapore port, and through PSA International, the global container terminal operator taking advantage of the global container trade growth.

As history evolved, Singapore has been able to keep her position as a predominant player in the maritime trade, surely but not solely, due to its strategic geographical position. Government policies, the clever use of innovation, and a reliable and high quality workforce, all constituted the basis to guarantee Singapore's role in global maritime transportation.

This chapter will provide a perspective of how Singapore could maintain and enhance its competitiveness until the present and which challenges will face in the future. We will focus on the role of technology and engineering solutions in the design of container terminals, the equipment used and the management of the port activities as the main factors behind Singapore's success as an international maritime centre.

2. Development of The Port of Singapore

"[…] The Island of Singapore is 88 miles north of the equator and contains an area of 217 square miles, the population numbers some 418,000 persons and is probably the most cosmopolitan one in the world, about three quarters of the people are, however, Chinese, and an eighth are Malays, indigenous to the country, while the Europeans total about 6000.[…] The approach to the Harbour of Singapore is one of the most beautiful in the world [...]. The Harbour is practically land–locked by islands and these afford such protection that until the reconstruction of the wharves was put in hand some fourteen years ago (1910, [ndr]), the berthing accommodation consisted only of wooden wharves on wooden piles. [...] 5,674 merchant vessels, representing a tonnage of 8,538,853 tonnes, entered the Port of Singapore in the year 1929, and of these vessels, 2,899 were British, 1,324 Dutch and 638 Japanese, the remaining 903 being American, French, Siamese and other nationalities. Singapore is therefore one of the greatest sea ports in the world and, according to Whitaker 1922, in the matter of tonnage entered and cleared, ranks next below Liverpool[…]."

(From "A Short History of the Port of Singapore". March 1922,
Raffles Library).

More than 90 years have passed from the day the monograph "A Short History of the Port of Singapore" was published, and Singapore remains among the busiest ports in the world. How has the port been able to evolve in order to maintain its competitive advantage? The effective use of engineering solutions and information-communications technology, as well as far-sighted government policies and port management strategies have given to the Port of Singapore, throughout these years and especially over the last 50 years since the establishment of Singapore as an independent nation, the special capability to evolve, react and stay in the top ranks, providing various strategies and always being innovative and providing effective services and infrastructures.

Fig. 5.   Port of Singapore at night.

Table 2.   Singapore terminals development.

Port name	Date of commencement	Company	Government authority
Keppel Harbour	1850	The Tanjong Pagar Dock Company (until 1905) KCL (1968)	SHB (1905/1964) PSA (1964/1968)
Telok Ayer	1879 (until 1932)	The Tanjong Pagar Dock Company (until 1905)	SHB PSA (until 1997) MPA (since 1997)
Tanjong Pagar Terminal	1972 (arrival of the first container ship on the 23rd of June)	PSA (until 1997) PSA Corporation	SHB PSA MPA
Keppel Terminal	1991 (split from Tanjong Pagar Terminal)	PSA (until 1997) PSA Corporation	PSA MPA
Brani Terminal	1988	PSA (until 1997) PSA Corporation	PSA MPA
Pasir Panjang Wharves	1976	PSA (until 1997) PSA Corporation	PSA MPA
Pasir Panjang Terminal Phase 1	1998	PSA (until 1997) PSA Corporation	PSA MPA
Pasir Panjang Terminal Phase 2	2006	PSA Corporation	MPA
Sembawang Wharves	1971	PSA (until 1997) PSA Corporation	PSA MPA
Jurong Port	1960	JTC	MPA

Table 2 reports the historical development of the main Singapore harbours. In its early settlement, the Singapore port was managed by the Tanjong Pagar Dock Company. The government took over the private company in 1905 constituting the Singapore Harbour Board (SHB), which was eventually replaced by the Port of Singapore Authority (PSA) in 1964. In 1997, PSA was corporatized, and renamed as PSA Corporation Limited, to focus on commercial container handling operations. With the corporatization of PSA, a new regulatory authority, the Maritime and Port Authority (MPA) was established to take over the regulatory functions of PSA. Currently, PSA Corporation Limited, MPA, the Jurong Town Corporation (JTC), and the Keppel Corporation Limited (KCL) are the main players in the Singapore maritime activities.

In the next section, we will go through some momentous developments in the history of Singapore with the objective to outline the challenges faced and strategies adopted by Singapore to develop its port in order to maintain and improve the port's competitiveness in terms of (1) productivity, and (2) provided services.

2.1. *Early establishment of the Singapore port*

The Kingdom of Singapura rose in importance during the 14th century under the rule of Srivijayan prince Parameswara and Singapore became an important trading port, until it was destroyed by Acehnese raiders in 1613. The modern history of Singapore began in 1819 when Englishman Sir Stamford Raffles arrived as an agent of the British East India Company and established a British port on the island. Free trade and petroleum then became the two keys to foster the maritime business growth in Singapura, the Lion City, as Singapore was then known.

By the 15th century, Singapore had declined as an international trading port due to the ascendance of the Malacca Sultanate, and the local trade was the main business on the island. Singapore used to provide regional ports with local products demanded by international markets. For instance, blackwood (a generic term used by Europeans to refer to rosewood) was exported from Singapore to Malacca, and was in turn purchased by Chinese traders and shipped to China for furniture-making (Heng, 2007).

In the early 17th century, Singapore's main settlement and port were destroyed by a punitive force from Aceh, one of the most powerful Indonesian sultanates. After that, there was no significant settlement or port at Singapore until 1819 when Sir Stamford Raffles, excited by the deep and sheltered waters in Keppel Harbour, established for Britain a new settlement and international port on the island. Raffles was convinced that Singapore had all the requisite potential to attract Asian as well as European traders. The construction of the most ancient ports in Singapore had started with Keppel Harbour and Tanjong Pagar Harbour. Keppel Harbour was put in operation in 1850 by the Tanjong Pagar Dock Company.

"It began as two holes in the ground at the water's hedge. The world floated into these holes as ships from all the corners of the globe came to be serviced and repaired. [...] Today, there are still the holes in the ground — in fact, a lot more of them are bigger too — but instead of just the world coming in, it has also gone out into the world." (Lim, 1993).

Raffles directed that land along the banks of the Singapore River, particularly the south bank, be reclaimed where necessary and allocated to Chinese and English traders to encourage them to establish a stake in the port-settlement. Moreover, roads and railways were developed, enhancing communication with Malaya. As a result of these constructions and policy efforts, by the late 19th century, Singapore had become an entrepôt port servicing the geographical hinterland of the Malayan peninsula, becoming its administrative capital due to the action of the British Forward Movement. Primary materials, such as crude oil, rubber and tin, were transported from the Malayan peninsula to Singapore to be processed into staple products, and then shipped to Britain and other international markets. During the colonial period, this was the most important role of the Port of Singapore.

Nevertheless, Singapore kept an eye on international trading. In particular, the port relied on three main international trade networks: (1) the Chinese network, which linked Southeast Asia with the southern Chinese ports of Fujian and Guangdong; (2) the Southeast Asian network, which linked the islands of the Indonesian archipelago; and (3) the European and Indian Ocean network, which linked Singapore to the markets of Europe and the Indian Ocean littoral. These networks were complementary, thus positioning Singapore as the natural transshipment point of regional and international trade (a critical role that the island nation has kept until today), to the point that, by the 1830s, Singapore had overtaken Batavia (now Jakarta) as the centre of the Chinese trade, and also became the centre of the English country trade in Southeast Asia.

These achievements in regional and international trade should not be attributed exclusively to the infrastructural development and the geographical position. In fact, during those times (and largely also today), one of the major competitive factors for Singapore were the few business restrictions and regulations characterizing the business activities, common with the port of Hong Kong (Heng, 2007). As a result, the Southeast Asian traders preferred the free port of Singapore to other major regional ports which had cumbersome restrictions.

Petroleum refining and distribution, one of the main businesses for the island, began at the start of the 20th century. Pulau Bukom (also spelt as Bukum), an island located 6.5 kilometres southwest of the main island of Singapore, was its first petroleum tank depot and oil refinery. Syme & Co., the company managing the depot, opened a petroleum tank depot with a tank capacity of 4,500 tonnes, the first of its kind in the East. It was a storage and distribution centre for kerosene. By 1902, Pulau Bukom was the oil supply

Source: Collection of the National Museum of Singapore, National Heritage Board.

(a) A view of the Port of Singapore 1830.

(b) A view of the Port of Singapore 2013.

Fig. 6.   Views of the port of Singapore in the history.

centre for the Far East, marking the start of a long history of leadership of Singapore in the oil trade and bunkering. From the 1840s, Singapore became an important coaling station for steam shipping networks that were beginning to form.

The sentiment of Raffles was right. As the volume of its maritime trade increased through the 19th century, Singapore became a key port of call for sailing and steam vessels in their passage along Asian sea routes. Figure 6 depicts the evolution of the Port of Singapore. From a sweep of land along the Singapore River within the city, the Singapore harbour has now evolved into an open-sea deep-water port. Singapore is one of the few ports in the world that manages to keep its leadership throughout the history of regional as well

as global maritime trade. Surely, the location of the harbour played a fundamental role. Nonetheless, the quality of service, the actions of the government
(from the nation's independence in 1965 onwards), and the investment policy,
were more important than the geographical preconditions in determining the
leadership position of the port.

2.2. *1940–1960 The reconstruction, Singapore as entrepôt port*

The Pacific War and reconstruction posed the accent on one of the most important 'assets' for the island for recovering and regaining its past competitiveness:
the people of Singapore. As stated in the Singapore Harbour Board annual
report (1949):

> "[…] The key to its achievement and to the handling of so large a trade while
> it all went on was the spirit in which the men of the board's traffic and ship
> repair departments co-operated with those colleagues whose work it was to
> restore the physical damage and to reconstruct the port's organization.[…]".

The joint effort on labour efficiency improvement as well as system and
equipment renovation were also one of the key factors for the Singapore Port
success:

> "[…] The port of Singapore is not only the principal outlet and point of entry
> for Malaya, but is also the centre of a large entrepot trade. Through Singapore
> pass manufactures from all parts of the world through neighboring countries
> and through it these countries export their produce. Singapore is used by lin-
> ers which link it directly with every part of the world and the facilities of the
> Singapore Harbour Board are versatile enough to meet the demands of the
> multitude of different cargoes which so world-wide a trade embraces. […]"
> (Singapore Harbour Board, 1949).

Government policies represent another key aspect. The effort exerted in
the War Recovery Plan was one of the strongest actions for the success of the
Singapore Port. Eventually, we will see the first signals of one of the most
impactful events in the history of the Port of Singapore: the creation of the Port
of Singapore Authority or PSA.

2.2.1. *The war recovery plan*

The Pacific war represents one of the most critical turning points in the
history of the Port of Singapore. The Japanese surrendered and Singapore
was handed over to the British on 5th September 1945. The Port then

Table 3. The system of wharves owned and operated by the Singapore harbour board, 1949 (from the Singapore harbour board, 1949).

Terminal	Length [ft]	Depth of water [ft]
West Wharf	4,525	33
Empire Dock	3,522	27
Main Wharf	3,152	33
East Wharf	700	24
Sheers Wharf	325	24

presented a scene of complete desolation. The role of the government is key at this defining moment of the port history: when the Singapore Harbour Board again took control, on the resumption of the Civil Government in April 1946, there were several problems that had to be solved simultaneously (SHB Annual Report, 1947). The physical reconstruction of warehouses was completed, the administration was restored, and the labour force was strengthened. The demand driving and justifying the enormous effort and investments in the port reconstruction during this period came from Malaya, mainly for food and material needs. As a result of the reconstruction, five main wharves were operating in 1949. Table 3 characterizes these wharves according to the length and depth of water.

Under this restoration flow, it was possible to begin the execution of the Singapore Harbour Board's long-term restoration plan for the improvement of the port's facilities. The modernization of the port started, and to gain back its competitiveness, three main tasks had to be accomplished: (1) renovation of the available equipment, (2) enhancement of labour efficiency and system productivity, and (3) increase in the volume of port services.

Equipment renovation

The most recent technologies on the mechanical equipment were introduced for the handling of cargo between ship and shed. Cargoes were mechanically discharged and expeditiously handled by motorised elevated platform trucks and a fleet of mobile cranes of capacity ranging from two to 10 tonnes. The SHB bought the innovative "Nimrod", a floating pontoon crane with a capacity of 80 tonnes. Thanks to the fully equipped salvage tug, the "S.T. Griper", with a range of approximately 5,000 miles, difficult salvage operations could be successfully performed.

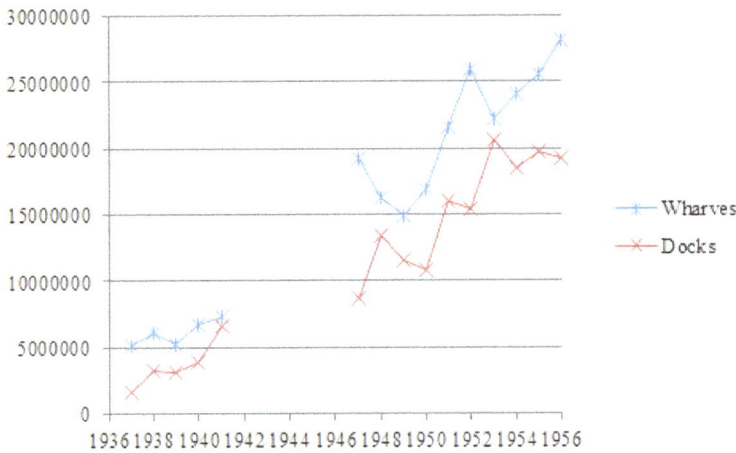

Fig. 7.   Gross revenues from wharves and docks in the years 1937–1956.

Labour efficiency

In light of this objective, radical reforms were introduced in the terms and conditions of employment, resulting in the introduction of a scheme for the decasualization of the men. As a result of the government action on labour conditions and renovation of the facilities in 1949, the rate of work at the Port of Singapore was higher than before the war, and the damage caused to cargo during the course of handling was lower. Figure 7 shows this incredible growth through the gross revenues from wharves and docks activities. Note that between 1941 and 1947, no data was collected due to the Japanese Occupation.

In 1950, the Port of Singapore was already one of the fastest port of call, largely due to the efficiency of port operations.

Increased volume of services — ship repair, oil refinery and distribution

The enhanced technological level and labour efficiency provided the port with the capability to considerably speed up the time for handling ship-to-shore operations. In addition to the increased efficiency in port operations and productivity, Singapore was known as one of the main repair locations for all types of ships in the late 1940s. Table 4 shows the capacities of the terminals as of 1949.

With her five dockyards, Singapore was recognized worldwide for the quality of ship repair services (primarily engine and hull repairs), and also as an oil supply centre. The repair yards were renowned for the modern oxygen plant for use during ship repair work. Also, the equipment for engine repair was

Table 4. The system of docks owned and operated by the Singapore harbour board, 1949 (from the Singapore harbor board, 1949).

Dock	Max length of vessel [ft]	Max breadth of vessel [ft]	Max draft of vessel [ft]
King's Dock	850	92	32
Albert Dock	455	55	18
Victoria Dock	450	55	16
No. 1 Dock	430	51	15
No. 2 Dock	380	43	13

constantly kept up-to-date with the most modern facilities, such as overhead cranes with capacity of up to 30 tonnes, were made available. Engine lathe, vertical borer and horizontal borer were of the latest generation. Services in engine repair included casting, forging and pipe work in iron and copper. Modern plating shops were available for hull repairs, where all classes of riveted and welded work were executed.

In oil supply and refinery, the installation at Pulau Bukom, the oldest kerosene depot centre in Singapore, was taken over by Shell Company. Shell had built Singapore's first oil refinery in 1961 (today, the oil giant has its largest refinery at Pulau Bukom, with a capacity of more than 400,000 barrels per day). This was a momentous year in the economic history of Singapore, and has since developed to become one of the major bunkering locations worldwide. Oil constituted the main trade for Singapore (excluding the trade of goods with peninsula Malaya), and its role as a main oil distribution centre in Asia made the island an entrepôt port as well as a major bunkering port for vessels calling at Singapore during this period.

2.2.2. End of the Singapore harbor board: Towards a new era for the port

In the "Report of the Commission of Inquiry into the Port of Singapore" (1957), commissioned by the Government of Singapore to evaluate the port's performance, we find:

> "[...] It is no accident that Singapore has achieved its present stature. It is true that the island has considerable natural advantages, in its favoured geographical position in relation to large areas of production and therefore to important trade routes, in its possession of a safe anchorage, and in having a waterway through the centre of the city. To these natural advantages, the industry and hard work of the people of Singapore have added man-made port facilities, modern transit methods and handling equipment. [...] As the general trading prosperity of the post-war years, there has grown up and

added trade in oil. Singapore is not only a bunkering port, or even merely an entrepôt port for oil. It ranks as one of the largest storing, blending and distributing centres in existence […]."

Another key aspect in this period was the creation of the Port of Singapore Authority. Towards the end of the 1950s, with the container era at its door step, the Singapore Government was aware that the volume and type of maritime trade would soon change radically, and the island had to anticipate these changes to keep its leadership in the east. A Commission of Inquiry was launched in 1957 to evaluate the port activities as well as the port equipment and port organization as the key elements. The Commission suggested an important move in the late 1950s — the establishment of a separated port authority. The Commission considered the establishment of the Port of Singapore Authority (PSA) as a necessary step towards the solution of noticeable inefficiencies hindering the desired development of the Port of Singapore. A clear example was the lack of cohesion and unity in the many facilities which the port utilized in its services to the shipping industry and trade. In order to favour the synchronization and sharing of the facilities and increase the port efficiency, a single authority (which will become PSA) was proposed. This authority partially replaced the Singapore Harbour Board with a view at the whole port. This represented a fundamental change in the organization and the management of the Port of Singapore, and thus initiated the beginning of a new era.

2.3. *1960–1970 PSA and Singapore as a transshipment hub*

As pointed out in the previous section, the establishment of PSA was a necessary step to further develop the Port of Singapore, enhancing its competitiveness with respect to other port cities not managed by a single central entity. The decision to have a unique overall controller was apparently brought on by the opportunity to maximize both the port efficiency and the utilization of port infrastructures.

On 27 December 1963, Singapore's Head of State, granted formal assent to the legislation to establish the Port of Singapore Authority to replace the Singapore Harbour Board. Provision was also made for the establishment of the dockyards as a separate entity. A new era in the history of the Port of Singapore had thus begun.[a] PSA was officially nominated on 1st April 1964. This was another momentous period for the State of Singapore, as the state gained independence on 9th August 1965. PSA and her newly established

---

[a]From "Singapore Port History 1819–1963". (1964). Maritime Museum, Port of Singapore Authority (PSA).

country were close, to the point that we read the following in the incipit of the PSA Annual Report in 1965:

> "[…] There has been a general increase in shipping activity, and the number and tonnage of vessels entering and leaving the port in 1965 surpassed all previous records. This is a most encouraging trend and it is indicative of the drive and resilience of the people of Singapore to push forward as an independent republic in this part of the world. […]".

(*PSA Annual Report 1965*).

The decade from 1960 to 1970 was intense not only in the Lion City. In 1967, the Association of Southeast Asian Nation (ASEAN) was formed. The Vietnam War and the closure of the Suez Canal in 1967 represented two key events for the bunkering industry in Singapore, which experienced a period of high growth resulting in a number of plans for the construction of more plants and acquisition of new equipment. In December 1968, the national shipping line, Neptune Orient Lines (NOL), was founded, giving a substantial push to the new Republic's economy.

2.3.1. *PSA: The development plan*

The key elements for maintaining competitiveness had not changed: (1) *system performance*, (2) labour efficiency, (3) equipment improvement, and (4) services offered. These were still the key considerations guiding the port development. However, different from the previous decades, the management policies became more relevant. Since PSA was now the "global eye" on the Port of Singapore, optimal policies to improve the overall terminal operations, crucial to increase the port's efficiency, were now applicable. Furthermore, the new organization and the nation's independence (determining the loss of Singapore's "hinterland", i.e., Malaysia) considerably fostered the development of the Port of Singapore as a transshipment hub, a key role that the port city has firmly maintained up to the present days.

Improved system performance

The maximization of efficiency of the wharves in cargo handling was the highest priority for PSA since the settlement as stated in the PSA Annual Report (1964):

> "[…] strenuous efforts were made to improve the gang-hour rate for the handling of cargo alongside the authority's wharves […]"

(*PSA Annual Report, 1964*).

As a pioneer of the management of cargoes, PSA introduced the Advance Berth Allocation scheme which enabled the scheduling of vessels up to 72 hours prior to their scheduled arrival time. In addition to this new system, the Automatic Cargo Removal System was also installed with the objective of clearing transit sheds seven days after the carrier's completion of discharge.

Transshipment is always a special capability and a priority at PSA, and an efficient system avoiding the double handling of cargoes was offered to users of PSA wharves. Cargoes were automatically moved by coasters to the transshipment berths upon discharge by ocean-going vessels to await on carriage. This advantage in avoiding double handling generated a noticeable trend of movement of cargoes towards PSA wharves in 1965 as a result of PSA committing massive investments in telecommunication and, in 1967, announcing a 24-hour working control tower, the first in the world.

Enhanced labour performance

Labourer's shifts underwent a significant modification: a two-shift system for cargo handling at the wharves was introduced for the first time on 5th October 1964. As a result of this innovative system, PSA succeeded in: (1) eliminating the long over-time hours worked under the old system, and (2) supplying fresh labour for each shift, working a maximum of 7 hours. The effect of such a change was apparent and the monthly average gang-hour rate went up by 8.49% (PSA Annual Report 1964).

Improved equipment and facilities

Upon the establishment of PSA, a modernization programme was brought on through the supply of new high-tech forklifts and mobile cranes. PSA pioneered the installation of electronic control devices in forklifts enabling the equipment to operate in ships' hatches. Training courses for operators in the use of the new equipment was an essential expense for PSA.

PSA departments were also constantly innovating. In 1979, the Mechanical Engineering Department (MED) designed innovative modifications for handling special cargoes such as drums. The improvement in the forklift was derived from a practical need: forklifts were less effective in lifting drums or cylindrical cargoes as the gap between the forks was too wide. With the new proposed rounded forks, drums could be easily accommodated with the narrowed gap between forks. It was a period of remarkable development for the Port of Singapore which was well aware of the upcoming growth in trade volumes due in part to the spread of containerization.

In 1963, Jurong Port was set up by the Singapore Economic Development Board (EDB) to support the growth of Singapore's first and biggest industrial estate, the Jurong Industrial Estate. In 1965, Jurong Port officially commenced operations. In 1968, Jurong Town Corporation (JTC) was set up to drive the industrial estate development in Singapore and Jurong Port became a business division under JTC.

New services

Due to the increasing trade volume and the sharing of port facilities, it was becoming critical for PSA to keep to high standards of customer service for the warehouses. The late 1960s represented the early start of the containerization era. PSA was well aware that storage yard space and technologies would become the key elements for maintaining its leadership. PSA organized a one-week training course focusing on warehouse management and operation activities due to the critical importance of warehousing.

The Ship Repair and Electrical Division was the core for the Port of Singapore and consequently, for PSA. Investments in new infrastructures and innovative electrical installations were both performed since the first year. Nevertheless, despite almost 150 years and ships using the docks have changed completely, the Victoria Dock (one of the main ship repair docks) was very much the same in the 1960s as when it was formally opened in 1868, proving the far-sighted dock design of early Singapore engineers.

The year 1968 was a significant year in the history of Singapore's ship repair business as PSA handed over the management of the dry docks to Keppel Shipyard Company, presently responsible for ship building and repairing in Singapore (Loh and Tey, 1995).

In addition to the traditional services, dangerous goods and pilotage become protagonists in the PSA portfolio. The PSA was competent in the management of dangerous goods; by the end of 1969, more than 600 vessels with dangerous goods (e.g., explosives) were cleared in the Port of Singapore, showing how safety was rising as a distinguishing feature for PSA. Another important addition was the incorporation of the pilotage service within PSA. The industry understood that pilotage was one of the key activities determining the service efficiency and effectiveness at the port. Hence, in order to maximize the port's efficiency, PSA chose to incorporate the pilotage service and invest in its improvement. A new watch system was then introduced in October 1964 ensuring that pilots would be available all the times.

It was in the combination of these four key elements that PSA found the optimal balance to maximize port efficiency. A clear evidence of the success of this strategy resulted in the increase of cargo handling capacity and the decrease of terminal congestion, on the one hand, due to the new cargo location scheme, and on the other, as a result of the reorganization of the port Traffic Department and the introduction of employee incentives based on personal exerted effort, bringing the port operations to the maximum efficiency.

2.3.2. The "Bet" on containerization and jurong industrial estate

The end of the 1960s represented a strategic moment for PSA, which confirmed its trust in the development of the container as the future marine transportation. As a result of this sentiment, huge investments were committed to enable the Port of Singapore to be the most efficient container hub in the world. The investment plan for the first 5 years from the establishment of PSA amounted to $137 million, including the second phase of the development of the East Lagoon wharf leading to four new berths for ocean-going vessels.

In a period of labour-intensive industrial development in Singapore, the Economic Development Board (EDB) was the driving force behind the establishment of Jurong as an industrial estate. From a swamp in the western part of Singapore, Jurong was transformed into Singapore's first and biggest industrial estate to initiate the country's industrialization programme with factories producing garments, textiles, toys, wood products and hair wigs. Further, in 1963, EDB set up Jurong Port to support the growth of the Jurong Industrial Estate. In 1965, Jurong Port officially commenced operations. In 1968, Jurong Town Corporation (JTC) was set up to drive the industrial estate development in Singapore, and Jurong Port became a business division under JTC. JTC would soon acquire a key role in one of the core businesses of the Lion City: petrochemicals. In 1969, the entire port area and the Jurong wharves were declared a Free Trade Zone, and traders were provided with cabinet entrepôt facilities to commence a new phase of operation in the port.

2.4. 1970–1980 Beginning of the container era

The most significant change in global trade and, consequently, in the development of the Port of Singapore in the 1970s was a result of containerization. The port authority envisioned a remarkably growing business and a consequent dramatic increase in container traffic volume. In fact, such innovation in trade contributed to deep changes in maritime endeavours. The 1970s represented a difficult period in the shipping industry when it faced one of the most severe shipping crisis ever experienced. The crisis affected shipping companies with high investments in tankers and tramp/bulk ships, whereas the liner trade was not affected due to its cost-plus rate setting system (UNESCAP, 1978). On the other hand, the economic reform in China (1978) represented a spectacular push against the decline of the maritime trade.

The challenge for the growing Singapore was clear: to be able to exploit a growing market and a completely new transportation means keeping the competitiveness coming from past core assets. The response of Singapore was decisive: investments were necessary (1) to enhance the port's capacity to handle the growing containerized transportation, and (2) to improve the port efficiency to reduce waiting time and turnaround time. PSA saw in the containerized trade a major business for the future. The authority also understood the challenges the new technology would have brought and the consequent need to innovate to efficiently run the new business.

2.4.1. *The challenges: diversified demand and the increased port traffic*

Table 5 shows the growth of containerized trade in Singapore in the period 1972–1977. Six months from the first container operation of Tanjong Pagar Terminal on 23rd June 1972, 184,855 tonnes of cargo were handled in 11,810 containers. A total of 25 container vessels berthed at the container port, of which 10 were on their maiden voyage from Europe to Singapore. The container handling rate was comparable to those being achieved at established container terminals in other parts of the world. The best handling was achieved with a workout of an average of 27 containers per hour (this record was surpassed in 1977 when 50 containers per hour were handled).

For the first time in 1973, the containerized cargo volume through Tanjong Pagar passed the 1 million tonnes mark (*PSA, Annual Report 1973*). The congestion at container ports became a problem that guided the port

Table 5. Growth in container traffic 1972–1977.

Year	Average monthly no. of ship calls	No. of shipping lines	No. of containers handled
1972	4	3	14,042
1973	14	5	95,905
1974	24	8	153,411
1975	40	14	191,981
1976	59	17	296,671
1977	83	19	351,140

Source: PSA BERITA (1977).

into two main development directions: (1) to renovate the equipment, and (2) to increase the storage space.

Containers were not the only source of marine trade for the country. In 1974, Pasir Panjang Wharves received the first LASH (Lighter Aboard Ship) vessel visiting Singapore, introducing a new mode of international cargo shipping. The LASH vessel discharged empty barges which were subsequently loaded with rubber shipments for American ports. The loaded barges were collected by the LASH vessel on its second visit. The Phase I of Pasir Panjang Wharves was completed in 1977 and, in the same year, the authority embarked on the development of Phase II with a $40 million investment in anticipation of (1) the expected increase in containerized regional trade, (2) the possibility to convert a part of Keppel Wharves into container feeder berths, and (3) the improvement of facilities in order to receive larger vessels in the future. The first Ro-Ro vessel visited the Keppel Wharves, kick starting the new Roll-On Roll-Off vessels business, while the Telok Ayer terminal was in operation providing additional general cargo capacity in 1979.

Together with the growth of containerization, the 1970s had brought challenges of both diversified demand and increased port traffic. In response, port expansion and information technology were the means to answer these challenges.

2.4.2. The response: Port expansion and information technology

Due to high capital investments required for the container business development, PSA moved towards the privatization of the general cargo business, a strategy which would be followed by several port operators in the world. Berths for general cargo were then assigned to private shipping companies. This allowed PSA to lower the investments in the general cargo handling, leaving this expensive development to the companies under the agreement.

The development of East Lagoon (Tanjong Pagar) was visionary: Singapore was building its container ports before a single shipping line had indicated that a containership would call (*Port of Singapore, the maritime hub Series 1999–2002*): "[...] We are moving towards the ultimate global system of container routes and container ports. [...]" (*PSA Chairman, Howe Yoon Chong, 5th August 1970*). Apparently, PSA had well understood the importance of containers in the marine trade, as we read in the *PSA Annual Report (1966)*:

> "[...] the new technique of cargo handling by containers and container ships was given a great deal of prominence in shipping circles during the year and the Port Authority, being aware of its role in this latest intermodal movement

of goods, made a preliminary announcement regarding its plans for the construction of a container complex at the East Lagoon. [...]".

In 1966, PSA had planned to provide wharf and shore facilities for the handling of container ships and containerized cargo. It is interesting to note that the original investment budgeted by PSA was $77 million. Nonetheless, being well aware of the impending arrival of third-generation container ships, the authority decided to increase the investment in order to be able to accommodate them.

At the beginning of the 1970s, the project for the container port at Tanjong Pagar started to be laid down, and the Tanjong Pagar Terminal started its first operations in June 1972. The new development plan implied several points: (1) investments in fixed assets; (2) system efficiency increase; (3) use of Information Technology, and (4) growth of the service businesses.

Fixed assets investments

The late 1960s represented a momentous period of preparation for the coming container era. The project for 120 acres of land dedicated to container storage was proposed, and investments for container equipment were planned. Plans were also made for training port labour force to handle containers and to use the container handling equipment.

Together with the East Lagoon terminal the Sembawang basin was opened for the handling of low-value high-volume homogeneous cargoes in 1971. Indeed, given the spectacular growth in the volumes, PSA had to manage the allocation of the traffic through appropriate policies as well as increasing its capacity.

Table 6 reports the incremental value of the PSA fixed assets in the years 1969–1979 (*PSA Annual Report, 1980*). Notice the increased value in the

Table 6.   Incremental value of PSA fixed assets for 1969–1979.

Additional fixed assets ($Million)	'69	'70	'71	'72	'73	'74	'75	'76	'77	'78	'79
Freehold Land	—	—	0.25	0.31	—	18.6	1.42	12.13	1.19	2.68	2.43
Leasehold Land	—	—	1.78	4.35	—	14.4	7.04	0.03	41.1	17.4	16.2
Wharves	—	—	—	33.2	—	39.1	25.02	6.08	29.6	7.14	41.4
Floating crafts and equipment	1.1	0.3	4.7	1.89	0.75	6.11	2.18	10.18	10.2	7.59	3.93
Plant machinery and equipment	0.8	1.4	1.08	12.63	0.83	3.24	16.82	10.24	29.4	26.8	13.7

wharves as well as the value of equipment-related assets. This value increase can be largely attributed to the growth on the container terminal infrastructures.

The large investments in the container business were justified: (1) containerized cargo through Singapore was increasing at a rate of 5% a year in the period 1972–1977 (see Table 5); (2) the Singapore Container Terminal was deemed a pivotal port in South Asia and Southeast Asia and, as a result, inexpensive feeder vessels used Singapore as a distribution/collection centre; and (3) the re-opening of the Suez Canal. The analysis proved to be right, and on 23rd June 1972, the first fully-cellular container vessel (M. V. Nihon) called at the Container Port at East Lagoon (Tanjong Pagar Terminal) and the Tanjong Pagar Terminal ended the landmark year of 1972 by handling 24,515 TEUs of containers, using two quay cranes.

In 1978, offshore docks were developed to extend Tanjong Pagar and Keppel terminals. These berths had no backup space since it was not possible to construct a deck for both terminals, without the risk of generating congestion in the movements of containers (*PSA BERITA, 1978*). Composite decking and yard capacity increase through the use of new technologies (Rubber-Tyred Gantry (RTG) cranes) in Phase II of Tanjong Pagar made it possible to control the congestion while not affecting efficiency.

In 1978, the first state-of-the-art yard gantry cranes were acquired by the container terminal. The transtainers were equipped with quick-change lifting spreaders to straddle 20-foot or 40-foot containers in six rows and stack them 5-high. Operations on each crane were controlled by a single operator in a cabin-on-trolley which travelled with the load.

Improve system efficiency through port operations management

A policy innovation was adopted by PSA to improve the port performance. In particular, a new tariff structure was introduced at the beginning of 1972 to incentivize port users to make efficient use of the port facilities. The revised tariff was based on the following main principles (*PSA Annual Report, 1972*): (1) to pay higher premium when using more expensive facilities in return for a reduction in delay, waiting time and turnaround time; (2) to divert high-volume low-value cargo to Sembawang Wharves, where simpler, inexpensive and basic facilities were available; and (3) to accelerate the development of Jurong Port as the bulk cargo port of the Republic. Consequently, the turnaround time improved by 18% and the waiting time of ships at the Keppel Harbour decreased from 4.5 hours to only 1.7 hours (*PSA Annual Report, 1972*). As a result of the redirection of port traffic to Sembawang, the highly developed facilities at Keppel Harbour were relieved of congestion and could then be devoted to the handling of high-value consumer goods and machinery with minimal delay to ships calling at the wharves. The improvement in the packaging methods of general cargoes also fostered the increase in the mechanization of Keppel Wharves, leading to a growth in productivity of 12 % in 1973.

Information technology

Information technology started to become one of the key elements in Singapore's port development strategy and a new computer system was installed in 1972 (*PSA Annual Report, 1972*). The authority was then able to undertake the computerization of the accounting procedures. The computer system also kept track of the many movements involved in the loading, discharging, transfer, stacking and delivery of containers through the Container Handling Information System (CHIS). The CHIS was designed to operate in real-time mode aided by visual display units. A teletypewriter terminal, installed within the East Lagoon Container Terminal Building, monitored these movements and subsequently used the information stored for operational and billing purposes. In 1973, the Computer Integrated Marine Operation System (CIMOS) and a first version of the Computer Integrated Terminal Operation System (CITOS) were introduced by PSA, again ahead of time with respect to all the major ports in other parts of the world.

Growth of the service businesses

The management of the container terminal went beyond the sole container handling. PSA was aware of the developed competencies in the service and repair market, and the step to become a renowned container repair service provider was short. In August 1972, the repair yard of the Container Port commenced operations on the repair of damaged containers, pre-trip inspections and monitoring services on refrigerated containers. In particular, the "refrigerated" technology would become one of the key expertise provided by PSA. In 1979, the development of cold storages for reefer containers was one of the key investments of the port, paving the way to become a service for which Singapore is globally known for presently.

The ship-repair industry experienced a remarkable growth at the point that, by the end of the 1970s, Singapore was one of the main centres providing a complete range of services. Table 7 shows the increase in the revenues coming from the services provided by the Port of Singapore.

In 1977, Singapore was the third busiest port in the world with 128.3 million nrt (behind Rotterdam with 189.3 million nrt and Yokohama with 154.5 million nrt (PSA BERITA, Vol. 12, 1979).

Table 7.   PSA revenue from port operations for 1975–1979.

Source	1975 S$'000	1976 S$'000	1977 S$'000	1978 S$'000	1979 S$'000
Container Terminal	36,533	68,998	79,080	102,643	133,217
Cargo Handling Service	56,580	70,541	68,073	70,611	74,403
Wharf Services and storage	56,295	66,389	67,906	75,419	91,314
Pilotage tugs port and garbage dues	40,436	57,016	67,265	75,191	75,596
Sundry revenue	23,526	33,696	37,691	40,848	51,615

Source: PSA Annual Report (1979).

Despite the general market downsize due to the Middle East oil crisis, reflected in a decrease of traffic in 1975, Singapore experienced an unprecedented growth in the second half of the 1970s, also in the oil industry. With six international oil companies operating refineries or installations in Singapore, the country was the third oil refining centre in the world. In 1977, the volume of mineral oil in-bulk handled in Singapore was 45 million tonnes. In 1977, a Bunkering Terminal with a total capacity of 25,240 tonnes, was directly controlled by PSA from ESSO. The reason of the acquisition was strategic: due to the location of the facility, vessels calling at Keppel or at the Tanjong Pagar Terminal could directly be serviced by underground common oil facilities pipelines.

The increase in the bunkering activities led to the need to directly face the issue of pollution of the seas, beaches, shore establishments and territorial waters. In February 1971, the Prevention of Pollution of the Sea Act was promulgated and PSA was charged with the responsibility of enforcing the Act. Slop facilities with the latest technologies in oily water treatment, waste treatment and sludge disposal were constructed on Pulau Sebarok so that tankers and other vessels could discharge their slops or waste oil before proceeding to the repair yards. Slop facilities were a fundamental service that Singapore Authority was able to supply with high efficiency, due to the ability to provide centralized facilities, compared to the solution used in other container terminals in the world where separate facilities were provided by refineries and shipyards at high capital costs. Also, besides their pollution reduction functionality, the slop facilities gave further momentum to the ship repair industry providing again an edge over the other ship repair centres in the world.

As a result of the significant increase in port traffic, another challenge was land occupation. The land reclamation for the port expansion and warehousing purposes became the primary focus in the plans of PSA throughout the decade (and they still are nowadays). From 1971, 'Warehousing' Service became a standalone section in the PSA Annual Report. It was in this year when the service started to be provided by PSA out of the port perimeter fence. The need for new land dedicated to storage space was justified not only by the increasing demand in cargoes and the predicted increase in containerized goods. Indeed, PSA was aware the urban renewal plan which was about to start along the Singapore River would have resulted in a dramatic space shortage. Documentation, delivery, transportation, storage, re-packing and shipment were all provided and available to users of PSA wharves from October 1971 (*PSA Annual Report, 1971*).

The warehouses in Nelson Road, Dover Road, and Clementi Road provided an overall storage space of 37,400 square metres. The success of the service was proven by a 100% utilization of the Nelson Road and Clementi Road warehouses by the end of the year. Space continued to be developed and mechanization improved in order to increase the efficiency in the material handling. An offshore supply centre strategically close to the Container Terminal at East Lagoon and a multi-storey complex at Pasir Panjang both became operational by the end of 1979.

2.5. 1980–1990 Port modernization, the answer to the crisis

The 1980s were another period of crisis, especially for European and American economies: the OECD (Organization for Economic Co-operation and Development) countries achieved a growth of about 1% in their gross national product, but high interest rates in the USA, aimed at dampening inflation, caused a slowdown in industrial production. The United Kingdom experienced a negative growth rate while West Germany saw her growth decelerate from 4% in the previous year to 2% in 1980. The Iran–Iraq war started in September caused an increase in oil prices. However, the ASEAN economies were resilient and registered growth ranging from 6%–8%. Japan, one of the major trading partners for Singapore, registered a growth of 5% and Singapore itself registered a growth of 10%.

Transshipment volumes rose to 26%, mainly due to tariff concessions and rebates. The ability to manage these containers more efficiently and quickly was a key factor for success. The continuous growth in containerized demand also constituted a relevant aspect for the port business. Figure 8 shows the share between containerized, general and oil in-bulk cargoes.

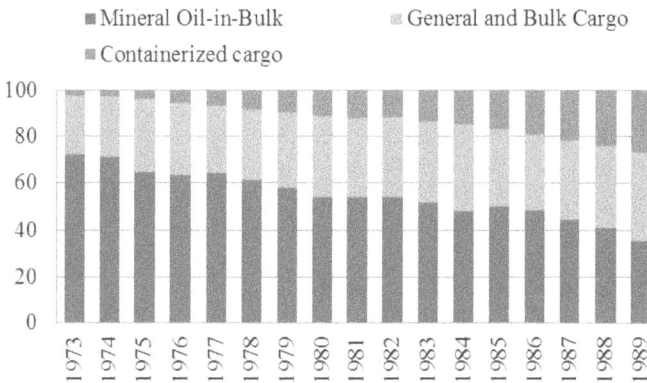

Fig. 8.   % Share of sold tons of containerized, general and oil in-bulk cargoes for 1973–1989.

Despite the relevance of oil which remained the main source of income for the Singapore port, the increased demand in general and containerized cargoes required a modification of the investment structures and the innovation used.

Manpower remained a key element in the success of PSA and the 1980s was an important decade for manpower development at the Port of Singapore. Quality Circles (QCs), were created in the early 1980s in order to spread port operation management skills and enrich the capabilities of PSA employees. By the end of 1988, PSA declared its value added per employee had rose by 26% to $96,270 (PSA Annual Report, 1988). The enhanced capability of the manpower was reflected in the decrease of daily rated employees, as Figure 9 suggests.

2.5.1. *Modernization through improved facilities*

PSA showed again how flexibility and the ability to adapt to external changes were the key factors for success. Despite the global demand downturn, PSA saw $1.5 billion worth of capital projects approved at the end of 1981. Orders amounting to $60 million for mechanical handling equipment were placed for delivery and the capital expenditure for the year 1982 alone amounted to some $500 million. The strategy was clear:

> "[…] The years ahead will pose difficult challenges with prospect of reduced world trade as nations struggle to improve their economies between alternate bouts of inflation and recession. For PSA to cope with these challenges, it must sharpen its competitive edge by investing in new improved port facilities.[…]"

> (Lim Kim San, PSA Chairman) (*PSA Annual Report, 1981*).

■ Senior Officers ■ Junior Officers ■ Daily-Rated Employees

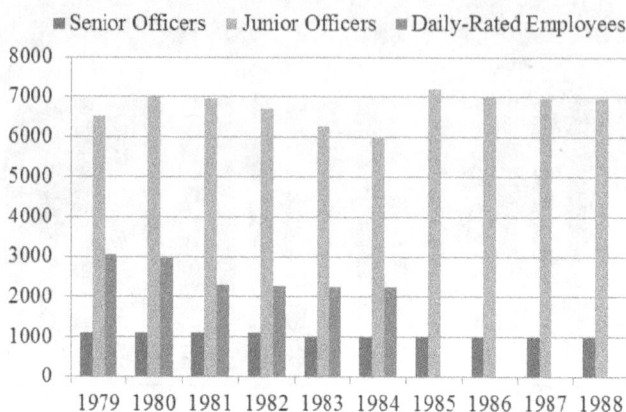

1979 1980 1981 1982 1983 1984 1985 1986 1987 1988

Source: PSA Annual Report (1998).

Fig. 9.   Staff strength as the end of the year.

The capacity to understand the trend and the investments before the demand was manifested made Singapore a leading country.

"[...] The evolution of the port is the result of adapting to changing realities. Singapore flourished as an entrepôt for the first 140 years of its modern history (1819–1959, n.d.r). Then faced with a growing population entering the employment market and the increasing pace of the economic expansion in Japan, Europe and the United States of America, Singapore had to opt for an industrialized, export-oriented economy. As external trade flourished, so did the port. With more than 600 ships in the port at any one time, and about 30,000 ships calling each year, Singapore today is truly a global port. [...]".

(*PSA. Singapore: Portrait of a Port, 1984*).

The late 1980s marked a success story for the Port of Singapore and PSA. For the 4th Asian Freight Industry Awards, Singapore was voted the Best Seaport in Asia. PSA was also voted the Best Terminal Operator (Seaport) and the Best Warehouse Operator for the third successive time.

At the end of 1989, PSA operated 5 cargo terminals and Tanjong Pagar was one of the largest container terminals in the world with a record of 4 million TEU and was the main gateway in Singapore for handling containers. Keppel Terminal (which split from Tanjong Pagar in 1991), Pasir Panjang Wharves, Sembawang Wharveswere were owned by PSA. Jurong Wharves were owned by the Jurong Town Corporation but managed by PSA. At the end of this decade,

Table 8. Quay and yard equipment for 1986–1990.

Year	Quay Cranes			Yard Transportation		
	Type	Total	New	Type	Total	New
1985	Panamax	18	—	Transtainer	41	—
1986	Panamax	18	—	Transtainer	41	—
				Van carriers	19	—
1987	Panamax	18	—	Transtainer	47	6
				Van carriers	19	—
1988	Panamax	18	—	RTG	67	67
	Post Panamax	6	6	Van carriers	19	—
	Total	24	6	Transtainer	48	7
1989	Panamax	18	—	RTG	74	7
	Post Panamax	9	3	Van carriers	19	—
	Total	27	3	Transtainer	48	—
1990	Panamax	21	3	RTG	115	41
	Post Panamax	11	2	Van carriers	19	—
	Total	32	5	Transtainer	48	—

Source: PSA Annual Reports.

over 50 port projects worth $200 million were under construction and more than $700 million of new equipment were purchased or were under approval.

Table 8 gives an idea of the distribution of investments in the innovation and expansion of equipment in Tanjong Pagar Terminal. Singapore was a pioneer in the installation of Rubber-Tyred Gantry (RTG) cranes. Concerning the quay cranes, it is noteworthy to consider that, despite the first historical post-Panamax vessel can be dated back to 1944, the development of post-Panamax container ships was dated exactly to 1988 (Rodrigue, 2014). The Port of Singapore showed the ability to be in pace with changes leading the country to be the first call for the super post-Panamax vessel Emma Maersk in 2006.

As the container traffic grew, space was increasingly becoming an issue. In light of the forecasted growth of container demand in 1985, PSA invested in container berths and storage yard space. This was carried out through the conversion of general berths in Keppel Wharves to container berths and the increase of yard space as well as the construction of new multi-story warehouses such as Alexandra Distripark detached from the port area.

In 1988, the construction of container terminals with the latest technologies in Palau Brani was approved at a cost of $870 million. It became operational in 1992 as one of the most modern terminals worldwide. The expansion came with the creation of storage yard space. It was the year in which the Keppel Distripark development was approved and Singapore became one of the pioneers in warehouse automation. The complex could occupy about 470,000 square metres at the cost of $399 million.

2.5.2. IT as facilitator for improved port operations, local and international trade

The policies for effective port management and investment in information technology for the optimal use of always-increasing data from several port users were important for Singapore to maintain her leadership position. From 1982 to 1984, PSA spent some $6 million to provide on-line terminals to the operations and engineering departments.

The complexity of port activities, exacerbated by the multitude of provided services, the heterogeneity of cargoes and the complex warehousing services supplied, all required computer applications to be able to manage

"[…] all facets of port operation […]"

(Lim Kim San, PSA Chairman) (*PSA Annual Report, 1980*).

The control of container ship operations, the tracking of dangerous goods, and packing and unpacking activities at the container freight stations were computerized in 1985. At the Tanjong Pagar Terminal, the vessel crane split information system was an important innovation: the system was able to estimate the number of moves of each sensitized crane when working on a particular vessel with a particular stowage pattern (*PSA Annual Report, 1988*).

The IT development was planned in detail to improve both port-to-customer relationships as well as port-to-port communication.

Developments in IT was fundamental to ease and speed up the gate activities and, in general, the bureaucracy behind marine shipment. In 1982, various batch-processing requests were made online to reduce the incidence of error and to speed up the flow of information. The on line applications were extended to the Sembawang port, the Warehousing Services department and the Supplies department. The Tanjong Pagar container gate, commissioned in 1986, was equipped with four electronic weighbridges, laser scanners to read bar-codes on documents, and a document-transfer system.[b] With this new gate system, the processing time of documents was reduced from three minutes to about a minute. As a result, more time was given to shippers to send their Less-than-Container-Load (LCL) export cargo, since PSA could accept cargoes for up to 18 hours (instead of the usual 24 hours) prior to the vessel arrival. With the new system, 93% of the container hauliers were served within the first hour (*PSA Annual Report' 1986*). To further improve service at Tanjong Pagar Terminal, the closing time for receiving export containers was reduced from 12 to 8 hours before the ship berthing time. This reduction was possible due to improvements in work methods and further computerization in the ship stowage planning. The introduction of advance clearance of customs permits for import containers also allowed the ships to leave the terminal faster. Another strong contribution in the reduction of the port time was due to the reform of the PSA Act in 1986. A remarkable modification resided in the mandatory requirement for the vessel owner, agent or master to provide the information relating to the principal dimensions, including the height, of the vessel. These data were fundamental for PSA to satisfactorily plan in advance the berthing and stowage actions.

Developments in IT proved to be a strategic asset to improve shipping line companies port experience. PortNet was first released in 1984, with an initial group of 120 subscribers. It was the world's first nation-wide business-to-business (B2B) port community solution connecting shipping lines, hauliers, freight forwarders and government agencies, helping them to manage information better and synchronise their complex operational processes. The same year, TRADENET started operations and it was the first nation-wide electronic data interchange (EDI) platform for trade administration ever implemented worldwide.

In 1988, CITOS (Computer Integrated Terminal Operations System) was implemented in Tanjong Pagar Terminal and it became one of the major assets for the Port of Singapore as an Enterprise Resource Planning system that coordinates and integrates every asset from prime movers, yard cranes and quay cranes to containers and drivers. The DataBox system became active with 37 subscribers in 1986. The features of the system included berth applications and direct input data related to containers.

---

[b]The increase of the gate operation efficiency will be of primary importance for the port of Singapore and several projects will culminate in the well-known Flow Through Gate System introduced in the 1997.

The development of the Computer Integrated Marine Operation System (CIMOS) started in this period, leading to the increased efficiency and safer use of the anchorage space and simultaneously improved the services provided to the port users.

Port-to-port relationships is especially important for one of the major transshipment hubs in the world. In 1988, PSA established computer-to-computer links with the ports of Hong Kong and Bremen in order to facilitate berth and yard planning. The teleport links with Bremen became active in 1989. With such a development, training was necessary, and a scheme of service was introduced in 1982 in order to train employees to become computer programmers.

2.5.3. *Flourishing of the ship repair and construction market*

Major changes also took place in the service business. After 100 years, the Victoria dock and Albert dock were closed down in 1983 to make way for the expansion of the Tanjong Pagar Terminal. This was in line with the environmental policies to relocate the shipyards in Singapore. However, the relocation apparently took a long time. Shipyards were still generating $60 million revenue a year, thus owning an important decision power (CLBa, 2013). The years that followed the establishment of Keppel Shipyard saw feverish activities in the search of offshore oil, and several smaller shipyards were established to construct tugs, workboats, barges, tankers and offshore supply vessels. Rig builders went into the production of jack-ups, semi-submersibles and drill-ships.

After the 1985 recession, the petroleum industry saw a particularly critical period. There was a strong need for Singapore to move towards value-added industries and to diversify the economy. EDB identified the chemicals cluster (petroleum, petrochemicals and specialty chemicals) as an industry it would strongly pursue. Meanwhile, Jurong Town Corporation (JTC) was facing a constraint of industrial land, and thus began to reclaim land off Tuas. JTC soon realized that the relatively low cost of reclamation meant that this was an economically viable option to explore. Soon, the idea of combining the seven southern offshore islands in the Jurong archipelagos through land reclamation was substantiated.

Jurong Island was a clear illustration of how JTC built infrastructures "just in time". EDB had begun its intensive promotion effort with the top players in the international chemicals industry while the land reclamation was still in progress. Many companies made a bold commitment to Jurong Island based on EDB's and JTC's successful track record [CLB, 2013].

As a result, 20 dry docks and floating docks with a total capacity of over 1 million DWT and ship repair berths extending over 9,000 metres offered their facilities for building and repairing ships in the mid-1980s.

2.6. 1990–2000 Singapore — the fastest call

"[…] 1990 marks the beginning of an exciting decade for the maritime and trade industries. The Port of Singapore Authority has also embarked on a new programme to develop infrastructure to meet the increasing demands of the shipping and distributions communities.[…]"

(Ng Kiat Chong, Executive Director, PSA, 1990)
(*Singapore Port Services Index, 1990*).

The 1990s opened in a spectacular way: celebrating its 25th anniversary as an independent republic, Singapore became the world's busiest container port, in terms of handled TEUs, for the first time (5.22 million of handled TEUs were registered in 1990). In 1997, Singapore was the first call for the largest vessel in the world (the first post-Panamax container ship), the Regina Maersk. The same year, the preparation for the upcoming Pasir Panjang Terminal (where all the containerized cargoes were planned to be moved to) was becoming more and more feverish. The latest innovations such as the innovative remotely-operated over-head bridge cranes (OHBC) were tested. The Virtual Terminal Agreement was another fundamental innovation of the time. Thanks to this agreement, shipping lines were guaranteed a certain standard of quality service at a predetermined cost, thus enabling an easy planning of their operations.

A clear sign of the acknowledgement of the island as a key global shipping hub was the set-up of their presence in Singapore for organizations such as the Baltic International Maritime Council. Assisted by the Trade Development Board (TDB) and the Economic Development Board (EDB), the maritime community expanded enormously in this decade (*Port of Singapore, the maritime hub Series, 1999–2002*) and Singapore became an International Maritime Centre (Huff, 1997).

On 2nd February 1996, the Maritime and Port Authority (MPA) was created to take over all the regulatory functions of PSA. This was the first step towards the corporatization of PSA which became effective on 1st October 1997. PSA was renamed as PSA Corporation Limited turning PSA from a statutory board into a commercially-run organization.

The 1990s had also their challenges: massive over-capacity in the container shipping industry drove freight rates down, and the mission of the shipping lines narrowed down onto cost minimization forcing port to increase their efficiency and lower their costs to remain competitive.

Also, the biggest single crisis for Singapore as a port for many years came on the night of 15th October 1997. The VLCC Orapin Global collided with the oil tanker Evoikos, causing the Republic's worst ever oil spill (Yeo and Teo, 1997).

Facility enhancement, port services, Information Technology and employee training were the port's answers to the challenges of increasing market demand. Also, the increased port traffic and the 1997 VLCC incident required an enhancement in the Traffic Management Systems to increase the safety of the busiest strait in the world.

2.6.1. *Enhancement of facilities for service level maximization*

PSA answered to these needs with advanced technologies and improved its competitiveness with the customization of the provided services to the different shipping lines' needs. Major innovation focused on the yard technology enhancement, as well as the warehousing activities.

In 1990, a new gate was put under construction in order to cater for the increase in container traffic in Tanjong Pagar Terminal and the newly-developed container terminal in Pulau Brani. The gate had 14 lanes installed with high-tech equipment such as transponder readers, Closed Circuit Televisions (CCTVs) and the weight pad system which quickly cleared vehicular traffic.

The quay side was also among the most advanced worldwide: by 1991, all the terminal quay cranes, yard cranes and yard equipment were equipped with computers and wireless data terminals to receive operating instructions from the yard control computer. Singapore pioneered the world's first full remote control of cranes operation.

The Keppel Terminal was added to the Keppel Harbor in 1991 as a dedicated container terminal. In particular, four conventional berths in the harbor were converted into container berths. Empire Dock was closed for constructions to convert it into a stacking area of 32,000 TEU. The same year, the construction of one new container terminal started on Pulau Brani, and was officially opened in 1992.

In 1994, Keppel Terminal was one of the first yard with Rail-Mounted Gantry (RMG) cranes. This new crane system enabled stacking more containers across and higher than the RTG yard cranes. In 1996, the first 16 driverless over-head bridge cranes (OHBC) were ordered for the new Pasir Panjang Terminal. The Keppel and Tanjong Pagar terminals were also renovated with RTG cranes the same year.

The first step towards the automation of the terminal yard was announced in the PSA Annual Report in 1993. The software Automated Container Operations System (ACOS) was indeed a key step towards the introduction of automated guided vehicles (AGV). With the ACOS system, it was envisioned that the prime movers would be replaced by AGVs. In 1994, PSA awarded contracts worth $17.7 million for the testing of AGV systems.

The Anchorage Management System (AMS), implemented in 1999, was another technological breakthrough. The system, the only one of the type in the world, selected the anchorage position for ships based on the overall demand for a particular anchorage, the vessel's destination, type and physical dimensions and provides the optimal distance between anchors.

The improvements to warehousing and storage areas were the answers to the continuously increasing traffic (Table 9).

Figure 10 proves the 1990s effort in development: it can be seen that 1993 was a spectacular year in terms of investments and capital expenditures.

Table 9.   Distriparks area in 1996.

Distriparks	Area [m²]
Alexandra	212,000
Pasir Panjang	196,000
Keppel	112,000
Tanjong Pagar	65,000

Source: PSA Annual Reports (1996).

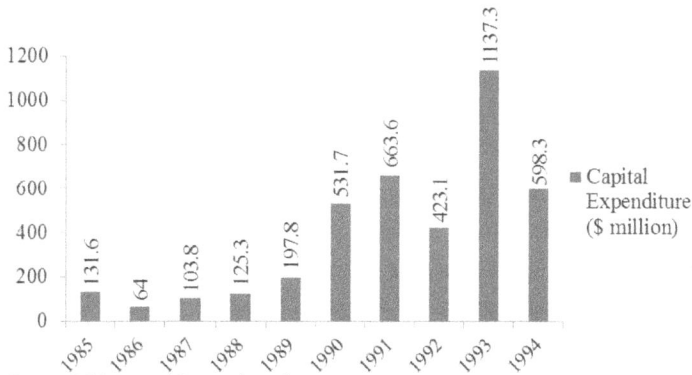

Source: PSA Annual Report (1994).

Fig. 10.   Capital expenditures for 1985–1994.

Table 10.　Main capital expenditures for 1992–1993.

Capital expenditure ($Million)	Year		
	1992	1993	% Change
Container Handling Equipment	135.5	381.7	181.7
Container Berth Facilities	125.4	542.3	332.5
Computer Systems	19.5	35.5	82.1

Source: PSA Annual Reports (1993).

As Table 10 shows, these investments were on handling, facilities as well as IT. This is reasonable as improved IT systems were increasingly important due to the increased technological complexity.

2.6.2. Optimising efficiency and enhancing safety through information technology

1990 represented a spectacular year for IT evolution at the Port of Singapore. For its effort towards IT development at the corporate level, PSA received the National Information Technology Award and became the first partner in the Strategic Corporate Partnership Programme of the National Computer Board (NCB). In 1994, PSA was the first organization outside USA to receive the Salzberg Memorial Medallion Concept Award (Syracuse University) acknowledging PSA's excellence in port operations and its efforts in pioneering ideas, practices and concepts in transportation and distribution.

IT for efficient port operations

By the end of 1990, 65% of shipping documents were electronically submitted through PORTNET, PSA's electronic link with the port's users. Shipment and delivery of cargoes were sped up through Freight Auto Service Terminals (FAST), which were installed at PSA's freight stations and warehouses. These terminals enabled port users to get faster shipment and delivery of their cargoes at the gateways.

Technology was also fundamental in the employee training programmes. PSA was acutely aware of the importance of training and, with $8.8 million in training budget, simulators were bought for training quay crane operators in 1994. This type of training placed Singapore ahead of time in the maritime industry. In fact, port authorities and port operators from all over the world started to send employees to the country for training.

CITOS underwent a remarkable development process. In particular, CITOS Phase I enabled the full automation of document processing for vehicles entering and leaving the container terminal. It also permitted to capture containers' locations in the yard in real time. In fact, its expert-system based Ship Planning System (SPS) was implemented to plan the best stowage execution for loading and unloading containers by factoring in the quay crane workload, hydrostatic stability of the ship, next port of call, and stacking operations at the yard. Due to the reduced time for shuffling container, container handling became much more efficient. CITOS was designed to send job directives via wireless connection so that the crane operators could receive, at any one time, the instructions to perform the needed yard/quay activities.

The Crane Maintenance and Monitoring System (CMMS) was another fundamental enhancement in CITOS. It was a powerful man-machine communication system that gave the operations and maintenance personnel a constant view of the operation of the crane control system. All quay crane functions were monitored on-line from a remote site. This meant that technicians need not hunt for a fault on the quay crane when it was reported by the crane operator. Instead, the system facilitated the resolution of the fault ultimately reducing the disruption to crane operations. The CMMS also provided early warnings of component failures, hence triggering the maintenance teams to rectify the problem before it caused further damage (*PSA–IT Innovations, 1997*). Another functionality of CITOS was the Gate Automation System comprising a network of CCTV allowing containers to be registered, weighed and assigned off-loading location in 45 seconds leading to a 50% reduction of the document processing time, and hence doubling the gate capacity and contributing to a further reduction of turnaround time. In the same direction, the Flow-Through Gate System (FTGS) was first implemented in 1996. The system allowed hauliers to drive through the container gate without stopping. The enabler of this application was an advanced vision system called Container Number Recognition System (CNRS). The time to flow through the gate was reduced to less than 25 seconds

"[...] a record which no other container terminal can match [...]"

(*PSA Annual Report, 1996*).

In 1995, together with a group of shipping lines, PSA re-engineered the container transshipment process. FASTCONNECT enabled shipping lines to select the earliest connecting carrier, reducing the processing of transshipment containers.

Safety as a new key element for technology development

It was apparent how safety was a challenge for the busiest port in the world. In particular, two new challenges required fast actions from the Port of Singapore: (1) the management of high port traffic volume, (2) the inspection of containers.

At 1990, Singapore was one of the few ports in the world with a radar system for tracking and managing ships in the port. CIMOS continued to be improved, aiming at the integration of the radars and expert systems to manage, monitor and control PSA's marine resources. In particular, the Vessel Traffic Information System (VTIS), a CIMOS component, constituted the ready response to the first issue. VTIS Phase I was completed in July 1990. Its radar system immediately tracked ships coming to Singapore. The system can then prevent collisions in the busy port water as a result of the monitoring of ships movements and the detection of hazardous behaviours.

As a result of the VLCC collision in 1997, STRAITREP was proposed as a mandatory ship reporting system supporting VTIS and covering the Straits of Malacca and Singapore (the agreement to start this reporting was established among Singapore, Indonesia and Malaysia). This system requires all ships to communicate their position and other information to VTIS.

CIMOS was fully implemented in 1995. Upon its completion, CIMOS improved services to vessels and enhanced navigational safety in the port. With an average of 800 vessels in port at any one time, radars in the port were able to track their real time movements which were displayed in the control room. VTIS locates and tracks in real time smaller marine and harbour craft through the innovative technology of Differential Global Positioning System. Another expert system assigns pilots and tugs to arriving ships in the shortest possible time.

The inspection of containers was also highly relevant for the sake of safety (*MPA Annual Report, 1999*) for the port with the largest traffic in terms of containers (Fig. 11). According to the MPA policy for inspection, vessels were randomly selected for control based on a specific selection matrix. If a ship was detained after inspection, the embassy or diplomatic representative of the flag administration and the International Maritime Organization (IMO) would be informed of the vessel's detention and the reason(s) for the detention.

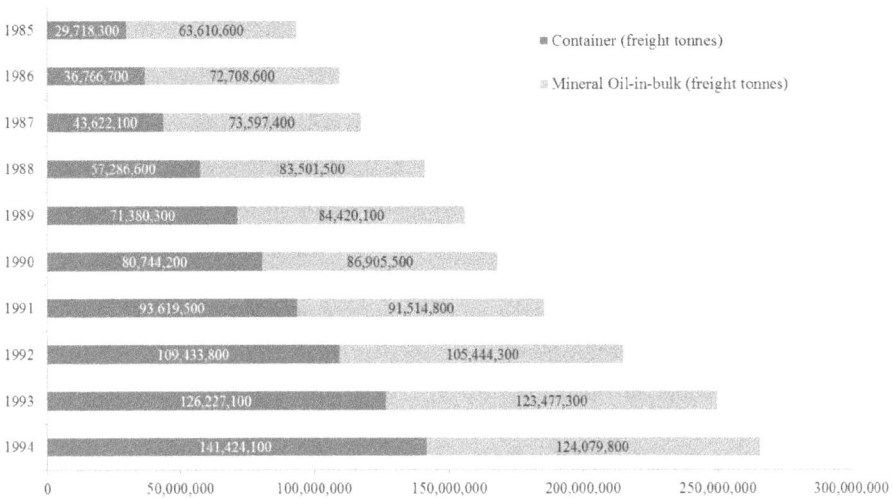

Source: PSA Annual Reports (1994).

Fig. 11.   Cargo handled by the port of Singapore.

2.6.3. *Leveraging on port services*

The most impressive statistics that Singapore can boast is the sheer number of services available at the port. Singapore maintained its key role in the ship repair and bunkering market.

In fact, substantial success was derived from the ship repair and ship re-building for which Singapore remains one of the main centres worldwide with its output of S$3.8 billion in 1998. Over the years, Singapore's ship repair industry grew in tandem with the port as vessel traffic increased. The industry was supported by the government to drive the industrialization of Singapore. The knowledge, skills and technology from ship repair activities enabled the island to expand into areas like ship conversion, ship building and rig building. The success in this business was not only based on the competitive pricing, but even more so, on the shipyard's track record for quality service, timely delivery and ability to handle sophisticated and strategic projects. From the traditional market of the class VLCC Tankers, Singapore proposed its services to more complicated ships such as Floating Production Storage and Offloading (FPSO), offshore platforms and passenger ships.

Singapore was also confirmed as the top bunkering centre in the world, supplying some 11 million tons of bunker fuel in 1990. In August 1992,

Singapore was the first country to formally establish a Bunkering Procedure (CP:60) and the Government established the bunker quality certification standard ISO:8217. Singapore's dominance in the bunker market was guaranteed by the combination of its massive maritime trade and its role as a top-ranking global refining and cargo centre. In the refining sector, Shell Eastern Petroleum was by far the largest player, basing its position on its 450,000 barrels/day at the Pulau Bukom refinery (*Port of Singapore, the maritime hub. Series, 1999–2002*).

As a relevant aspect of this decade, insurance, as well as complex financing arrangements, became one of the cores in the Singapore service portfolio. A clear demonstration of this novel entry into the port-related services was the number of companies in financing and brokerage industry establishing office in Singapore. Some examples are SSY, Clarkson, Lorentzen & Stemoco and RS Platou, as well as maritime lawyers such as Ince & Co, Cliffors Chance, Sinclaire Roche and Temperley and Homan Fenwick & Willan.

The increasing demand in these ancillary services gave the chance to Singapore to attract shipping banks such as Nedship Bank, Mees Pierson, ANZ Bank, Hong Kong and Shanghai Bank and Bank of Nova Scotia to either set up base in the Republic or expand their presence here in this region.

2.7. *2000s New millennium challenges: Competition, safety & sustainability*

The 2000s represented a particularly critical period as the competition in the Asian territory became increasingly intense. Nowadays, due to both regional as well as global trends, regions and countries are all ultimately competing in one market. The clear consequence of this is the increased competition in international trade.

The growth of China played a key role: several ports started to operate under the "hub-and-spoke" paradigm and several transshipment-dedicated ports were created as well (Fig. 12).

Figures 12(a) and 12(b) confirm how transshipment was the fastest growing segment of the containers port market. It also supports the fact that during the mid-2000s, several pure transshipment hubs at strategic locations were put in operation. In fact, Asia saw the consolidation of major ports as well as the growth of new competing hubs (Song, 2003; Yap, and Lam, 2006a; Yap *et al.*, 2006b; Yap *et al.*, 2014 and Yap, 2009). Another reason for the growth of larger ports resided in the changed policies of shipping lines that, in order to

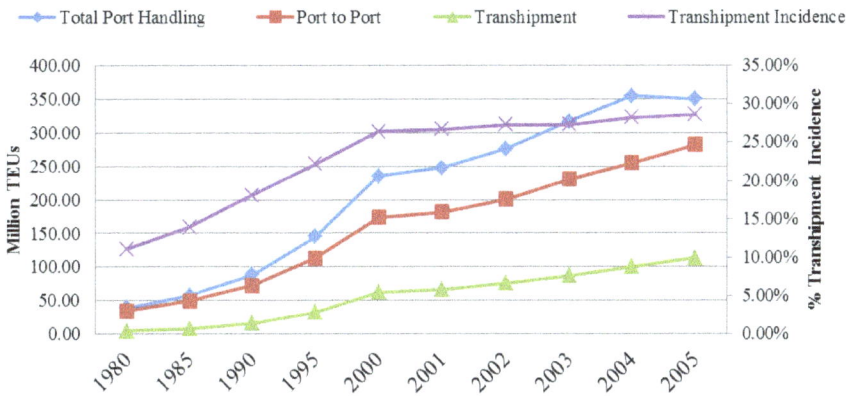

(a) Estimated world container traffic and transshipment incidence, 1980–2005.

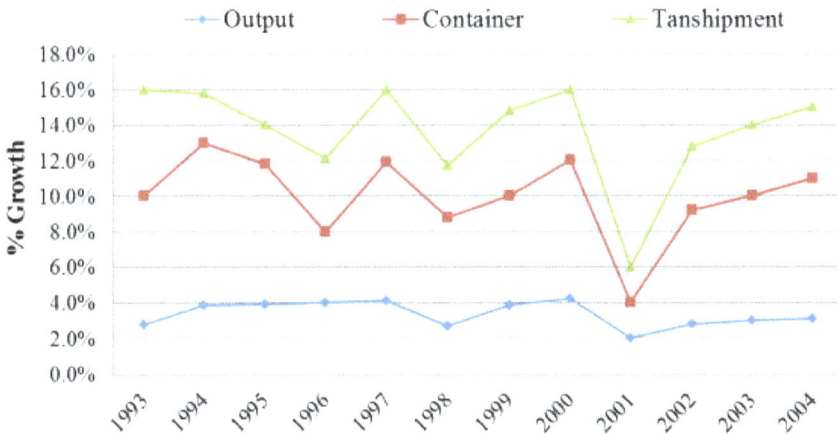

(b) World economic, container and transshipment growth for 1991–2004.

Fig. 12.   World economy and transshipment growth (from Wang (2007)).

minimize costs, increased the ship size and, to maximize yield, were keen to include fewer ports in their service routes and transship containers to other vessels for the regional transportation.

Already in the early 1980s (*UNCTAD, 1980*) the amount of transshipment at the Singapore port accounted for more than 40% of its throughput, when the global percentage of transshipped cargo was around 10%. Currently more than 80% of the Singapore port's handled containers are for transshipment [Yap, 2014] resulting in an almost pure transshipment port, which effectively serves a hinterland including the whole of Southeast Asia and even parts of the Indian subcontinent and Australasia (*Cullinane et al., 2006*). To keep its role as a leader in transshipment, Singapore needs to provide capacity and lower port times: application of information-communication technology and consolidation of activities to the larger Pasir Panjang Terminal represent a clear sign of its resolve to keep the pole positions.

> "[...] To compete effectively, we have to become a complete maritime cluster by adding our core group of ship owners and operators, maritime support services such as maritime finance, insurance and legal services [...]".
>
> (Lee Kuan Yew, Minister Mentor)
> (*Inaugural Singapore Maritime Lecture, 2007*).

Another key event in this decade was undoubtedly the World Trade Center terrorist attack on 11 September 2001. The safety concerns, which started to be one of the main points in the port's agenda in the past decade, became top priority.

The concept of sustainability was also one of the main driving factors for engineering and IT development. However, not only economic sustainability was of concern, but environment protection was now the key requirement from the public bodies.

In this section, with the usual focus on engineering and technology, we will review how the Singapore port responded to these new challenges.

2.7.1. *Responding to competition*

Pasir panjang — the terminal of the future

The increase in competition required Singapore to maintain excellent performance levels in order to protect its position as the best transshipment port despite its port being located in a particularly expensive area in the heart of the city. In fact, the majority of hubs being constructed in the same period were located well away from high-cost land, in order to be able to charge their customers, the shipping lines the lowest possible price.

Singapore had to bank on higher performance: the increase in automation level, qualification of the operators, and the intense use of Information Technology were the key elements of Singapore's reaction to the fierce price-based competition. The following factors were of fundamental interest and guided the development of Pasir Panjang Terminal whose Phase II ended in 2007 (Chu and Huang 2005): (1) The handling system; (2) Crane dimensions and stacking height of cranes; (3) Container yard size; (4) Average dwell time of containers and transshipment ratio; and (5) Characteristics of terminal operator in expediting the planning procedure of the terminal. The consideration of these aspects led to the development of one of the most advanced container terminals in the world pioneering the Remote Yard Crane control, making a fundamental step towards the minimization of the human presence in the terminal yard (*MPA Annual Report, 2001*).

In 2008, Singapore was the first off the block on the world track in providing mobile WiMAX within its port at Pasir Panjang. The Wireless-broadband access for SEaPORT (WISEPORT) network provided wireless broadband access to maritime users along the coast and at sea. Crew onboard ships and harbour craft, and users on offshore islands could ride on this information communication infrastructure to access the Internet, online value-added applications, and for voice communications.

The year 2009 was momentous for Pasir Panjang. Having started its operations in 1974, the port reached its peak, handling 18 million tonnes of cargo. In 2013, container throughput handled by the Port of Singapore hit 32.6 million TEUs, a 2.9% increase over the 31.6 million TEUs in 2012, according to the Maritime and Port Authority of Singapore (MPA). Total cargo tonnage handled increased by 3.6% in 2012 to reach 560.9 million tonnes.

Enhancement of information technology and investment in research & development

> "[…] For the world's busiest transshipment hub, technology is vital in ensuring that its terminal customers are served seamlessly, efficiently and effectively round the clock. Through this Programme, MPA hopes that Singapore will continue to be at the forefront of technology in port operations and services adding to its development and growth as an international maritime centre […]"

> (MPA Chief Executive Lam Yi Young) (*MPA Annual Report, 2011*).

The MPA kept the philosophy of PSA, understanding that the main success key of Singapore is

> "[…] to anticipate change, be adaptable, and constantly innovate to stay ahead in the game […]"

> (Peter Ong, Chairman MPA) (*MPA Annual Report, 2007*).

> "[…] We have to remember that the rate of acceptance of technology developments depends to some extent upon the intellectual development needed to sustain them. […]"

> (*Leech, 2000*).

It was in the direction of promoting technologies that the Integrated Simulation Center was created in 2002, the Foundation of Maritime Innovation and Technology was founded in 2003, and the Centre For Maritime Studies was started within National University of Singapore in 2005. In the PSA Annual Report of 2006, the following simulators were listed:

1) Full mission ship handling simulator
2) TUG/Small vessel simulator
3) Vessel Traffic System simulator (VTS)
4) Electronic Chart Display Information System simulator
5) Crisis Management simulator

Table 11 reports the number of assessment and courses in simulation for the period 2002–2006. We can notice a substantial increase in the use of these advanced tools as a training means for Singapore port operators.

A further clear sign of the effort invested by Singapore in Research and Development, was the Memorandum of Understanding (MOU) signed in March 2000 with the National Research Council in Norway, establishing a framework to encourage co-operation in maritime research, development, education and training (Sorensen, 2000).

The service industry: Delocalization, government cooperation and e-services

Competition was not only affecting the container market. The ship repair industry also had to face the competition coming from countries with lower labour

Table 11. Number of simulation assessments and simulation courses for 2002–2006.

	2002	2003	2004	2005	2006
	Number of assessments				
Simulator	Number of courses				
Full mission ship handling	172	278	214	205	323
	139	394	363	414	430
TUG/Small vessel	380	619	653	629	649
	352	1096	884	1083	907
Vessel Traffic System	6	11	36	24	—
Electronic Chart Display Information System	—	—	58	43	170
Crisis Management	83	33	55	32	15

Source: MPA Annual Report (2006).

costs. Keppel Shipyard and Sembawang Shipyard, beside maintaining a strong presence in the traditional niche markets of tankers (VLCC), were renowned for the repair, upgrade and conversion of high value-added sophisticated ships such as FPSO (Floating Production Storage and offloading). In addition, cruise ship conversion and refurbishment quickly gained importance in the local ship repair industry (*Port of Singapore, the maritime hub. Series, 1999–2002*). In order to specifically address the competition matter, since the early 1990s, major Singapore shipyards have strategically expanded their repair activities and expertise into the region in order to leverage on cheaper land and labor resources. In parallel, shipyards collaborated with EDB and National Science & Technology Board (NSTB), the Association of Singapore Marine Industries (ASMI) and the Ministry of Manpower (MOM), for the implementation of training and skill upgrading programmes for workers and supervisors in order to seek more cost effective solutions (*Port of Singapore, the maritime hub. Series, 1999–2002*). In fact, until today, technology and training remain as the fundamental aspects in Singapore's development.

Singapore continues to shape the business environment in the bunkering industry. The SSCP60 bunkering standard was always applied rigorously. The introduction of new business strategies also made Singapore the best bunkering port in the world. An example was the Special Bunkering Anchorage (SBA) scheme, which allowed qualified vessels to enjoy port dues concession of over 40%, as one of the incentives for ships to lift bunker in Singapore's port. Its popularity with ship owners and bunker suppliers meant that most of the SBA slots were taken up every day. Hence, the MPA added four new SBA slots in February 2006 so that more vessels could benefit from it. The energy market in Singapore changed during the 2000s. In fact, Singapore's role in Asia's oil and petrochemicals trade can be divided into four phases: (1) the 1960s and 1970s were the two decades for building the production facilities and infrastructures; (2) in the 1980s, it consolidated its role as Asia's leading supplier and trader; (3) in the decade leading to the 2000s, this role picked out, but Singapore succeeded in adding a new and important status as the centre for the region's paper oil trade. (4) Singapore is nowadays the region's e-commerce hub for energy trade. This role is of strategic relevance for Singapore as more Asian companies hedge their physical trades here. The MPA developed BUNKERNET, a network linking all parties within the bunker supply chain for the exchange of information and to automate and streamline processes of bunker purchasing and delivery. The application enabled oil majors, tank storage operators, bunker traders, surveying companies, shipping lines, bunker vessel operators and testing labs to interface seamlessly. To develop this service, several platforms were made available to the MPA stakeholders to engage in dialogue and provide feedback. Examples of such platforms included the Bunker Quality Advisory Panel providing a high-standard, systematic procedure for sample testing, review of test results and provision of technical advice to the shipping community in cases of off-specification fuel delivery. Another example was the Bunker Working Group evaluating suggestions and feedbacks from the bunker industry in order to identify and propose measures to further enhance bunker quality and quantity in the Port of Singapore. Finally, the Bunkering Forum was initiated as a series of MPA-sponsored forums organized on a regular basis offering the opportunity for industry-wide networking (*from MPA, www.mpa.gov.sg*).

On 1 January 2001, Jurong Port was corporatized and became a fully owned subsidiary of JTC. This decade saw the development of Jurong Island as a crowning achievement of a success story of the Singapore port in the petrochemical scene. JTC initiated a process which would eventually lead to a world-class petrochemical hub with 150 oil companies and an investment value of $40 billion creating 15,000 jobs for Singaporeans. The harbour covered 800,000 square metres of land, directly linked to the chemical plants through common pipeline corridors in order to avoid traditional isotank conveyance, reducing logistic costs, time and human effort. Following the Singapore port development strategy of providing multiple services, the Jurong Harbour handled bulk liquid and solid chemicals, including hazardous chemicals. It provided logistics services such as storage tanks, chemical warehouses, tank filling, cleaning and maintenance, drumming and waste treatment facilities. IT was a main guideline also for this petrochemical hub: before starting the constructions, a broadband optical fiber was installed on the island. A platform for the integrated management of the port services was developed in a form similar to PORTNET, guaranteeing an Internet-based service to efficiently connect the different stakeholders in the supply chain of chemicals (Kim, 2000).

2.7.2. *The security challenge*

The 2001 terrorist attacks in New York, USA, had a tremendous impact all over the world. The research on technologies for increasing container control without slowing down port activities became primary. Cargo x-rays, Automatic Identification Systems (AIS), covert alarms, electronic cargo seals, crane-mounted radiation detectors and supply chain software were just some of the wide range of technologies being utilized and developed in its efforts to beef up maritime security (*Abelman, 2002*).

The safety issue was exacerbated by the Singapore traffic conditions. In 2004, the VTIS system was improved as a consequence to the connection to the Automatic Identification System. The AIS enabled the automatic identification, tracking and transmission of data between ships and coastal authorities as well as between ships. It was mandated by the International Maritime Organization's Safety of Life at Sea (SOLAS) Convention.

In 2007, Singapore established a momentous milestone being the first in the world to implement a Harbourcraft Transponder System (HARTS). HARTS is a transponder tracking system for harbour and pleasure craft plying within the island's port waters. Around 750 crafts were tracked daily by the system. With 98% of all vessel traffic monitored automatically by HARTS, attention could be focused on the remaining 2%, which comprised mainly small coastal vessels not subjected to SOLAS regulations (*Port View, 2007*).

Also engineering solutions were fundamental to improve control over the strait's traffic. In the 2000s, the introduction of satellite-based positioning system and multi-beam sounding system were fundamental for hydrographic survey, enabling the precise evaluation of vessel's draft, essential information for the port operators when assigning the berth to the incoming ship. Forward-looking echo-sounders were another fundamental technology offering increased safety to survey vessels in reef areas. Geographical Information System (GIS) was the main reference platform to access vessel data and a strong effort was devoted in this decade towards fostering the use and enhancing of GIS data services (*Port of Singapore, the maritime hub. Series, 1999–2002*).

The Information Sharing Centre (ISC) for the Regional Cooperation Agreement on Combating Piracy and Armed Robbery Against Ships in Asia (ReCAAP) was officially launched in Singapore on 29 November 2006 as an international organization.

2.7.3. *Towards a sustainable port*

From 2010, the environment was a key consideration in port development projects. MPA worked on the implementation of a proactive feedback Environmental Monitoring and Management Plan (EMMP) for the development of Pasir Panjang Terminal (PPT) Phases III and IV. The feedback EMMP operations were carried out on a daily basis making use of various tools and resources. Such innovative use of continuous feedback of information within the EMMP provided MPA with a responsive and reliable system that allowed for unexpected impacts to be mitigated prior to them becoming significant, and allowing the development of PPT Phases 3 and 4 to proceed despite its close proximity to sensitive marine habitats and facilities.

Marine pollution was another threat in which Singapore, as the main bunkering port in the world, led the research in increasing responsiveness and lowering the risk of maritime accidents, as well as developing technologies to reduce oil spills and imrpove ballast water management. IMO adopted a quality standard for ships to install a Ballast Water Management System to treat ballast water before discharge in 2004. Singapore again proved to be ahead of its time and, while the standard was still on the way to be implemented, the National University of Singapore developed a Blue Seas Ballast Water Management System to disinfect ballast water, giving proof of the country's prime engineering efforts towards the concerns of marine pollution.

The Maritime Singapore Green Initiative marked the year 2011: a comprehensive initiative to promote environmental-friendly shipping, launched by MPA in April 2011. The Initiative comprised three programmes: (1) Green Ship Programme, (2) Green Port Programme, and (3) Green Technology Programme. MPA will invest up to $100 million over a period of five years in the Maritime Singapore Green Initiative. Under the Green Ship Programme, MPA will provide incentives to ship owners who adopt energy-efficient ship designs that reduce fuel consumption and carbon dioxide emissions. Singapore-flagged ships which go beyond the requirements of IMO's Energy Efficiency Design Index will enjoy a 50% reduction of Initial Registration Fees (IRF) and a 20% rebate on Annual Tonnage Tax payable. Ship owners will also be recognized through certificates and a new "SRS Green Ship of the Year" award. The Green Port Programme aims to reduce the emission of pollutants like sulphur oxides and nitrogen oxides by ocean-going ships calling at the Port of Singapore. Ships that use type-approved abatement/scrubber technology or clean fuels with sulphur content of less than 1% within Singapore port limits can enjoy a 15% reduction on port dues payable. Finally, the Green Technology Programme aims to encourage local maritime companies to develop and adopt green technologies through co-funding of up to half of qualifying costs.

2.8. Summary remarks

Figure 13 proposes a brief overview of the main milestones in the history of the Port of Singapore with respect to the main historical events in the maritime industry in other major countries (in terms of sea trade) in the world in the period between 1956 to present. When Singapore welcomes her first container ship, Felixstowe, which was the first dedicated container port worldwide, it had been in operation for only five years, despite the demand for containers in Europe and USA that started way before it did in Asia. Singapore proved an incredible sense of timing, as after the mid-1970s, the number of container

April 1st: PSA
August 9th
1965:
Singapore
Independence

NOL

June 23rd:
1st
containership

$1,500M for
Port
expansion

TRADENET

Singapore 1st
CITOS introduced

PPT ph1

1956   1964   1967   1968   1972   1980   1981   1985   1986   1990   1993

April 26: First Call
for the Container
IdealX in NYNJ

Global Recession

HK 1st

Japanese
Asset price
crisis and
Gulf War

First Dedicated Container Port
Felixstowe (GB).
ASEAN union is established

Thamesport

Service Hub   Independence   Containerization TP & IT Leader
               & Expansion

(a) Period 1956–1993

PSA corporatized
Flow Through Gate System

Singapore 1st call for
Emma Maersk

Singapore is
the 1st call
for Regina
Maersk
MPA

Port Net
Remote Crane
Operations and Control
(RCOC) introduced in
PPT

Pasir Panjang
Phase II fully
operational

PSA and SMI
launch NGCP

WISEPORT

1996   1997   2000   2002   2006   2008   2010   2011   2012   2014

ECT
1st Post
Panamax:
Regina Maersk

ASIA
Financial Crisis

ASEAN –
China FTA

Global
Recession

Triple-E
by Maersk

China publishes the
arctic shipping routes

1st Super Post Panamax EMMA
Maersk 15,000 TEU

Shanghai is the Busiest
Port in the World

Singapore
Fastets Call

Pasir Panjang and the consolidation Plan

(b) Period 1996–2014

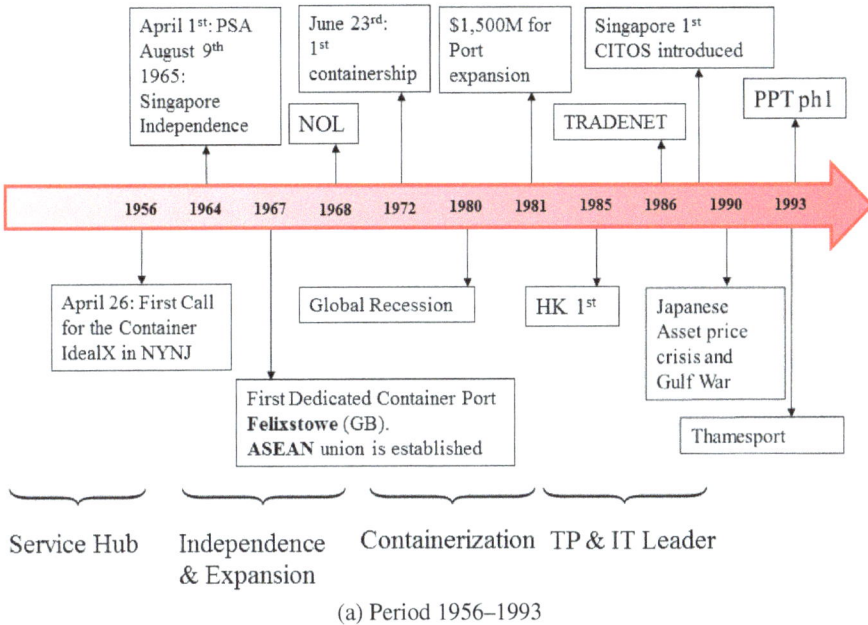

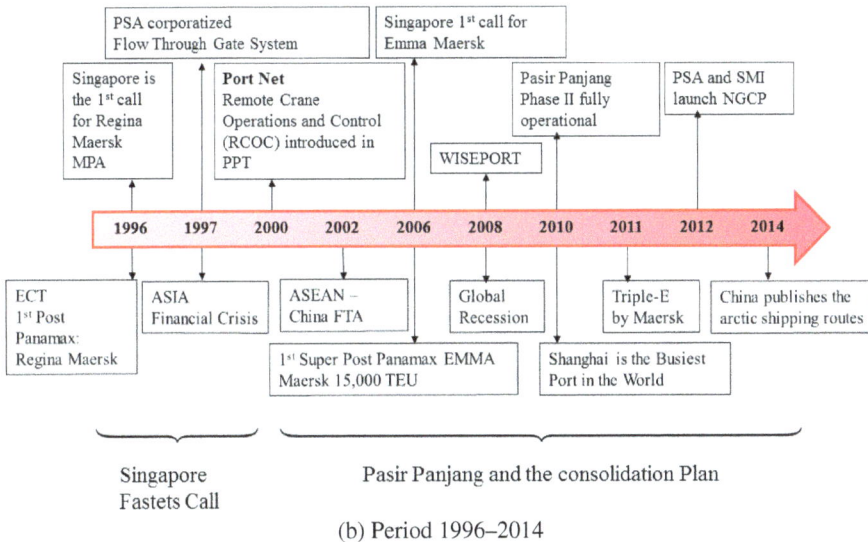

Fig. 13.   Main Time line for maritime events in the world and Singapore for 1956–2014.

terminals in the world grew exponentially, putting Singapore in a particularly favourable situation of being among the few providing efficient container service.

Hong Kong soon became be the historical rival of Singapore: the container terminal was developed a few months before Tanjong Pagar, and the competition between the two ports might partially justify the PSA investment in 1981 for port development in terms of expansion and automation, despite the global crisis (Section 2.5). Nevertheless, Hong Kong became the busiest container port in the world in 1985, surpassing Rotterdam. In this period, Singapore pioneered the use of IT, putting a lot of effort in the realization of CITOS and TRADENET. While IT developed and Pasir Panjang Terminal started its Phase I, in Europe the automation started to be the priority for port authorities which started to develop Automated Container Terminals (ACTs) in the early 1990s. In 1993, the innovators at Thamesport in London, UK, began automating its terminal operations, a development which was quickly followed by the world's first fully automated terminal, the European Combined Terminals (ECT) at the Port of Rotterdam, followed by the Container Terminal Altenwerder (CTA) in the Port of Hamburg. Back when Thamesport began operations, the technology available to the design team was very different to that which is available now, and the plan required in-house development of a whole range of new, purpose-built technologies to integrate the automated operation.

PSA was not unaware of these developments and thus, one of the largest tests for AGV installation was promoted in 1994. Nonetheless, the technology was apparently still not suitable for the Singapore infrastructures and performance requirements. In fact, it has been the choice of the most suitable technologies which led the Port of Singapore to gain the position of "busiest container port" in 1996 (Section 2.6).

The subsequent decade was strongly innovation oriented in the equipment, exploitation and exploration of new efficient IT solutions. In the meantime, China grew spectacularly despite the crisis affecting the world economy in the 21st century. Despite the efforts of PSA, the growth of the Chinese ports was such that in 2010, Shanghai captured the position as the busiest port.

The influence of the Singapore port on the local economy (as we will see in Section 2.8.1.), as well as its global impact (as we will see in Section 2.8.2.), should not be underestimated. The maritime industry is fundamental in contributing to the overall Singapore economy. Port-related businesses contributed about 7% of Singapore's annual GDP and employed over 170,000 people. Singapore is home to more than 5,000 maritime companies including over 100 international shipping groups (*Maritime Gallery, Singapore*).

2.8.1. *Singapore port local impact*

PSA has been showing a great capacity to capture the trade trend over the years, avoiding disruption for the incumbent changes (e.g., containerization, IT).

Already in 1973, PSA was estimated to generate about 7% of Singapore's GDP (*from Chow Kit Boey. "A Report on the Survey of Shipping Activities". 1975*).

The effort exerted by the Singapore government to the port's development highly reflected the impact of the port on the local economy. Table 12 shows a comparison of the distribution of major industry groups for the years 1960 and 1976: the number of establishments increased by four times, the number of workers increased by six times, output increased by 10 times and value-added increased by 22 times. In 1960, the most important industry group was the manufacture of rubber products and rubber processing which registered 73% of total industrial output and it was also the most important for value added and number of workers. In 1976, however, the most important was the chemical and petroleum industry which accounted for 41% of the total throughput and 23% of total value added. It is not surprising that, while in the late 1950s, tin and rubber were the main products to be traded, in the late 1970s Singapore became one of the major bunkering centres. The chemical and petroleum products replaced the rubber products and processing industry as the most important, and the latter became relatively insignificant (*Chew and Lee, 1991*). The industry showing the fastest growth between 1960 and 1976 according to the two factors of number of establishments and output, was the transport equipment group (mostly ship building and repair). The electrical machinery industry ranked first for employment generation and second for growth in the number of establishments and value added.

Over the years, PSA and, subsequently, MPA have attracted many international ship owners and operators to set up operation in Singapore. This has helped boost efforts to grow a wide range of services, including shipbrokers and charterers, marine insurers, maritime lawyers and ship financiers.

Today, Singapore is the world leader in the building of jack-up drilling rigs, and it has more than two-thirds of the global market share of the business. It is a major builder of semi-submersible rigs and also among the world's top centres for ship repair and ship conversion. Singapore is also a global leader in the conversion of Floating Production Storage and Offloading (FPSO) vessels. It is a niche player in the construction of customized and specialized vessels, and hosts leading marine equipment manufacturers and classification societies from all over the world.

2.8.2. *Singapore port global influence*

The influence of the Port of Singapore is not confined to the small island: as a prominent marine centre, Singapore has a global influence. Indeed, at the global level, all this expertise makes Singapore particularly renowned in the port

Table 12. Percentage distribution of major industry groups by number of establishments, number of workers, output and value added for 1960 and 1976.

Industry group	1960				1976			
	Establishments	Workers	Output	Value added	Establishments	Workers	Output	Value added
Food Beverage	21.3	19.1	9.2	23.8	10.2	5.9	7.7	6.5
Paper, Printing & Publishing	18.5	13.7	2.9	14.3	11.9	6.1	2.5	4.6
Chemical & Petroleum Products	5.8	4.4	3.8	5.2	4.8	3.8	41.1	22.5
Rubber products and processing	6.6	19.3	73.0	25.0	1.8	2.0	5.4	1.8
Fabricated Products and Machinery	7.3	4.4	1.0	3.5	19.1	12.9	6.7	11.5
Electrical Machinery	2.3	3.8	1.0	4.3	6.9	22.4	13.1	7.1
Transport Equipment	6.8	6.9	1.9	4.5	7.3	13.2	8.4	15.7

Source: Report on the census of industrial production, 1960 and 1976.

community and officials from other port authorities often join training classes provided by PSA Institute. An example is 12 officials from Tanjong Priok Port Authority Indonesia who completed a training by PSA in 1976 (PSA BERITA, Vol. 10-11, 1977). In 1977, visitors from the Port of Longview (located on the north bank of California River, ranked among the busiest port in USA at that time) commented in this way on their visit:

> "[...] At this time, the Port of Longview is making a sincere attempt to increase its share in the Transpacific Traffic and is preparing a comprehensive port facilities development programme. We are impressed with the facilities and operations here. We believe our programme should be compatible to the port of Singapore as she is considered one of the best run ports in the world [...]".

> (Mr. R.E. Mc Nannay, Manager Port of Longview)
> (*PSA BERITA, Vol. 10, 1977*).

A further example of the island's global impact dates back to the 1980s when Singapore's Ambassador-at-large, Professor Tommy Koh, was elected president of the 3rd United Nations Conference on the Law of the Sea (UNCLOS) from 1981 to 1982. During his term of office, UNCLOS was adopted. Besides being a council member of the International Marine Organization (IMO) since 1993, Singapore has held various leadership positions including Chairman of the IMP Council and vice-chairman of two main IMO committees — the Marine Environment Protection Committee and the Maritime Safety Committee. Outside IMO, MPA also actively participates and has held key positions in several international maritime bodies.

Singapore's port development has been influenced by global port cities as well. European ports constituted the reference for Singapore, because of their intense investment in new port technologies. The PSA Zone Manager V. Thirupathy, when asked about his impressions on a visit to the major European ports, answered:

> "[...] There are many things we can learn from these advanced ports, like their high productivity, systems of work and technological innovation in cargo handling [...]"

> (*PSA BERITA, 1977*).

Besides the European ports, the port of Hong Kong had a predominant position in the 1970s in the containerized trade. In 1976, Hong Kong ranked as the fourth in the list of top ports in the world for container traffic after New York, Rotterdam and Kobe. At that time, Singapore was 21st on the list. PSA understood the potential of Hong Kong operations and handling. The strong policy of training and investment in equipment renewal would eventually lead Singapore to be the top port in 1982 for the first time.

Currently, Singapore is leading in bunkering, petrochemicals and sustainability areas due to the efforts of MPA and PSA in terms of investments, hence guaranteeing the leadership position to the island nation despite competition and challenges.

3. Sailing Into Tomorrow: Future Challenges in the Marine Industry

"We are limited in terms of land, but I think we are not limited in terms of imagination and creativity."

(Grace Fu, Senior Minister of State for National Development
and Education, February 2010).

Statistics clearly show how the main ports in the world are seeking to keep their positions in the maritime market through intensive investments in innovation to increased efficiency in ship handling a key element for attracting shipping lines. In other words, we can recognize a "global trend" in port development as well as a regional trend.

Stimulated by the removal of national borders and by the ever-increasing interaction among different regions, globalization has remarkably changed the role traditionally played by ports as centre of transport activities. Containerization and intermodal transportation have also significantly contributed to this change. As a result, the world gateways, i.e., the nodal points through which intercontinental containers are transshipped onto continental axes (*Lee et al., 2008*), could become hub port cities (*Fleming and Hayuth, 1994; Slack, 1993*). Nonetheless, specificities of the different regions cannot be neglected. Inland transportation is of central concern in both European and American ports, but the same is not observed in Asia. Furthermore, Asian ports show a limited inland penetration. On the other hand, while America and Asia share the characteristic of coastal concentration of cities, the same cannot be said for Europe which shows inland centrality of big cities.

Given the fierce competition, it is clear how Singapore port has to maintain the pace and take the necessary actions to control and increase its competitiveness (Chew *et al.*, 2011). The container trade is foreseen to keep on increasing and the land scarcity has become a major problem for Singapore and keeping its competitiveness requires always increasing effort, as we will see in the next section.

3.1. *Future challenges*

We can recognize five main challenges posed to Singapore as a major trans-shipment hub in order to face the competition and keep its leading position: (1) Increase capacity and productivity, (2) Optimizing land use, (3) Sustainability, (4) Flexibility, (5) Creation of an integrated logistics hub and value-added services provider (Tan *et al.*, 2014; Yim, 2014).

Increase capacity and productivity

The challenge is to cater to higher volumes of cargo and vessel frequency without compromising the port's schedule reliability and the functioning of other port activities. In fact, the main repercussion of high traffic will be increased and possibly highly variable demand for space and handling operation leading to potential congestion peaks. This reflects not only a need in physical expansion or intensive use of IT, but also the need to maximize labour productivity.

Optimizing land use

Singapore, like many other main ports in the world, has gone through terminal restructuring, expansion and development of capacity in new locations (e.g., Tuas). Nonetheless, gradual enlargement of the city centre leads container ports to compete increasingly with activities such as commercial and high-end residential real estate developments which are able to generate higher returns (e.g., through rentals) per square foot compared to container handling activities. As shown in *Notteboom and Rodrigue (2005)*, port activities are gradually relocated to the peripheral areas of the city as urban sprawl takes place. In some cases, the new location of the container port may be beyond the statutory confines of the city as seen in the case of Shanghai with the new offshore development at Yangshan.

Sustainability

Sustainability is becoming a major challenge for industrialized, as well as developing, countries. As a result, the port development has to consider new constraints and new technologies that need to be developed to favour the abatement of carbon dioxide emissions. This aspect is apparently related to the optimization of land use: the issue of conflict over use of land for port development and urban development is a topic of interest for many research communities.

De Langen (2007) analyzes the environmental protection aspect. The need to minimize the negative externalities generated by the port activities such as noise, air and water pollution is always more predominant. At the present moment, with a common stake and interest in sustainable development, container ports are engaging environmental groups on an increasingly frequent basis in formulating development strategies.

Sustainability has also an additional interpretation which is not related to environmental considerations, that is, the capacity to respond to the changes smoothly in terms of cost and performance. Singapore is currently facing two major events that could change the trade profile: China is building alternatives to oil shipping via the Straits in the form of pipelines, and are continuing to expand its refining capacity and direct shipping option in the intra-Asia market offers options to bypass Singapore entirely. Likewise, Chinese mega ports are offering shipping lines good incentives to do direct services, being able to fill their large container vessels by calling just a few Chinese ports. The arctic route, proposed in 2014, is an example of faster routes being offered.

Flexibility

The fourth challenge is the need to manage a diverse range of containerized cargoes. Shipments mainly differ in customers: shippers determine the goods to be transported, seasonality and direction of the trade. In their effort to maximize their profitability, shipping lines gravitate towards ports with large traffic volumes that go direct from origin to destination (*Yap et al., 2006b*).

Another aspect in flexibility is related to the diversity in the geographical reach, such as the continental area of origin and destination for cargo traffic flowing through the port. In this merit, the expansion of the reach is reflected in the share of transshipped containers handled.

A further aspect is related to the ability to serve always diverse vessels in size and service type. Flexibility leads to the challenge for the ports of facilitating interchange of containerized cargoes for a greater variety of customers from increasingly diverse sources.

Integrated logistics hub & value added services provider

In today's market, the commercial success of a port does not only rely on the advantage in the capability of its "primary services" such as cargo handling and operations. Eventually, shippers and carriers will make their choice based on the benefits obtained from the services that the port can deliver. Providing diversi-fied services has always been one of the major targets for the Singapore port. Nevertheless, in order to afford the always increasing requests, a new business model should be adopted. The whole port, from a primary maritime centre, must become an integrated logistics hub.

Currently, Singapore's Changi Airport is one of Asia's largest cargo airports and is served by over 6,500 weekly flights connecting to 250 cities in 60 coun-tries, handling close to 2 million tonnes of cargo annually. The island is also close to the world's major markets, being situated within a seven-hour flight radius to half of the world's population in the Asia-Pacific.

The port connects Singapore by 200 shipping lines to 600 ports in 123 countries, with daily sailings to every major port of call in the world. The port will then be the centre of this integrated logistics hub. The relevance of a strong logistics connection is even more apparent when the consolidation to Tuas is considered. The connection to the urban network will require new developments in the city infrastructure to avoid congestion. As a result, the city infrastructure, the sea-port and the air-port will constitute a unique efficient hub in the city of the future with maximum concern for the abatement of the emissions in order to guarantee the city sustainability. When the Tuas project is completed in 2027, the port land at Tanjong Pagar, Keppel and Pulau Brani will become a new waterfront district. With a size comparable to Marina Bay, the area would allow the business district to grow and accommodate homes, hotels and other recreational facilities. The Singapore urban background will be vastly changed and the urban logistics needs to be planned. From 2027, Singapore will not only have one of the biggest port in the world at Tuas; in the event that the Central Business District area is expanded to Tanjong Pagar, Keppel and Brani, Singapore will also have the largest business district in the world.

In addition to the basic services such as massive warehouses for cargo transshipment and distribution operations directly connected to various mul-timodal transport facilities for transit shipment, value-added services will be a key competitive factor for the new port. One of the clear evidences of the effort of the government in making Singapore a leader in this direction was the initia-tive for free port, 30,000 square metres of storage such as showrooms, photo studios and wine cellars to facilitate the storage and movement of delicate, exclusive cargo.

The significance of value-added services in logistics centres has been widely recognized by both logistics companies and shippers as it specifically emphasizes on the development of the customized logistics products, and thus

Table 13.   Value added services for industry.

Petrochemical	Cold products	Healthcare	General value-added cargo services
Blending Heating Chilling Weighing Trucking Circulation	Tracking and detailed temperature information; Customized packaging solutions	Warehousing, Distribution, Inventory management and control, and an express lane for urgent products	Receiving cargoes, Breaking shipments, Preparing for shipment

plays a very important role in supply chain management. Singapore as a logistics hub will have to provide not only traditional activities but also value-added services mainly due to the high degree of global production and the demands from customers (see Table 13 for examples by industrial sector). The logistics hub will be established as a free trade zone to consolidate and centralize the logistics management where goods can be stored without any customs duties or excise tax charged within prolonged periods, and goods may also be broken up, repacked, assembled, distributed, mixed with other goods or labeled by the logistics operator.

Remarks

All these challenges deeply affect the Port of Singapore and the related growth strategies needed to consider several aspects. The good news for Singapore is that there seems to be no economically feasible alternative to a big-scale replacement of the Straits for shipping between east and west. Compared to local ports with a smaller market, Singapore has a broader demand base which should give less volatility in terms of the markets where the cargo comes from. Furthermore, oil demand in East Asia (in particular China) is estimated by several sources to continue increasing, and the supply is expected to increasingly come from the Middle East and to some extent, Africa.

The massive investments in the oil sector done by oil majors and others will create some stickiness for Singapore over the short to medium term. Moreover, Intra-Asia trade (and in particular trade between China and India) is increasing, and generally, so is trade in merchandise goods in general to and from Asia. As a consequence, Singapore can benefit from the increased volume of trade.

Singapore has to be always mindful that, in its nature, transshipment can be done from a wider area of ports, as it is not dependent on hinterland connections and its position to hinterland markets in the same way as captive market ports. Malaysia seems keen on competing with Singapore as transshipment hub for the Straits and has already captured more than six million TEU of the transshipment market, while Indonesia is clearly contemplating similar moves. China is building alternatives to oil shipping via the Straits in the form of pipelines and continues to expand its refining capacity and direct shipping in the intra-Asia market, offering options to bypass Singapore entirely. Likewise, the Chinese mega ports are offering shipping lines good incentives to do direct services, being able to fill their large container vessels by calling just a few Chinese ports.

3.2. *Shaping the future: Next generation container port challenge and SINGA port*

To tackle the challenges catering for the anticipated growth in container volumes through Singapore, the Pasir Panjang Terminal (PPT) Phases III and IV development project kicked off in 2007, and will contribute an additional annual handling capacity of more than 14 million TEUs upon completion leading the capacity of the port to 50 million TEUs. For the development of phases III and IV of Pasir Panjang, automation and environmental consideration are key pillars guiding the terminal development. The upcoming phases at PPT will feature an automated container yard with proprietary intelligent planning and operation systems, and unmanned rail-mounted gantry (RMG) cranes. By 2020, both phases will add an annual capacity of 14 million TEUs, bringing the port's total annual handling capacity to 50 million TEU of containers. In order to harness the most innovative and effective technologies, MPA and PSA signed an MOU to establish the Port Technology Research and Development Programme. The five-year programme co-funded by MPA's Maritime Innovation and Technology (MINT) Fund and other local research institutes and industry partners will cost up to $20 million. A key project identified under the MOU is the development and test-bedding of Automated Guided Vehicles (AGVs) for future container terminals to improve port efficiency and productivity. Other projects seek to develop intelligent systems to streamline port planning processes and operations, and increase the value-add and productivity.

The Phase III and Phase IV expansions of PPT are part of Singapore's plan to relocate its container terminals. In October 2012, Mr Lui Tuck Yew, Minister for Transport, announced a long-term plan to consolidate Singapore's container port activities at Tuas (*MPA Annual Report, 2012*). Tuas is a suitable location because of its sheltered deep waters and proximity to both major industrial areas and international shipping routes. To maintain Singapore's hub port leadership position, the new Tuas Terminal will be progressively developed to handle up to 65 million TEUs per annum to meet the long-term needs of the industry (*MPA, Annual Report 2012*).

Before the announcement of the port relocation plan, MPA and the Singapore Maritime Institute (SMI) jointly organized an international competition, the New Generation Container Port (NGCP) Challenge, with a grand prize award of US 1 million, to solicit participation from professionals and researchers all over the world to submit innovative and novel proposals and ideas on how to plan, design and operate the next generation of container ports that exemplified performance, productivity and sustainability.

The judging criteria of the NGCP Challenge were seemingly extensions of the future challenges that the Port of Singapore will have to tackle in order to keep its current leadership. Figure 14 shows the key performance indicators (KPIs) for the design of the port of the next generation. As observed, these indicators largely correspond to those in Section 3.1.

If existing systems and operating norms were to be employed in the quay and yard of the next generation container port, the operating specifications posed in the NGCP Challenge would not be achievable, especially considering the limited available land area of only 2.5 square kilometres. Focusing on the quay side productivity, about 75 quay cranes operating at the target productivity of 35 moves/hour would be required. On yard side analysis, the

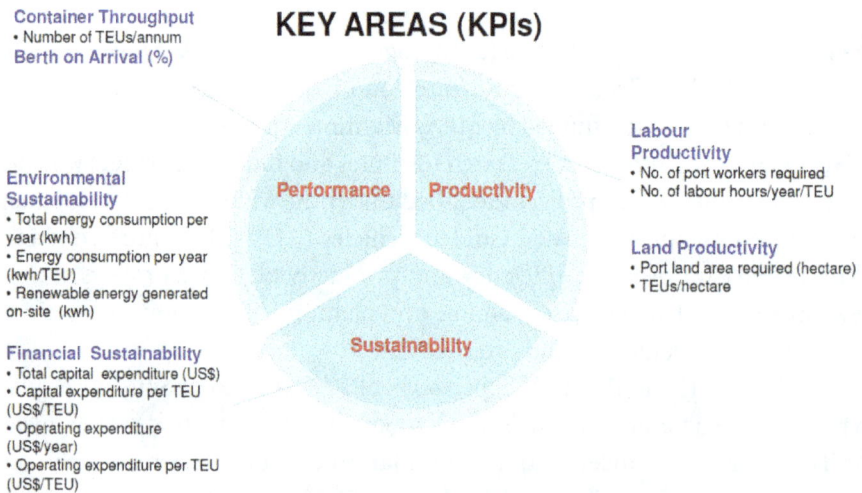

Fig. 14.   KPIs in NGCP challenge (*Source: NGCP project proposal, 2012*).

conventional yard layout would not be able to provide sufficient storage space to accommodate the given annual target container throughput without compromising the overall terminal efficiency. Hence, what is crucially needed is a novel and innovative design that is able to address the following issues:

1) New quay crane technology to meet the quay crane handling capacity requirement of more than 35 moves/hour.
2) Efficient and congestion-free supporting operations at the yard side essential for the seamless operation of the quay cranes.
3) Elimination of unnecessary handshakes between the various components of the equipment in order to minimize congestion and unproductive waiting time;
4) Optimal allocation of vehicular buffer spaces at both quay and yard sides to optimise the operation efficiency of the quay cranes.
5) Multiple accesses for quay cranes to load and unload containers are needed in order for quay cranes to work with minimal interruption.
6) An innovative yard layout design concept to increase the land productivity for storage is needed to ensure seamless operations at the yard side.

Winning design — SINGA port

The US$1 million grand prize was announced in April 2012. The winning proposal, entitled "SINGA Port" (for Sustainable Integrated Next Generation Advanced Port) was from the team composed of engineers and researchers from the National University of Singapore (NUS), Shanghai Maritime University (ShMU) and Shanghai Zhenhua Heavy Industries Company (ZPMC).

To address the challenges at both the quay and yard sides, the team of "NUS-ShMU-ZPMC" proposed an innovative and novel double-storey container terminal design concept, which is illustrated in Fig. 15. The double-storey terminal provides not only additional space for storage and transportation, but also more access points for the quay cranes, since containers can be loaded onto and unloaded from vessels at both the upper and lower levels. To take advantage of this double-storey design, a new innovative triple-hoist tandem lift quay crane design was proposed by the ZPMC engineers. There are also cutouts at strategic locations in the floor of the second storey to allow fast and efficient container movements between the first and second levels. Moreover, the second floor provides a natural shelter for the first floor, which will help in the reduction of energy usage for storage locations on the first floor that are used to stack refrigerated containers.

Fig. 15.   Double-storey container terminal.

To reduce the construction costs of the second floor, an innovative indented storage yard is proposed for the first floor. This will help to reduce the height of the second floor. The columns which support the second floor will also form a natural support structure for the possible use of overhead bridge cranes (OHBC) on the first floor. With the double-storey structure, the terminal will have more than sufficient land to meet the NGCP Challenge, and this surplus land can be used to develop an integrated logistics centre within the terminal premises.

At the quay side, the design proposes a triple-hoist quay crane with tandem lift capability to improve the productivity for the quay side. This triple-hoist quay crane is specially designed for the double-storey structure which allows the quay cranes to load or unload containers on the first and second floors simultaneously. Moreover, the proposed quay crane can be further improved to allow multiple quay cranes to work on consecutive bays simultaneously, further improving the quay side productivity by reducing interference between quay cranes.

In order to reduce the waiting time of the transportation equipment at both quay and yard cranes, the design proposes the use of automated lifting vehicles (ALV) as the transportation equipment to eliminate the handshakes between the container handling equipment (quay and yard cranes) and the transportation equipment.

The double-storey container terminal can be fully automated and it is one of the determining factors for the winning team. A spokesperson for the winning team said, "The innovative double-storey container port concept, named the SINGA (Sustainable Integrated Next Generation Advanced) Port, was made possible through the synergistic and team effort of NUS, ShMU and ZPMC. The team, which includes leading researchers and highly experienced practitioners, was able to generate innovative ideas that are achievable, namely a port concept that promises high productivity and throughput, through the integration of the unique features of a double-storey structure with the latest technologies." (*source*: *http://www.maritimeinstitute.sg/*).

The view of the port as an integrated logistics hub discussed was one of the main ideas in the SINGA Port design. Specifically, the logistics hub is located within the container terminal which allows containers to be speedily taken back to the system after stuffing/unstuffing, and the cargoes to be efficiently transported to and from the terminal. The multi-story logistics hub will include blocks of storage space and office space. An open yard for heavy machinery storage and heavy lift operations will be built with very high ceiling and high-floor loading deck to increase the efficiency of the storage space and support high-rack automated storage and retrieval systems. Conventional warehousing services, such as cargo storage and transshipment are provided by the logistics hub. Latest communication technology and highly skilled work-force will also be employed to meet the needs of customers.

4. Conclusions

The Port of Singapore has been, for the past two centuries, one of the most important ports in Asia and, ultimately, of the world. This leading position can only be partially credited to the geographical location of the country, and to a greater extent, the smart and strategic moves of the government authorities and companies that managed the port. In fact, the role of the government has been fundamental to foster the growth of the port, the busiest in the world for more than 20 years.

Clever investments and far-reaching planning, the attention for equipment innovation and port expansion and information technology, all contributed in distinguishing the port as being fully ahead, ready for the new generation vessels and showing great flexibility in the management of the ever-increasing volumes and changing market demand.

Singapore never focused on a unique business as a whole; instead, it was able to keep its identity as a bunkering and ship repairing port renowned for the quality of the services provided in addition to the efficiency in the handling activity in the most popular business, such as containerized trade. Singapore could become a fundamental hub in the shipping network thanks to the strong connections built through the decades. Examples of the effort of the country are the initiatives for strategic cooperation with other governments with big sovereign funds, to major innovations in oil and industrial hubs in cooperation with oil majors.

The plans for the future port promise to bring Singapore into the next generation highly capable to coping with diverse and critical challenges. The appropriate adoption and innovative integration of automation in container terminal operations will be one of the main novelties in the next generation container port which might be the first fully automated in the region.

Appendix A. Appendices

Appendices should be used only when absolutely necessary. They should come before the References.

A.1. List of Acronyms

HIT	Hong Kong International Terminals
PSA	Port of Singapore Authority
MPA	Marine Port Authority
SHB	Singapore Harbour Board
MTL	Modern Terminals Limited
SCT	Shanghai Container Terminals Ltd
HKC	Hutchinson Korea Terminal
KECT	Korea Express Container Terminal Co Ltd

A.2. Milestone in the Port Development

The following table reports the main milestones in the global and Singaporean facts relevant to the maritime global development.

Year	Location: Facts
1819	**Singapore**: Sir Stamford Raffles arrives in Singapore. Start of British dominance.
1869	**Suez**: Suez canal opening.
1905	**Singapore**: The Tanjong Pagar Dock Company is expropriated by the British government.
1912	**Singapore**: The Singapore Harbour Board is constituted
1914	**USA**: *Panamax* Vessel specifications are defined.
1928	**Denmark**: First Maersk line voyage.
1941/1945	**Asia**: Pacific War.

(Continued)

(Continued)

Year	Location: Facts
1944	**Japan:** 1st *Post Panamax* vessel Yamato — class battleship constructed by: (1) Kure Naval Arsenal, (2) Yokosuka Naval Arsenal, and (3) Mitsubishi Nagasaki Shipyard.
1948	USA: Establishment of the International Container Bureau
1951	**Denmark:** First Container ships designed in Denmark by Denmark's United Shipping Company EU.
1953	**USA:** Alaska Steamship company uses containers for wood and steel on the service route Alaska — Seattle.
1954	**USA:** Dravo Transportation (Pittsburgh, USA) created *Transportainer* a steel box 7 feet x 9 inches long. More than 3000 in use in one year.
1956	**USA:** *IdealX* is the first container ship. April 26 first container tanker *Ideal X* calls at Newark, New Jersey.
1959	**USA:** January 9, First Land Quay Crane Design by PACECO USA in operation at Alameda Terminal (San Francisco).
1960	**USA:** Los Angeles Port Starts installing land cranes.
	USA: Sea Land corporation is founded by Malcom McLean.
1961/1964	**Singapore:** First development policy proposed by the new Singapore government.
1961	**China:** The China Ocean Shipping (COSCO) company is funded.
1964	**Singapore:** 1st April 1964 — PSA is established by the Singapore Government.
1964	**Japan:** NYK (Nippon Yusen Kabushiki) line merge with Mitsubishi Shipping Co. Ltd. (Japan).
1965	**Singapore:** 9th August 1965 — separation from Malaysia (represents the transition from import–substitution trade to export–oriented strategy. **USA:** 24 September The Standard for Container Fittings is release by ISO. The containerization era begins.
1966	**Singapore:** Committees are appointed in Singapore to consider implications of containerization. Prior to the commissioning, conventional vessels carrying containers have to be managed at existing berths.
1967	**Asia:** ASEAN is founded.
	Europe: First European Container Shipping Line (Overseas Container Limited).
	England: First Container Port in Felixstowe (London, England).
1968	**China:** Evergreen Marine Corporation is funded in China.
1969	**Hong Kong:** Orient Overseas Container Line (OOCL) is funded in Hong Kong.
1969	**Singapore:** The entire port area is declared Free Trade Zone.

(Continued)

(Continued)

Year	Location: Facts
1970	**USA**: The Port Authority of New York and New Jersey is officially constituted with full control on NYNJ port.
	Asia: First dedicated Containership start to be produced (fully cellular containerships – FCC – second generation).
	Germany: Hapag Loyd container shipping Company is funded.
1972	**Singapore**: Taskforce to increase the mechanization and improve work procedures start. These followed the Government move in the late 70s to restructure the economy, to phase out foreign workers and to usher in a high–wage, high skilled economy.
	Hong Kong: International Container Terminal is put in operation.
	Singapore: June 23rd, first container berth opened at Tanjong Pagar. Containership Nihon calls at Singapore. First container port in Far East Asia. Port and Maritime services society is created.
1973	**Middle East**: Middle East War and Oil crisis.
	Singapore: Computer integrated terminal operations system (CITOS) and computer integrated marine operations system (CIMOS).
1978	**China**: Shanghai First Container Terminal (China).
1980	**Netherlands**: First container Terminal in Maasvlackte (Rotterdam).
1981	**Global Recession**: the immediate effect was on the oil and repair industries for which the reduction in capacity was dramatic. Singapore's dockyard capacity dropped that year to 2.4 million dwt.
1982	**Singapore**: First time Singapore is the first port for shipping tonnage overtaking Rotterdam.
1984	**Singapore**: The first version of PortNet named DataBox is developed to coordinate the PSA port services. It links port users to the port computer system online.
1985	**Europe**: *Panamax Vessels* are introduced with a capacity up to 4000 TEUs.
	Recession: a decrease in the growth rate (8.3% in 1984) of −1.8% in 1 year.
1988	**Singapore**: 1988 — PSA introduces CITOS® system to integrate all aspects of port operations. Singapore is the world's top bunkering port. PSA puts the Computer Integrated Terminal Operation System. Pioneer of the integrated control systems.
	USA: APL takes delivery of the first *Post Panamax* vessel.
1989/1991	**USA**: Savings Loans crisis.
1990	**Japan**: Collapse of the Japanese Asset price bubble.
	Singapore: overpasses Hong Kong in terms of containerized traffic.

(Continued)

Year	Location: Facts
1990/1991	Gulf War.
1992/1993	**England**: Black Wednesday financial crisis.
1993	**Singapore**: The Pasir Panjang Container Terminal Phase I is operational. **England**: First Automatized Container Terminal in Thamesport (London, Europe).
1994/1995	**Mexico**: Economic crisis.
1996	**Singapore**: MPA is created. PSA is corporatized.
	USA: APL pioneers shipment transactions via the Internet.
	Singapore: Regina Maersk, the largest vessel in the world (1st Post Panamax), has Singapore as first port of call.
	Rotterdam: ECT automated container terminal.
1997	**Singapore**: NOL and APL merge. Establishment of the national standard for bunkering with SS CP60. **Asia**: Asian financial crisis. **China**: China Shipping Container Lines (CSCL) is funded.
	Singapore: The Flow Through Gate System is introduced reducing the truck identification time from 5 minutes (1988) to 25 seconds.
1998	**Southeast Asia**: Singapore, Malaysia and Indonesia jointly introduce STRAITREP.
	Russian Financial Crisis
1999	**Singapore**: APL introduces HomePort®, the container shipping industry's first web portal. PSA Introduces Port Care.
1999/2002	Argentine economic crisis.
2000	**World**: Economic recession. **Singapore**: PSA introduces the new Port Net® world's first nation-wide e-commerce system. The Remote Crane Operations and Control Technology (RCOC) is introduced in PPT.
2001	Internet Bubble.
2002	**Asia**: ASEAN–China Free Trade Area (ACFTA) is formed.
2004	**Hong Kong**: Hong Kong International Company pioneers the use of Rail-Mounted Gantry Cranes in Hong Kong. Hong Kong International Terminal (automated) 6/7 starts operation.
2004	**Singapore**: Total vessel arrivals for the year, in shipping tons, is 1.04 billion GT. For the first time Singapore crosses the billion GTs.

(Continued)

Year	Location: Facts

(Continued)

Year	Location: Facts
2006	**Panama**: the Project for the expansion of the Panama canal is approved. The extension will be completed in 2015.
	Denmark: Maersk releases the first mega vessel Emma Maersk with 15000 TEU of cargo capacity (Denmark) 145,000,000+[US$].
	Singapore: Singapore is the first port hosting Emma Maersk. November 29th ReCAAP is established in Singapore against robbery and piracy in Asia.
	Singapore: APL introduces OceanGuaranteed®, the industry's first date-definite delivery service for Less than Container load (LCL) shipments.
2007/2008	Global Financial crisis.
2008	**Singapore**: MPA–IDA initiative: first port to adopt wireless broadband system allowing instant and real time communications between ships and shore-based locations: important communication facilities are provided and access to real time vessel data becomes possible. Infocom@SeaPort Programme and WISEPORT programme.
2010	**Europe**: European sovereign debt crisis.
	Singapore: The Pasir Panjang semi-automated terminal 1/2 becomes fully operational. **South Korea**: The Pusan Newport International (automated) Terminal opens.
	China: Shanghai is the first container terminal for container throughput for the first time.
	Europe: Fully Automated Terminal Euromax becomes operational in Rotterdam (the Netherlands).
2011	**Denmark**: Maersk launches Triple E new generation vessel capacity of 18,300 [TEU] (Denmark).
2012	**Singapore**: PSA announces the development of Pasir Panjang Terminal Phase 3 & 4 which will be fully operational by 2020.
	Singapore: NGCP Challenge.
2013	**Singapore**: APL introduces TEMASEK, its first and largest continership (14,000 TEU).
	China: Hong Kong International installs remote control operations.
2014	**China**: June 21st — China publishes arctic shipping guide.

References

Journal Publications:

[1] Abelman, R. (2002). Creating a world-class chemical hub. Port of Singapore. *The Maritime Hub Series*, 7, pp. 10–12.

[2] Chan, C.T., and H.H., Lee. (2000). *Containers, Containerships and Quay Cranes, A Practical Guide*. (VT Editorial Consultancy, Singapore).

[3] Chew, E.C.T., and E., Lee. (1991). *A story of Singapore*. (Oxford University Press, New York).

[4] Chew, E. P., L.H., Lee, J., Jianlin, and G.C., Chun. (2011). Models for Port Competitive Analysis in The Asia-Pacific Region. In Advances in Maritime Logistics and Supply Chain Systems, 4, pp. 91–115.

[5] Chu, C.Y. and W.C., Huang. (2005). Determining container terminal capacity on the basis of an adopted yard handling system. *Journal of Transport Reviews*. 25(2), pp. 181–99.

[6] Cullinane, K., Yap, Y.W., and J.S., Lam. (2006). The port of Singapore and its governance structure. *Research in Transportation Economics*, 17, pp. 285–310.

[7] De Langen, P. W. (2006). Stakeholders, conflicting interests and governance in port clusters. *Research in Transportation Economics*, 17, pp. 457–477.

[8] Fleming, D. K., and Y. Hayuth. (1994). Spatial characteristics of transportation hubs: centrality and intermediacy. *Journal of Transport Geography*, 2(1), pp. 3–18.

[9] Heng, D. (2007). Continuities and changes: Singapore as a port-city over 700 years. *Biblioasia (Singapore: National Library Board)*, 1, pp. 12–16.

[10] Huff, W. G. (1997). *The economic growth of Singapore: Trade and development in the twentieth century*. (Cambridge University Press).

[11] Kim, O. C. 2001. Stepping Up maritime security and the technologies available. *The Maritime Hub Series*, 9, pp. 12–15.

[12] Lee, S. W., D. W., Song, and C., Ducruet. (2008). A tale of Asia's world ports: the spatial evolution in global hub port cities. *Geoforum*, 39(1), pp. 372–385.

[13] Leech, J. (2000). Future Technology Developments in Hydrography. (Port of Singapore the maritime hub, Singapore).

[14] Levinson, M. (2010). *The box: how the shipping container made the world smaller and the world economy bigger*. (Princeton University Press).

[15] Li, Y. (2011). Current logistics situation of container transport in Shanghai Port. (Bachelor Thesis, Shanghai).

[16] Lim, R. (1993). *Tough Men, Bold Visions. The Story of Keppel*. (The Keppel Shipyard Company, Singapore).

[17] Loh, G. and S.H., Tey. (1995). *What's behind the Name Jurong Shipyard Limited*. (Times Academic Press, Singapore).

[18] Ma, H., and Z., Feng. (1997). *Ying-yai sheng-lan: the overall survey of the ocean's shores*. (J. V. Mills (Ed.). White Lotus).

[19] Merk, O. (2013). *The Competitiveness of Global Port-Cities: Synthesis Report (No. 2013/13)*. (OECD Publishing).

[20] Notteboom, T. E., and J.P. Rodrigue. (2005). Port regionalization: towards a new phase in port development. *Maritime Policy & Management*, 32(3), pp. 297–313.

[21] Rodrigue, J.P. (2014). *The Geography of Transportation Systems*. (Dept. of Global Studies & Geography, Hofstra University, New York, USA).

[22] Slack, B. (1993). Pawns in the game: ports in a global transportation system. *Growth and Change*, 24(4), pp. 579–588.

[23] Song, D. W. (2003). Port co-opetition in concept and practice. *Maritime Policy & Management*, 30(1), pp. 29–44.

[24] Sorensen, P.F. (2000). Memorandum of Understanding between Norway and Singapore. 2000. *Port of Singapore. The Maritime Hub. Series*, 5, pp. 20–22.

[25] Wang, J. J. (Ed.). (2007). *Ports, cities, and global supply chains*. (Ashgate Publishing, Ltd).

[26] Yap, W. Y., and J.S., Lam. (2006a). Competition dynamics between container ports in East Asia. *Transportation Research Part A: Policy and Practice*, 40(1), pp. 35–51.

[27] Yap, W. Y., J.S., Lam, and T., Notteboom. (2006b). Developments in container port competition in East Asia. *Transport Reviews*, 26(2), pp. 167–188.

[28] Yap, W. Y. (2009). *Container Shipping services and their impact on container port competitiveness*. (Antwerp, Belgium: University Press Antwerp).

[29] Yap, W. Y. (2014). Container Port Development in the 21st Century: Challenges and Trends. In *Managing Logistics and Supply Chain Challenges*. (Tan, Y. W., Sim, T. and de Souza, R. eds.), 18, pp. 341–370.

[30] Yeo, G., and G., Teo. (1997). The big spill: Waters turn black while big clean-up keeps slick away. *The Straits Times, p. 1. Retrieved December, 19, 2014, from NewspaperSG.*

Reports:

[31] Center for Livable Cities (CLB). (2013). *Industrial Infrastructure, growing in tandem with the economy*. (Singapore Urban System Studies Booklet Series, Singapore).

[32] Center for Livable Cities (CLB). (2013a). *Sustainable Environment, Balancing growth with the environment*. (Singapore Urban System Studies Booklet Series, Singapore).

[33] *Containerization International Yearbook* 1991–2012 (London, National Magazine Company).

[34] Maritime Port Authority. *Annual Reports* 1997–2014.

[35] Maritime Port Authority. (2007). *Port view*. (MPA, Singapore).

[36] Port of Singapore Authority. IT Innovations 1990s. (Port of Singapore Authority, Singapore).

[37] Port of Singapore Authority. (1964). *Singapore Port History 1819–1963*. (Maritime Museum, Singapore).

[38] Port of Singapore Authority. (1990). *Singapore Port Services Index 1990*. (FETP Business Publications Pte. Ltd, Singapore).

[39] Port of Singapore Authority, (1978). *BERITA*, (PSA, Singapore).

[40] Port of Singapore Authority, (1979). *BERITA*, (PSA, Singapore).

[41] Port of Singapore Authority. (1999). The maritime hub Series 1999–2002, (PSA, Singapore).

[42] Port of Singapore Authority. (1965). *Annual Reports* 1965–2014. (PSA, Singapore).

[43] Port of Singapore Authority. (1984). *Singapore: Portrait of a Port*. (PSA, Singapore).

[44] Report of the Commission of Inquiry into the Port of Singapore. (1957). (Singapore Harbor Board, Singapore).

[45] The Singapore Harbor board. (1922). A Short History of the Port of Singapore, (Raffles Library, Singapore).

[46] The Singapore Harbor Board. 1949. *Annual Report* 1949–1963, (SHB, Singapore).

[47] UNCTAD (United Nations Conference on Trade and Development. (1980). *Review of Maritime Transport* 1980–2013. (UNCTAD)

[48] United Nations Economic and Social Commission for Asia and the Pacific (UNESCAP). (1978). *Report of the workshop on shipper's cooperation* (country-level) 23–25 May 1978. (UNESCAP, Singapore).

[49] *Singapore Nautilus*, 2008. Issue 2, Q1 2008.

Websites:

[50] http://www.worldshipping.org/

[51] http://unctadstat.unctad.org/_

[52] http://www.worldshipping.org/

[53] www.jp.com.sg (Jurong Port)

[54] www.singaporepsa.com (port connectivity)

[55] www.internationalpsa.com

[56] www.mpa.gov.sg

[57] http://www.mpa.gov.sg/sites/pdf/singapore_nautilus_issue2.pdf

Chapter 8

Development of Singapore's Global Air Transport Hub

CHIA Lin Sien and Karmjit SINGH

For the island-state of Singapore that has a mere land area of 716.2 square kilometres and a population size of 5.47 million in 2014, to have reached the status of a developed nation with a GDP of $390 billion and a per capita income of $71,318 is an improbable situation. Singapore's external trade at $983 billion is 2.5 times its GDP, indicating a heavy dependence on the rest of the world. Singapore has little choice but to be connected to the rest of the world by sea and air transport. Total international arrivals[1] number 15 million giving a ratio of 2.9 times to the population, while the volume of sea cargo handled was 581 million tonnes compared to 1.8 million tonnes of air cargo — a mere fraction of the volume of sea cargo handled at our seaports. However, the value of air cargo is far greater than that of sea cargo, which tends to be heavy and bulky.[2]

From early 19th century, British administrators clearly saw the value of building a seaport that adopted the *laissez faire* policy, and opened to all comers with minimum intervention. As for air transport, the vision of Singapore as a major air hub was due to the realisation of the need for a seamless international mobility of people, businesses, finance and investments, and tourism. Singapore's success is the result of operating an unmatched international airport combined with a premium world-class airline in the form of Singapore Airlines. This can only be achieved with the support of a superb domestic transport system, telecommunications system and facilities such as aviation business parks, all with the backing of a strong and efficient financial system.

---

[1] Excluding Malaysian arrivals by land but includes arrivals by sea and air.
[2] Latest Data, Dept of Statistics, Singapore. Available at http://www.singstat.gov.sg/statistics/latest-data#7. Accessed 8 March 2015.

Singapore has also aspired to be a preferred workplace and home to both locals and foreign residents and visitors with a conducive environment that is clean and green. Without the hand of an efficient and far-sighted government, the creation of a global air hub would not have been possible.

1. An Overview of Air Transport Development

According to the International Aviation Transport Association (IATA), international air passenger traffic grew at double-digit rates from its earliest post-1945 days until the first oil crisis in 1973.[3] Annual growth of global air travel has averaged approximately 5% over the last three decades, increasing at about twice the annual growth in GDP. It is expected to double over the next 10 to 15 years.[4] Airport Council International (ACI) reported that worldwide air passenger numbers increased by 4.6% in 2013 to 6.3 billion, registering increases in all six regions. The two fastest growing regions were Asia Pacific (2.06 billion up 8.7% over 2012) and Middle East (278 million, up 7%) compared with Europe (1.73 billion, up 3.2%), North America (1.57 billion, up 1.1%), Latin America-Caribbean (501 million, up 5.5%), and Africa (164 million, up 0.5%).[5]

The steady expansion of international air travel worldwide since the middle of the 20th century was the result of several factors: (1) technological advancements in aircraft and engine manufacturing; (2) liberalisation of international aviation regulatory regime that enabled (3) the emergence of new airline operators, and, more recently low-cost carriers; and (4) the rapid post-war economic prosperity of especially the Asia-Pacific region.

Much of the impetus for air traffic growth came from technical innovation from the dominant Western aircraft manufacturers, notably Boeing in the United States and the Airbus European consortium. Turbo-propeller aircraft made its appearance in the early 1950s. By the late 1950s, the jet age with many technical advances had arrived, ushering in the transatlantic jet planes. In 1958, the Boeing 707 could carry up to 181 passengers and travel at speeds

---

[3] International Air Transport Association, see http://www.iata.org/about/Pages/history_3.aspx. Accessed 3 January 2015.

[4] Eric Henckels, Airline industry overview. Available at http://www.columbia.edu/cu/consulting-club/Resources/Airlines_Eric_Henckels.pdf. Accessed 3 January 2015.

[5] ACI releases 2013 World Airport Traffic Report: Airport passenger traffic still going strong; air cargo inches along after third year of weak growth — Sep 16, 2014. Available at: http://www.aci.aero/News/Releases/Most-Recent/2014/09/16/ACI-releases-2013-World-Airport-Traffic-Report-Airport-passenger-traffic-still-going-strong-air-cargo-inches-along-after-third-year-of-weak-growth. Accessed 5 January 2015.

of 550 miles per hour. In 1969, the revolutionary Boeing 747 aircraft made its appearance. It was the first wide-bodied jet that featured two aisles and four engines. It has a seating capacity of 450 passengers, 80% bigger than the largest jet up until that time, the Douglas DC-8. Other aircraft manufacturers such as Douglas and Lockheed followed suit, creating their own versions of wide-bodied planes.[6] The Boeing 787-8 Dreamliner can carry 242 passengers up to 14,500 kilometres, while the longer B787-9 can carry 280 passengers for up to 15,372 kilometres. Dreamliners offer more than 20% fuel efficiency compared to similar existing aircraft. The still larger aircraft are a series of Airbus A380s that provide 40% more capacity with lower seat-kilometres cost in its class. These newest aircraft offer higher speeds and longer range, lighter frames, greater capacity and payload, reduced aircraft maintenance, improved operating performance and lower unit operating costs. Other aircraft such as the A330 series and Boeing's B777 that came onto the market in 1995 cater to long-range trans-continental routes.

In 2005, Changi Airport upgraded its infrastructure and unveiled a A380-compatible gate holdroom and a third passenger loading bridge (aerobridge) at all 19 A380-compatible gate holdrooms. Later in November, Changi Airport became the first airport outside Europe to welcome the A380, when the super jumbo double-decker aircraft arrived for airport compatibility verification tests. The world's first A380, commercial passenger flight took off from Changi Airport in October 2007.[7]

1.1 *Importance of the aviation industry*

In the words of the Minister of State for Transport, Mrs Josephine Teo, "Singapore thrives by being a global city, connecting ideas, people and goods across the region and the world. Aviation is a vital enabler of our global connectivity, which helps many other sectors, such as financial services, manufacturing, and tourism to flourish. Aviation is in fact a fairly diverse sector, comprising airlines, airport, aerospace manufacturing and aviation services. These different activities provide jobs for an estimated 40,000 people."[8]

---

[6] See Avjobs, History of Aviation — First flights, available at http://www.avjobs.com/history/. Accessed on 26 December 2014. See Doganis, 1991 for discussion on aircraft development.

[7] See CAAS website: http://caas.gov.sg/caas/en/About_CAAS/Our_History/?__locale=en. Accessed 26 December 2014.

[8] Speech by Josephine Teo, Minister of State for Transport, on Air-Sea Transport, for Committee of Suppl Debate 2012, 07 March 2012 in Parliament. Available at http://www.mot.gov.sg/News-Centre/News/2012/Speech-(Part-3---Air-Sea-Transport)-by-Mrs-Josephine-Teo,-

The aviation industry is a significant contributor to the Singapore economy. The available data for 2009 show that it contributed S$14.2 billion (5.4%) in direct and in-direct value-added to the Gross Domestic Product (GDP). This included S$8.7 billion of direct contribution from the sector, S$3.1 billion through indirect contributions from the sector's supply chain and S$2.4 billion from consumer spending by employees of the aviation sector and its supply chain. 58,000 direct jobs were made available in the industry, and up to 119,000 jobs if the entire supply chain was included. Out of the 58,000 direct jobs, aerospace manufacturing companies employed approximately 18,000 people, airline companies employed 16,000, while 24,000 worked in the airports and ground service companies. The sector also directly contributed S$1.5 billion in taxes in 2009. Overall, the airlines contributed over S$5.5 billion to the economy and supported 34,000 jobs in Singapore.[9] The industry is indeed an engine of growth and a major source of employment for Singapore.

1.2 *Changing international aviation regime*

A major factor in the rapid expansion of the airline industry in the Asia-Pacific has been the substantial regulatory reforms at both the domestic and international levels during the decade from the mid-1980s. During this period, many countries progressively liberalised the bilateral air service agreements to allow for multiple designations, i.e., more than a single international airline from each side, thereby enabling greater flexibility in meeting the growing air transport demand.

According to CAAS, Singapore has concluded air services agreements with more than a hundred countries and territories, as well as 40 open skies agreements.[10] Open skies agreements allow carriers to operate any number of flights between and beyond both signatory states, enabling them to tap on traffic from third countries to improve the commercial viability of scheduled flights. Singapore has signed open skies agreements with Thailand, the United Kingdom, the United States, United Arab Emirates, Chile, New Zealand, Denmark, Iceland, Sweden, Portugal, Kuwait, Oman, Sri Lanka, and many others. In other cases, a more liberal bilateral has been signed with, countries

---

Minister-of-State-for-Transport,-for-COS-2012/. Accessed 12 January 2015
[9] Wikipedia, Aviation in Singapore. Available at: http://www.ask.com/wiki/Aviation_in_Singapore?o=2802&qsrc=999&ad=doubleDown&an=apn&ap=ask.com. Accessed on 12 January 2015.
[10] CAAS website http://www.caas.gov.sg/caas/en/About_CAAS/Our_Strategic_Thrusts/Air_Hub_Development/Air_Services_Policy/. Accessed on 10 March 2015.

such as, Malaysia and Australia. In November 2001, Singapore became a signatory to the Multilateral Agreement on the Liberalisation of International Air Transportation (MALIAT) with Brunei, New Zealand, Chile and the United States, which allows for unlimited traffic rights among participating countries under third, fourth, fifth and sixth freedoms, as well as unlimited seventh freedom traffic rights for cargo-only flights. New Zealand, Chile, Singapore and Brunei have gone even further, and granted each other seventh and eighth freedom rights for passenger flights.[11] Singapore is a signatory to the ASEAN Roadmap for the Integration of the Air Travel Sector (RIATS). In November 2007, Singapore concluded a landmark Agreement with the United Kingdom that included unlimited "hubbing" and cabotage rights.

Bilateral Aviation Service Agreement (ASA) reforms gained momentum in the 1990s, and were given a boost in 1997 when the USA signed new 'open skies' agreements with Singapore, Malaysia, Taiwan and Brunei, though in practice, Singapore Airlines' attempt to open up new destinations in the United States was restricted to just a handful of destinations. The reason was that the route proposed was already experiencing over-capacity by existing airlines. Elsewhere, restrictions to increase air services were constrained by the unavailability of landing and departure time slots and parking space in already congested airports, as in the case of Tokyo's Narita Airport and more recently, in Jakarta's Soekarno–Hatta Airport. While Singapore has signed a large number of liberal ASAs, a number of major markets still have significant restrictive agreements including Malaysia, Indonesia, India, China, Hong Kong, Japan and Australia.

1.3 Performance of Asia–Pacific aviation sector

Data from the International Air Transport Association (IATA) in the December 2014 review of the performance of the global air transport industry shows passenger departures to be 3,134 million and 3,306 million for 2013 and 2014 respectively, and is projected to reach 3,530 million in 2015, giving 5.4%, 5.7% annual growth and an estimated stronger growth for the three respective years. Air freight for these three years had been growing slowly since 2010, but a moderate cyclical upturn is expected in 2015. Internationally, freight tonnes handled came to 49.3 million, 51.3 million and 53.5 million for 2013, 2014 and 2015 respectively, giving an annual growth rate of 2.3%, 4.1% and 4.3%

---

[11] See website for the Multilateral Agreement on the Liberalization of International Air Transportation. http://www.maliat.govt.nz/. Accessed on 12 March 2015.

respectively. This compares with the world GDP growth of 2.5%, 2.6% and 3.2% for the three years respectively; and for world trade growth, of 2.7%, 3.0%, and 4.0% respectively.

IATA revealed that there would be 1% of world GDP to be spent on air transport in 2015, totaling up to US$820 billion. Expected revenue growth of 7.0% was to be expected, which is a significant improvement since 2010, and above the 5.5% level of the previous two decades.[12]

Four of the world's busiest airports are located in the Asia-Pacific region. These are, in 2012, Beijing (China) with 82 million passengers and an increase of 4.1% of the number over the previous year. It is fast catching up with the world's busiest airport in Atlanta, USA that handled 95 million passengers over the same period. At fourth place, Tokyo's Haneda Airport handled 67 million and had a growth of 6.7%. Jakarta (Indonesia) came in 9th position with 58 million passengers and growth of a high 12.1%, followed by Dubai (United Arab Emirates), a relatively new mega airport that handled 58 million passengers and a phenomenal growth of 13.2%.[13]

As for cargo airlines that carried the largest volume of international air freight in terms of tonne-kilometres, SIA Cargo ranked 6th among the world's top 10 cargo airlines in 2013. The top two, Emirates and Cathay Pacific Airways, respectively carried 10.5 and 5.0 billion tonne-kilometres ahead of FedEx at third place with 7.7 billion tonne-kilometres, and followed by Korean Air, Lufthansa ahead of SIA Airlines with 6.2 billion tonne-kilometres of cargo.[14] SIA Cargo operated 900 flights a week from its hub at Changi Airport. It has links with more than 65 cities in over 30 countries across the globe with its fleet of dedicated B747-400 freighters and belly-hold capacity in its hundred wide-body passenger aircraft.[15] Among the top ten air cargo carriers are five other Asian carriers — Emirates Skycargo (ranked 3rd), Cathay Pacific Cargo (ranked 4th), Korean Air Cargo (ranked 5th), China Airlines Cargo (ranked 9th), and EVA Air Cargo (ranked 10th). Clearly, Asia is the world's

---

[12] IATA, Economic performance of the airline industry, 2014 year-end report. Available at www.iata.org/economics. Accessed 20 December 2014.

[13] Airports Council International (ACI), ACI Annual World Airport Traffic Report, Available at http://www.aci.aero/Data-Centre/Annual-Traffic-Data/Passengers/2010-final. Accessed 3 January 2015. Accessed 4 January 2015.

[14] International Air Transport Association (IATA), 2015. Available at http://www.iata.org/publications/Pages/wats-freight-km.aspx. Accessed 5 April 2015.

[15] SIA Cargo is ranked 8th by Supply Chain Digital. See Sam Jermy, Top 10 Air Freight Companies, dated 31 July 2014. Available at http://www.supplychaindigital.com/top10/3556/Top-10-Air-Freight-Companies. Accessed on 7 January 2015.

producer of air cargo and this will likely continue to be the case for the future.[16]

2. Aim of This Study

This chapter attempts to trace the development of Singapore into a premier air transport hub from the time of the nation's birth 50 years ago. The roots of the improbable flowering of such a strong international air hub goes back to the nation's pre-independence days, especially with respect to the development of our airports that led to the construction of the Changi Airport which was completed in 1981. This chapter covers both air passengers and air freight that have closely-linked factors that helped to create the Singapore hub both for air passengers as well as for air cargo. The approach adopted is essentially a chronological one, but each step taken explains the measures implemented on the part of the management leaders of key agencies, not only to react to the quick changing circumstances but also to take bold decisions based on long-range and forward-looking planning. This study also hopes to indicate the difficulties encountered and the adverse conditions that have hampered Singapore's growth as an international air hub, including the strong rivals both in competing airports within the surrounding region as well as the firmly established and powerful airlines that rival Singapore Airlines.

3. Making of An International Air Transport Hub

Changi Airport has become a beacon of Singapore's success. According to the Civil Aviation Authority of Singapore (CAAS), the airport has garnered more than 470 accolades. For over 30 years since its inception, Changi Airport has become the world's most awarded "Best Airport". In 1988, Changi Airports Terminal 1 received its first recognition as the world's best airport from Business Traveller (UK). Changi Airport has been named the World's Best Airport at the 2013 Skytrax World Airport Awards held in Geneva, ahead of other world-class airports such as Incheon International Airport in Korea and Hong Kong International Airport. It was the fourth time in a row that Changi Airport has won this prize. The success story of Changi Airport has been oft reported.

Changi Airport has firmly established itself as a major air hub in Asia and ranks as the region's sixth busiest international airport. It serves more than 100

---

[16] Wikipedia, World's Largest Airlines. Available at http://www.ask.com/wiki/World%27s_largest_airlines?qsrc=3044&lang=en. Accessed on 5 January 2015.

international airlines flying to some 300 cities in about 70 countries worldwide, Changi Airport handled more than 53.7 million passengers in 2013, roughly 10 times the size of Singapore's population. A flight takes off or lands at Changi Airport roughly once every 90 seconds.

The steep growth of air passengers passing through Changi Airport, and decades earlier at Kallang and Paya Lebar airports from 1960 to 2013, is shown in Fig. 1 and Table 1. The number of flights — for both passenger and freight — rose from 6,233 in 1960 to 343,765 in 2013. The latter figure represents an average take-off and landing every 1.53 minutes from Changi Airport in 2013. Starting from just a mere 225,000 air passengers in 1960, the volume of air passengers grew steadily until about 1970. The then Paya Lebar Airport could not cope with the rising traffic — annual passenger volume rose rapidly from 300,000 in 1955 to 1.7 million in 1970 and jumped to 4 million in 1975. Passenger traffic continued to rise in 1997, when it was temporarily halted by the Asian Financial Crisis when the Thai, Malaysian, Indonesian, the Philippines, South Korean as well as Singapore an economies were severely affected. It resumed its rise until the attack on the World Trade Center on 11 September 2001. From thence, the volume began to climb at a fast rate but stalled in 2003 due to the outbreak of Severe Acute Respiratory Syndrome (SARS) in 2003.[17] The 2007–2008 global financial crisis and the consequent global recession that lasted till 2010 appears only to have had a temporary effect on the underlying upward trend, as the volume of air passengers continued to power through the period and continued thereafter. The volume of air passenger movements through Changi Airport reached a remarkable 53.7 million passengers in 2013 (Fig. 1). However, Table 1 shows that the number of air passenger movements tapered off between 2012 and 2013, growing at 5.0% compared to growth rates of 10.7% and just below 10.0% for 2011–2012 and 2010–2011 respectively. The slowdown in 2013 is likely to be the result of the restrictions of travellers from China. For 2014, air passenger numbers were affected by the disappearance of the Malaysian Airlines MH370 on 8 March 2014, with 239 individuals on board the ill-fated plane of which 152 were Chinese nationals. Chinese travellers reacted by staying away not only from Malaysia, but also from the region including Singapore. The crash of Indonesia AirAsia Flight QZ5801 on 28 December 2014 in the Java Sea en route from Surabaya to Singapore with a loss of 162 lives would not help the confidence of air travellers in the region. Indonesia

---

[17] SARS first made its appearance in Guangdong, China, in November 2002 then spread to many countries including Singapore. It came to an abrupt stop by mid-2003.

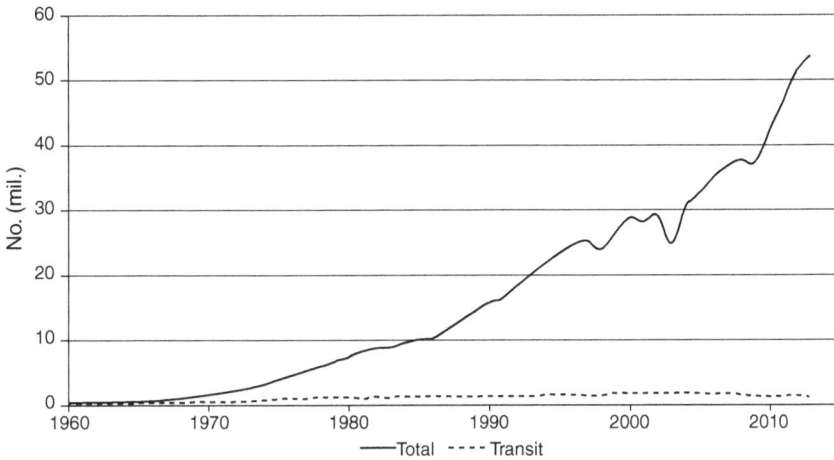

Source: Singapore Yearbook of Statistics, Ministry of Trade and Industry. Data from CAAS.

Fig. 1.   Singapore: Air Passenger Movements, 1960–2013.

also has had a poor record of air safety, and the European Union barred 62 Indonesian carriers from flying to Europe.[18]

Table 1 shows the number of flights at Changi Airport growing rapidly from 1960 to 2013. Initial growth was spectacular up till 1980, when the number of flights more than tripled between 1960–1970, and rose further by 4.4 times over the following decade. Thereafter growth rates were more moderated. In terms of air cargo, its growth during the period was rapid, though was less impressive than the increase in the volume of air passengers. Total air cargo handled at the airport was a miniscule 1,639 tonnes in 1960, but rose spectacularly by 8.6 times between 1970 to 1980 then by 3.4 and 2.7 times respectively, in the following two decades. By 2013, the volume of cargo handled had risen to 1.8 million tonnes, with an average growth rate of 4.7% for the past 15 years. Transshipment cargo accounts for about 40% of total throughput — taking advantage of Singapore's well-established connectivity to the world. Singapore is thus both an air hub for passengers and cargo.

CAAS reveals that as a global logistics hub, Singapore has been able to attract leading brand names across various industries such as Zuellig Pharma, Moet Hennessy, Louis Vuitton (LVMH), Avaya, Merck and Schering-Plough, and many have located their Regional Distribution Centres (RDC) in Singapore. Singapore is party to an extensive network of Free Trade

---

[18] See, Indonesia's troubling aviation safety record, ST 2 January 2015.

Table 1. Singapore: Civil Aircraft Arrivals/Departures, Passenger Movements, Cargo and Mail, 1960–2013.

Year	1960	1970	1980	1990	2000	2010	2011	2012	2013
Flights (No.)	6,233	17,113	75,971	97,645	173,947	263,593	301,711	324,722	343,765
Passengers ('000)									
Total	225	1,338	7,295	15,621	28,618	42,039	46,544	51,182	53,726
Arriving	93	512	3,141	7,237	13,546	20,486	22,778	25,055	26,500
Departing	91	521	3,151	7,166	13,419	20,437	22,651	24,854	26,275
In Transit	40	304	1,003	1,217	1,654	1,115	1,115	1,272	951
Cargo (million Tonnes)									
Total	11.3	21.1	181.8	623.8	1,682.5	1,813.8	1,865.3	1,829.1	1,837.7
Discharged	7.2	12.8	90.7	324.2	848.3	941.4	983.1	975.8	995.8
Loaded	4.4	8.3	91.1	299.7	834.2	872.4	882.1	853.3	841.9
Mail (Tonnes)									
In-coming/Out-going	1,639	3,081	8,051	8,693	22,881	27,192	33,596	35,633	35,743

Source: Yearbook of Statistics, Singapore, 1970–2014. Data from Civil Aviation Authority of Singapore.
Note: Figures refer to Changi Airport only.

Agreements, Avoidance of Double Taxation Agreements and Investment Guarantee Agreements that, together with its excellent air, sea and IT infrastructure, have enabled a smooth flow of goods and services from Singapore to markets around the world.[19]

The distribution of air passengers by origin and destination is shown in Table 2. There is a consistent pattern with few changes in the trends over the period from 1963 to 2013. The data clearly shows the dominance of Singapore's neighbouring South East Asian region as the main source of air passengers accounting for between 38% (2003) and 46% (2013). The pattern was likely distorted by the year 2003, the year affected by SARS when travellers limited their travel due to the fear of being infected. The dip in 2003 is most noticeable for North East Asia for the same reason, but rebounded from 22% in that year to 25% in 2013. The other two significant sources of air passengers are travellers from Europe and Australasia — likely to use Singapore as a stopover destination before going on to other destinations.

Figure 2 illustrates the growth of air travel between Singapore and the top seven origins/destinations. The highest growth rate between 2003 and 2013 was the Middle East at 387.3% due to the huge investments in airports in Abu Dhabi, Dubai and Doha, and the success of a number of Middle Eastern airlines such as Emirates, Etihad, and Qatar Airways. Like the Middle East, Vietnam and the Philippines also recorded high rates of growth at 381% and 290% over the same period due to the low base but augmented by the operation of a number of successful Low-Cost Carriers (LCCs) taking advantage of the ensuing more liberalised aviation regime. Air passengers to and from other South East Asian countries continued to grow, giving a rate of 169%, as did North East Asia (161%) and South Asia (160%).

Traditional sources of air travellers from Europe, North America and Oceania grew more modestly. Penetration into the North American market remains weak with only 617, 000 accounted for. The pattern of travel confirms that Singapore is very much an international air hub with a strong market in Northeast Asia and increasingly India and the Middle East, while Europe and Oceania remain as major markets that promise huge potential for future growth. Expectedly, China is featured as a growing market (Fig. 2), and will continue to grow strongly in the future. It is anticipated that with the adoption

---

[19] CAAS website. Available at http://caas.gov.sg/caas/en/About_CAAS/Our_Strategic_Thrusts/ Aviation_Industry_Development/Air_Cargo_x_Logistics_Hub/?__locale=en. Accessed 5 January 2015.

Table 2.   Air Passenger Volume by Region and Selected Countries, 1963–2013 ('000).

Region/Country	1963	1973	1983	1993	2003	2013	% incr. 2013/2003
Total	261	2,031	6,133	18,796	23,143	52,775	128.0
South East Asia	—	—	—	8,724	8,805	23,680	168.9
Indonesia	33	528	1,075	2,380	2,712	7,417	173.5
Malaysia	29	153	443	3,474	2,113	5,451	158.0
Philippines	8	53	261	404	645	2,515	289.9
Thailand	47	264	781	2,013	2,589	5,266	103.4
Brunei	—	—	—	253	198	—	—
Vietnam	—	—	—	—	375	1,805	381.3
North East Asia	52	377	1,571	4,482	5,070	13,246	161.3
China	—	—	—		1,309	4,643	254.7
Hong Kong	38	227	701	1,465	1,315	3,408	159.2
Taiwan	—	—	—	720	494		
Japan	3	77	516	1,757	1,375	2,421	76.1
South Asia	25	85	377	939	1,681	4,357	159.2
India	—	—	—	531	1,179	—	—
Middle East	—	—	—	221	252	1,228	387.3
Oceania	39	284	733	1,920	3,567	5,485	53.8
Australia	—	—	—	1,641	3,064	—	—
New Zealand	—	—	—	247	479	—	—
Europe	25	284	740	1,871	3,002	3,960	31.9
France	—	—	—	126	329	482	46.5
Germany	—	—	—	322	633	784	23.9
Netherlands	—	—	—	181	220	—	—
United Kingdom	18	148	268	664	1,213	1,262	4.0
North America	—	—	—	453	532	617	16.0
USA	—	—	—	402	491	—	—
Other Regions	4	3	154	187	234	206	−12.0

Source: Yearbook of Statistics Singapore, 1970–2014. Data from Civil Aviation Authority of Singapore.
Notes: Refers to Changi Airport only. Data exclude transit passengers who continued their journey on the same flight.

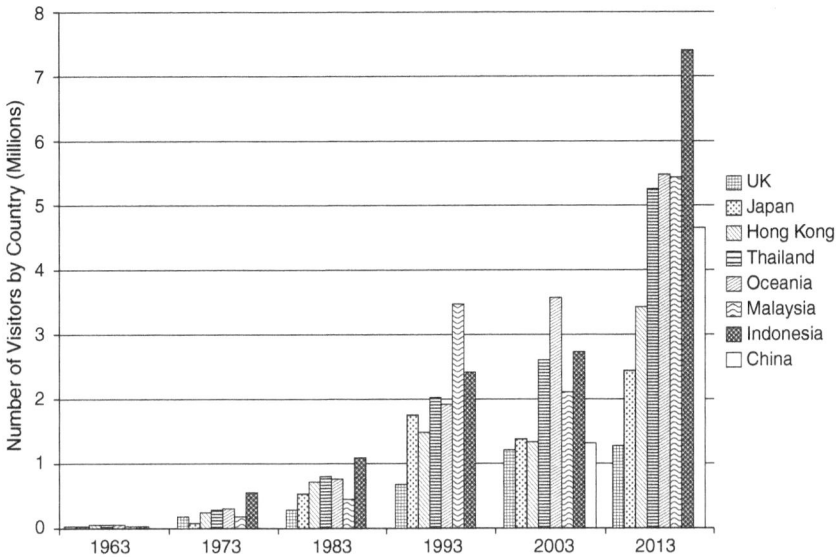

Fig. 2.   Singapore's Top Seven Origins/Destinations, 1963–2013.

of the ASEAN Open Skies Agreement in 2015, the number of air passenger movements would increase by 5% per year.[20]

Table 3 shows Singapore's top 13 destinations by flight frequency, confirming the above observations that Changi Airport serves the Southeast Asian region well. Hong Kong as the gateway to Southern China is ranked third with 149 flights per week as of April 2014, after Jakarta (282 flights/week) and Kuala Lumpur (249 flights/week), which ranked at first and second. Taipei, Shanghai (Pudong), Chennai and Sydney ranked 10th to 13th, with Chennai featuring strongly since June 2010.

The Asia–Pacific region is home to some of the busiest air transport routes. The Singapore–Jakarta route is expected to double within the next 10 years. The largely maritime nature of the region, with two of the world's largest archipelagic nations, Indonesia and the Philippines, lends itself most amendable to air travel domestically as well as internationally. Air travel between Peninsular Malaysia and Sarawak/Sabah, separated by the South China Sea, represents growth opportunities for Singapore, though this has been strongly challenged by Kuala Lumpur. SIA has a share of 35% while

---

[20] Wikipedia, Singapore Changi Airport. Available at http://www.ask.com/wiki/Singapore_Changi_Airport?o=2802&qsrc=999&ad=doubleDown&an=apn&ap=ask.com#cite_note-changiairportgroup.com-11. Accessed 5 January 2015.

Table 3.   Singapore's Top Destinations by Flight Frequency (weekly one-way), 2010, 2013 & 2014.

Rank	Destination	Jun2010	May2013	Apr2014
1	Jakarta, Indonesia	211	247	282
2	Kuala Lumpur, Malaysia	201	285	249
3	Hong Kong, China	95	159	149
4	Bangkok, Thailand	131	165	128
5	Manila, Philippines	85	129	119
6	Denpasar, Indonesia	68	94	111
7	Penang, Malaysia	66	80	97
8	Ho Chi Minh City, Vietnam	67	82	85
9	Phuket, Thailand	64	73	76
10	Taipei, Taiwan	—	—	66
11	Shanghai (Pudong), China	60	50	60
12	Chennai, India	49	55	59
13	Sydney, Australia	—	—	53

Source. Wikipedia, Singapore Changi Airport, Available at http://www.ask.com/wiki/Singapore_Changi_Airpo rt?o=2802&qsrc=999&ad=doubleDown&an=apn&ap=ask.com#cite_note-changiairportgroup.com-11. Accessed 5 January 2015.

Garuda has 17% share for the Singapore–Jakarta route. Currently, acute airport infrastructure limitations in Soekarno–Hatta Airport in Jakarta are hindering further growth of traffic on this route.[21] Singapore has also developed itself into a hub for an increasing number of other Indonesian airports served by a number of active LCCs.

Singapore and Malaysia have plans to build a high-speed train to connect Singapore to Kuala Lumpur. Work is expected to begin sometime in 2015 with a target completion date of 2020. Travel time could take just 90 minutes on the train, while the total time by air including immigration clearance, could take a total of 2 1/2 hours.[22] High speed rail between Singapore and Kuala Lumpur will be likely to substantially reduce the growth of air traffic between the two cities, judging by a study on the 342-kilometres Tokyo–Nagoya Shinkansen

---

[21] Zubaida Nazeer, Garuda adds briyani, more flights on hot Jakarta-Singapore route, ST 18 November 2014.

[22] Adrian Lim, Malaysia confirms its Singapore-Kuala Lumpur high-speed rail stations, ST 23 October 2014; Chong Pooi Koon, Singapore-Kuala Lumpur bullet train may miss 2020 deadline, Bloomberg, October 29, 2014.

route that found 71% of its Japanese commuters chose to travel by train, 23% by car, 5% by bus and only 1% by plane.[23]

3.1 *Role and contribution of low-cost carriers (LCCs)*

The contribution of LCCs to the development of Changi Airport as an air hub has been significant. Changi Airport opened its Budget Terminal in March 2006, the second in Asia after Kuala Lumpur Airport opened a dedicated terminal catering to budget airlines. The terminal underwent expansion work costing $10 million and added facilities to it in September 2008. With the expanded terminal, it was able to handle seven million passengers a year, up from the original 2.7 million However, the Budget Terminal was closed in September 2012 to make way for Terminal 4 with a larger design capacity of 16 million passengers and is expected to be completed in by 2017.

LCC operations through Changi Airport have expanded rapidly since the opening of the Budget Terminal. The terminal handled about 657, 000 passengers by October 2006, six months after its opening in March. The number of passengers using LCCs at Changi has grown dramatically at the rate of 35% per year over the five years from 2006. LCC flights in Changi Airport constituted 11.3% of total flights in October 2006, compared to 9.6% in April the same year. In 2011, LCCs accounted for about 25% of passenger movements (compared to 22.4% in 2009) and 28.6% of flight movements (26.3% in 2009). In more mature markets such as the United States and Europe, LCCs make up 30–35% of total passenger volumes. The profile of LCC travellers has become more diverse, and now includes business travellers who demand better quality airport and airline services and facilities.

Minister Josephine Teo revealed that in 2012, LCCs served 50 cities in 16 countries. Several Changi city links are served solely by LCCs. These include Yogyakarta, Shantou, Krabi, and Tiruchirapalli in India, and more new destinations from the Changi Airport will be added. Scoot, a budget subsidiary of Singapore Airlines that began commercial operations in 2012, offered long-haul services to Australia's Gold Coast. As a result, LCCs now take up a 25% share of Changi Airport total passenger volume, compared to just 7.7% in 2006, when the Budget Terminal was first opened.[24]

Partly in response to the competition from Malaysia's AirAsia, Singapore-based LCCs only began to operate from 2004 when Valuair, a Singapore based

---

[23] Adrian Lim, Singapore-Kuala Lumpur distance 'ideal for high-speed rail', ST 24 Dec 2014.
[24] Speech by Josephine Teo, 7 March 2012, op cit.

LCC, began scheduling flights in May 2004. This was quickly followed by Singapore Airlines' Tiger Airways (Tigerair) and Qantas' Jetstar Asia Airways, which began commercial flights at the Changi Airport in September 2004 and November 2004 respectively. A planned Singapore-affiliated budget airline by AirAsia was not realised when it failed to obtain an air operator's certificate from the Singaporean aviation authorities.

For Singapore, as a small island nation, all low-cost air routes have to be international in nature, and this imposes higher risks on Singapore-based airlines as there is a greater dependence on aviation negotiations between Singapore and its markets. With relatively limited air rights on offer in the Southeast Asian region, airlines were able to fly to only a limited number of destinations where air rights are available. This resulted in intense, direct competition on specific routes. Nevertheless, airlines were able to take advantage of the subsequent liberation of air rights between Singapore, Brunei and Thailand in the late 2004, although the high density Singapore-Kuala Lumpur route remained closed to all low-cost airlines. However, in spite of the fears of adverse impacts on Malaysia Airlines, Singapore and Malaysia agreed on 23 November 2007 to allow for up to two flights a day for LCCs from each country from 1 February 2008. The route was fully liberalised in December 2008, with the possibility of opening up other routes between Singapore and secondary Malaysian cities.[25] As a result, Singapore Airlines and Malaysia Airlines terminated the shuttle agreement on their service between Singapore and Kuala Lumpur with effect from 1 June 2008. The two national carriers had started the shuttle in 1982 in response to demands for more frequent flights between the two cities.

4. Beginnings of Singapore's Aviation

Commercial passenger carrying air services began in the 1920s preceded by services that carried mail to link European centres to the far-flung Dutch and British Empires. Imperial Airways Ltd (IAL), the state airline of Britain, began in 1926 and linked the United Kingdom to Cairo, the Middle East and Karachi, and thence in 1929 from Delhi to Calcutta and Rangoon, carrying mail and a few passengers. The Royal Dutch Airlines (KLM) also had services that spanned the Dutch empire to connect to the East Indies and the Far East. On 12 September

---

[25] Wikipedia, Aviation in Singapore. Available at http://www.ask.com/wiki/Aviation_in_Singapore? o=2802&qsrc=999&ad=doubleDown&an=apn&ap=ask.com#cite_note-55. Accessed 10 March 2015.

1929, a service was operated fortnightly carrying mail from Amsterdam to Batavia (now Jakarta). On 1 October 1931, KLM started a weekly service with four passengers and mail.

Then came KLM's subsidiary, Koninklijke Nederlandsch-Indische Luchvaart Maatschappij (KNILM), which flew its first aircraft from Batavia and landed in Seletar Aerodrome on 11 February 1930. This was followed by Imperial Airways which began a service in February 1933 jointly operated by IAL and Qantas Empire Airways (QEA), and from December 1934, it connected a network that included India, Burma (Myanmar), Thailand through to Malaysia, Singapore, and Indonesia to destinations in Australia.

The Dutch were ahead of the British in providing international air services from Amsterdam, Singapore and Jakarta since May 1933. A weekly service was started by the British between London and Singapore on 31 December 1933. Qantas operated the Darwin–Singapore route on the Australia–London sector starting on 13 April 1935. The rudiments of an international network with Singapore as its hub had thus already taken shape, with services connecting the metropolitan cities of Europe to the British and Dutch controlled territories in South Asia and Southeast Asia to Australia and New Zealand. The key operators then were state airlines of the United Kingdom, the Netherlands and Australia.

The credit of the creation of Singapore as a modern global air transport hub belongs to the Republic of Singapore, and is part of Singapore's narrative of its 50 years of nation-building. The success of the Singapore hub could not have come about without the vision, drive and commitment of the leadership of those in charge the development of a new nation and its vibrant economy. The growth of the air transport industry is intimately tied to the superb physical planning of the island as well as the vital interlinks with the network of roads, telecommunications, utilities as well as the support of manufacturing industries and attendant logistics and financial services that relied on an efficient air transport system.

5. Path to Creating A Premier International Airport

The Governor of Singapore, Sir Cecil Clementi declared in August 1931 that:

> "Looking into the future, I expect to see Singapore become one of the largest and most important airports in the world, and it is therefore essential that we should have close to the heart of the town, an aerodrome equally suitable for landplanes and for seaplanes" [Allen, 1990].

Clementi's vision was fully justified though understandably, he could not have possibly imagined the creation of a nation-state and the huge changes to the demographic and economic status of the country that ensued, and the creation of such a huge network of today's air hub.

This section of the chapter briefly traces the airport's evolution from Singapore's first airports from the completion of the Seletar Aerodrome in 1929 and the Kallang Airport built in 1937, then the Paya Lebar Airport in 1957, and to the opening of Changi Airport in 1981. The airport has subsequently added three other terminals excluding the Budget Terminal which has itself given way to the planned Terminal 4 now under construction. Table 4 captures the sequence of developments or earlier airports leading to the present Changi Airport as well as the planned airport developments in the future.

5.1 *Predecessors of Changi Airport*

This section traces the early development of commercial airports in Singapore beginning with the Seletar Airport under the British Administration. The facility was built in 1929 to serve as a military airport but operated as a civilian airport during 1930–1937. The Kallang Airport was constructed in 1937 before the Paya Lebar Airport was established in 1955. Singapore then firmly took on the development of Changi Airport that opened in 1981, while modernising the Paya Lebar Airport in the years preceding its move to Changi.

5.1.1 *Seletar airport*

The earliest airport in Singapore that operated commercially is the still existing Seletar Airport, which operated as a commercial airport during 1930 to 1937, performing the function of a general aviation airport, mainly for chartered flights and training purposes. It was managed by the Civil Aviation Department under the Ministry of Transport prior to its change to CAAS. Subsequently, the Changi Airport Group took over the control of the airport from CAAS on 1 July 2009.[26]

Currently, the airport is open 24 hours a day, has a single runway with 27 aircraft stands and can handle 840 tonnes of freight per day. In 1998, the airport received a total of 7, 945 scheduled flights, handled 23, 919 passengers and 6,025 tonnes of cargo. In 2013, the airport handled 22,000 passengers and in the first nine months of 2014, 18,000 passengers were handled. About 80%

---

[26] Seletar Airport. Retrieved from http://www.seletarairport.com/. Accessed 3 December 2014.

Table 4. Evolution of Singapore's Airports and Future Expansion.[27]

Airport	Date Opened	Remarks	Capacity (pax/annum)
Seletar Aerodrome [Allen, 1990]	1929 Work began in 2015 to expand and upgrade the airport	Built in 1928 as part of the naval base. Currently handles mainly private and pilot training flights, as well as chartered flights. A new control tower was planned and the runway to be lengthened from 1,600 metres to 1,840 metres	—
Kallang Airport	12 June 1937	Covered an area of 105 hectares (grass surface) with a slipway for flying boats.	8 million
Paya Lebar Airport	20 August 1957	Decision to built made in 1951. Built under British rule by Public Works Department. One runway 2,438 metres long and 61 metres wide.	1.0 million
Changi Airport Terminal 1	29 December 1981	Phase I Runway length 4,000 metres Phase II development from 1981 comprised a second 4,000-metres runway, 45 aircraft parking bays, airfreight complex, cargo agent buildings and a 78 metres high control tower, fire station, and a third cargo agent building. Phase I cost $1.3 billion Together with the Jewel Project, redevelopment of T1 will add passenger handling capacity by 3 million passenger movements per annum. Completion by 2018.	Original 21 million Increased to 24 million

(*Continued*)

[27] Information on Changi Airport in this table is drawn from a number of sources including Wikipedia, Changi Airport Singapore http://en.wikipedia.org/wiki/Singapore_Changi_Airport#cite_note-73; and CAAS website http://caas.gov.sg/caas/en/index.html.

Table 4. (*Continued*)

Airport	Date Opened	Remarks	Capacity (pax/annum)
Terminal 2	1 June 1991	Second runway completed in 1983. Construction of T2 began in 1985 and completed in 1989 and opened for operation in Nov 1990. 2006 — Official completion of S$240 million upgrading of Terminal 2.	23 million
Budget Terminal	31 October 2006 Closed in September 2012 to make way for T4.	Dedicated terminal for low cost carriers is the first in Asia. It became operational on 26 March 2006. In September 2008, expansion works costing $10 million was undertaken before it was closed.	7 million expanded from 2.5 million
JetQuay CIP Terminal	Opened March 2002 as Changi Business Aviation Centre before changing to its present name.	Privately operated, offers private check-in, baggage handling, and immigration clearance services. The dedicated CIP (Commercially Important People) terminal caters to private jets or in any class, on any airline, through other terminal.	
Terminal 3	July 25, 2008	Construction started in 1999 and began operation on 8 January 2008 after a delay of two years. Cost $1.75 billion	22 million
Terminal 4	2017 Projected	To be built on the site of the Budget Terminal. Cost $985 million(4). Construction commenced in November 2013 with a focus on LCCs.	16 million
Terminal 5	Mid-2020s Projected	Covers 10,000 hectares, located in Changi East. A third runway, cargo complexes and other supporting infrastructure will be built.	30–50 million

of the total number of flights operating at the Seletar Airport are accounted for by the various flying schools, while the remaining flights are due to aircraft charters, repairs and maintenance.

5.1.2 *Kallang airport*

The Kallang Airport was officially opened on 12 June 1937 and operated till 1955. It had a single runway and a small terminal. It was described as "the finest airport in the British Empire". The airport was built on reclaimed land. Improvements were made by the Japanese Occupation Forces (1942–1945) and subsequently, after the war, it was upgraded by the British. By that time, the airport could not handle the larger and heavier aircraft that demanded better infrastructure and facilities.

Due to the lack of suitable space for extension and Kallang Airport abutted the already heavily built-up precincts, it was decided that the new airport will be located elsewhere. It is interesting that the then Public Works Department (PWD) had explored the possibility of locating it in Changi but had then abandoned the idea due to unsuitable soil conditions and by technology that was then not available to the construction of a modern airport. Finally, the decision was made in 1950 to build the new airport in Paya Lebar, located just 8 kilometres to the east of the city centre.

5.1.3 *Paya Lebar airport*

Paya Lebar Airport was Singapore's third main civilian airport. It was opened in August 1955 with a single runway and a small passenger terminal. With rapid growth in global air transport, the airport started to encounter congestion problems. Its inability to cope with the fast-rising passenger traffic became critical by the 1970s; annual passenger numbers rose dramatically from 300,000 in 1955 to 1.7 million in 1970, and to 4 million in 1975.

Following the recommendations by a British aviation consultant, a decision was made in 1972 to keep the Paya Lebar Airport. The plan was to build a second runway together with an extensive redevelopment and expansion to the passenger terminal building. However, concerned that the existing airport was located in an area with the potential for housing development and urban growth, which would physically hem it in on all sides, the government subsequently decided in 1975 to build a new airport at the eastern tip of the main island at Changi, at the site of the former Changi Air Base, where the new airport would be easily expanded by means of land reclamation. In addition, airplanes could fly over the sea, avoiding noise pollution issues within residential areas and help to avoid disastrous consequences on the ground in the event of an air mishap. The Paya Lebar airport was then converted for use as a military air base and was closed to civil traffic.[27]

5.1.4 *Changi airport — birth, expansion and upgrading*

The decision to build the Changi Airport was made in 1975, where more space to accommodate a modern international airport that can handle larger aircraft was needed. The policy has been, and continues to provide timely incremental capacity ahead of demand, to avoid congestion problems as well as to maintain high service standards and attractiveness to the users of the airport.

Terminal 1 went through its first major refurbishment in 1995 at a cost of S$170 million, prior to the commencement of expansion works three-year later to add 14 aerobridges at a cost of S$420 million, which was completed in 1999. When Terminal 2 was opened two years later, it was well ahead of traffic demand at the time. An additional two piers of aerobridges costing S$330 million were completed in 1996. In 2002, work commenced on the Changi Airport Skytrain as well as the Terminal 2 building, and in September 2006, the airport marked the completion of an extensive upgrade costing S$240 million in Terminal 2, which included an updated glass-fronted facade, interior decor, and terminal layout modifications.

Terminal 3 began construction in 1999 and opened in July 2008. The terminal increased the airport's passenger capacity annually by 22 million, bringing the total annual capacity up to 70 million passengers. The nine-storey Crowne Plaza Hotel adjacent to Terminal 3 was also constructed.

The Budget Terminal, with a capacity of 8 million, was opened in October 2006 as a dedicated terminal for handling LCCs. However, it was closed in Sept 2012 to make way for the new Terminal 4. Due to be completed in 2017, Terminal 4 would double the original Budget Terminals' capacity to 16 million passengers. In terms of facilities, the Changi Airport Group (CAG) has indicated that Terminal 4, like the present Budget Terminal, will not have aerobridges. This will keep open the option of a no-frills model preferred by most LCCs. The new Terminal 4 would better position Changi Airport to take advantage of the continued strong growth of LCCs in support of regional tourism. Some LCCs now also fly long-haul, offer transfer services, and even business class options, therefore requiring much improved facilities and services.[28]

---

[27] Wikipedia, Singapore Changi Airport. Available at http://www.ask.com/wiki/Singapore_Changi_Airport?o=2802&qsrc=999&ad=doubleDown&an=apn&ap=ask.com. Accessed 23 November 2014.

[28] Josephine Teo, speech, 7 March 2012. Op cit.

Changi Airport's present three passenger terminals, operating with two runways can handle 66 million passengers a year. By around 2020, with the addition of a fourth terminal, together with a third runway, the airport will be able to handle a total of 85 million passengers per year. Plans are also in place to further expand Terminal 1.

Of special interest is the construction of a spectacular facility named the Jewel that will be connected directly to Terminals 1, 2 and 3. Work on the project began in December 2014 and is expected to be completed by 2018. It will feature a distinctive dome-shaped façade made of glass and steel. The five-storey structure will be built over the 3.5 hectares open-air Terminal 1 carpark and will have a 13,000 square metres (1.4 million square foot) mixed-use complex with retail outlets, airport services and a hotel. The feature will transform the airport and will further differentiate Changi Airport from other major air hubs. Project Jewel is a joint-venture between the Changi Airport Group (CAG) and a local developer, CapitaMalls Asia.[29]

Well beyond that, Terminal 5 when completed will combine to handle a total of 135 million passengers. In addition, space will also be set aside for future satellite terminals if additional capacity is required. Terminal 5 will be linked to other airport terminals via underground tunnels so the expanded Changi Airport can be operated as a single, integrated airport with easy transfers between terminals, and provide maximum convenience for passengers and efficient airfield operations.

The expansion and upgrading plans were set in motion by a high-level multi-agency Changi 2036 Steering Committee set up in March 2012 to look into the airport's future needs.[30] The Committee comprises eight government agencies and the CAG, the Changi Airport operator. The purpose of the Steering Committee is to provide strategic directions for the future developments of the Changi Airport and its surrounding area. The aim is to guide Changi Airport's growth over the next quarter of a century, and take it to the next major milestone, and to greatly strengthen Changi Airport's premier air hub position for the future. The Committee was tasked to develop a concept plan that addresses key issues which include: The timeline and development of

---

[29] See Adrian Lim, Work on $1.57b Jewel project to kick off by year end, ST 29 October 2014; Karamjit Kaur, Changi's Jewel: Not just another mall, ST 6 Dec 2014; Speech by Mr Lui Tuck Yew, Minister for Transport, at the groundbreaking ceremony of Jewel Changi Airport on 5 December 2014. Available at http://www.mot.gov.sg/News-Centre/News/2014/Speech-by-Mr-Lui-Tuck-Yew,-Minister-for-Transport,-at-the-groundbreaking-ceremony-of-Jewel-Changi-Airport-on-5-December-2014/. Accessed 28 January 2015.

[30] Karamjit Kaur, Budget 2013: Third runway for Changi Airport by 2020, ST 13 March 2014.

a new passenger terminal; measures to augment Changi Airport's infrastructure, in particular the civilian use of Changi's third runway; and the phasing of development for related infrastructure, facilities and transport links to support Changi's future growth. Relevant stakeholders such as airlines, airport service-providers, and aviation industry players would be consulted.[31]

Improvements and upgrading of physical facilities, services and procedures have been implemented continuously at the Changi Airport. Apart from upgrading and expansion works on the airport terminals, other projects include, the launch of the first A380-compatible gate holdroom in August 2005, F31, and the installation of a third Passenger Loading Bridge (aerobridge) at all 19 A380-compatible gate holdrooms at the Changi Airport. On 11 November 2005, Changi Airport became the first airport outside Europe to welcome the A380. In the same year, CAAS awarded a site at the airport for lease for the development of an airport hotel adjoining Terminal 3 and connected to the Skytrain from Terminal 1 and Terminal 2. The process of renewal and implementing new innovative schemes is sustained, and the Jewel Project would be the next exciting feature of the airport, pushing Changi Airport to the forefront of the international airport scene.

Changi Airport has garnered more than 470 accolades since it was opened in 1981. In 2015, Changi Airport was awarded by Skytrax the best airport ahead of Incheon Airport of Korea and Munich Airport in Germany for the third year in succession.[32] It ranks as the sixth busiest international airport today, and is a global air hub serving more than 106 international airlines flying to some 300 cities in about 70 countries and territories worldwide. More than a mere airport, it has also about 350 retail and services outlets and 120 food and beverage outlets covering over 70,000 square metres of space. It has become a favourite destination for Singapore residents, especially families offering a welcomed outlet for people to gather for an experience.[33]

A major source of revenue for the Changi Airport is retail sales, which came in at more than $2 billion in both 2013 and 2014 according to the CAG. Visitors from China accounted for a quarter of the airport's retail business,

---

[31] Ministry of Transport, Press Release, 7 March 2012. Fact Sheet on Changi 2036 Steering Committee http://www.mot.gov.sg/News-Centre/News/2012/Fact-Sheet-on-Changi-2036-Steering-Committee/. Accessed 28 January 2015. The Steering Committee is chaired by Minister of State for Finance and Transport, Mrs Josephine Teo. Members comprise senior officials from Ministries of Transport, Defence, and Finance, and statutory boards, CAAS, Economic Development Board, Land Transport Authority, Singapore Tourism Board and Urban Redevelopment Authority.

[32] Karamjit Kaur, Changi Airport tops global survey again, ST 13 March 2015.

[33] See Changi Airport Singapore website http://www.changiairport.com/our-business. Data accurate as on 14 August 2014.

followed by Singapore an air passengers who contributed one-fifth of the sales. Other top spenders were from Indonesia, India and Australia.[34]

5.2 *Providing excellent ground handling services* [35]

While it is crucial to have a truly world-class airport infrastructure, the success of the Changi Airport is also due to the superior quality of the ground handling services in providing passenger handling, baggage handling, ramp and cargo handling as well as inflight catering services. It is acknowledged as one of the world's best international airports for superior standards of ground handling. Ground handling services are provided by three companies, i.e., Singapore Airport Terminal Services (SATS), Dnata Singapore, Aircraft Service International Group (Asig). As of 2014, SATS controls over 70% of Changi Airport's ground handling business and serves 51 passenger carriers at Changi Airport.[36]

SATS provides cabin service, catering, ramp service, passenger service, cargo handling as well as aviation security. SATS uses a baggage reconciliation system (BRS) to automatically track baggage, and that significantly reduces the number of mishandled bags.[37] In addition, SATS also provides 'technical ramp' services such as aircraft marshalling and parking and tow-in/push-back services. The company has invested over $500 million in infrastructure at Changi, comprising six air freight terminals, two inflight catering centres, maintenance centres and its head quarters. By the mid-1980s, SATS was able to handle about 20,000 passengers a day at the Changi Airport, a 60% increase over the 12,700 passengers handled daily at the Paya Lebar Airport in 1980. Air cargo handled also registered double-digit growth rates.

In 1985, SATS was restructured into four companies so that it could better manage demand for its services. SATS' created four subsidiaries to cater to apron services, cargo services, passenger services, and catering. In September 2009, SATS was divested from SIA to form a separate entity. The company changed its name then from Singapore Airport Terminal Services Limited to SATS Ltd. SATS unveiled a new brand identity in June 2011. The company has

---

[34] Adrian Lim, $2b in retail sales at Changi despite drop in China visitors, ST 2 February 2015.
[35] Section on ground handling drawn mainly from Wikipedia, Singapore Changi Airport. Available at http://www.ask.com/wiki/Singapore_Changi_Airport?o=2802&qsrc=999&ad=doubleDown&an=apn&ap=ask.com. Accessed on 5 January 2015.
[36] SATS. (September 2010). Gateway Services. Available at http://www.sats.com.sg/AboutUs/WhatWeDo/Pages/WhatWeDo.aspx. Accessed 18 November 2014.
[37] Karamjit Kaur, Fewer travellers' bags lost, delayed at Changi. *Straits Times*. 25 April 2012.

transformed itself into Asia's largest gateway services and aviation food solutions company, with a network of operations at 35 airports in 10 countries principally in the Asia-Pacific region.

Renowned for its product and service innovations, SATS launched its $16 million Coolport @ Changi, in 2010 — Asia's first on-airport perishables handling centre within the Free Trade Zone. The facility is the first dedicated on-airport facility in Asia for handling terminal and transit perishables cargo including flowers, meat and pharmaceuticals. Located within SATS Airfreight Terminal 2, Coolport @ Changi has a handling capacity of around 250,000 tonnes, with scope for expansion from the current 8,000 sq m to 14,000 square metres. [38]

The Changi International Airport Services (Private) Limited (CIAS) was formed in 1977 by the Port of Singapore Authority (PSA) and five equity partner airlines — Air France, China Airlines, Garuda Indonesia, KLM and Lufthansa. The company handled the remaining market share. CIAS was restructured when its shareholding was bought over by the Emirates Group's Dnata, a subsidiary of Emirates Airlines in 2004. It was rebranded as Dnata Singapore in August 2011. Dnata too, subsequently opened a 1,400 square metres temperature-controlled perishable handling centre in 2013, Coolchain, handing 75,000 tonnes of perishable cargo each year.[39]

In the early 2000s, CAG believed that the addition of the new entrant would boost competition between ground handlers, and lower the cost of ground handling for all airlines operating into and out of Changi Airport.[40] The government offered a third license. It was won by Swissair's Swissport and they commenced operations in March 2005. The entry of a third ground handler resulted in reduced handling rates then by more than 10%.[41] By September 2007, Swissport had secured seven customers at the Changi Airport, including Northwest Airlines, Tiger Airways, AirAsia, Adam Air, Swiss World Cargo and Cardig Air. In 2009, due to the severe global economic recession and intense competition, Swissport Singapore suffered a hefty loss of more than $50 million and it had to withdraw operations from Singapore.[42] Following the

---

[38] Wikipedia, SATS Ltd. Available at http://en.wikipedia.org/wiki/SATS_Ltd. Accessed 13 January 2015.

[39] Natasha Ann Zachariah, All in a Chilly Day's Work, ST 14 February 2015.

[40] Leo, D. Dated 25 July 2011. ASIG's success as third ground handler is important to Changi. Available at http://www.aspireaviation.com/2011/07/25/asigs-success-as-third-ground-handler-is-important-to-changi/ Accessed 20 November 2014.

[41] Karamjit Kaur, Third ground-handler gives boost to airport's competitiveness. Straits Times 10 September 2007.

[42] Karamjit Kaur, Swissport to pull out of Changi in March. Straits Times 13 January 2009.

withdrawal of Swissport, the US-based Aircraft Service International Group (Asig) was granted a 10-year ground handling licence starting in January 2012.[43] However, Asig was only able to secure its first customer, Jetstar, in October 2014.

With the strong growth of LCCs, SATS decided to set up a subsidiary, Asia-Pacific Star Pte Ltd, in March 2009 to cater to mainly LCCs operating at the then-Budget Terminal. The company offered differentiated ground handling services with a lower cost structure for LCC airlines.[44] Its clients included Tiger Airways, Jetstar Asia, Valuair, and Lion Air.

5.3 *Role of the government*

The Changi Airport came under the care of the Civil Aviation Authority of Singapore (CAAS) which became a statutory board in 1968. It grew out of the Department of Aviation under the Ministry of Transport. CAAS remains as the regulatory authority for civil aviation in Singapore and continues to provide air navigation services. It has a broader role of fostering the growth and development of Singapore as an air hub and the aviation industry as a whole, while overseeing and promoting safety in the industry. As part of its strategic thrusts, CAAS also fosters aviation knowledge and expertise by becoming a centre of aviation learning and thought leadership. The Singapore Aviation Academy (SAA) was set up in 1958 as the training arm of CAAS. CAAS also contributes and engages actively in the international aviation community.

On 1 July 2009, Changi Airport was corporatised and since then has been managed by the Changi Airport Group (Singapore) Pte Ltd (CAG) which was formed on 16 June 2009. CAG undertakes key functions focusing on airport operations and management, air hub development, commercial activities and airport emergency services. Through its subsidiary Changi Airports International, the Group invests in and manages foreign airports to leverage on the success of managing Changi Airport.[45]

[43] Changi Airport Group (CAG) dated 9 June 2011. Changi Airport Group awards third ground handling license. Available at http://www.changiairportgroup.com/export/sites/caas/assets/media_release_2011/9_June_2011.pdf. Accessed 20 November 2014.

[44] SATS, Media Release March/09, SATS Launches low-cost ground handling unit. 10 March 2009. Available at http://www.sats.com.sg/Media/NewsContent/PR-10_Mar09.pdf. Accessed 13 January 2015.

[45] Changi Airport Group (CAG), A record 51 million passengers for Changi Airport in 2012, Press Release 31 January 2013. Available at http://www.changiairportgroup.com/export/sites/caas/assets/media_release_2013/Media_Release_-_A_record_51_million_passengers_for_Changi_Airport_in_2012_xwebx.pdf. Accessed on 26 January 2015. MOT, Corporatisation of Changi

5.3.1 *Promoting air hub development*

CAAS continually strives to position Singapore as a dynamic international air hub. In November 2002, CAAS announced a three-year $210 million Air Hub Development Fund (AHDF) to help airlines and the aviation industry tide over the uncertainty in the global aviation industry. In 2006, CAAS extended the Air Hub Development Fund 2 (AHDF2) of $300 million for a further three-year period. AHDF2 is an enhanced extension of the AHDF which had expired. In February 2009, CAAS further increased relief packages for aviation partners to $200 million. Following that, CAG announced the Changi Airport Growth Initiative on 25 November 2009, offering customised incentives to airlines and other airport partners to boost passenger and cargo traffic at Changi, and to strengthen Changi's air traffic network.

An Airport Productivity Steering Committee (APSC) was formed in 2013 comprising key Changi Airport stakeholders. The Committee was tasked to provide strategic directions, coordinate airport-wide collaboration, and rally airport agencies to support productivity improvement efforts. The committee has developed productivity roadmaps for passenger handling, cargo handling, baggage handling and apron processes. These roadmaps outline initiatives to improve labour productivity through technology adoption and automation, infrastructure changes, and process and job redesign. The APSC has also set an annual airport productivity improvement target of 3–4% until 2020.

On 18 July 2014, CAAS announced that it would set up an Airport Productivity Package, with S$100 million up till 31 March 2018. Two new programmes will be rolled out under this package. In the first programme, S$20 million has been set aside to encourage airport stakeholders such as ground handlers, line maintenance companies and airlines, to adopt off-the-shelf equipment to improve the efficiency of their operational processes and reduce reliance on manpower. The second programme, Aviation Challenge, is a competition-based programme that seeks to tap on the intellectual capacities of the industry, academia and others to develop innovative solutions for the strong and sustainable development of Singapore aviation.

In March 2015, CAAS announced that it would renew the Aviation Development Fund (ADF) for five years from 1 April 2015, with an expanded size of S$160 million to drive innovation and productivity at Changi. The renewed ADF aims to drive transformative developments to ensure the

---

Airport and restructuring of Civil Aviation Authority of Singapore (CAAS) — Corporatisation model announced. Available at http://www.mot.gov.sg/DATA/0/docs/Corporatisation_news_release.pdf. Accessed 30 December 2014.

Singapore aviation hub's continued growth and competitiveness, with airport innovation and productivity as the key focus.

In order to relieve airline cost pressures in a period of slowing traffic growth, CAAS offered to absorb S$50 million of the air navigation services charges to Changi Airport Group (CAG) for two years. CAG has committed to passing on the savings to all airlines operating at the Changi Airport. This will translate to a 10% rebate in landing charges for airlines at the Changi Airport from 1 September 2014 to 31 March 2016.[46]

5.3.2 *Providing supporting infrastructure*

For a small, densely populated and highly developed country, urban planners need to adopt a long-term physical development plan in order to minimise the problem of conflicting and incompatible land use, and be able to fulfill the necessary targets efficiently. The Urban Redevelopment Authority (URA) has adopted the approach of Long-Range Comprehensive Concept Plans backed by the Statutory Master Plans with this in mind. In the case of the Changi Airport, the 1971 concept plan known as the 'Ring Plan' had already envisaged a large airport to be located in the Changi area. The decision to build the Changi Airport in 1975 was made easier for a number of reasons, including ample available land that could be augmented by land reclamation, an existing military airfield in the area, minimum conflict with existing alternative users such as public housing or seaport development, and a good distance from densely urbanised areas [Olszewski and Chia, 1991]. It was revealed by the chairman of CAG, Mr Liew Mun Leong, that the decision made in mid-1975 to drop the plans to expand the Paya Lebar Airport in favour of building a new airport at Changi was made by former Prime Minister, the late Mr Lee Kuan Yew. The enormous responsibility of having the airport successfully constructed by 1981 was given to the then Public Works Department (PWD), while the task of land reclamation was assigned to the Port of Singapore Authority (PSA).[47]

Within the Changi Airport complex, a Mass Rapid Transit (MRT) station was officially opened in February 2002. The $850 million project

---

[46] CAAS Media Release dated 18 July 2014. Available at: http://www.caas.gov.sg/caasWeb2010/ opencms/caas/en/Media/news_details.html?newsURL=http://appserver1.caas.gov.sg/caasme-diaweb2010/opencms/Journalist/Press_Releases/2014/news_0008.html accessed 7 December 2014. See also Karamjit Kaur, Talk under way to inject more money into aviation fund, ST 17 December 2014.

[47] Liew Mun Leong, Lee Kuan Yew — Truly the father of Changi Airport, Business Times (Singapore), 31 March 2015. Available at http://www.businesstimes.com.sg/opinion/lee-kuan-yew-dies/lee-kuan-yew-truly-the-father-of-changi-airport. Accessed on 14 April 2015.

involved a 6 kilometre MRT line extension from the Tanah Merah MRT station on the East–West line. Travel within Changi and to other parts of Singapore will be made easier when the Downtown Line, currently under construction, opens by 2017. Serving the Upper Changi area and connected to the Changi Airport Line (CAL) at the Expo MRT station, commuters in the Changi area would have ready access to the Central Business District, Marina Bay and the eastern regions of Singapore.

The proposed Cross Island Line, expected to be completed by 2030, will provide an alternative east-west MRT route connecting the Changi Airport to the surrounding public housing and industrial estates, as well as recreation areas in the east and north-east to the centre and west of Singapore. There will also be improvements along Loyang Avenue and Nicoll Drive, a realigned Changi Coast Road and the widening of Tanah Merah Coast Road.[48]

Land transport around Changi will be improved to ensure convenient access to the future Terminal 5, which will also be connected to the MRT network. The adequacy of bus services to the airport would also be reviewed. The existing Changi Coast Road will be realigned to make space for a continuous and integrated airfield.[49]

As an international aviation, shipping, logistics and financial services hub, Singapore is well supported by advanced telecommunications technology, and is rated the world's second 'most network-ready' country for five years and running. Singapore has one of the world's leading national ultra-high speed fibre infrastructure and services direct to homes and offices. Singapore is one of the most advanced telecommunications hubs worldwide with a well-established infrastructure for international connectivity. It is among the first in the world to implement a nationwide wireless broadband network Wireless@ SG, with about 5, 000 hotspots across the nation. Singapore's mobile penetration stands at 156% — one of the highest in the world.[50]

---

[48] Urban Redevelopment Authority (URA), Changi – Draft Master Plan. Available at http://www. ura.gov.sg/MS/DMP2013/regional-highlights/~/media/dmp2013/ Planning%20Area%20 Brochures/Brochure_Changi.ashx/. Accessed on 30 December 2014.

[49] Royston Sim, Changi T5 to land by mid 2020s, capacity to handle 50 metres passenger movements a year. ST 30 August 2013. Available at http://www.straitstimes.com/breaking-news/ singapore/story/changi-t5-land-mid-2020s-capacity-handle-50m-passenger-movements-year-#sthash.UZxxtzvG.dpuf

[50] IE Singapore, The Singapore Advantage. Available at http://www.iesingapore.gov.sg/~/media/ IE%20Singapore/Files/Publications/Brochures%20Foreign%20Companies/The%20Singapore%20 Advantage/IEThe20Singapore20AdvantageJul2014.pdf. Accessed on 13 January 2015.

5.3.3 *Promoting aerospace industries*

Singapore is a regional leader in aerospace maintenance, repair and overhaul (MRO) and manufacturing. It has become the most comprehensive MRO hub in Asia. Since 1990, the industry has grown at a compounded annual growth rate of 13% since 2009 and accounted for an output of over S$7 billion in 2010. There were over 100 international companies employing some 20, 000 workers carrying out MRO activities in Singapore offering a wide range of capabilities that include airframe maintenance, engine overhaul, structural and avionics systems repair, as well as aircraft modifications and conversion. These core competencies, coupled with our commitment to quality and safety, have made Singapore a recognised one-stop solutions provider for airlines' maintenance and repair needs (Yeo and Tan, 2014). CAAS reveals that Singapore's MRO cluster accounts for more than 20% of the Asia–Pacific MRO market. In short, Singapore's aerospace industry has become an integral economic driver of Singapore.[51]

5.3.4 *Establishing aviation industrial and business parks*

As an integral part of developing Singapore as a global air hub, the government has, since the second half of 1990s, developed business and industrial parks in the vicinity of the Changi Airport that cater to MRO and related activities such as warehousing and logistics services.[52] This requires the whole-of-government (WOG) approach involving the Economic Development Board (EDB), Urban Redevelopment Authority (URA), Ministry of Transport (MOT), and the support of a number of other government agencies. EDB has been the lead government agency driving the growth and development of Singapore as an aerospace hub in encouraging OEM part manufacturers to locate and trade out of Singapore.

JTC Corporation (JTC) has also played an important part in the development of Singapore's aviation hub in the design and construction of all the aviation industrial parks specifically catering to the aerospace industry and air logistics. Leading industry players who have established a firm presence in Singapore include Goodrich Aerostructure, Thales Solutions Asia, Nordam Singapore, Hamilton Sundstrand, Diethelm Keller and Windsor Airmotive in

---

[51] CAAS website. Available at http://caas.gov.sg/caas/en/About_CAAS/Our_Strategic_Thrusts/ Aviation_Industry_Development/Aerospace_Hub/?__locale=en. Accessed on 5 January 2015.

[52] Yeo, D. and Tan, S. Dated Feb 2014, Bridging Skies: The Aviation Hub of Choice. Economic Development Board. http://www.caas.gov.sg/caasWeb2010/export/sites/caas/en/PDF_ Documents/Others/BS_singles.pdf Accessed on 5 January 2015.

Changi North; and GE Aviation Service Operations, Airfoil Technologies International, Pratt & Whitney, Combustor Airmotive Services, International Aerospace Tubes–Asia, Honeywell Aerospace Singapore and Messier Services Asia in Loyang; and Rolls-Royce, ST Aerospace, Bombardier, Pratt & Whitney, Jet Aviation, Hawker Pacific, Fokker Services, MAJ Aviation, Air Transport Training College, Eurocopter, Bell Helicopter and Cessna Aircraft Company and Wah Son Engineering at Seletar Aerospace Park.[53] As of end-2014, there were more than 120 aerospace companies including home-grown companies, ST Aerospace and SIA Engineering Company, offering comprehensive MRO services.[54]

Table 5 summarises the six aviation industrial and business parks in the vicinity of Changi Airport. These are the Changi Airfreight Centre, the Airport Logistics Park of Singapore (ALPS), the Changi Business Park (CBP), the Seletar Aerospace Park (SAP), the Changi International Logispark (South) and the Changi International Logispark (North).

5.3.4.1 Changi Airfreight Centre

When the Changi Airport was opened in 1981, the Changi Airfreight Centre was built at a site on the western edge of the airport within the Free Trade Zone (FTZ). Facilities at the centre include six SATS Air Freight terminals and SIA's aircraft hangars. There is also a hangar operated by SASCO/ST Aerospace.

The key tenants within the airport free trade zone include SIA Airline House, SATS Cargo, SATS Maintenance and Dnata (formerly CIAS) and SIAEC–SIA's Engineering subsidiary company. SingPost's AirMail Centre is also located in the area. The Cargo Agents Building and Changi Megaplex were completed in November 2002. The presence of the Immigration and Customs Authority (ICA) and AVA's Animal and Plant Quarantine facility help to facilitate the processing of animals, livestock and plants as well as perishables.

CAG has worked with DHL Express since 2013 to invest in a $140 million 24-hour sophisticated automated facility at the Changi Airfreight Centre to be completed in 2016. DHL had been able to handle up to 225 tonnes a day of 2,400 shipments an hour. The planned automated system would provide the necessary capabilities to support the fast pace of growth of trade within the

---

[53] JTC, Aerospace Industries (n.d.). Available at: http://www.jtc.gov.sg/Industries/Aerospace/Pages/default.aspx. Accessed 13 January 2015.

[54] Speech by Josephine Teo, Senioir Minister of State for the Ministry of Transport, during MRO Asia 2014 on 5 November 2014. Available at http://www.mot.gov.sg/News-Centre/News/2014/Speech-by-Senior-Minister-of-State-Mrs-Josephine-Teo-at-MRO-Asia-2014-on-5-November-2014/. Accessed on 15 January 2015.

Table 5.   Aviation-related planned industrial and business parks.

Park	Launched	Notes
Changi Airfreight Centre & Vicinity	Opened in1980 together with T1. Added more buildings and facilities since then.	Located at the Western end of the Changi Airport. Accommodates SIA Airline House, SATS AFT, AFT5, AFT6, SIA Engineering hangar, Dnata Cargo Terminal, Cargo Agents Building, Singapore Post, warehouses, office blocks, Immigration and Customs Authority Import Office, and Agri-Food & Veterinary Authority (AVA) Animal and Plant Quarantine (CAPQ) Office.
Airport Logistics Park of Singapore (ALPS)	26-hectares, opened in 2003, located within a Free Trade Zone	Dedicated to third-party logistics(3PL) operators and home to world-class players such as Menlo Worldwide Forwarding Inc, Schenker (Singapore) Pte Ltd, Nippon Express (Singapore) Pte Ltd, SDV Logistics (Singapore), UPS SCS (S) Pte Ltd and DHL Supply Chain Singapore Pte Ltd.
Changi Business Park (CBP)	S$1.5 billion investment. Launched in July 1997. Completed in 2012. 71-hectares in area and located South of Changi Airport	Comprises a mix of high-tech business, data and software enterprises, research and development as well as knowledge-intensive facilities. Home for major MNCs such as Honeywell, IBM, Invensys, Ultro Technologies, and Xilinx. Also host to financial institutions such as DBS Bank, Citibank, Credit Suisse and Standard Chartered Bank within the park.
Seletar Aerospace Park (SAP)	Completion of first two phases by 2018.	Caters to aerospace industries. Located in Seletar, Cost S$60 million to develop 140 hectares of land adjacent to Seletar Airport to include the upgraded Seletar Airport within the Park. The Park is home to 100 leading global and local aerospace companies including engine-maker Rolls Royce, helicopter manufacturer Europcopter, ST Aerospace and Jet Aviation, Bell Helicopter, Cessna, Hawker Pacific and Fokker Services.
Changi International Logispark (South)	Developed by Jurong Town Corporation (JTC) on 40.4 hectres area	Located close to Changi Airport, it is dedicated to logistics companies, it is fully allocated. Major tenants are: Schenker Singapore, DHL Global Forwarding, UPS Singapore.
Changi International Logispark (North)	Developed by Jurong Town Corporation (JTC) on 16.5 hectares area	Located close to Changi Airport, Changi International LogisPark (North) has land dedicated to logistics and warehousing usage. Tenants include Zuellig Pharma, Agility International Logistics.

region and to support up to five times more flights in Singapore. This development would further bolster Singapore's position as top logistics hub in Asia as ranked by the World Bank in 2014. [55]

5.3.4.2 Changi Business Park (CBP)

Following the International Business Park in Jurong East, the Changi Business Park (CBP) is the second business park planned by the Singapore government. The construction of the CBP started in 1997 with an investment of more than S$1.5 billion and land occupancy of 71 hectares in 2014. One initial objective of this business park was to attract a new category of business involving inventory control, freight management, technical support and training.[56] With time, the CBP has now evolved into a major info-communications technology and banking hub in Singapore. It is also a centre for higher-value added industry, such as technology business and research and development divisions not necessarily related to the aviation industry.[57] Its proximity to the airport significantly lowers the cost of transportation and logistics for companies located in CBP.

5.3.4.3 Airport Logistics Park of Singapore (ALPS)

Opened in 2003, the Airport Logistic Park of Singapore is strategically located in the Free Trade Zone. The 26 hectares of logistic park is fully occupied by 3rd-party Logistics Players (3PLs) providing *the quick turnaround of value-added logistics and regional distribution services. ALPS is currently hosting world-class logistics players including some of the largest air express and air freight companies, DHL, UPS, Fedex, TNT, and Nippon Express.[58]

5.3.4.4 Seletar Aerospace Park (SAP)

In 2007, JTC Corporation together with the Economic Development Board (EDB) announced its master plans to convert the ageing Seletar Airport into a new aviation hub, the Seletar Aerospace Park (SAP). This move was intended to separate the functions of the Changi Airport and the nearby the Seletar Airport. Unlike the uniformity and scale of aviation services that Changi Airport provides, the focus of the Seletar Airport is to host "a multitude of specialist and industrial players, operators and companies that have different needs and

---

[55] Ariel Lim, DHL to build $140m hub in Changi Airport, ST 6 March 2015.

[56] Oon, D. Government to spend S$1.5b on Changi Business Park. Business Times Singapore 26 June 1997.

[57] Chng, G. Changi Business Park Shaping up as Infocom Tech Hub. Straits Times. 6 March 2014.

[58] JTC. (July 04 2014). Airport Logistics Park of Singapore. Available at http://www.jtc.gov.sg/Industries/Logistics/Pages/Airport-Logistics-Park-of-Singapore.aspx. Accessed on 3 January 2015.

business models".[59] SAP's success is based on the agglomeration of industrialists, suppliers and service providers in the same value chain in a single location. For companies located in SAP, it yields economies of scale and creates opportunities for collaboration and synergy. It also allows JTC and the other agencies to engage the industry very closely to provide customised infrastructure solutions to specifically cater to the industry's needs.

Before 2007, the Seletar Airport area was already home to 30 aerospace companies. The expansion and renovation follows years of requests for improved facilities from the aerospace industry. The land for Seletar Aerospace Park (SAP) was allocated in January 2007. JTC has now started infrastructure works for Phase 3, which will provide another 60 hectares of industrial land when completed. SAP has become an integrated aerospace hub hosting a wide range of high value-added activities including MRO, manufacturing and assembly of aircraft engines and components, business and general aviation, training, and research and development.[60] The target for SAP was initially to create 10, 000 new jobs and contribute to 1% of Singapore's GDP.[61]

The Seletar Aerospace Park covers an area of 160 hectares and is hosting aerospace MRO, the design and manufacturing of components and systems, as well as the assembly and manufacturing of small aircraft. It will complement Changi and Loyang with developments in the narrow-body aircraft market. SAP is also the location of an aviation campus for the training of pilots, aviation professionals and technical personnel to meet the demand of the Asia–Pacific region in the near future.

Table 5 also lists the major tenants located in SAP. Up till February 2014, 45 aerospace companies are already situated in the park. In 2010, the British engine manufacturer Rolls-Royce launched its $700 million project to set up factories, training facilities and technology centres in SAP.[62] It offers a diverse range of services including maintenance, repair and overhaul (MRO), technical support, procurement, sales and marketing as well as head quarter functions for research and Development (R&D) in Singapore. Scoot will fit its new Boeing 787 aircraft with Rolls–Royce's Trent 1000 engine which will be

---

[59] Sreenivasan, V. Rumblings over Seletar Hub's Goals. Business Times 29 September 2008.

[60] Minister of State for the Ministry of Trade and Industry (MTI), Teo Ser Luck, Keynote Address by Mr Teo Ser Luck, Minister of State for Trade and Industry, at Launch of Aviation Two @ Seletar Aerospace Park and Expansion of JTC aeroSpace @ Seletar Aerospace Park on 7 February 2014, 3.55pm, at Seletar Aerospace Park. Available at http://www.mti.gov.sg/NewsRoom/Pages/Mr-Teo-Ser-Luck-at-the-.aspx. Accessed 14 January 2015.

[61] Lee, C. W. Seletar Airport to Become Aerospace Park. Today (Singapore). 11 May 2006.

[62] Ramchandani, N. Rolls-Royce Plans Plant at Seletar. Business Times Singapore. 29 July 2009.

manufactured in Singapore. Similarly, Singapore Airlines has selected the same engine to power its Boeing 787-10X when these enter into commercial service from 2018. Rolls-Royce will undertake to maintain, repair and overhaul the engines, as well as provide spare engine support to both airlines.[63]

Singapore is also a hub for corporate aircraft sales and servicing. Seletar offers corporate aircraft owners access to Singapore and to manufacturer-approved service centres such as Jet Aviation (for Gulfstream Aerospace, Bombardier and Cessna) and Hawker Pacific (Hawker Beechcraft and Dassault Falcon). Seletar is the home base for several smaller charter aircraft operators serving regional corporate travel and aeromedical evacuation services, as well as charter brokers such as Singapore Aviation and aviation consultancies specialising in regional aircraft and corporate jet operations.[64]

It should be noted that although Singapore has been the leading aerospace hub, there is emerging competition in the nearby Senai International Airport in Johor, Malaysia that has been developing its own aviation hub.[65]

6. Singapore Airlines and Other Carriers Operating out of Singapore

The success story of the Singapore Airlines including its emergence from a small regional player has been well documented by publications of the airline itself (Singapore Airlines, 1988; Allen, 1990), and others including Sabhlok (2001), Baranwal (2013), Heracleous and Wirtz (2009), Heracleous *et al.* (2009) and Chan (2000). Heracleous and Wirtz (2010) undertook an analysis of SIA's management strategy. There is also an abundance of information on the airlines' website as well as a full account of the airline provided by the airline's own website.[66] There is therefore no necessity to repeat the contents of these studies except to provide information on the key events on the evolution of the airline, and to provide updates of the more recent developments affecting the airline. The emphasis would be to highlight the crucial role of SIA in fostering and moulding Singapore into a global air hub.

---

[63] Karamjit Kaur, Scoot selects Rolls Royce engine for B787, ST 7 June 2013. Available at http://www.straitstimes.com/breaking-news/singapore/story/scoot-selects-rolls-royce-engine-b787-20130607#sthash.kOcq1aO5.dpuf/

[64] Wikipedia, Aviation in Singapore. Available at: http://www.ask.com/wiki/Aviation_in_Singapore?o=2802&qsrc=999&ad=doubleDown&an=apn&ap=ask.com. Accessed on 12 January 2015.

[65] Han, K. 7 February 2014. Aerospace Park Centerpiece of Singapore's Aviation Dream. DPA: Deutsche Presse-Agentur.

[66] See http://en.wikipedia.org/wiki/Singapore_Airlines; and Singapore Airlines — Our History, available at http://www.singaporeair.com/en_UK/about-us/sia-history/. Accessed 4 December 2014; and Heracleous, *et al.* [2009].

For decades now, SIA has been held up as the benchmark for a highly efficient global aviation enterprise as well as a strong global corporate brand. SIA has also evolved its own family of airlines, namely the wholly-owned subsidiary full-service SilkAir, and Scoot, which is a medium-haul LCC, and is a majority shareholder of Tigerair, a short-haul LCC. SIA had also previously reached out of its home base through limited equity stakes in other airlines such as Swissair and Virgin Atlantic, and subsequently Virgin Australia. SIA has since exited these equity-based partnerships. Most recently SIA added Vistara Airlines (49% owned by SIA in the Tata-SIA joint venture), an endeavour which finally came to fruition after its failed attempts to enter the Indian market in 1994. This will be followed by looking at the various strategies employed by SIA and Singapore to expand and gain market share in various parts of the world.

6.1 *Emergence of Singapore Airlines*

Singapore Airlines was formed in 1972 as a result of the restructuring of the Malaysia–Singapore Airlines (MSA) after Singapore's separation from the Federation of Malaysia. SIA's beginnings started from the establishment of a domestic airline, Malayan Airways, on 17 August 1936. The intention for the airline was to serve the countries in the surrounding region. Malayan Airways was initially incorporated in Singapore on 21 October 1937, with Britain's Imperial Airways holding 50%, two private shipping companies, Ocean Steam Ship and Straits Steamship each holding approximately 17% and 32%, respectively. Meanwhile, an Australian company, Wearne Brothers, had launched Wearne's Air Services (WAS) in June 1937 to provide services linking Singapore, Kuala Lumpur and Penang. With a direct competitor operating on the key route, Malayan Airways decided to hold back until May 1947, before operating scheduled services due to support traffic growth (Allen, 1990).

In 1947, Malayan Airways Limited operated thrice weekly with services linking Singapore, Kuala Lumpur, Ipoh and Penang. The airline grew and by 1955, it began to open up international services. When Singapore merged with the Federation of Malaysia in 1963, the name of the airline was changed to Malaysia–Singapore Airlines Ltd (MSA). Singapore broke away from Malaysia in August 1965, and in the following year, the governments of Malaysia and Singapore acquired joint control of the airline. It was not long after that, in 1971, that the company was restructured into two independent entities: Singapore Airlines and Malaysian Airlines System (MAS) [Chan, 2000]. This was the formal beginnings of SIA as it became a publicly listed company with limited liability and a wholly-owned subsidiary of Temasek Holdings (Private) Limited on 28 January 1972. The Government of Singapore through its investment arm, Temasek Holdings (Pte) Ltd, owns 55.95% of the shares of the company as of 3 June 2014 [SIA Annual Report, 2014].

6.2 Success of Singapore Airlines

Singapore Airlines is among the most admired international airlines in the world. The airline has garnered many industry and travel awards over the years, including the Conde Nast Traveller's "World's Best Airline" Award for the 23rd time, Travel and Leisure Magazine's "World's Best International Airline" Award for 16 consecutive years, and Wall Street Journal's "Asia's most admired Company" Award for 18 consecutive years. However, it does not currently rank among the World's top 10 in any of the indicators, whether by revenue, passengers carried, passenger-kilometres flown, freight tonne-kilometres flown or by fleet size for either passenger or cargo airlines, nor by the number of destinations according to a compilation of Wikipedia for the year 2013. However, the airline's freighter subsidiary, Singapore Airlines Cargo, is ranked sixth in 2013 as measured by freight tonne-kilometres flown.[67]

6.2.1 Financial performance

It is remarkable that the Singapore Airlines as a group, in all its years of operations, has always been profitable (Table 6). In its earlier years, between the financial years of 1972–73 and 1983–84, SIA's Group revenue increased eight-fold from S$340 million to S$2.7 billion, which worked out to an average growth rate of 21% per annum [SIA Annual Report, 2008]. SIA Group's revenue reached a peak of S$16 billion in 2008–2009 thereafter, maintaining its revenue at about that level till 2013–2014.

SIA Group itself has consistently made a profit each year, reaching a record of S$2.1 billion profit and making its lowest profit of S$63 million in 2010, though bouncing back to register a profit of S$851 million the following year (Table 6). Profits were squeezed since then in the following three years (S$181 million, $187 million and $256 million respectively) due to a number of factors — high fuel costs, strong competition from Middle Eastern airlines, and the drag due to the offer of low airfares by LCCs. Since 2005, the global airline industry has been struggling to show consistent profitability. It was only

---

[67] http://en.Wikipedia.org/wiki/World%27s_largest_airlines/. Accessed 11 November 2014.

Table 6. Financial statistics of Singapore Airline group, 1988–2014.

Year	1988	1993	1998	2003	2008	2009	2010	2011	2012	2013	2014
Income Statement											
Total Revenue (S$mil)	4,272	5,656	7,213	9,762	15,973	15,996	12,707	14,525	14,858	15,098	15,244
Total Expenses (S$mil)	3,407	5,111	6,617	9,093	13,848	15,093	12,644	13,254	14,572	14,869	14,985
Operating Profit (S$mil)	865	544	596	669	2,125	904	63	1,271	286	229	259
Productivity and Employee Data											
Value Added	2,052	2,412	3,125	3,899	7,082	5,571	4,276	5,419	4,344	4,500	4,370
Value Added Per Employee (S$)	221,891	195,276	228,254	131,126	235,380	174,995	159,151	246,361	192,960	194,040	184,268
Revenue Per Employee (S$)	465,142	457,462	526,859	328,308	530,859	502,491	472,918	660,308	659,936	651,093	642,769
Average Employee Strength	9,246	12,363	13,690	29,734	30,088	31,834	33,222	21,997	22,514	23,189	23,716

Source: SIA annual reports, 1987/1998–2013/2014.

by the end of 2014 that the industry began to improve its profitability, owing to the sharp decline in fuel prices.

6.2.2 *The Strategies of SIA*

In order to be the best among the world's top airlines, Mr J. Y. Pillay, the company's first chairman (1972–1996), "…. best known as the man who led Singapore Airlines from a small startup with just 12 aircraft to a global industry leader"[68], had laid down clear business and operating guidelines for SIA. Right from the start, SIA did not receive any financial subsidy from the government and was set to be a profit-driven company able to face competition from the rest of the world's airlines. Decisions made have been bold and ground-breaking while encouraging a corporate culture of flexibility, innovation and enterprise. The company upholds its management philosophy based on being cost competitive; focusing on its core business; being customer-oriented and providing distinctive and impeccable service, supported by its innovative ability to offer creative products and services; maintaining communications excellence; and consistent standards in all areas. SIA chose to deploy state-of-the-art aircraft, engine and inflight entertainment equipment, and has striven to create a world-class brand of distinction. The airline's logo is instantly recognisable and the "Singapore Girl" has become the personification of SIA's tradition of friendly service, warm hospitality and a unique travel experience.

6.2.2.1 Providing superior services

SIA prides itself by assuming product leadership in a multitude of ways including providing excellent cabin comfort by way of lie-flat beds in the First Class and the Business Class. The airline imposes stringent employee training regimes to ensure quality in-flight service from its cabin crew while supporting high staff morale. The airline continually upgrades its In-flight Entertainment System and serves world-class cabin cuisine, including wine selected by world-acclaimed wine consultants. SIA adheres to high standards of operational excellence in terms of insisting on safety and security as a matter of top priority. It also places high importance on punctuality and on-time departures.

SIA has taken advantage of the government's proactive policy in developing the Changi Airport for fast and comfortable transit passenger and cargo traffic, as have other airlines operating out of the airport. Changi Airport's automated facilities have raised efficiency through faster clearance, shorter

---

[68] Quotation from Lee Kuan Yew School of Public Policy, National University of Singapore website http://lkyspp.nus.edu.sg/faculty/pillay-j-y/.

waiting time, and its wide range of excellent airport facilities and services. The symbiosis of the national airline and an outstanding airport supported by SATS' premium ground-handling services has bolstered Singapore to become a preferred international air hub.

SIA aspires to be a full-service industry leader pioneering innovative products and services. It has featured many firsts, including offering free head-sets, free drinks and a choice of meals in the economy class (1970s); to fly non-stop on the London-Singapore route (1984); to provide in-flight tele-phones (1991); to launch audio/video on demand (1997), the in-flight trial of e-mail (2001); to operate world's longest non-stop flights (2004); to be the launch customer for the Airbus A380 aircraft (2007); to be the first to have iPod and iPhone connectivity in the Economy Class (2009).

Consequent to SIA joining Star Alliance, the airline has had to be more creative in maintaining its competitive position by further improving its services. SIA's lounge service is highlighted here. SIA continually upgrades its luxurious premium lounge — the SilverKris lounges — not only at the Changi Airport but also worldwide, to showcase the distinctive high standards of services that the airline provides and to project the image of always thinking of its consumers. The new Silver Kris Lounge at Terminal 2 was updated and opened in August 2000. In October 2004 and January 2008, two more lounges — the Silver Kris First Class Lounge and the Silver Kris Lounge (in the then newly-opened Terminal 3) were opened. Up till October 2014, the Sliver Kris lounges were opened in 15 other airports worldwide. From 2012, a $20 million investment programme was proposed to further enhance customers' travel experience by adopting a new design concept for all of the airlines' VIP lounges worldwide. A five-year programme that started in 2014, featuring a "Home Away From Home" design concept for the Silver Kris Lounge was introduced.[69]

6.2.2.2 SIA's family of airlines

Table 7 provides a summary of characteristics of airlines that operate out of the Changi Airport, other than Singapore Airlines. SIA remains focused as a full service long-haul international airline in a global route network. A key strategy of SIA is to focus on its core airline business of operating a profitable premium

---

[69] Singapore Airlines Website, SilverKris Lounges Worldwide. Available at https://www.singa-poreair.com/images/Travel_Info/Lounge1/SilverKris_Lounges_Worldwide.pdf/. Accessed on 20 December 2014.

Table 7. Airlines operating out of Changi Airport.

Airlinew	Year began	Fleet	Network of destinations
SilkAir (formerly Tradewinds the Airline)[70]	1989	Operates a fleet of 25 Airbus aircraft, comprising 6 A319s,14 A320s and 5 Boeing 737-800s. In August 2012, the airline announced a huge order of 68 Boeing 737 aircraft, with firm orders for 23 Boeing 737-800s and 31 Boeing 737 MAX 8s. Average age 5 years.	Operates 350 weekly flights to 48 destinations in 12 countries including Indonesia (11), China (8), India (8) and Malaysia (5). In November 2014, it added a service to Cairns.
Singapore Airlines Cargo[71]	2001	The airline operates 12 B747-400 freighters, and employs the belly-hold space of all passenger aircraft of SIA and Scoot aircraft as well as an interline arrangement with SilkAir.	Combined Scoot Freight and SIA Cargo's network covers more than 65 cities in over 30 countries across 6 continents with over 900 flights scheduled weekly.
Tigerair[72]	2003	Operates a fleet of Airbus A320 aircraft. Tigerair Singapore operates flights to 38 destinations across 13 countries in Asia/Australia	Network covers 13 countries and 38 destinations in mainly India, China, Australia, Indonesia, Malaysia, Thailand and Vietnam.

Jetstar Asia Airways[73] and Valuair	2004	Merged with Valuair in July 2005. Operates 18 A320-200 aircraft. As of January 2014, Jetstar Asia Airway's fleet comprised 19 Airbus-A320-200 aircraft with one more on order.	Regional destinations — Myanmar, Indonesia, Cambodia, Malaysia, Thailand and Vietnam — and to Japan, Taiwan, and China including Macau and HK. Jetstar offers more than 400 flights per week. Jetstar Asia (incl. JetStar Airways, Valuair) flies from Singapore to 25 cities in 13 countries as on 31 May 2014. Has codeshare agreements with Myanmar Airways International, Qantas, and Valuair.
Scoot	2012	The fleet comprised Boeing 777-200ERs, acquired from its parent company. These will be phased out to make way for 20 Boeing 787 Dreamliners starting from January 2015.	Flies to Australia (Gold Coast, Perth, and Sydney), China (Qingdao, Shenyang, Tianjin, Nanjing), Hong Kong, Tokyo (Narita), S. Korea (Seoul), Taiwan (Taipei), and Thailand (Bangkok Don Mueng).

[70] SilkAir Airlines website: http://www.silkair.com/SAA-flow.form?execution=e1s1. Accessed 25 November 2014.
[71] SIA Cargo website: http://www.siacargo.com/index.asp. Accessed 26 November 2014.
[72] Tigerair website: http://www.tigerair.com/sg/en/about_us.php. Accessed 11 February 2015.
[73] Jetstar Asia Airways, Wikipedia http://www.ask.com/wiki/Jetstar_Asia_Airways?qsrc=3044#Merger_with_Valuair. Accessed 23 November 2014.

airline. As the company expanded, and faced by low-cost regional airlines, it became necessary to establish subsidiary airline brands to cater to new market segments and grow the medium- and short-haul services to destinations. The rise of LCCs in Asia following the success of airlines such as Ryan Air in Europe and Virgin Air that started as a trans-Atlantic operator, has not only generated new travel demand but also had the effect of providing stiff competition to established full-service airlines. LCCs are a force to be reckoned with, and these airlines constitute almost 30% of the total traffic at the Changi Airport. In Southeast Asia, LCCs are expanding very fast, representing almost 60% of the regional traffic.

SilkAir: Labelled as the regional wing of Singapore Airlines, SilkAir operates as a full service short-haul airline using a narrow body fleet and extends the Group's network by developing popular regional tourist destinations in the Asia-Pacific region. The airline began its operations in February 1989 under the name of Tradewinds before being rebranded as SilkAir in 1992. The subsidiary airline works closely with SIA and operates more than 350 weekly flights to 48 destinations in 12 countries. In 2013 alone, the network was expanded to include Semarang, Makassar and Yogyakarta. The airline's strategy is to expand to secondary destinations in key markets such as China, India and Indonesia. Starting from February 2014, the airline also operated services to cities inclusive of Kuala Lumpur, Penang, Phuket and Medan, as well as Siem Reap, Danang, Davao, Cebu and Kochi from March 2014.

SIA Cargo: In July 1992, Singapore Airlines created a cargo division to be a standalone profit centre to drive growth in airfreight. In July 2001 however, Singapore Airlines Cargo (SIA Cargo) was incorporated, taking over the airfreight operations of Singapore Airlines as a separate subsidiary. SIA Cargo also entered into an alliance with Lufthansa Cargo and SAS Cargo Group to form WOW Alliance in October 2001.

The company had a fleet of about a dozen B747-400F all-freighters in 2003, increasing to 16 B747 in 2006. But its freighter fleet was reduced to nine aircraft as a result of the global slowdown in the cargo markets. The average age of the freighter fleet has been rising to 12 years and three months by March 2014.[74]

SIA Cargo has operated via cargo hubs at the Sharjah International Airport (United Arab Emirates), the Brussels Airport and a secondary hub at the Schiphol Airport, Amsterdam. SIA Cargo deploys a fleet of dedicated B747-400 freighters and markets belly-hold space of 100 wide body SIA passenger

---

[74] See Singapore Airlines, 2013/2014 Annual Report.

aircraft. The pure freighter fleet services 24 destinations in 19 countries. Sharjah and Amsterdam, and to a less extent, Brussels, are hubs for the Middle Eastern and European regions respectively. A round-the-world freight service was introduced in May 2003, providing direct service between China and the United States.[75]

Tigerair: Tigerair Singapore was incorporated on 12 December 2003, with Singapore Airlines holding 32.84% and Dahlia Investments Pte Ltd holding 7.37% of the shares. Tigerair Singapore is wholly-owned by Tiger Airways Holdings Limited, a holding company set up in 2007 to manage both Tiger Airways and its Australian subsidiary Tigerair Australia. Tiger Airways Holdings Limited was listed on the Singapore stock exchange in 2010. The airline rebranded itself as Tigerair in mid-2013, dropping its leaping tiger logo.

In mid-2007, Tigerair ordered 30 aircraft worth US$2.2 billion with another 20 on option. Towards the end of 2007, Tigerair increased its fleet size to 70 A320 aircraft in total. However, the airline reported a loss of $182.4 million for the third quarter of 2014 due to high fuel costs and intense competition from other regional LCCs in a market riddled with excess capacity. In order to reduce expenses, the airline took the drastic step of cancelling its order of nine aircraft, grounded eight A320 planes and withdrew from its unprofitable joint-venture airline operating services in Australia, Indonesia and the Philippines.

In October 2014, Singapore Airlines took majority control of Tigerair as part of the budget carrier's restructuring and turnaround plan, and increased its stake to 55%.[76] Tigerair subsequently sold off its 40% share in Tigerair Australia to Virgin Australia for just AUD$1.[77]

Jetstar Asia: Jetstar Asia Airways Pte Ltd (Jetstar) was incorporated in Singapore and was launched in 2004 as a partnership between Qantas, holding a 49% stake in the airline, with two Singaporean shareholders held 22 and 10%

---

[75] Round-the-world route begins in Singapore flying to Hong Kong, Dallas, Chicago, Brussels, Sharjah, and back to Singapore on Wednesdays, and on Fridays, the route includes Singapore, Hong Kong, Dallas, Chicago, Brussels, Mumbai, and back to Singapore. Information on SIA Cargo's operations drawn from Wikipedia: http://www.ask.com/wiki/Singapore_Airlines_Cargo?o=2802&qsrc=999&ad=doubleDown&an=apn&ap=ask.com. Accessed 26 November 2014.

[76] Pangarkar [2014] reports that the stake was raised from 40% to 71%.

[77] Karamjit Kaur, Loss-making Tigerair seeks turnaround by clipping own wings, ST 3 May 2014. http://www.straitstimes.com/news/business/companies/story/loss-making-tigerair-seeks-turnaround-clipping-own-wings-20140503#sthash.FdmlxzG5.dpuf; and SIA to take majority control of Tigerair as budget carrier continues to bleed cash, ST, 12 Feb 2015. http://www.straitstimes.com/news/singapore/transport/story/sia-take-majority-control-tigerair-budget-carrier-continues-bleed-cas#sthash.9cqOTiQo.ynV4IUEJ.dpuf. Accessed on 11 February 2015.

respectively, and the Singapore government's investment company, Temasek Holdings (Private) Limited, owned 19% stake. Jetstar Asia operates as an LCC based in Singapore, and is one of the Asian offshoots of its parent company, Jetstar Airways Australia, itself a low-cost subsidiary of Qantas Airlines. It operates a dense network of services throughout the Asia-Pacific. It is the main feeder airline for its parent company Jetstar Airways for budget passengers flying to Australia. Its sister airlines include Jetstar in New Zealand, Jetstar Pacific and Jetstar Japan.

The merger between Jetstar and Valuair in 2005 represented the first major consolidation of Southeast Asia's already crowded and competitive low-cost airline industry. The new company, Orange Star, received a cash injection of more than $50 million in fresh capital into the new entity, largely to be provided by Qantas. Qantas owns 42.5% of both airlines. Newstar Holdings is the holding company that operates and manages Jetstar Asia in Singapore.

Jetstar and Valuair have a codeshare partnership with parent Qantas, effectively making them the regional feeder LCC airline for Qantas. This is similar to the feeder role SilkAir plays for Singapore Airlines, though the latter is a full-service carrier.[78]

Valuair: Valuair was the pioneering Singapore-based LCC that was launched in 2004. The airline was the first of the LCCs to begin operations in Singapore offering services initially to just Bangkok and Hong Kong. Valuair sought to differentiate itself from its LCC competitors such as Jetstar Asia and Tiger Airways by offering free hot meals, wider legroom and assigned seating, while marketing itself as a low-fare airline. The airline flew beyond the five-hour stage length typical of LCCs to destinations such as Perth, Xiamen and Chengdu.

However, it was unable to hold up against its competitors. Valuair was acquired by Jetstar Asia Airways in July 2005, and its brand was retained only for Jetstar's scheduled services to destinations in Indonesia, namely Jakarta, Medan, Surabaya and Denpasar.

Up until 2014, Indonesia prohibited the LCCs of Singapore and Indonesia from adding more flights. The merger suited Jetstar as it had not yet established routes into Indonesia. The merger then served as a way to enter the burgeoning Indonesian market. Subsequently in October 2014, the Indonesian Government

---

[78] Wikipedia, Valuair. Available at http://en.wikipedia.org/wiki/Valuair. Accessed on 8 March 2015.

lifted operational restrictions on foreign-owned LCCs into Indonesia. Valuair was dissolved and its flights were taken over by Jetstar Asia.[79]

Scoot: In November 2011, Singapore Airlines launched Scoot Pte Ltd, a low-cost wholly-owned subsidiary for medium and long-haul routes. The following year, Scoot began its first flight to Sydney, the Gold Coast (Queensland) and Perth (Western Australia). Since then, the airline has expanded its route network to cover destinations in the Asia-Pacific region and India. In October 2012, it provided an interline service with Tigerair.[80]

Demonstrating the vitality of the Singapore Airlines, in December 2013, Scoot announced its new long-haul low-cost carrier based in Thailand, NokScoot, a joint venture between Scoot and Nok Air, Thailand's second largest LCC. The new airline, starting with three Boeing 777-200ERs, started commercial flights from Bangkok's Don Mueang airport in the second half of 2014. It had plans to operate international services hubbing at Bangkok's Don Mueang Airport. Using its base in Bangkok, the plan is to provide low-cost services across Pacific Asia as well as to reach some 20 destinations within Thailand.[81]

Vistara: Vistara is the latest member in the SIA family of airlines. It is a full-service airline based in New Delhi. It was established with a total investment of US$100 million in 2014 as a joint venture between India's Tata Sons and Singapore Airlines. Tata Sons owns 51% and Singapore Airlines (SIA) owns the remaining stake in the company. Two earlier attempts by the two companies at opening up the Indian air travel market had failed. In 1994, it was blocked due to a change in civil aviation policy that barred foreign carriers from holding a stake in domestic airlines. Five years later, the two bid for a stake in the loss-making state-owned Air India but the offer was later withdrawn. Vistara received approval from India's Foreign Investment Promotion Board in October 2013. The airline now has five Airbus A320s with eight more on order to be delivered by April 2016 and seven A320-neo to be delivered thereafter till 2018. The airline began operations in January 2015, with 164 weekly domestic flights linking its hub in Delhi to Mumbai, Pajim in Goa, Guwahati in Assam, Hyderabad, Ahmedabad, and Silgiri in West Bengal. It has ambitious plans to open up more hubs and to fly to more destinations across

---

[79] *Source: Wikipedia, Valuair. http://en.wikipedia.org/wiki/Valuair.* Accessed 5 January 2015.

[80] *Source: http://www.ask.com/wiki/Scoot?qsrc=3044.* Accessed 10 December 2014.

[81] Thailand's second-largest LCC, Nok Air, owns 51 per cent of NokScoot, with Scoot Pte Ltd, part of Singapore Airlines, owning 49%. See http://www.ask.com/wiki/NokScoot?qsrc=3044. Accessed 10 December 2014.

India as well as to operate international long-haul flights to Europe and North America. Vistara has codeshare agreements with SIA and SilkAir.[82]

6.2.2.3 SIA's fleet management strategy

Singapore Airlines has maintained its strategy to operate a young and modern fleet that will yield operating cost savings due to lower fuel and maintenance costs. A young fleet will also be an important advertising feature to attract customers. The strategy also helps to utilise fleet commonalities so that aircraft, technical and cabin crew resources can be optimised when it comes to shared maintenance equipment, smaller range of stocked spare parts, optimising maintenance of aircraft, and cabin crew services. SIA's fleet strategy maximises aircraft fleet utilisation to match the global route network and to optimise use of expensive capital assets.

In 1973, just one year after incorporating the airline, SIA boldly took delivery of two 400-seat Boeing 747s to become the first Asian airline outside of Japan, to operate the first wide-bodied aircraft! SIA ordered ten more B747s and four B727s, the package cost a total of US$1billion — the biggest single order in aviation history ever made by any airline at that time. A year later, SIA ordered six Airbus A300s. In 1981, just three years after the unprecedented large order, the airline placed yet another huge order worth US$1.4 billion for eight B747-300s and two more A300s. Once again, there was another US$1.4 billion order for six Airbus A310s, four B757s and six more B747-300s in 1983. The huge aircraft orders were made in 1983 under global economic recession conditions, thus SIA was able to extract very attractive terms from both Boeing and Airbus Industrie. In addition, SIA was able to refine and phase out its older B747s and other air planes as part of the packaged aircraft orders.[83]

Table 8 provides data on SIA's fleet of aircraft type from 2002/2003 to 2013/2014. The airline shed 11 aircraft from 96 in 2002/2003 to 85 in the following year, in response to the drop in demand but thereafter the fleet size gradually increased to 103 in 2013/2014. There was a pause in the growth of the fleet size between 2009/2010 and 2010/2011, dropping to a total of 100 aircraft in 2011/2012 — the drop in the number of aircraft was made up by the inclusion of the larger capacity Airbus A380s that began service from 2007/2008 while phasing out the ageing B747s. It should be noted that SIA has deployed a fleet of B777s, which have smaller seating capacity to serve long-range destinations but subsequently preferring the B777-300ER (Extended

---

[82] Source: Wikipedia, Vistara. http://en.wikipedia.org/wiki/Vistara. Accessed 6 March 2015.
[83] Singapore Airlines annual report 1988, p. 21–22.

Table 8.   Singapore Airlines fleet and aircraft types, 2002/2003–2013/2014.

	Seating capacity	2002/ 2003	2003/ 2004	2004/ 2005	2005/ 2006	2006/ 2007	2007/ 2008	2008/ 2009	2009/ 2010	2010/ 2011	2011/ 2012	2012/ 2013	2013/ 2014
B747-400	375	39	30	27	27	22	18	12	7	7	1	0	0
B777-200	288	11	12	14	14	14	14	13	12	8	5	2	0
B777-200A	323	12	17	17	17	17	17	17	16	7	7	7	5
B777-200R	266	0	0	0	0	0	0	0	1	11	11	11	11
B777-200ER	285	14	15	15	15	15	15	15	15	9	9	9	4
B777-200ERR	271	0	0	0	0	0	0	0	0	0	0	2	9
B777-300	332	8	8	11	12	12	12	12	11	11	1	0	0
B777-300R	284	0	0	0	0	0	0	0	1	1	7	7	7
B777-300ER	278	0	0	0	0	9	14	19	19	19	19	19	22
A310-300	183	9	0	0	0	0	0	0	0	0	0	0	0
A340-300	265	3	0	0	0	0	—	0	0	0	0	0	0
A340-500	181	0	3	5	5	5	5	5	5	5	5	5	0
A380-800	471	0	0	0	0	0	3	6	10	11	11	11	11
A380-800A	409	0	0	0	0	0	0	0	0	0	5	8	8
A330-300	285	0	0	0	0	0	0	4	11	19	19	20	26
Subtotal	—	96	85	89	90	94	98	103	108	108	100	101	103
SIA Cargo													
B747-400F	—	12	13	14	16	14	14	12	11	11	13	12	9

Source: SIA annual reports, various years.

Range) planes after the mid-2000s. The data reveals the shift from smaller to larger aircraft types with advanced features thus reaping greater operational efficiencies and customer satisfaction.

In the earlier years, the average age of SIA's fleet was just 3½ years but by the end of March 2014, it had increased to six years and nine months. Up until 2014, SIA had placed orders for as many as 118 aircraft worth US$17 billion. At the same time, the retirement of the ageing B747-400 was done progressively, starting with 39 in year 2002 and finally bidding farewell to the "Queens of the Sky" in April 2012. These planes were leased, traded-in or disposed of. Perhaps one SIA's proudest achievements was to purchase the A380-800 and B777-300ER series aircraft for commercial usage. SIA thus made history for being the first airline in the world to take this step. The A380 with a seating capacity of 471 is considered to be the largest aircraft in the skies. The strategic move reinforced SIA's continued leadership role in the global airline industry.[84]

6.2.2.4 Enlarging route network

As a tiny country with no hinterland to speak of, SIA — and Singapore as a whole — had to think big and spread its wings internationally. The airline systematically developed the Singapore hub through the expansion of routes and by advocating the 'open skies' policy in bilateral and multilateral fora. Leveraging on its traditional ties with Western countries and ethnic ties with China, India and Southeast Asia, the airline has successfully opened up routes to many destinations in these regions.

As the international aviation business opened up through deregulation in the US and through progressive liberalisation in Europe, it became imperative to explore and supplement organic growth through strategic stakes in other airlines and related businesses to tap into markets that had been hitherto closed to gain additional sources of income. Hence, there is the need to join airline alliances, enter into joint ventures with other airlines and to open up more destinations through third-party code sharing agreements. With the success of LCCs, it prompted SIA to establish one of its own, Scoot, and now has a majority stake in Tigerair.

In March and August 2000, Singapore bought 49% stake in London-based Virgin Atlantic and 25% of Air New Zealand (owned by Ansett Holdings of Australia). The arrangement was one of complementarity rather than overlapping networks. The arrangement also potentially opened up the huge UK,

---

[84] Singapore Airlines, Annual Report 2013/2014, p. 209.

Transatlantic and Australian markets, and offered sizeable new possibilities for networking partnerships (Morrish & Hamilton, 2002].

Singapore's position as a global air hub has been also due to the country's push for a liberal open skies air transport regime. The result is the high degree of connectivity thereby consolidating Changi as a global air hub. The table below shows that at the end of January 2015, the airport was connected to 306 cities in 71 countries worldwide. The routes are served by more than 100 airlines operating out of Changi and those that fly in and out of the airport. Between 24 June 2014 and 29 Jan 2015, there had been a net increase of 10 cities linked to Singapore. Connections to destinations in the us increased by 12 (though six of the city links are for freighter airlines). There had been an addition of six destinations to 53 for Southeast Asian destinations — the increase was for cities other than those in the four more popular countries. It is expected that, with the joint venture of Nok-Scoot, many more destinations in Thailand will be connected through codesharing (see section below). Similarly with Vistara's expansion in future, there will be more destinations in India.

As a result of the persistent efforts on the part of Singapore and SIA over several decades, Changi Airport is connected with 71 countries and 306 cities across the globe as in January 2015 as shown by Table 9. This is truly a demonstration of Singapore's status as a global air transport hub.

Table 9. Destinations connected to Singapore (as on 29 January 2015).

Region	Countries	Cities
Africa	9	14 South Africa (5)
Central Asia	1	1
Europe	27	76 — France (5) Germany (11) Norway (5) Turkey (10)
Middle East	6	8
North America	2	38 — United States (35)
NE Asia	6	44 — China (28) Japan (10)
South America	1	1
South Asia	6	17 — India (12)
Southeast Asia	10	54 — Indonesia (13) Malaysia (9) Philippines (6) Thailand (6)
SW Pacific	3	53 — Australia (45) New Zealand (7)
Total	71	306

Source. Changi Airport Group.

6.2.2.5 As member of Star Alliance

Aviation deregulation in the U.S. in 1978, created a new regulatory environment that led to the establishment of open skies continental blocs which included the US and Canada. The new regime benefitted airlines operating in Canada, Western Europe, Asia and the rest of the world. Given the serious regulatory constraints of the bilateral air services obliging airlines to negotiate the various airline freedoms for capacity and route expansion, airline alliances became a good alternative for growth, hence the proliferation of airline alliances — one-world, Start Alliance, Skyteam, and Mercator. Airlines were allowed to set up efficient hub-and-spoke networks, and have led to trans-border consolidation within the airline industry. For example, Star Alliance provides ample opportunities to go beyond mere co-ordination of passages but also to generate new synergies, products, markets, and levels of service and organisation. The first benefit is the expansion and optimisation of route network through code sharing arrangements with alliance partners.[85]

There will still be constraints on market entry, capacity, and pricing in the future. These barriers are applied more stringently on foreign carriers. Foreign ownership by airline restrictions continues to constrain the industry. The alternative is to acquire a minority equity stake in an indigenous carrier to gain access to each other's market, assuming aviation regulatory authorities approval is first obtained.

The Singapore Airlines was among the first to enter into an international alliance. It was a small organisation, the Excellence Alliance in 1989 that included Delta from the US and Swissair from Europe. This was then regarded as a pioneer of global air alliances. Alliance members took small equity stakes in each other's company to undergird and strengthen the commercial relationship. Such an alliance then opened up opportunities to expand each partner's market without losing their commercial and brand independence or the right to develop its own markets. The alliance enabled SIA and Swissair to operate flights to more destinations in the US. On the other hand, Delta secured access to the international market. However, this pioneering alliance ended in 1997 due to conflicts among the partners' goals and priorities [Vasigh, 2008].

---

[85] Details of Star Alliance are drawn from Wikipedia, Star Alliance. Available at http://en.wikipedia.org/wiki/Star_Alliance. Accessed 28 December 2015. For benefits of alliances and joint ventures, see IATA, Economics Briefing, The Economic Benefits Generated by Alliance and Joint Ventures. http://www.iata.org/whatwedo/Documents/economics/Economics%20of%20JVs_Jan2012L.pdf. Accessed on 19 December 2015.

There are currently three dominant alliances, namely Star Alliance, Oneworld and SkyTeam, that accounted for 77% of the world airline capacity in 2013 (Wang, 2014]. Star Alliance is the first truly global alliance which was formed in 1997 by United, Lufthansa, Scandinavian Airlines, Air Canada and Thai Airways. By 2014, it became the largest airline group which included 27 members[86] with a combined total of 4,651 aircraft. Its member airlines operate in 193 countries, carrying some 654 million passengers. Its network connects 1321 airports with 18,521 daily departures.[87] Singapore Airlines (SIA) joined the Star Alliance in April 2000.

6.2.2.6 Entering into joint ventures

Mention of joint-ventures with other airlines have already been made in an earlier section. Apart from joining an air alliance, SIA also acquired stakes in other airlines or sought to establish joint-venture companies with other airlines to strengthen its relations including entering into codeshare agreements with them. This helped improve operational collaboration to raise efficiency (IATA) [see Bowen, 1997].

Among the Star Alliance partners, SIA bought a 25% stake in Air New Zealand in April 2000 to increase its penetration into both the New Zealand and Australian markets. As Air New Zealand owned Ansett Airline, a domestic airline in Australia, this enabled SIA to indirectly hold a 25% stake in Ansett Airline. By October 2004, it still held 19.7% of Air New Zealand, but eventually lost its entire stake as a result of losses and capital restructuring. Among non-alliance airlines, SIA bought a 49% stake in Virgin Atlantic in March 2000 to gain access to the UK market. Similarly in October 2012, SIA acquired a 10% stake in Virgin Australia[88] to increase its presence in the booming tourist market of Australia.

With Virgin Atlantic, SIA gained trans-Atlantic codeshare flights between London and Boston, Washington and Miami, Manchester and Orlando in

---

[86] Current members of Star Alliance are Adria Airways, Aegean Airlines, Air Canada, Air China, Air India, Air New Zealand, All Nippon Airways, Asiana Airlines, Austrian Airlines, Avianca Holdings, Brussels Airlines, Copa Airlines, Croatia Airlines, EgyptAir, Ethiopian Airlines, EVA Air, LOT Polish Airlines, Lufthansa, Scandinavian Airlines, Shenzhen Airlines, Singapore Airlines, South African Airways, Swiss International Air Lines, TAP Portugal, Thai Airways International, Turkish Airlines and United Airlines. Avianca Brazil, as a subsidiary of Avianca joined in 2014.
[87] Star Alliance website. http://www.staralliance.com/en/about/member_airlines/. Accessed 20 December 2015.
[88] Wikipedia, Virgin Australia. Available at http://en.wikipedia.org/wiki/Virgin_Australia. Accessed on 14 Apr 2015.

October 2004; from London to Los Angeles, San Francisco and Dubai in 2006; and also in the Asia-Pacific between London and Sydney via Singapore in 2007.

The cooperation between SIA and Virgin Australia Airlines (formerly Virgin Blue Airlines) started before the acquisition of a 10% stake in the latter company in 2012. The two airlines had established a long-term alliance not only covering codeshare of each other's flights but also the reciprocal frequent flyer benefits and airport lounge access, among other initiatives. The alliance also connected SIA's network with Virgin Australia's network of Australian and the Pacific destinations. The success of this cooperation is shown by SIA having gained access to 32 Australian cities and the number of flights per week to Australian destinations had grown by 45% compared to 2010.[89]

6.2.2.7 Code sharing

Since joining the Star Alliance in 2000, Singapore Airlines has entered into a large number of codeshare agreements with other Star Alliance members as well as with airlines outside of the Alliance (Tables 10 and 11). By 2015, SIA had signed such agreements with over 15 Star Alliance members for 177 flights and

Table 10.   Singapore Airlines' codeshare flights with other Star Alliance members (2015).

Star Alliance members	No. of Flights
Aegean Airlines	7
Air Canada	3
Air New Zealand	59
Asiana Airlines	4
Brussels Airlines	3
Egyptair	3
Ethiopian Airlines	1
EVA Airways	7
LOT Polish Airlines	6
Lufthansa	28
Scandinavian Airlines	19
Shenzhen Airlines	1
South African Airways	4
Swiss International Airlines	5
Turkish Airlines	27
Total: 15	Total: 177

---

[89] See Singapore Airlines annual reports, various years.

Table 11.   Singapore Airlines' codeshare flights with
non Star Alliance airlines (2015).

Non Star Alliance members	No. of Flights
All Nippon Airways	15
Garuda Indonesia	2
JetBlue Airways	16
Malaysia Airlines	2
SilkAir	47
Transaero Airlines	4
Virgin Australia Airlines	85
Virgin Atlantic Airways	14
Virgin America	24
Total: 9	Total: 209

9 airlines with non-Star Alliance for 209 flights.[90] These codeshare agreements have enabled SIA to expand its network both in terms of the number of destinations and flight frequency. In this way, SIA has been able to feed passengers into Changi Airport either as a destination or to transit to other destinations, hence strengthening the hub position of the Changi Airport.

6.2.2.8 Managing fuel costs

Aviation fuel constitutes a major component of an airline's operating cost accounting for as much as 40% of total operating costs. Crude oil prices had been rising since 2009 to as high as S$110 in July 2014. However Brent oil fell to below US$50 per barrel in January 2015, recovering slightly in the two months thereafter. Spot jet fuel at US$70 per barrel in early 2015 had declined by 44% since mid-June 2014. According to IATA, Jet fuel prices have been averaging US$99/bbl in 2015 compared with US$116/bbl a year before. This would have a beneficial effect on the financial performance of the airline industry as a whole. However, with global revenues generated at US$975 billion for the world's airlines and a profit of US$19.9 billion in 2015; fuel savings will improve industry profit margins slightly to just 3.2%. However, SIA has hedged its fuel needs for six months at US$116/bbl till March 2015 which means the benefits

---

[90] Singapore Airlines, Codeshare Destinations. Available at http://www.singaporeair.com/en_UK/about-us/psh-codeshare-psh/psh-codeshare-dest/ http://ec.europa.eu/competition/sectors/transport/reports/airlinecodeshare.pdf. Accessed on 20 December 2014.

of declining fuel costs would only be realised on the expiry of its hedging. SIA is likely to continue with the practice of hedging to take advantage of the low fuel costs. Other airlines, such as China Eastern Airlines that do not hedge, will reap substantial savings earlier as a result of lower fuel costs.[91]

6.3 *Promoting Singapore as a tourist destination*

Passengers travelling to Singapore by air constitute premium visitors who are likely to spend more in the country. Realising the benefits of tourism as a source of revenue and employment, the government endeavoured over the decades to develop Singapore as a tourist destination. This included an allocation of $10 million in 1972 to develop Sentosa as a tourist attraction. At the same time, a total of $393 million was made available to build the Changi Airport and roads.[92] Also, Singapore Airlines on its own or jointly with organisations such as the Singapore Tourism Board (STB) had contributed much to promoting air passengers to visit or stopover in Singapore. The Singapore Government has taken the initiative to move ahead of rival destinations by developing many iconic structures aimed at attracting international air visitors, as well to provide venues for top-class sporting, cultural and other world-class events including the Formula 1 night Singapore Grand Prix now sponsored by SIA.

Global travel company, Lonely Planet, named Singapore the world's top country to visit in 2015 over Namibia (ranked second), Lithuania (ranked third), and the nearby Philippines (ranked eighth). Singapore scored high on its three criteria, namely topicality, 'wow' factor and broad appeal.[93] STB reported that the nation's Business Travel and Meetings, Incentive Travel, Conventions and Exhibitions (MICE) industry hosted, a total of 3.5 million business visitors in 2013, an increase of 3% from 2012. The benefits to the economy are considerable as visitors spent an estimated S$5.5 billion excluding sightseeing, entertainment and gaming expenditure.[94]

In recent decades, the Government successfully gained the support of private initiatives to invest in many outstanding attractions including the

---

[91] See Nisha Ramchandani, IATA forecasts record profits for airlines as fuel costs dive, Business Times, 11 December 2014. See also Karamjit Kaur, Some carriers burnt by fuel prices hedging, ST 6 December 2014. Also see Grace Leong, SIA, Tigerair stocks up on falling oil price, ST 12 February 2015. Available at http://www.straitstimes.com/the-big-story/oil-price-plunge/story/sia-tigerair-stocks-falling-oil-price-20150108#sthash.slGiGL46.dpuf.

[92] Memorable S'pore budgets, ST 7 January 2015.

[93] Melissa Lin, Lonely Planet picks Singapore as top travel spot in 2015, ST, 22 Oct 2014.

[94] Singapore Promotion Board (STB), Press release, Meetings, Incentive Travel, Conventions and Exhibitions. Available at https://www.stb.gov.sg/industries/mice. Accessed on 12 January 2015.

Esplanade, the six-star Fullerton Hotel, the Singapore Flyer, the Marina Bay Sands, the Gardens by the Bay and the Helix Bridge. All of these attractions were built around the man-made Marina Bay (now a lake). Part of the area has also become the new financial centre extending from the old financial centre within the Central Business District. Other attractions include the Singapore International Cruise Centre located just outside of the bay area, the recently completed Singapore Sports Hub, the earlier enhanced Sentosa Island with many new attractions including the Resorts World Sentosa (RWS). These and the refurbished and upgraded Central Area precincts contain numerous attractions including museums, art galleries, theatres, ethnic quarters with conserved buildings and structures that house appropriate activities all give the country a new life and buzz in order to attract international visitors as well as to develop the city-state for local residents. In addition, large convention and exhibition facilities have been put in place to support large international business conventions and meetings.

A part of this success can be attributed to collaborations between the Singapore Airlines and the Singapore Tourism Board (STB). In 2013, they jointly promoted the Singapore Stopover Holiday (SSH) package with the aim of encouraging visitors to enjoy their stopover in Singapore. The programme successfully targeted travellers from Europe who are attracted by Singapore's unique cultural mix, and a clean and green environment. The SSH programme showcased the Republic's Asian Civilisations Museum and Peranakan Museum in its package for Europe.

This was followed in April 2014 by the SIA and STB joint programme (till 30 June 2015) to boost visitorship to Singapore. The programme involved a joint global marketing effort between SIA and STB that targeted a wide range of leisure, cruise and business visitors. The two partners invested $4 million in joint worldwide campaigns focusing on Australia, China, Germany, India, Japan, the UK, and the US. The collaboration between SIA and STB bolstered Singapore's position as a preferred destination for leisure, business and cruises and, consequently, as an international air hub.[95]

As part of its efforts to celebrate Singapore's 50th birthday (SG50), SIA announced a programme to encourage visits or stopover holidays in Singapore in January 2015. The airline offered a one-night Singapore Stopover Holiday (SSH) package or a one-day Singapore Explorer Pass (SEP)

---

[95] Singapore Promotion Board (STB), Press release. Singapore Airlines and Singapore Tourism Board Enter into Wide-Ranging Partnership to Boost Visitorship to Singapore, 30 April 2014. Available at https://www.stb.gov.sg/news-and-publications/lists/newsroom/dispform.aspx?ID=504/. Accessed on 1 February 2015.

at a special promotional rate of $1. The SSH package was made available to SIA or SilkAir customers who are stopping over in Singapore before flying to their next destination. The offer was valid for new bookings and ticket sales made from 27 January to 28 February 2015 and was valid for travel for two until mid-2015.[96]

It was reported in April 2015, that CAG and STB entered into a joint-venture, making available $35 million over the following two years to attract more visitors to Singapore. This was in response to a drop of 4.1% of visitors from 15.1 million in 2014, compared to 15.6 million in the previous year. The aim of this scheme was to achieve an average annual growth of 3–4% in visitor arrivals over the next decade. Joint global marketing campaigns would be targeting key markets such as Australia, Indonesia, China and India. The scheme would include SIA and other carriers and cruise liners, travel agents. In addition to promoting Singapore, it would also attempt to sell the Changi Airport as a destination in its own right. In 2014, Changi Airport handled a record 54.1 million passengers but the growth was only 0.7% over the previous year.[97]

7. Challenges and Responses

The half a century of high growth and development of Singapore as a global air transport hub, in tandem with the economic growth of the country, has been remarkable. Its success can be attributed to the commitment on the part of the government in adopting a long-term airport infrastructure development approach commencing with the Paya Lebar Airport and followed by the superb Changi Airport, together with the necessary supporting infrastructure to create a strong aviation hub and to enable air transport activities to thrive. Singapore champions a liberal and open skies policy that encourages airlines to operate out of the Changi Airport, using it as a gateway to the surrounding region and the world. Singapore has been a beneficiary of an increasingly deregulated international air transport regime since the start of the 1900s that has opened up markets allowing access to more destinations, higher frequencies of flights, and greater capacity. As a member state of ASEAN, Singapore will gain much from the expected adoption of the ASEAN Open Skies Agreement by the end of 2015.

---

[96] Singapore Airlines, news release. SIA Shows Commitment To Singapore's Tourism Sector With SG50 Promotions. http://www.singaporeair.com/jsp/cms/en_UK/press_release_news/ne150126.jsp. Accessed on 6 January 2015.
[97] Karamjit Kaur, $35 million tie-up to attract more visitors to Singapore, ST 2 April 2015.

The success of Singapore's air hub can be attributed also to the Singapore Airlines, our national carrier that has come to symbolise the country for its efficiency and meticulous attention to quality in all aspects of its operations and services. In order to keep up with the rapid increase in demand for air travel, SIA has spawned a family of airlines — SilkAir to serve regional destinations, Singapore Airlines Cargo and Scoot to compete against other LCCs in both the medium and long-distance routes. Scoot has joined Thailand's Nok Airlines to form Nok-Scoot. More recently, SIA's joint-venture airline in India, Vistara Airlines, finally took to the skies. SIA is also a majority shareholder of TigerAir, a regional LCC that has helped to strengthen the Changi Airport hub. More recently, SIA was reported to be in talks to invest in South Korea's largest LCC, Jeju Air. This would be SIA's first North East Asian venture to enable the airline to tap on to the rising number of Chinese passengers in the region.[98]

A key SIA strategy has been joining the global Star Alliance thereby enabling the airline to gain access to destinations served by fellow members through the practice of codesharing with other airlines. It has also helped to gain greater efficiencies and to share airport facilities (and costs) with other Alliance members. Through this means, SIA has been able to expand its network of routes served, and generate more passengers from hitherto untapped markets.

Pangarkar [2014] suggests that "... selected partnerships with occasional small equity stakes or joint ventures..." is the way forward for SIA. He questions the wisdom of taking up a large 71% stake in the loss-making Tigerair. The earlier failures of SIA's joint-ventures with Virgin Atlantic and Air New Zealand have been due to factors beyond its control.

7.1 *Impediments to growth*

The Changi Airport has held on as one of the best airports worldwide for the third year in a row. On the other hand, in terms of passengers handled, Changi is ranked only 18th among the world's busiest airports based on data from January to September 2014 by the Airports Council International (ACI). It had in fact dropped from 13th position the previous year.[99] This indicates that, in spite of the awards its receives other airports had gained in relative to the Changi

---

[98] Joyce Lee and Siva Govindasamy, Singapore Air in talks over stake in Korea budget carrier Jeju Air, Reuters, 16 March 2015. Available at http://www.reuters.com/article/2015/03/17/us-singapore-air-jeju-air-equity-idUSKBN0MD03X20150317. Accessed on 8 April 2015.

[99] Wikipedia, Airports Council International, World's busiest airports by passenger traffic

Airport. Among the number of impediments to its further growth as an air hub, the following are most notable:

7.1.1 *Competition*

The aviation industry is intrinsically a high risk enterprise due to the high capital and operating costs and its stringent international regulations governing the industry. The task of running a successful airline requires unusual industry leadership as well as both internal high level management skills and competence, and external factors favouring Singapore to achieve unhindered development of its air hub status. Competition to both the Changi Airport as well Singapore Airlines, the two pillars of Singapore's air hub, has been intense from the 1990s, since the emergence of strong air hubs within the region, and airlines that are already existing as well as the newly established Middle East airlines such as Emirates, Etihad Airways and Qatar Airways.

It is ironic that the great liberalisation of the international aviation regulations restricted by the 1944 Chicago Convention that was led by the advanced countries and then followed by a countries in the Asian region has benefitted and, at the same time, opened up strong competition in allowing foreign airlines to operate out of the Changi Airport. Further liberalisation and deregulation would likely provide more intense competition while opening up new opportunities. Changi has become an international and regional gateway, and welcomes more foreign airlines thus triggering strong competition to SIA at its home base. One report noted that three Middle Eastern airlines are bolstering its already formidable position in Changi Airport, posing a strong challenge to SIA. In 2014, more than 90% of passengers who fly the three Middle Eastern airlines from Changi used their Middle East hubs as a stopover for flights to Europe, the US, and elsewhere, thereby scavenging on passengers that would otherwise use Changi as a key stopover point. All three airlines are targeting Singapore's corporate market dominated by SIA and hence weakening Singapore's hub position.[100]

Competition against SIA and its subsidiary airlines is also presented by regional airlines such as full-service airline Garuda Indonesia, Thai International and Malaysia Airlines, and, the aggressively expanding LCCs — Air Asia, Lion Air, and Jetstar. These and other regional airlines have high ambitions to break into the top league of international airlines and will also pose strong challenges to SIA-owned SilkAir and Scoot as well as TigerAir.

---

[100] Karamjit Kaur, Mid-East carriers muscle in on SIA's home ground, ST 16 February 2015.

It is sufficient to highlight two of the region's fast growing airlines, Garuda Indonesia — Indonesia's flag carrier — and Indonesia's privately-owned Lion Air. Garuda experienced a series of accidents and mishaps that led to it being banned in June 2007 from flying to European Union destinations. In July 2009, the airline launched a five-year expansion plan known as the Quantum Leap, during which time its fleet doubled from 62 to 116 aircraft. The airline also planned to boost passenger annual numbers from 10.1 million to 27.6 million through increasing domestic and international destinations from 41 to 62.[101]

Garuda operates a fleet of 134 aircraft with an order of 88 (as of January 2015). It has a low-cost subsidiary, Citilink, that was hived off in 2012. Garuda entered into a partnership agreement with Etihad Airways in October 2012. The partnership includes a codeshare agreement for a total of 36 flights between the two airlines. In November 2010, Garuda Indonesia joined Sky Team Alliance and became a member in March 2014. The airline offers flights to 28 additional international destinations through codeshare agreements.

In March 2013, Lion Air placed an order to purchase a total of 234 A320 aircraft worth US$24 billion — recorded as the most valuable commercial order booked in history. The second biggest order was also made by Lion Air earlier in 2011, in a US$22.4 billion order for 230 Boeing jets. The airline operates a fleet of 108 aircraft and had on order 528 aircraft. As of January 2014, the airline serves a total of 120 destinations — 100 domestic and 20 international.[102]

The consolidation of airlines into alliances has brought airline competition to a new dimension. Now airlines compete more intensely as part of an alliance rather than as individual entities. Among fellow alliance members, close collaboration in joint efforts to promote air routes has had the effect of helping less efficient and newer airlines to learn and find ways to match or even surpass the quality of facilities and services provided by SIA. This puts pressure on SIA to maintain its market share as well as holding on to its premium air fares.

The arrival of low-cost carriers (LCCs) has brought along a new dimension to competition, due to their strategy of adding new routes and capacity in order to gain market share by taking advantage of the greatly liberalised aviation policies of many countries within the region and worldwide. The proliferation of LCCs operating out of Changi Airport was bolstered by the

---

[101] Wikipedia, Garuda Indonesia. Available at http://en.wikipedia.org/wiki/Garuda_Indonesia#2009.E2.80.93Present:_Rebirth. Accessed on 15 January 2015.

[102] Wikipedia, Lion Air. Available at http://en.wikipedia.org/wiki/Lion_Air#cite_note-inquirer1-3. Accessed 16 January 2015.

opening of the Budget Terminal in March 2006 that would later be replaced by Terminal 4. As mentioned earlier, LCCs have generated a substantial proportion of passenger traffic and benefitted Changi Airport and Singapore as a whole, while travellers can now choose from among many more regional destinations. However, aggressive expansion strategies of many regional LCCs as shown above have led to severe excess capacity, leading to deep discount fares that put pressure on both LCCs and full-service carriers.

7.1.2 A question of location and size

The locational advantage of Singapore in supporting the Changi Airport as an international air hub due to our East–West and North–South pivotal position has been challenged by other hubs such as Hong Kong and Bangkok that are better positioned along the East–West corridor. Likewise, Singapore and SIA are facing competition from the Middle Eastern airlines and from hubs such as Dubai. The availability of long-range aircraft starting with the Airbus 340–500 that can fly 16, 600 kilometres followed by the Boeing 747s and then the Boeing 777LR and 777ER that operates a range of 20, 000 kilometres has not disadvantaged Singapore. The high-capacity wide-body planes have made it possible for routes that connect Singapore to major Australian destinations as well as European hubs such as Frankfurt, making Singapore a logical international air hub serving these destinations. However, these same aircraft have made it possible for routes serving the same Australian destinations to partly bypass Singapore to fly directly to the Middle Eastern hubs that also provide connections to numerous European destinations. The close cooperation between Australia's Qantas and Dubai Government-owned Emirates that hubs in Dubai International Airports has opened up more than 50 destinations in Australia and New Zealand, and a host of European and nearby destinations benefitting the two airlines.

The limitation of Singapore's small size puts the country in a disadvantageous position of not having more than a single destination to offer during the process of negotiations for bilateral air traffic rights with other countries. This factor explains Singapore's adoption of a liberal aviation regime that has led to the country making numerous bilateral air service agreements. Together with many countries adopting a more liberal policy on air transport, it has allowed foreign airlines to even use Singapore as their hub to serve destinations other than the capital cities. LCCs such as Jetstar and those controlled by major airlines like Qantas exploit Singapore's hub position to fan out to many regional destinations to feed regional passengers to their trunk routes operated by their parent airlines. Jetstar, with its 7th freedom traffic rights, is able to base its fleet and operate flights out of the Changi Airport.

The small population size of Singapore with a mere 5.4 million people as well as a modest (non-oil) manufacturing sector also means that Singapore's base for airfreight is limited. Our manufacturing sector is dependent on electronics and pharmaceuticals for generating air cargo. Electronics has been on the decline for some years. The growth of the MRO and aerospace industry has proven to be a success providing synergy among aviation sector players as well boosting air cargo traffic.

7.2 External factors

Air transport is sensitive to the vagaries of global and regional economic and socio-political developments, and the fluctuation of air travellers handled through the Changi Airport has testified to this. In addition, outbreaks of contagious diseases such as SARS and the annual occurrence of influenza and its variants, as well as natural disasters and aviation accidents have had their occasional adverse effects on air traffic. With the end of the 2008–2009 financial crisis, the cost of aviation fuel had risen sharply thereby restricting the profitability and growth of the air industry. However the substantial reduction in the cost of aviation fuel since January 2015, has improved the profitability of the aviation industry as well as the Singapore Airlines Group's profitability, especially if aviation fuel prices continue to remain soft over a longer term.

External factors can also impact on air travel affecting passenger arrivals in the Changi Airport. The 3% reduction in international visitor arrivals in the first half of 2014 was attributed to the impact of China's newly, introduced tourism law enacted in October 2013, prohibiting cheap air travel packages with hidden costs. In addition, the disappearance of the Malaysia Airlines flight MH370, the abduction of Chinese visitors in Sabah and political unrest in Thailand, have all had a dampening effect on Chinese travel to the region. From January to June 2014, Chinese visitor arrivals fell 30% year-on-year. However, during the same period, Chinese visitors that stayed for at least two days outnumbered those who spent a day or less in Singapore.[103]

As a whole, the Asia-Pacific region has remained vibrant economically. The GDP of East Asia is expected to rise 6.2% in spite of the slowing growth of China's economy.[104] In recent years, the burgeoning Indonesian economy has

---

[103] Singapore Tourism Board website, Quarterly Tourism Performance Report Q2 2014. Available at https://www.stb.gov.sg/statistics-and-market-insights/marketstatistics/singapore%20 tourism%20board_tourism%20sector%20performance%20q2%202014.pdf. Accessed 2 January 2015.

[104] Fiona Chan, East Asia to remain world's fastest growing region, ST 8 April 2014.

resulted in a massive growth of domestic and international air travel, including the emergence of a good number of Indonesian airlines. The growth of Jakarta's Soekarno-Hatta Airport was fueled by a large increase in air passengers flying between Jakarta and Singapore, making it the world's second busiest international route. Capacity constraints of Jakarta's airport have adversely affected it while awaiting the completion of its third terminal expected in August 2015. It was in 2013 that Jakarta became the world's 10th busiest airport handling 60 million passengers which was around three times its designed capacity. There has been a rapid increase in the number of flights to destinations in Indonesia both from Jakarta and Singapore. Clearly, Jakarta's congestion has also benefited the Changi Airport where passengers would prefer to transit through Changi for other destinations in Indonesia, thus bypassing Jakarta.

The congestion of the Soekarno-Hatta Airport and Bangkok's new Suvarnabhumi Airport was due to the slow progress in expanding their capacity to meet fast rising demand. This has not been the case for Singapore where additional capacity had always been put in place well before full capacity has been reached, thereby giving Singapore an edge over other hub airports. Inadequate airport facilities and lack of professionalism have also surfaced as a result of the disappearance of Malaysia Airline's MH 370 and the crash of AirAsia's QZ8501 on 28 December 2014 en route from Surabaya to Singapore. Thus it can be surmised that Changi Airport would grow even faster had the capacity of regional airports and aviation facilities and services been more adequate.

An unexpected problem arose in early April 2015 when China, Japan and South Korea stopped Thailand-based airlines from operating charter flights due to safety concerns highlighted in an international audit by the UN's International Civil Aviation Organisation (ICAO). The consequences were said to be 'unimaginable', affecting nearly 120,000 people over the following two months. Airlines affected were LCC Thai AirAsiaX and SIA subsidiary Scoot's JV airline, NokScoot, and Asia Atlantic Airlines.[105]

While many countries in the Asia-Pacific region have liberalised their air service agreements (ASAs), airline ownership and control restrictions still remain and these continue to prevent free and open trade in air transport. Bilateral ASAs specify limits on pricing, capacity, designated airlines and

---

[105] Reuters, Thai-based airlines in for turbulence after bans, ST 1 Apr 2015. See also MailOne, Thai Airways has been banned from flying to China, Japan, and South Korea over safety fears — so why does it still fly to Australia? Available at http://www.dailymail.co.uk/news/article-3023086/ Thai-Airways-banned-flying-China-Japan-South-Korea-safety-fears-does-fly-Australia.html posted on 8 Apr 2015. Accessed 8 April 2015.

routes operated. Often, changes desired by airlines would need to be submitted to the relevant authorities for approval which typically takes time for processing, following several rounds of prior lengthy negotiations between the airlines. For a number of reasons including strategic, safety and defence reasons and the need to control airlines in times of national emergency, most countries continue to impose foreign ownership and control restrictions on their airlines. While Singapore has signed a large number of liberal ASAs a number of major markets within the Asia-Pacific region still have restricive agreements.

At the 20th ASEAN Transport Ministers' meeting held in Mandalay, Myanmar on 27 November 2014, the ministers signed a Joint Ministerial Statement wherein member countries agreed to the implementation of the ASEAN Single Aviation Market (ASAM), including the ASEAN air services agreements which are set to be fully implemented by 2015.[106] When the agreement does come into force, ASEAN airlines, not limited to a single airline for each country, ideally would be able to fly freely to all destinations within the entire ASEAN community, but not including domestic traffic. However, Tan (2010, 2013) notes that the agreement is limited to opening up the 3rd, 4th and 5th freedoms. Cabotage has not been included to enable a foreign airline to add-on a second domestic destination beyond its first entry point. The Philippines has still not agreed to open up its capital city for the agreement in order to protect its weak national airline, while Indonesia has recently relented and opened up several international airports to ASEAN airlines. Indeed, the uneven development of the aviation industry of the member states has been the main reason for the unwillingness of some countries to open its air transport sector to foreign competition.

Tan [2013] further shows that the inability of ASEAN to fully adopt the open skies regime actually strengthens the competitive position of non-ASEAN airlines. This is demonstrated by the ASEAN–China Air Transport Agreement adopted in 2010. It is in force among China, and five ASEAN countries, namely Singapore, Malaysia, Thailand, Myanmar and Vietnam. The agreement allows airlines from participating countries to enjoy 3rd and 4th freedom and more recently also 5th freedom rights from any point in China to any point in the five ASEAN countries. However, the airlines of each of the five ASEAN member countries can only fly from their domestic destinations to any point in China. Thus, unless there is full integration within the ASEAN countries, open skies agreements with India, Japan and Korea, the ASEAN airlines

---

[106] ASEAN Multilateral Agreement on the Full Liberalisation of Passenger Air Services. Available at http://www.asean.org/archive/transport/Agreement-101112.pdf. Accessed 26 February 2015.

including the Singapore Airlines will suffer from the same disadvantage as China.

7.3 *Future growth prospects*

ASEAN has a combined population of over 600 million. A fast-expanding middle class coupled with the economic prosperity of the two Asian giants of China and India will mean that regional air travel will grow rapidly.[107] In addition, the rise of LCCs making air travel affordable to a large segment of the population will mean a healthy growth of air passengers. The availability of modern advanced aircraft with vastly increased passenger loads as well as enhanced fuel efficiency, lower maintenance requirements has significantly reduced the operating costs of running a fleet of aircraft. Should fuel costs remain low, it should provide a much needed stimulus to the aviation industry providing new avenues for growth for both airports and airlines. Severe competition for both the Changi Airport and the SIA group will remain a reality that must be overcome by better and more innovative strategies.

In 1972, the then Minister for Foreign Affairs, S. Rajaratnam, in a speech expounded the concept of Singapore as a "Global City" considering Singapore to be more than a regional city and then outlined the strategy for transforming Singapore into a global city. He argued that the traditional role of Singapore as a trading hub in Southeast Asia and as a marketplace of the region had gradually diminished in importance. Singapore would, instead, become part of the world and its global economic system.[108] The concept of a global city resonates well with Singapore as a global air hub that he anticipated citing the rapid increase in air passengers. A strong air transport hub supported by superior logistics services and a thriving aerospace hub in effect bolsters Singapore's position as a global city.

But there is more to our being only a global air hub as Singapore is already known as a powerful shipping hub as well as world-class centre for ship/rig building-repairing, petroleum refining and petrochemicals, pharmaceuticals, as well as being a centre for finance, legal and professional services,

---

[107] Speech by Senior Minister of State Mrs Josephine Teo at Inter Airport South East Asia, 21 January 2015. Available at http://www.mot.gov.sg/News-Centre/News/2015/Speech-by-Senior-Minister-of-State-Mrs-Josephine-Teo-at-inter-airport-South-East-Asia-on-21-January-2015/. Accessed 28 January 2015.

[108] S Rajaratnam, 'Global city' success for Spore: Raja — speech on Singapore as a global city. ST 7 February 1972. Available at http://eresources.nlb.gov.sg/newspapers/Digitised/Article/straitstimes19720207-1.2.5.aspx. Accessed 23 February 2015.

communications, healthcare and education services. Each of these nodes of activities reinforce each other as a dynamic global city.

Recognizing our severe size limitation, the lack of natural resources including scenic sites, Singapore has had to create man-made facilities and attractions to draw visitors. The pursuit of a well-planned and organised city, and a clean and green country has made Singapore a highly desirable city for local residents and foreigners alike. Added to this is the rich and varied multi-ethnic and cultural character that adds to the country's attractiveness. Singapore continues to gain its way to be a tourist destination as shown by the recent longer period of stay as a stopover point for visitors from China. There is much optimism for Singapore's continued quest to be a strong and resilient global air hub.

Acknowledgement

The authors are grateful to our three very able research assistants, Yong Kuan Chen, He Sichuang, and Chen Zhuhe for their efforts.

References

Allen, R. (1990). *Take-off to Success*. (Singapore Airlines Ltd., Singapore).

Baranwal, H. (2013). Singapore Airlines — Sustainable Advantage through their Dual Strategy, (Amity Global Business School, Singapore).

Bowen, J. T. (1997). *Air Pacific Air Transport: Challenges and Policy Reforms*, eds. Findlay, C. C., Chia, L. S. and Singh, K., Chapter 8 "The Asia-Pacific airline industry: prospects for multilateral liberalisation," (Institute of Southeast Asia Studies, Singapore) pp. 123—153.

Chan, D. (2000). The story of Singapore Airlines and the Singapore girl. *Journal of Management Development*, 19 (6), pp 456–472.

Changi Airport Group. *Annual Reports*, (Changi Airport Group, Singapore).

Civil Aviation Authority of Singapore. *Annual Reports*, (Civil Aviation Authority of Singapore).

Doganis, R. (1991). *Flying Off Course: The Economics of International Airlines*, 2nd Ed. (Routledge, New York).

Pangarkar, N. (2014). SIA's strategy in tough airline business, *The Strait Times*. Dated 22nd October 2014.

Heracleous, L. and Wirtz, J. (2009). Strategy and organisation at Singapore Airlines: achieving sustainable advantage through dual strategy, *Journal Air Transport Management*, 15, pp. 274–279.

Heracleous, L. and Wirtz, J. (2010). Singapore Airlines' balancing act, *Harvard Business Review*, Jul-Aug Issue, pp. 141–149.

Heracleous, L., Wirtz, J. and Pangarkar, N. (2009). *Flying High in a Competitive Industry*, 2nd Ed. (McGraw Hill, Singapore).

Morrish, S.C and Hamilton, R.T. (2002), Airline alliances — who benefits? *Journal Air Transport Management*, 8 (6), pp. 401–407.

Olszewski, K. F. and Chia, L. S. (1991). *The Biophysical Environment of Singapore*, eds. Chia, L. S., Rahman, A. and Tay, D. B. H., Chapter 7 "National development, physical planning and the environment in Singapore", (Singapore University Press, Singapore) pp. 185–206.

Sabhlok, A. (2001). The evolution of Singapore business: a case study approach, *Institute of Policy Studies Working Papers*, 10(1).

Singapore Airlines. (1988). *Perspectives*, (Singapore Airlines Ltd., Singapore).

Singapore Airlines. *Annual Reports*, (Singapore Airlines Ltd., Singapore).

Tan, A. K. (2010). The ASEAN multilateral agreement on air services: En route to open skies?, *Journal Air Transport Management*, 16(6), pp. 289–354.

Tan, A. K. (2013). ASEAN's single aviation market: many miles to go, *Airline Leader*, 16, pp. 12–15.

Wang, S.W. 2014. Do global airline alliances influence the passenger's purchase decision? *Journal Air Transport Management*, 37, pp. 53–59.

Yeo, D. and Tan, S. (2014). *Bridging Skies — The Aviation Hub of Choice*, (Civil Aviation Authority of Singapore, Singapore).

Index

www.ingramcontent.com/pod-product-compliance
Lightning Source LLC
Chambersburg PA
CBHW080546270326
41929CB00019B/3214